T0325331

ATTRITION IN THE PHARMACEUTICAL INDUSTRY

ATTRITION IN THE PHARMACEUTICAL INDUSTRY

Reasons, Implications, and Pathways Forward

EDITED BY

ALEXANDER ALEX
C. JOHN HARRIS
DENNIS A. SMITH

Published by John Wiley & Sons, Inc., Hoboken, New Jersey
Published simultaneously in Canada

For general information on our other products and services or for technical support, please contact our Customer Care Department within the United States at (800) 762-2974, outside the United States at (317) 572-3993 or fax (317) 572-4002.

Wiley also publishes its books in a variety of electronic formats. Some content that appears in print may not be available in electronic formats. For more information about Wiley products, visit our web site at www.wiley.com.

Library of Congress Cataloging-in-Publication Data:

Alex, Alexander, editor.
  Attrition in the pharmaceutical industry : reasons, implications, and pathways forward / edited by Alexander Alex, C. John Harris, Dennis A. Smith.
    pages cm
  Includes index.
  ISBN 978-1-118-67967-8 (cloth)
1. Pharmaceutical industry–Management.   2. Pharmaceutical industry–Capital productivity.
3. Industrial efficiency.   4. Drug development.   I. Harris, C. John, editor.   II. Smith, Dennis A., editor.
III. Title.
  HD9665.5.A38 2015
  615.1068′5–dc23
                                                                                            2015024772

Set in 10/12pt Times by SPi Global, Pondicherry, India

Printed in the United States of America

10 9 8 7 6 5 4 3 2

1 2016

CONTENTS

CONTRIBUTORS

Alexander Alex, Evenor Consulting Ltd, Sandwich, Kent, UK

Thomas A. Baillie, School of Pharmacy, University of Washington, Seattle, WA, USA

Andrew Bell, Institute of Chemical Biology, Department of Chemistry, Imperial College, London, UK

Scott Boyer, Swedish Toxicology Sciences Research Center, Södertälje, Sweden

Clive Brealey, AstraZeneca R&D, Mölndal, Sweden

Kelly Chibale, Department of Chemistry, University of Cape Town, Rondebosch, South Africa

Robert T. Clay, Highbury Regulatory Science Limited, London, UK

Andrew M. Davis, AstraZeneca R&D, Mölndal, Sweden

Wolfgang Fecke, VIB Discovery Sciences, Bio-Incubator, Leuven, Belgium

Peter Gedeck, Novartis Institute for Tropical Diseases Pte Ltd, Singapore

Rosalia Gonzales, Hit Discovery and Lead Profiling Group, Department of Pharmacokinetics, Dynamics and Metabolism, Pfizer Inc., Groton, CT, USA

C. John Harris, cjh Consultants, Kent, UK

Cornelis E.C.A. Hop, Department of Drug Metabolism & Pharmacokinetics, Genentech, South San Francisco, CA, USA

Wilma W. Keighley, WK Life Sciences, Kent, UK

Christian Kramer, Roche Pharmaceutical Research and Early Development, Molecular Design and Chemical Biology, Roche Innovation Center, Basel, Switzerland

Geoff Lawton, Garden Fields, Hertfordshire, UK

Richard Lewis, Novartis Pharma AG, Basel, Switzerland

J. Richard Morphy, Lilly Research Centre, Surrey, UK

Peter Mbugua Njogu, Department of Chemistry, University of Cape Town, Rondebosch, South Africa

Marie-Claire Peakman, Hit Discovery and Lead Profiling Group, Department of Pharmacokinetics, Dynamics and Metabolism, Pfizer Inc., Groton, CT, USA

Anne Schmidt, Hit Discovery and Lead Profiling Group, Department of Pharmacokinetics, Dynamics and Metabolism, Pfizer Inc., Groton, CT, USA

Dennis A. Smith, Department of Chemistry, University of Cape Town, Cape Town, South Africa; The Maltings, Walmer, Kent, UK

Matthew Troutman, Hit Discovery and Lead Profiling Group, Department of Pharmacokinetics, Dynamics and Metabolism, Pfizer Inc., Groton, CT, USA

Christine Williams, Ipsen BioPharm Ltd, Global Project Management and Analytics, Slough, UK

INTRODUCTION

ALEXANDER ALEX[1], JOHN HARRIS[2] AND DENNIS A. SMITH[3]

[1] *Evenor Consulting Ltd, Sandwich, Kent, UK*
[2] *cjh Consultants, Kent, UK*
[3] *Department of Chemistry, University of Cape Town, Cape Town, South Africa;*
The Maltings, Walmer, Kent, UK

Taking on this very complex and important topic and putting together a book seemed a large but rewarding task for individuals who have spent their careers discovering and developing drugs. Having completed the task, there is still the feeling of not quite answering the problem. What the book represents is a detailed analysis of what is largely failure and some important directions that can be followed. At the time of publication, the industry is moving from blockbuster drugs to patient-targeted entities. These have the potential to lower attrition and may change the commercial process. In assembling the volume, the editors felt more and more the massive importance and urgency to find solutions for the issue of attrition in the pharmaceutical industry, which has been an ever-growing threat to the entire industry for at least 20 years. The editors have themselves experienced significant changes designed to increase productivity, reduce cost, and tackle attrition in the sector. These range from the implementation of a "more is better" philosophy with compound library synthesis and high-throughput screening to the "genome revolution" through all the way to alliances, collaborations, mergers, and acquisitions. However, it seems that none of these approaches have really worked since drug discovery productivity, as measured by number of new chemical and biological entities (NCE and NBE), has essentially stayed flat since the 1980s, despite exponential increases in research spending throughout the industry until investment started to stagnate in the last few years. Many questions have been raised, and many attempts have been made to resolve this conundrum, but it appears that a long-term, sustainable solution has yet to be found and recent events with yet more reorganizations and takeovers on the horizon seem to confirm this.

A strong cohort of new drug approvals by the FDA toward the end of the year increased the total to 41 for 2014, the largest number in 18 years. Therefore, 2014 becomes the

Attrition in the Pharmaceutical Industry: Reasons, Implications, and Pathways Forward,
First Edition. Edited by Alexander Alex, C. John Harris and Dennis A. Smith.
© 2016 John Wiley & Sons, Inc. Published 2016 by John Wiley & Sons, Inc.

second highest year on record for the approval of new chemical entities since the record of 53 new drug approvals in 1996. This is good news for the pharmaceutical industry but also for patients in need of new medicines. It is noticeable that the number of NCEs has been highly variable over the last 5 years with a total of only 29 new drug approvals in 2013, which followed 39 approvals in 2012, although, by any measure, 2014 approvals outstrip those of recent years (average of 24 per annum in the first decade of the new millennium and 31 per annum in the 1990s).

Despite these encouraging numbers, the total number of drugs approved for the last 5 years is most likely still below the ideal in terms of the needed return on investment, particularly for large pharmaceutical companies. The challenges facing the pharmaceutical industry in terms of compound attrition in discovery and clinical phases all the way to postmarket withdrawals will be outlined in this book.

It would be presumptuous in the extreme for any book to claim to provide all the answers to a given problem, never more so than when dealing with attrition in the pharmaceutical industry. However, this book is intended to provide a perspective from a number of industry and academia experts in the field and to stimulate discussion on the topic that may even help to point in the direction of potential solutions. It is not intended to review every aspect of attrition in the pharmaceutical industry over the last three decades, but rather to provide some context in order to enable a measured attempt to look forward. Although it is not possible to predict the future, we hope that this book will provide some useful information and insights for a productive, collaborative, and positive discussion on attrition in the pharmaceutical industry. We hope that it will make a small but useful contribution to the debate on reducing attrition and increasing productivity. Above all, we should never lose sight of the ultimate goal of our efforts, which is to provide new and urgently needed medicines for patients across the world.

Attrition in the pharmaceutical industry has been a topic of intense discussion for at least three decades. As with most debates, the underlying facts are often complex and difficult to agree on by experts. One of the unarguable facts that have emerged over the last 30 years is that the number of new drugs coming to market has remained effectively flat since the early 1980s despite increasing research and development (R&D) budgets [1]. To a large extent, budgets have been essentially flat over the last 5 years, but productivity is still not in line with even the stagnant investments. However, in reality, the productivity of a pharmaceutical company is not measured, at least not by investors, by the output of new drugs but instead in terms of costs, sales, and profits; the market valuation of a company; and particularly the ability to pay dividends to its investors at an expected level. Remarkably, while innovation has remained relatively flat, profits and dividends have not actually fallen for decades. So what has been going on? As with most measures of success, productivity is relative. Many pharmaceutical companies expanded in the late 1990s in line with double-digit growth predictions for the decade ahead, which never materialized due to unforeseen economical circumstances and overoptimism, particularly but not exclusively around overinflated expectations in increasingly volatile stock markets and the impact of competition from emerging economies and severe challenges in the international patent landscape. This was despite the ever-increasing demand for existing and new medicines from those countries as well as the more established sectors.

There have also been severe challenges from economists to the wide claims that research to discover and develop new medicines entails the high costs and high risks outlined and published, primarily by the pharmaceutical industry, in a paper by the London School of Economics in 2011 [2]. A widely used figure for the cost of a new NCE is that of $802 million, which originates from a study done in 2003 [3]. However, it appears that in these numbers, factors like taxpayer subsidies have not been included, and accordingly, a corrected estimate would be $403 million per NCE [1]. Further adjustments as, for example, using a "cost of capital" rate called for by the US and Canadian governments in the calculations that is significantly lower than the one used in the 2003 study, leads to a further reduction of the actual cost to $180–$231 million [1]. In addition, it appears that one needs to be very careful when drawing firm conclusions about NCE costs from analysis of data, especially when it has been voluntarily submitted by the companies themselves and is confidential and therefore not verifiable [1]. Another way of calculating the cost of an NCE is by dividing the actual research budgets by the number of NCEs per company [4]. It turns out that from this analysis, the amount of money spent on a new NCE is simply staggering. For example, AstraZeneca would have spent $12 billion in research for every new drug approved, as much as the top-selling medicine (Lipitor, Pfizer) has ever generated in annual sales, whereas Amgen would have spent just $3.7 billion per new drug. It is probably fair to say that at around $12 billion per drug, inventing medicines would be considered an unsustainable business and at around $3.7 billion, companies might just about be able to make a profit [4].

Whatever the precise real costs for an NCE are and with the benefit of hindsight, the investments made in anticipation of overoptimistic growth rates led to a somewhat unsustainable economic situation across the entire pharmaceutical industry, especially in the R&D area. Indeed, companies had to adjust in an often drastic manner to the economic and social realities that pertained toward the end of the twentieth century, notably through a massive consolidation of the industry driven by both friendly and hostile takeovers and mergers on an unprecedented scale. The main objective for many of these acquisitions appeared to be either to access the revenue for already marketed drugs or to incorporate the most promising candidates from the respective R&D pipeline. It appeared that these actions were at least stabilizing for the profits of the remaining companies, although these measures could clearly only be a "fix" for a few years until the next wave of patent expiries were imminent. The first decade of the twenty-first century did not seem to help pharmaceutical companies to get back on track to achieve their desired profits and share-holders' expectations, with the stock market and housing market crashing around the world during that time. The inevitable consequences of these global crises, that is, stagnation of incomes, austerity measures by governments, and the increase of poverty across even many of the wealthy countries in the so-called developed world, also had a profound impact on the healthcare market, with prices for medicines being a particularly prominent target for governments and healthcare providers. In order to avoid government regulations in particular countries, some companies may even have withdrawn their products from those markets, and one can only assume that this was done in order not to put their pricing strategies in other, more profitable countries at risk.

The financial cuts, staff reductions, and general consolidation in the pharmaceutical sector have come at an enormous price, both economically and socially, for the people who rely on this industry for their income and prosperity, but even more importantly for patients who are getting fewer and fewer novel medicines at a time when the need for new therapies, especially in chronic diseases and increasingly resistant infections, is growing greater than ever before.

Covering the extremely wide theme of attrition in the pharmaceutical industry is a challenging endeavor, and this book claims neither completeness nor the provision of comprehensive answers to the many questions one might ask in relation to this topic. It does however attempt to provide not only a historical account that may help to facilitate learning but also, hopefully, to offer some stimulating and thought-provoking insights from a group of vastly experienced authors who have, despite the obvious challenges, kindly agreed to contribute. In order to make this book more forward looking, the editors strongly encouraged the authors to identify and incorporate new approaches and ways of thinking into their chapters and give their personal opinions and speculations about potential ways forward for reducing attrition. We hope that readers will find this approach appealing and useful and that this book will exert some positive influence through the vast expertise and considered opinions of their drug discovery research colleagues.

This book has been structured with the intention to guide the reader through the various stages of drug discovery and development in a systematic way, starting with an overview of attrition in drug discovery over the last 20 years in Chapter 1 and then focusing on more detailed analyses in Chapters 2–5 of the various stages from discovery through to phases I, II, and III and postlaunch. Following the chapters on the discovery and development pipeline, Chapter 6 investigates the influence of the regulatory environment, which has seen some major changes over the last 20 years. Chapter 7 then focuses on experimental screening strategies to reduce attrition, while Chapter 9 examines the influence of phenotypic and target-based screening strategies on compound attrition and project choice. Chapter 8 discusses the importance and evolution of medicinal strategies to reduce attrition in the early stages of the discovery process but also, as a consequence, reduce the risk of attrition later on in development. Chapter 10 focuses on *in silico* approaches to reduce attrition, highlighting the importance of the contribution of computational methods to modern drug discovery. Chapter 11 discusses current and future strategies for improving drug discovery efficiency, particularly on collaborations and interactions between industrial and academic drug research. Chapter 12 then looks at the impact of investment strategies, organizational structure and corporate environment on attrition, and future investment strategies to reduce attrition.

As might be expected, there is some overlapping content between chapters, primarily in the introductory parts but also on occasion in discussions and interpretations of the scientific literature. The editors have recognized this and considered it to be a very positive aspect of this book since it allows for diversity of views and opinions from all the authors.

The editors hope that this book will make a valuable contribution to not only the very intense ongoing discussion of attrition in the pharmaceutical industry but also to point out new approaches, productive critique and innovative thinking, as well as realistic and implementable ways forward to tackle this issue of such massive significance not only to the millions of people involved in the industry but also, most of all, to the billions of patients, who are still largely relying on the industry for the breakthrough medicines of the future.

REFERENCES

1   Schmid, E.F., Smith, D.A. (2005). *Drug Disc. Today, 15*, 1031.

2   Light, D.W., Warburton, R. (2011). Demythologizing the high costs of pharmaceutical research. *Biosocieties, 6*, 34–50.

3   DiMasi, J.A., Hansen, R.W., Grabowski, H. (2003). The price of innovation: New estimates of drug development costs. *J. Health Econ. 22*, 151–185.

4   http://www.forbes.com/sites/matthewherper/2012/02/10/the-truly-staggering-cost-of-inventing-new-drugs/ (accessed July 16, 2015).

1

ATTRITION IN DRUG DISCOVERY AND DEVELOPMENT

Scott Boyer[1], Clive Brealey[2] and Andrew M. Davis[2]

[1] Swedish Toxicology Sciences Research Center, Södertälje, Sweden
[2] AstraZeneca R&D, Mölndal, Sweden

1.1 "THE GRAPH"

If we had a confident grasp of the underlying reasons for attrition of projects and compounds in drug discovery and development, we would not need to write this book. But we are not confident, not confident at all. While attrition is a problem for both small and large molecules, and they share some common factors, it is small-molecule attrition that is currently crippling the industry. In some senses, the perceived greater success rates achieved with large-molecule drugs have increased the focus on large-molecule therapeutics.

With only 1 in 20 or fewer small molecules that enter clinical development reaching the market, greater than 95% of our innovation fails during the phases of clinical development [1]. A heated debate is currently raging in the scientific literature over the reasons for our dismal success rates. Many papers have been written concerning reasons for attrition, and many lectures given, often with contradictory messages. Substantial progress has been made in identifying new targets and rapidly designing small molecules active at these targets. However, converting these molecules into drugs has become more difficult [1]. Furthermore, to create value for patients and investors and to meet the health economic targets of those who pay for these drugs, let alone sustain a drug on the market for many years in the face of constant scrutiny and challenge, seems at times to be a superhuman task. Some limited progress has been made, but many great leaps in understanding are still to be taken. This books aims to help project teams and drug hunters in what is still a great endeavor.

One thing that everyone agrees on is that output from drug discovery industry is declining. "The graph" is a common first slide or figure in many public presentations.

Attrition in the Pharmaceutical Industry: Reasons, Implications, and Pathways Forward,
First Edition. Edited by Alexander Alex, C. John Harris and Dennis A. Smith.
© 2016 John Wiley & Sons, Inc. Published 2016 by John Wiley & Sons, Inc.

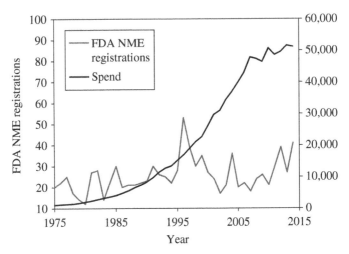

FIGURE 1.1 "The Graph"—Number FDA New medical entity registrations per year (gray curve) and total R&D expenditure/$ millions (black curve) [2, 3].

It shows the FDA new drug approvals and the costs of drug discovery and development per year [2, 3]. While investment in research and development (R&D) has dramatically increased, new drug registration has remained flat. It is shocking, we keep looking at it, we keep talking about it, and it is resulting in fundamental changes in the pharmaceutical industry (Figure 1.1).

The reasons for decreasing output are highly complex and poorly understood. Often cited reasons include, but are not limited to, higher regulatory hurdles required for drug safety, the requirement for adequate differentiation of new drugs versus existing therapies for reimbursement, inadequate choice of biological targets linked to disease, poor control of compound quality, and human decisions over which drugs to support through development and which to not support, so-called portfolio reasons.

The pressure is on; companies aspire to decrease attrition by implementing changes in the way they operate, but they do not just rely on their aspiration. They "manage" attrition by playing the numbers game. In order to "live" with attrition, you just need to run more projects. A recent 2010 review on R&D productivity[1] suggests that at a 7% success rate for small-molecule drugs reaching the market from a phase I entry and a 13.5-year development time, a company would need 11 phase I entries per year to yield 1 marketed drug per year. To sustain that level of availability of development compounds, a company would need a steady-state work in progress volume of 25, 20, and 15 projects in the target to hit, hit-to-lead, and lead optimization stages, respectively. Many large pharmaceutical companies have been attempting to maintain such a "volume" model. But this "volume" model is becoming unsustainable, for a number of reasons. First, the pharmaceutical industry cannot afford to sustain the volume model. While it was thought that the average cost of delivering a drug to market was $1.8B, Matthew Herper in *Forbes* magazine recently published the "real" costs of drug development [4]. By taking 10-year R&D costs of the top 100 companies and dividing by the number of drugs they delivered to market, the median cost for companies releasing more than three drugs was cited as greater than five billion dollars. For some companies, the figures were even worse. Topping the poll of worst performers were Abbott ($13B), Sanofi ($10B), and AstraZeneca ($9B). These staggering numbers are the result of higher than average failure in delivering drugs to market during the period of measurement despite somewhat similar overall levels of R&D investment. For companies that released only one drug in the 10-year period, the median costs were only $350M, but the attrition in this segment was likely in *companies* rather than projects. With the costs of delivery of drugs to market spiraling, the return from those few drugs that do reach the market needs to be higher; hence, the industry has continued its pursuit of blockbuster drug status (able to achieve >$1B/year sales). Where the number of treatable patients is limited by the disease, for example, for some cancers, increased prices are required to achieve commercial viability, with consequent issues in some health economic assessments. The industry's reaction to the failing output and increasing costs has been to experiment with changes to business models:

- Mergers to bolster weak portfolios and drive size and scale efficiencies, as exemplified by the 2014 attempted acquisition of AstraZeneca by Pfizer
- Closures or "virtualization" of "difficult" high-attrition disease areas, such as GSK's and AstraZeneca's minimized investment in neuroscience
- Outsourcing of synthesis and screening to lower cost base countries (although with demand, costs are increasing there)
- The scramble to develop a biologics business by partnerships, in-licensing, and acquisitions, based on perceived lower risks, higher returns, and lower generic competition with biologic drugs
- The move away from diseases apparently well controlled on standard therapy
- The hunt to build new markets in developing countries
- "Playing to company strengths" in discovery, clinical science, or sales and marketing expertise
- An increased focus on first in class drugs, as "innovative" drugs for new mechanisms are more likely to suffer less competition than follower drugs
- And lastly a focus on "quality" projects and "quality" compounds. How to achieve "quality" is perhaps the main aim of this book

However, many of these are essentially business operational strategies. What are we doing to address attrition head-on?

1.2 THE SOURCES OF ATTRITION

An early study by Prentis, Lis, and Walker in 1988 focused on reasons for attrition in the development pipelines on the then seven major UK pharmaceutical companies and categorized sources of attrition as shown in Figure 1.2a [5].

They highlighted 39.4% development compounds failed due to inappropriate human pharmacokinetics, with a further 29.4% failing due to lack of clinical efficacy. Pharmacokinetics are determined in phase I trials, while it is not until phase II that clinical efficacy results are uncovered. Anti-infectives comprised 30% of the database, and if they were excluded, clinical efficacy failure rose to 50%. At that time, drug metabolism and pharmacokinetics were not a part of preclinical optimization. Many companies began to invest in the discovery of drug metabolism and pharmacokinetic departments, where

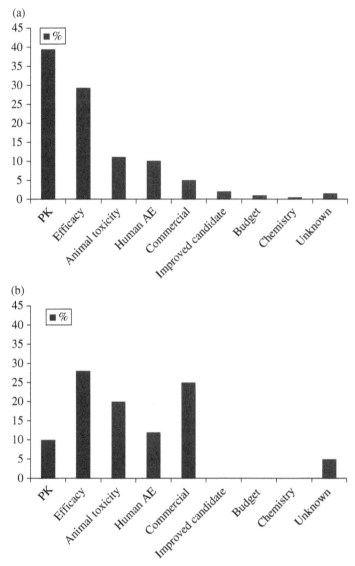

FIGURE 1.2 (a) Reasons for attrition. Data from Prentis and Walker [5]. (b) Reasons for attrition. Data from Kola and Landis [6].

compound weaknesses could be addressed during lead optimization. Reassuringly, it appeared that the investment was worthwhile, as in a 2004 follow-on review, attrition due to pharmacokinetics had apparently been reduced to around 10%. The major source of attrition remained lack of efficacy [6]. Poor pharmacokinetics was certainly a problem that needed fixing. But fixing it uncovered an unaddressed problem and moved attrition to phase II, a more expensive place to fail. The failure was that of translation of our mechanistic hypothesis into clinical efficacy. It had always been the major problem and remains the major challenge the industry faces. Attrition in phase II is now thought to be the highest of any phase, with some estimates putting it as high as 66% [1].

1.3   PHASE II ATTRITION

The problem of translation of mechanistic hypotheses into clinical efficacy is being tackled on a number of fronts. The choice of biological target on which to base a discovery program is receiving increased scrutiny at the earliest possible opportunity. Even before potent selective compounds are available, gene knockdown or gene editing can be conducted using siRNA knockdown, TALENs, or CRISPR-Cas technologies even using primary human cells. These experiments can probe the biological hypothesis and safety liabilities can be inferred [7, 8]. As potent selective compounds become available, experiments can be conducted with chemical probes that provide subtler control over the degree of modulation of the biological target than can be achieved with knockouts or generic mutations and indicative of the eventual candidate drug. As the discovery project progresses and compounds become closer to candidate drugs, further studies can be conducted, including *in vivo* testing. Although important questions are being asked about the value of animal models of disease [9, 10], such models can allow a more detailed pharmacokinetic–pharmacodynamic relationship to be explored, providing information on the concentration-time-biological mechanism relationship informing the design of clinical studies.

The definition of "patient populations to treat" is a further important focus, and the emerging paradigm is personalized healthcare. Identification of likely-to-respond patients maximizes the chances of observing a clinical efficacy signal without the dilution of nonresponding patients. It also avoids the risk of exposing nonresponding patients to possible drug-induced toxicity. Hence, personalized healthcare is of interest to patients, pharmaceutical companies, regulators, and payers alike. A recent PhRMA survey suggests that most clinical trials are now personalized [11], although very few diseases are understood at the genetic level.

Much of medical disease classification is empirical by nature, largely based on clinical manifestations, where a collection of similarly exhibited symptoms are used to classify indications. This is a major problem for drug development, which approaches disease from a molecular perspective. Where patients do not share a common molecular basis for disease, variability in drug response will, unsurprisingly, ensue.

Cystic fibrosis (CF) is a good case study to exemplify these points. CF was first described in 1938 by Dorothy Andersen, a pathologist, who noted the pancreatic lesions on a child who had presented with symptoms of celiac disease [12]. Prior to Andersen's description, there was increasing recognition that children with celiac disease were not uniform, and some of them presented with distinct pancreatic abnormalities, often identified post mortem. Up until this point, sporadic cases of infant deaths had been ascribed to pancreatic insufficiency, and some of the children were noted to have severe respiratory disorders also. At this time, infant death due to gastroenteritis and pneumonia, even in non-CF patients, was a relatively common occurrence, which had prevented the recognition of CF as a distinct disease. Andersen researched the post mortem records of similar patients to her own, which provided the evidential basis for her to classify CF as a distinct clinical entity.

The disease pathology was now understood at the level of clinical manifestations, but it would be years before a molecular understanding was provided. Andersen held on to the hypothesis that CF was caused by vitamin A deficiency, due to the similarities with celiac disease. We would now not be surprised that vitamin A supplementation was hardly likely to be effective. The hint to the underlying pathology can be traced as far back as 1857, to a passage in the "Almanac of Children's Songs and Games from Switzerland," which warned that "the child will soon die whose forehead tastes salty when kissed." This idea was proven in 1953 when Paul di Sant' Agnese revealed the increased salt content of sweat in people with CF, and this remains a cornerstone of CF diagnosis today. It was not until 1985 that Professor Lap-Chi Tsui, Dr. Francis Collins, and Professor Jack Riordan identified the first specific faulty gene mutation responsible for CF, ΔF508 in the gene that codes cystic fibrosis transmembrane conductance regulator (CFTR) [13]. CFTR normally transports sodium and chloride ions together with their waters of hydration. At least 1000 mutations to the CFTR are known to be part of the disease, and all affect the CFTR ability for ion transport. Vertex's recent drug registration for Kalydeco (ivacaftor), which improves function of mutant G551D CFTR, found in just 4% of patients, shows the success that can be achieved when the molecular basis of the disease is understood.

Crizotinib, an ALK kinase inhibitor, targets lung cancer patients with ALK mutations; likewise, AstraZeneca's gefitinib is most effective in mutated EGFR in non-small-cell lung cancers, although this was reportedly only discovered through subset analysis of clinical trial data rather than designed in during its discovery. The clinical use of these drugs is facilitated by the use of diagnostic tests to identify patients carrying the appropriate mutations [14, 15].

In most other diseases, where a genetic basis of disease has not been identified so far, patient selection is focusing at the level of biomarkers for disease classification, but you have to pick the right biomarker. A biomarker is defined by the FDA as [16] "measured in an analytical test system with well-established performance characteristics and for which there is an established scientific framework or body of evidence that elucidates the physiologic toxicologic pharmacologic or clinical significance of the test results." The FDA and European Medicines Agency (EMA) recognize "qualified biomarkers," which can be used for regulatory decision making, while the pharmaceutical industry will work with exploratory biomarkers, which they may use for internal decision making and for which they may seek to achieve qualification.

For example, subsets of asthmatics can be defined as eosinophilic, with high blood/sputum eosinophil counts, or with a high Th2 cell count phenotype. A working hypothesis is that these are biomarkers of a disease phenotype and that therapies targeting Th2 cells or eosinophils in these eosinophilic/Th2 high patient subsets would be expected to show increased efficacy over asthmatics with low eosinophil/Th2 cell counts. Lebrikizumab is a humanized IL-13 antibody; IL-13 is secreted by Th2 cells and apparently involved in eosinophil cell recruitment. In a phase II clinical study of lebrikizumab, the efficacy of lebrikizumab was compared in asthmatic patients segmented by high/low blood eosinophil counts and high/low Th2 cell phenotypes. But just prior to unblinding the study, a further subset was defined based on another biomarker, periostin. Periostin is also controlled by IL-13. The high/low eosinophil and high/low Th2 subsets did not produce any significant separation in clinical effect; similar effects were observed in high Th2 and low Th2 groups, but the periostin separation did show a significant difference with increased efficacy in the high periostin class [17].

In the absence of anything else, patient selection can be based on the lack of response to another drug, if preclinical evidence suggests the mechanism under question may be particularly efficacious. Through these steps of patient selection, we are aspiring to reduce phase II/III efficacy attrition for future programs, by how much we will succeed is difficult to say.

1.3.1 Target Engagement

Pfizer, through a systematic retrospective analysis of 44 of their phase II programs (with an overall success rate in achieving positive phase II readout of 33%), were able to define three pillars of survival success to reaching positive phase II decisions and phase III progression. The three fundamental elements that needed to be demonstrated early in development were:

- Exposure at the target site of action over a desired period of time
- Binding to the pharmacological target as expected for its mode of action
- Expression of pharmacological activity commensurate with the demonstrated target exposure and target binding

Only when they had confidence in both pharmacology and exposure were they confident of phase II success. Out of the 44 phase III projects studied, only 14 had experimental data providing confidence in both the pharmacology and exposure, and all 14 of these achieved a positive phase II decision, and 8 progressed to phase III. In comparison, 12 projects had no data demonstrating confidence in exposure and pharmacology, and all 12 were phase II failures [18].

Phase II is also the start of the investigation of the properties of the drug on wider groups of individuals and the context of its future uses as a drug, for example, in the presence of comedications. At this stage, the potential for drug–drug interactions is investigated in clinical pharmacology studies. Adverse findings can have an impact on the contents of the drug label, which might ultimately limit the scope for use of the drug and have an effect on market size. Such considerations must be weighed in the decision to progress to phase III and ultimately to the regulatory submission. Increasingly, multiple complications with the properties of a drug can undermine the commercial case, even if the drug demonstrates efficacy. Again, such trends will reduce the number of new drugs reaching the market, limiting the choice within a class for physician and patients.

1.3.2 Clinical Trial Design

As in other areas of biology dealing with populations, the clinical phases of drug discovery and development present the problem of signal to noise. Signals for efficacy and safety have to be detected against the noise from interindividual variability. The clinical development phase is by far the most expensive stage of the process of drug innovation such that decision making on the funding of studies is a significant source of attrition. Frequently, it is not possible to power early studies to deliver a statistical endpoint for a relatively weak signal, often leading to equivocal outcomes in phase II. Complex designs to compare subgroups of patients in phase II, which might be very beneficial in investigating the scope of a new target in disease, can be unattractive when viewed against the eroding patent life of a project. Furthermore, complex studies can be difficult to implement in practice, as clinical centers might not be available to deliver a biomarker, for example. Nevertheless, there are some encouraging trends in the design of phase I and II trials, which offer opportunities to reduce attrition or allow earlier decision points.

For a number of years, regulators have attempted to stimulate flexibility in phase I studies and in fact do seem to be open to novel and scientifically well-based study concepts. The exploratory IND is a clear example. The advantages are that it is possible to generate initial human data somewhat faster, requiring less preclinical data. Pharmacokinetics can be examined, and multiple compounds compared. However, the dose used needs to be sub-pharmacological for the target (less than 100 μg in most cases), and further progression requires a second stage with completion of a full IND.

More recently, microdosing studies using accelerator mass spectrometry are increasingly popular. The very low doses used (nanograms in most cases) are readily justifiable in terms of predicted biological effects. However, there are risks around nonlinearity of pharmacokinetics especially as this is a tool more likely to be used in cases where there is increased uncertainty over the prediction of human pharmacokinetics from preclinical studies. On balance, in many cases, a well-designed and rapidly executed normal phase I program probably takes less time and allows continuity into phase II. Most experienced project teams have good ideas how to reduce attrition at this stage, by thorough evaluation of dose to man predictions. For example, much time and cost can be saved by careful design of the toxicology program to attempt to avoid heroic doses in preclinical species, thus limiting the need for expensive drug substance at this stage.

Phase Ib studies where there is an attempt to demonstrate proof of mechanism or proof of principle in a small number of patients are increasingly popular, supported most commonly by biomarkers or less often by surrogate markers (simply as there are fewer of those well validated). Perhaps an overemphasis on the phase Ib aspect of a trial could become a source of attrition in itself—the purpose of phase I is to investigate clinical safety and set doses for phase II. Without a firm foundation at this stage, phase II can easily be compromised.

Adaptive designs for clinical trials (phase I, but possibly also phase II) where the dose selection and escalation are not fixed at the start of the trial but are modified during the trials in response to the results at the earlier stages (sometimes using Bayesian statistical methods) can be economical on subjects and drugs. However, such trials may be more complex and lengthy to conduct—there might be practical issues in the preparation of dose sizes, for example, or the rotation of subjects in the clinical pharmacology units. Specialist CROs and consultancies are experienced in these issues, so further progress can be expected.

Clinical trial simulation [30] is a powerful tool in the design of phase II trials—arguably the stage of clinical development responsible for most attrition. Computationally intensive stochastic simulations are now done relatively easily, so that the predictive power of different trial designs can be estimated before the trial design is finalized. For example, with a set budget for a trial, the number of subjects split between a number of doses or groups could be varied in the simulations. The signal to noise of a biomarker might be examined to assess its value in the trial, with the level of powering or measurement accuracy and precision available.

1.4   PHASE III ATTRITION

But what about failure in phase III? Historically, greater than 66% of phase III projects would be expected to reach the market. With the potentially large numbers of patients, and possibly long trials involved, failures here can be financially disastrous. While not all phase III trials are huge (patients can be around 100 per group in some indications), the commercial value of a company is based on the strength of its phase III pipeline. To a large pharmaceutical company, phase III failure can result in major share price fluctuations, and to small biotechs, it can be catastrophic. In 2012, the failure of Abbott's bardoxolone partnered with Reata wiped 3.5% off its share price in one day [19]. In 2011, AntiSoma closed in dramatic fashion after the failure of its phase III program for AS1413 and discontinuation of its phase IIb program for A1411 [20]. In 2008, it had already sold off its FDA-approved fludarabine to back its own development portfolio, with the loss of AS1413 there was little value left in the company.

So why do drugs fail in phase III, when efficacy failures appear to have been weeded out at such expense in phase II? In 2013, Eli Lilly's ramucirumab failed to meet its primary endpoint on progression-free survival among women with metastatic breast cancer (although it was successful in its phase III trial in advanced gastric cancer) [21]. Eli Lilly also stopped enzastaurin, a kinase inhibitor that failed to meet the main goal for boosting disease-free survival in a phase III study in patients with diffuse large B-cell lymphoma [22]. AstraZeneca's fostamatinib, an SYK kinase inhibitor, was stopped after 2 phase III trials as results did not "measure up to the promising results we saw earlier in development" [23]. GlaxoSmithKline and Prosensa announced that phase III clinical study of drisapersen, an investigational antisense oligonucleotide, for the treatment of Duchenne muscular dystrophy (DMD) patients with an amenable mutation, did not meet the primary endpoint of a statistically significant improvement in the six minute walking distance (6MWD) test compared to placebo [24]. Roche's PPARα/γ agonist aleglitazar was halted prior to the completion of its phase III program due to safety signals and lack of efficacy [25]. Roche was working in a high-risk area that has seen the failure of more than 50 other PPAR agonists in clinical development.

All of these drugs had positive clinical signals in phase II patient studies, which did not translate into phase III success. We appear to be failing on efficacy badly in both phase II and phase III. An analysis for 2011–2012 phase III failures found 56% of them failed for efficacy reasons (59/105 failures which reported reason for failure) [26], and most of these failed to demonstrate efficacy versus placebo. The mechanistic hypothesis was supposed to have been tested in phase II trials, and attrition in this phase was already the highest of any phase, at 66%. These phase II trials are conducted in 10s to 100s of patients and designed to give statistically significant, clinically meaningful indication of efficacy, and differentiation where there are preexisting therapies, on which to base decisions on the huge investments required in the phase III trials.

Phase III are the pivotal efficacy trials, which will be used to make the registerable claims for the drug. Phase III endpoints may be different from phase II trials. Phase II endpoints may be surrogate endpoints, biomarkers of clinical efficacy, or recognized endpoints that are thought to be indicative of clinically meaningful benefit such as blood pressure, cholesterol levels, bone density, or composite endpoints scoring systems, for example, ACR20, ACQ, and SLEDII, but phase III endpoints will generally be primary endpoints. They will be endpoints that directly measure how a patient feels, benefits, or survives [27]. Drugs granted accelerated approval may be registerable based on surrogate endpoints for life-threatening diseases with no treatment option. The lack of translation of positive phase II results into phase III trials may be the failure of the surrogates to translate to patient benefit, or a problem of sample size, or a problem of control over studies executed necessarily in many multiple centers across many countries. Phase III expectations from regulators are becoming more demanding, and not necessarily consistent across jurisdictions, and may also change during the conduct of phase III trials, and be applied retrospectively to the outcome. While Pfizer's JAK inhibitor tofacitinib met its primary phase III endpoints, and was approved by the FDA,[1] the EMA has so far refused registration [28]. The EMA Committee for Medicinal Products for Human Use had major concerns about the overall safety profile of tofacitinib relative to its efficacy, and while acknowledging tofacitinib resulted in a reduction in disease activity and physical function of patients, there was no consistent reduction in structural joint damage in the target patient population, who had failed at least two other disease-modifying antirheumatic drugs [29].

1.4.1 Safety Attrition in Phase III

In 2011–2012, 28% of phase III programs were stopped due to clinical safety/safety margin issues. Even after the extensive safety evaluation that drugs have undertaken before the final phase III trials, toxicity is still a source of late-stage attrition. For example, Takeda recently announced the termination of its phase III program for fasiglifam (TAK-875) due to concerns over liver safety [31]. Abbott and Reata's NRF2 activator bardoxolone trial in stage 3/4 chronic renal disease patients was stopped in 2012 due to "excess mortality" in the dosed groups [32]. Merck's with-drawn cholesterol drug Tredaptive, a combination of niacin and an experimental drug laropiprant, showed that one-quarter of patients in a new trial withdrew because of side effects including itching, rashes, and muscle problems. Bristol-Myers Squibb halted development of the hepatitis C nucleotide polymerase inhibitor BMS-986094 after 9 trial patients were hospitalized and one trial participant died of heart failure following drug administration [33].

Comprehensive safety packages are designed around our clinical programs to avoid harm to patients in clinical trials. *In vitro* and *in vivo* safety studies are valuable, have undoubtedly contributed to the avoidance of safety catastrophes, and have a critical place in our development framework. But rare or infrequent events are statistically unlikely to show up in any study in small numbers of patients, and it is only in late-stage studies, or even postmarketing where large numbers of patients are treated, that these events become significant. A safety finding of a single drug molecule in a unique mechanism may never be fully explainable, particularly if the finding is serious as further human studies would not be supported. But when multiple drug molecules exhibit common toxicologies, common patterns may be observed upon which hypotheses can be drawn and investi-gated. The identification that rare potentially fatal torsades de pointes were related to the use of certain marketed antihistamine drugs and that these drugs blocked the hERG cardiac ion channel enabled a life-threatening toxicology to be reduced to selective phar-macology [34]. We were then able to screen for hERG liabilities very early in the drug discovery. On a more positive note, observations of side effects of drugs are a common feature of new medical innovation. Viagra was originally trialed as an antiangina treatment, before its true value arose. The recently registered Tecfidera is rapidly becoming a blockbuster treatment for multiple sclerosis and is in fact a formulation of dimethyl fumarate, which had been used for many years as an antipsoriatic treatment. Drug repro-filing is becoming big business.

---

[1] http://www.fda.gov/NewsEvents/Newsroom/PressAnnouncements/ucm327152.htm

There are three particularly important points made in the previous paragraph that are not often stated overtly:

1. Safety issues, once you get past the obvious ones that appear in preclinical studies, are nearly impossible to predict with any useful degree of accuracy.
2. Basic mechanisms of an adverse event sometimes require not only many thousands of patients to be exposed but also that more than one drug in the class be developed such that a specific common mechanism can be identified or at least a "class effect" can be postulated.
3. An enormous amount of resource must be spent on characterizing a drug's therapeutic and side effects before other possible uses can be identified.

In the context of safety-related drug attrition, let us look at these three aspects in a little more detail.

1.4.1.1 Safety Issues are Nearly Impossible to Predict with any Useful Accuracy With this statement, one's mind automatically goes to the ability of preclinical species to reflect side effects in humans, which by some estimations is quite reasonable. Therefore, it should be reasonably easy to at least perform an adequate risk assessment for humans from preclinical species. This may be expecting too much and may in fact represent an experiment that is impossible to actually perform correctly. Expecting a readout in a preclinical species (or several) to translate to humans at the right exposure for the right duration in a vastly differing phenotypic background is asking quite a bit of studies consisting of the minimum number of fairly homogeneous animals as possible. But another aspect of our expectations of preclinical findings translating to the clinical setting is that the data used to determine whether preclinical studies actually do predict clinical outcomes are skewed by the fact that many drug candidates are abandoned after findings in preclinical species are deemed "unmanageable" or "unmonitorable" in the clinic and thus never make it to humans to test whether this relationship holds or not. While this is, in most cases, prudent, it must be acknowledged that our knowledge of how these examples would actually perform in the human population is poor. In some cases, judging a finding to be unmanageable or unmonitorable in the clinical setting is down to technical reasons, for example, no biomarker or imaging technique is available. However, in some cases, it comes down to a templated approach to clinical development in many organizations that does not accommodate research into side effects, either for resource reasons or for lack of early clinical investigators who are interested in the relatively unglamorous and complicated world of side effects. The consequence of this situation is that safety-related attrition in human trials will continue because the preclinical safety assessment of drugs is an oversimplification of the real human response to a drug and that some very useful therapies will be abandoned because we cannot adequately risk-assess their effects in humans due to an underdeveloped approach to research into human side effects.

1.4.1.2 Basic Mechanisms of an Adverse Event Sometimes Require not only Many Thousands of Patients to be Exposed, but also that more than One Drug in the Class be Developed such that a Specific Common Mechanism can be Identified or at least a 'Class Effect' can be Postulated With this statement, one must acknowledge the complexity of biological systems in general and these same systems under the influence of a pharmacological intervention in particular. The basic ambition of most therapies under development is that they are specific in two ways: first, that the intervention (small molecule, antibody, etc.) is specific to its target and, second, that the role of the target in the given disease is specific to the hypothesized disease mechanism. In the vast majority of cases, neither of these conditions is fulfilled and one is left managing the cornucopia of effects and carefully recording observable changes in the patient and, as discussed earlier, usually neglecting to thoroughly explain the mechanisms of these effects. However, as several projects attempt to target a specific biological mechanism and clinical (or preclinical) safety observations are accumulated, patterns oftentimes do emerge that can give indications whether side effects are related to the specific intervention or are a result of alterations in the target biology brought about by the intervention. This important learning exercise involving many clinical trials performed with many drug candidates in many patients becomes extremely expensive but is the only way to separate effects specific to a particular drug candidate and effects related to the alterations in the biological makeup of the patient. This situation, after we acknowledge it, may help to set expectations for future drug discovery/development projects. We have to resign ourselves to working with a "black box" and that understanding mechanisms will not happen until massive amounts of data from several attempts at therapies against a specific target are made. While this may be a rather pessimistic approach, it also points to the need to strengthen the research environment within regulatory bodies who will in the end be the only group to have a complete overview of all of the positive and negative effects of a group of drug candidates. One can speculate that a more open, balanced reporting environment may have spotted the relationship between hERG inhibition and torsade de points earlier, as this was neither a protein target class (on pharmacology) nor a chemical structural class safety problem, but rather a shared off-target pharmacology of diverse structural classes driven largely by bulk physical properties.

1.4.1.3 *An Enormous Amount of Resource must be Spent on Characterizing a Drug's Therapeutic and Side Effects before other Possible Uses can be Identified* The discovery and development of a safe and pharmacologically active substance is unquestionably a challenge. Once the drug candidate is given to human subjects, opportunities begin to appear, but without the previously mentioned "research-minded" clinical development, many projects are simply abandoned when they either show unacceptable side effects at the "therapeutic" dose for which they are being developed or simply show too little efficacy in the patient population chosen. At this point, the project is shelved and often quickly forgotten as other priorities take over. With a success rate in clinical development of less than 5%, one cannot help thinking that this "lost 95%" is a neglected resource. Of course, the issue is that it is essentially invisible, including the safety data, from which trends, patterns, and sometimes mechanisms could be derived. There is no current solution to this situation, partially because of the confidentiality (both patient and commercial), but perhaps more importantly because of the internal resources required to find, summarize, anonymize, and analyze the data. Hence, many opportunities are probably being lost to cursory data analysis and the inability to access these data for further analysis and merging with other relevant datasets. It is often said that many drugs have been discovered by serendipity. Serendipity can only occur when the right eyes view the right data. At the moment, this potential is severely handicapped by shelving this lost 95%.

The assessment of significance of an observed safety concern and the assessment of the relative safety risk patients are exposed to by a new treatment relative to the efficacy benefit a patient could potentially gain are the core discussions between pharmaceutical companies and regulatory bodies. When all the trials are completed, these discussions can lead to attrition, market opportunity, or delay. What is acceptable in one disease or patient population may not be acceptable in another and likewise with different regulatory authorities.

1.5 REGULATION AND ATTRITION

Earlier in 2013, the FDA rejected Novo Nordisk's Tresiba (long-acting insulin degludec), against the advice of its own advisory panel, as it asked for a further dedicated cardiovascular outcome trial to investigate potential cardiovascular risks associated with the treatment. A requirement for a cardiovascular safety study after the completion of phase III can torpedo many projects outright. The FDA as the tollgate to the largest pharmaceutical market in the world has been, and continues to be, a major source of controversy. The FDA is often accused of both hindering access of patients to potential lifesaving therapies and, at the same time, allowing harmful drugs to reach patients. The history of the FDA highlights that its regulatory framework was built on heart-breaking real-world experience of the harmful effects of inappropriate or unregulated drug use [35]. The 1912 Sherley Amendment to the Drugs Act prohibited not just false labeling of ingredients but also false claims and was in part a response to the widespread sale of dubious and outright dangerous tinctures, ointments, and treatments, like Mrs. Winslow's Soothing Syrup for teething and colicky babies, unlabeled yet laced with morphine, and thought to have been the cause of many infant deaths. The Federal Food, Drug, and Cosmetic (FDC) Act of 1938 contained many new provisions, not least in response to the marketing of Elixir of Sulfanilamide, which contained the poisonous solvent diethylene glycol, which it is claimed killed 107 persons, many of whom were children. In 1962, the Kefauver-Harris Drug Amendments were passed, in response to the public outcry caused by birth defects observed in many countries due to use of thalidomide, even though it was never approved for use in the United States. For the first time, drug manufacturers were required to prove to the FDA the effectiveness of their products before marketing them as well as safety. The FDA's mission then, and today, is to protect public health, and in increasing its vigilance, it protects patients from treatments whose risks outweigh their benefits but in doing so also delays useful medicines from reaching patients rapidly. Prior to 1962, the approval process in the United States on average took just 7 months but by 1998, with the impact of the Kefauver-Harris amendments, took almost 7 years [36].

During the HIV epidemic of the 1980s, organizations such as ACT-UP accused the FDA of unnecessarily delaying the registration of new HIV medications. In 1990, the then chairman of the presidential advisory panel on drug approval, Louis Lasagna, estimated that thousands of lives were lost each year due to delays in approval and marketing of drugs for cancer and AIDS. Partly in response to these criticisms, the FDA introduced expedited approval of drugs for life-threatening diseases and expanded preapproval access to drugs for patients with limited treatment options.

In 1992, the FDA instituted the *accelerated approval* regulations. When studying a new drug, it can sometimes take many years to learn whether a drug actually provides a real effect on how a patient survives, feels, or functions. Mindful of the fact that it may take an extended period of time to measure a drug's intended clinical benefit, the regulations allowed drugs for serious conditions that filled an unmet medical need to be approved based on a surrogate endpoint. Use of a surrogate endpoint enabled the FDA to approve these drugs faster. A surrogate endpoint is a measure that is thought to predict clinical benefit, but is not itself a measure of clinical benefit, such as blood pressure or cholesterol levels in cardiovascular disease.

Accelerated approval does carry an obligation that the sponsoring company continues postmarketing clinical studies to show clinical effectiveness of the treatment, which the surrogate endpoint was predictive of clinical effectiveness. Genentech's Avastin, a monoclonal antibody that inhibits angiogenesis by blocking vascular endothelial growth factor A, was approved for a breast cancer indication in February 2008 under the FDA accelerated approval process. But in 2011, on the basis of two additional clinical studies that showed only a small effect on tumor growth without evidence of an impact on mortality or improved quality of life, the FDA withdrew marketing authorization for this indication [37]. The drug remains on the market in the United States for other cancer indications and remains approved for breast cancer in other countries.

The FDA continues to evolve its processes to ensure beneficial drugs reach patients as rapidly as possible. In 2012, congress passed an amendment that allowed the FDA to base accelerated approval for drugs on a surrogate or an intermediate clinical endpoint. An intermediate clinical endpoint is an endpoint that is reasonably likely to predict clinical benefit (based on available data), even if it has not achieved such widespread acceptance, but following registration the company is required to conduct trials to demonstrate the clinical benefit, if it cannot be subsequently demonstrated, or if the company does not show due diligence in conducting such a trial, the drug may be removed from the market. The FDA recently approved a multiple sclerosis treatment on the basis of a large clinical effect on relapse rates after 13 months but with uncertainty of the durability of the effect. The sponsor was required to continue the trials to show durability of the clinical effect for a further 2 years [38]. In the oncology field [38], objective response rate was historically acceptable endpoint for drug registration, but in the 1980s, there was a move toward clinical endpoints of survival and patient reported quality of life outcomes. For accelerated approval, the FDA will accept progression-free survival, objective response rate, and complete response.

The year 2013 saw the introduction of a new breakthrough review status for drugs and biologics for serious life-threatening diseases, which brings the number of expedited programs to four: fast-track, breakthrough therapy, accelerated approval, and priority review. A drug may be accepted under one or more of these programs.

Fast-track review is granted upon request to facilitate development and speed review of compounds for life-threatening diseases. A fast-track designation allows the drugs to receive some or all of the following:

- More frequent meetings with the FDA to discuss the drug's development plan and ensure collection of appropriate data needed to support drug approval
- More frequent written correspondence from the FDA about such things as the design of the proposed clinical trials and use of biomarkers
- Eligibility for *accelerated approval and priority review, if relevant criteria are met*

- *Rolling review*, which means that a drug company can submit completed sections of its biologics license application (BLA) or new drug application (NDA) for review by the FDA, rather than waiting until every section of the application is completed before the entire application can be reviewed (BLA or NDA review usually does not begin until the drug company has submitted the entire application to the FDA)

Breakthrough status is granted upon request for drugs for life-threatening diseases where preliminary clinical evidence indicates that the drug may demonstrate substantial improvement over available therapy on a clinically significant endpoint and includes the benefits of fast-track review and in addition intensive guidance on an efficient drug development program, beginning as early as phase I, as well as organizational commitment involving senior managers.

Under the Prescription Drug User Fee Act (PDUFA), the FDA agreed to specific goals for improving the drug review time and created a two-tiered system of review times: *standard review* and *priority review*. A *priority review* designation means FDA's goal is to take action on an application within 6 months (compared to 10 months under standard review).

A *priority review* designation will direct overall attention and resources to the evaluation of applications for drugs that, if approved, would be significant improvements in the safety or effectiveness of the treatment, diagnosis, or prevention of serious conditions, documented evidence of patient compliance expected to lead to improvement in serious outcome or evidence of safety and effectiveness in a new subpopulation, when compared to standard applications.

One certainly feels that the FDA places science at the center of its decision making. In many areas, the required science does not exist, and the FDA has been a leading force in investment in the necessary research, for example, in the areas of pharmacovigilance and pharmacoepidemiology, clinical trial design, and integration of population pharmacokinetics, pharmacodynamics, and toxicokinetics in drug development. The regulatory bodies' role is to ensure patients benefit from treatments with an acceptable level of risk. They are independent and their decisions are driven by data, they are not in themselves sources of attrition, but they do implement it through their decisions both in premarketing and in postmarketing.

1.6 ATTRITION IN PHASE IV

Even after all development hurdles have been passed, and the drug reaches the market, the problems may not be over. The pharmaceutical company still has to be able to sell the drug and make a profit to return to shareholders. In 2011, Savient Pharma obtained marketing approval for its gout treatment Krystexxa (pegloticase). Savient was reputedly for sale after marketing approval was gained from the FDA, but due to a lack of suitors, the company took the drug to market itself. Sales remained stubbornly low, and with production problems, the challenge to increase sales to cover its rising debts proved too great and Savient was forced to file for bankruptcy in 2013. Too many marketed drugs fail to recoup their developments costs, but clearly the worst outcome of all is the withdrawal of a drug when it is on the market. This can have serious consequences for the manufacturer but also for patients—both for those who might experience adverse effects and for other patients (the vast majority) who lose the benefit of an efficacious medicine that they might have tolerated well (Table 1.1).

TABLE 1.1 Drug withdrawals By Year

Drug	Year	Country	Indication	Reason
Drotrecogin alfa (Xigris)	2011	Worldwide	Severe sepsis	Lack of efficacy as shown by PROWESS-SHOCK study
Propoxyphene (Darvocet/ Darvon)	2010	Worldwide	Mild to moderate pain	Increased risk of heart attacks and stroke
Gemtuzumab ozogamicin (Mylotarg)	2010	United States	Acute myelogenous leukemia	No improvement in clinical benefit; risk for death
Ozogamicin	2010	United States		No improvement in clinical benefit; risk for death; veno-occlusive disease
Sibutramine (Reductil/ Meridia)	2010	Australia, Canada, China, the European Union (EU), Hong Kong, India, Mexico, New Zealand, the Philippines, Thailand, the United Kingdom, and the United States	Weight loss	Increased risk of heart attack and stroke
Sitaxentan	2010	Germany	Pulmonary arterial hypertension	Hepatotoxicity
Efalizumab (Raptiva)	2009	Germany	Psoriasis	Withdrawn because of increased risk of progressive multifocal leukoencephalopathy
Aprotinin (Trasylol)	2008	United States	Inhibit bleeding in complex surgery	Increased risk of death
Rimonabant (Acomplia)	2008	Worldwide	Weight loss	Risk of severe depression and suicide
Lumiracoxib (Prexige)	2007–2008	Worldwide	Pain	Liver damage
Clobutinol	2007	Germany	Cough suppressant	Ventricular arrhythmia, QT-prolongation
Nefazodone	2007	United States, Canada, others	Antidepressant	Branded version withdrawn by originator in several countries in 2007 for hepatotoxicity. Generic

Drug (brand)	Country	Year	Indication	Concern
Pergolide (Permax)	United States	2007	Parkinson's disease	Risk for heart valve damage
Tegaserod (Zelnorm)	United States	2007	Irritable bowel syndrome	Risk for heart attack, stroke, and unstable angina. Ref. 2 was available through a restricted access program until April 2008
Alatrofloxacin	United States	2006	Broad-spectrum antibiotic	Liver toxicity; serious liver injury leading to liver transplant; death
Gatifloxacin	United States	2006	Respiratory tract infections	Increased risk of dysglycemia
Ximelagatran (Exanta)	Germany	2006	Anticoagulant	Hepatotoxicity
Natalizumab (Tysabri)	United States	2005–2006	multiple sclerosis and Crohn's disease	Voluntarily withdrawn from US market because of risk of Progressive multifocal leukoencephalopathy (PML). Returned to market in July 2006
Adderall XR	Canada	2005	Attention deficit hyperactivity disorder	Risk of stroke. The ban was later lifted because the death rate among those taking Adderall XR was determined to be no greater than those not taking Adderall
Hydromorphone (Palladone]]		2005	Opioid analgesic	High risk of accidental overdose when administered with alcohol
Thioridazine (Melleril)	Germany, United Kingdom	2005	Schizophrenia and psychosis	Withdrawn from UK market because of cardiotoxicity
Bezitramide	Netherlands	2004	Analgesic	Fatal overdose
Co-proxamol (Distalgesic)	United Kingdom	2004	Back pain	Overdose dangers
Dofetilide	Germany	2004	Maintenance of sinus rhythm	Drug interactions, prolonged QT
Rofecoxib (Vioxx)	United States	2004	Pain in osteoarthritis and rheumatoid arthritis	Risk of myocardial infarction and stroke
Valdecoxib (Bextra)	United States	2004	Pain in osteoarthritis and rheumatoid arthritis	Risk of heart attack and stroke

(Continued)

TABLE 1.1 (Cont'd)

Drug	Year	Country	Indication	Reason
Levomethadyl acetate	2003	United States	Opioid dependence	Cardiac arrhythmias and cardiac arrest
Kava Kava	2002	Germany	Social anxiety	Hepatotoxicity
Ardeparin (Normiflo)	2001	United States	Deep vein thrombosis	Not for reasons of safety or efficacy
Cerivastatin (Baycol, Lipobay)	2001	United States	Cardiovascular disease	Risk of rhabdomyolysis
Rapacuronium (Raplon)	2001	US multiple markets	Anesthesia	Withdrawn in many countries because of risk of fatal bronchospasm
Sparfloxacin	2001	United States	Antibacterial	QT prolongation and phototoxicity
Alosetron (Lotronex)	2000	United States	Irritable bowel syndrome	Serious gastrointestinal adverse events; ischemic colitis; severe constipation. Reintroduced 2002 on a restricted basis
Cisapride (Propulsid)	2000	United States	Gastrointestinal dysmotility	Risk of fatal cardiac arrhythmias
Phenylpropanolamine (Propagest, Dexatrim)	2000	Canada, United States	Decongestant	Hemorrhagic stroke
Troglitazone (Rezulin)	2000	United States, Germany	Antidiabetic	Hepatotoxicity
Trovafloxacin (Trovan)	1999–2001	European Union, United States	Broad-spectrum antibiotic	Withdrawn because of risk of liver failure
Amineptine (Survector)	1999	France, United States	Antidepressant	Hepatotoxicity, dermatological side effects, and abuse potential
Aminopyrine	1999	France, Thailand	Anti-inflammatory	Abuse; dependence; severe acne
Astemizole (Hismanal)	1999	United States, Malaysia, Multiple Nonspecified Markets	Antihistamine	Fatal arrhythmia
Grepafloxacin (Raxar)	1999	Withdrawn Germany, United Kingdom, United States, others	Broad-spectrum antibiotic	Cardiac repolarization; QT interval prolongation
Levamisole (Ergamisol)	1999		Antihelminthic	Still used as veterinary drug; in humans was used to treat melanoma before it was withdrawn for agranulocytosis

Drug	Year	Country	Use	Reason
Temazepam (Restoril, Euhypnos, Normison, Remestan, Tenox, Norkotral)	1999	Sweden, Norway	Insomnia	Diversion, abuse, and a relatively high rate of overdose deaths in comparison to other drugs of its group. This drug continues to be available in most of the world including the United States, but under strict controls
Bromfenac	1998	United States	Ocular inflammation and pain	Severe hepatitis and liver failure (some requiring transplantation)
Ebrotidine	1998	Spain	Antiulcer	Hepatotoxicity
Mibefradil	1998	European Union, Malaysia, United States, others	Hypertension	Fatal arrhythmia, drug interactions
Mibefradil (Posicor)	1998		Hypertension	Withdrawn because of dangerous interactions with other drugs
Proxibarbal	1998	Spain, France, Italy, Portugal, Turkey	Antianxiety	Immunoallergic, thrombocytopenia
Sertindole	1998	European Union	Antipsychotic	Arrhythmia and sudden cardiac death
Tolcapone (Tasmar)	1998	European Union, Canada, Australia	Parkinson's disease	Hepatotoxicity
Terfenadine (Seldane, Triludan)	1997–1998	France, South Africa, Oman, others, United States	Allergy	Prolonged QT interval; ventricular tachycardia
Dexfenfluramine	1997	European Union, United Kingdom, United States	Appetite suppressant	Cardiac valvular disease
Fen-phen (popular combination of fenfluramine and phentermine)	1997	European Union, United Kingdom, United States	Appetite suppressant	Cardiotoxicity
Pemoline (Cylert)	1997	Canada, United Kingdom	Attention deficit hyperactivity disorder	Withdrawn from United States in 2005. Hepatotoxicity Reason: hepatotoxicity
Phenolphthalein	1997	United States	Laxative	Carcinogenicity

(Continued)

TABLE 1.1 (*Cont'd*)

Drug	Year	Country	Indication	Reason
Chlormezanone (Trancopal)	1996	European Union, United States, South Africa, Japan	Anxiolytic and muscle relaxant	Hepatotoxicity; Stevens–Johnson Syndrome; Toxic Epidermal Necrolysis
Minaprine	1996	France	Antidepressant	Convulsions
Tolrestat (Alredase)	1996	Argentina, Canada, Italy, others	Diabetic complications	Severe hepatotoxicity
Alpidem (Ananxyl)	1995	Worldwide	Anxiolytic	Not approved in the United States, withdrawn in France in 1994 and the rest of the market in 1995 because of rare but serious hepatotoxicity
Bendazac	1993	Spain	Anti-inflammatory	Hepatotoxicity
Flosequinan (Manoplax)	1993	United Kingdom, United States, others	Congestive heart failure	Increased mortality at higher doses; increased hospitalizations
Ketorolac	1993	France, Germany, others	Anti-inflammatory analgesic	Hemorrhage, renal failure
Moxisylyte	1993	France	Benign prostatic hypertrophy	Necrotic hepatitis
Remoxipride	1993	United Kingdom, others	Antipsychotic	Aplastic anemia
Sorivudine	1993	Japan	Antiviral	Drug interaction and deaths. [citation needed]
Thiobutabarbitone	1993	Germany	Anesthetic	Renal insufficiency
Benzarone	1992	Germany	Peripheral venous disorders	Hepatitis
Temafloxacin	1992	Germany, United Kingdom, United States, others	Antibiotic	Low blood sugar; hemolytic anemia; kidney, liver dysfunction; allergic reactions
Temafloxacin	1992	United States	Antibiotic	Allergic reactions and cases of hemolytic anemia, leading to three patient deaths

Drug	Year	Countries	Indication	Adverse effect
Encainide	1991	United Kingdom, United States	Antiarrhythmic	Ventricular arrhythmias
Fipexide	1991	France	Senile dementia	Hepatotoxicity
Flunitrazepam	1991	France	Hypnotic	Abuse
Terodiline (Micturin)	1991	Germany, United Kingdom, Spain, others	Antispasmodic in urology	Prolonged QT interval, ventricular tachycardia and arrhythmia
Triazolam	1991	France, Netherlands, Finland, Argentina, United Kingdom, others	Severe insomnia	Psychiatric adverse drug reactions, amnesia
Dilevalol	1990	United Kingdom	Hypertension	Hepatotoxicity
Dinoprostone	1990	United Kingdom	Induction labor	Uterine hypotonus, fetal distress
Fenoterol	1990	New Zealand	Bronchodilator	Asthma mortality
Metipranolol	1990	United Kingdom, others	Glaucoma	Uveitis
Pirprofen	1990	France, Germany, Spain	Anti-inflammatory	Gastrointestinal toxicity
Broazolam	1989	United Kingdom	Hypnotic	Animal carcinogenicity
Etretinate	1989	France	Severe psoriasis	Withdrawn United States (1999). Risk for birth defects
Exifone	1989	France	Senile dementia	Hepatotoxicity
L-Tryptophan	1989	Germany, United Kingdom	Cognitive disorders	Eosinophilic myalgia syndrome
Proglumide	1989	Germany	Antiulcer	Respiratory reaction
Prenylamine	1988	Canada, France, Germany, United Kingdom, United States, others	Angina pectoris	Cardiac arrhythmia and death
a	1988	Germany	Anti-infective	Dermatologic, hematologic and hepatic reactions
Sulfamethoxydiazine	1988	Germany	Anti-infective	Unknown
Suprofen	1986–1987	United Kingdom, Spain, United States	Anti-inflammatory/ mitosis	Flank pain, decreased kidney function
Nikethamide	1988	multiple markets	Stimulant/tranquilizer overdose	CNS Stimulation
Cinepazide	1987	Spain	Vasodilator	Agranulocytosis

(Continued)

TABLE 1.1 *(Cont'd)*

Drug	Year	Country	Indication	Reason
Clometacin	1987	France	Anti-inflammatory	Hepatotoxicity
Cyclofenil	1987	France	Scleroderma/Raynaud's disease	Hepatotoxicity
Muzolimine	1987	France, Germany, European Union	Hypertension	Polyneuropathy
Vincamine	1987	Germany	Vasodilator/nootropic	Hematologic toxicity
Beta-ethoxy-lacetanilanide	1986	Germany		Renal toxicity, animal carcinogenicity
Bucetin	1986	Germany	Pain, fever	Renal toxicity
Canrenone	1986	Germany	Diuretic	Animal carcinogenicity
Difemerine	1986	Germany	Antispasmodic	Multiorgan toxicities
Sulfamethoxypyridazine	1986	United Kingdom	Antibacterial	Dermatologic and hematologic reactions
Cianidanol	1985	France, Germany, Spain, Sweden	Liver disorders	Hemolytic anemia
Indalpine	1985	France	Depression	Agranulocytosis
Perhexilene	1985	United Kingdom, Spain	Angina pectoris	Neurologic and hepatic toxicity
Phenylbutazone	1985	Germany	Pain, fever	Off-label abuse, hematologic toxicity
Suloctidyl	1985	Germany, France, Spain	Dementia thrombotic disorders	Hepatotoxicity
Oxyphenbutazone	1984–1985	United Kingdom, United States, Germany, France, Canada	Pain, fever	Bone marrow suppression, Stevens–Johnson syndrome
Nitrefazole	1984	Germany	Alcohol deterrent	Hepatic and hematologic toxicity
Althesin (=Alphaxolone amineptine + Alphadolone)	1984	France, Germany, United Kingdom	Anesthetic induction	Anaphylaxis
Antrafenine	1984	France	Pain, fever	Unspecific experimental toxicity
Fenclofenac	1984	United Kingdom	Rheumatoid arthritis	Cutaneous reactions; animal carcinogenicity
Feprazone	1984	Germany, United Kingdom	Pain	Cutaneous reaction, multiorgan toxicity
Glafenine	1984	France, Germany	Pain, fever	Anaphylaxis

Drug	Year	Country/Region	Indication	Reason/Comments
Isaxonine phosphate	1984	France	Lesions on peripheral nerves	Hepatotoxicity
Methaqualone	1984	South Africa (1971), India (1984), United Nations (1971–1988)	Insomnia	Withdrawn because of risk of addiction and overdose
Dimethylamylamine (DMAA)	1983	United States	Nasal decongestant dietary supplement	Voluntarily withdrawn from market by Lily. Reintroduced as a dietary supplement in 2006;[13]:13 and in 2013 the FDA started work to ban it due to cardiovascular problems [14]
Indoprofen	1983	Germany, Spain, United Kingdom	Pain, fever	Animal carcinogenicity, gastrointestinal toxicity
Isoxicam	1983	France, Germany, Spain, others	Pain, fever	Stevens–Johnson syndrome
Propanidid	1983	United Kingdom	Short-acting anesthetic	Allergy
Zimelidine	1983	Worldwide	Depression	Risk of Guillain–Barré syndrome, hypersensitivity reaction, hepatotoxicity [3, 45, 46] banned worldwide
Zomepirac	1983	United Kingdom, Germany, Spain, United States	Pain, fever	Anaphylactic reactions and nonfatal allergic reactions, renal failure
Benoxaprofen	1982	Germany, Spain, United Kingdom, United States	Pain, fever	Liver and kidney failure; gastrointestinal bleeding; ulcers
Clomacron	1982	United Kingdom		Hepatotoxicity
Methandrostenolone	1982	France, Germany, United Kingdom, United States, others	Female tonic	Off-label abuse
Pentylenetetrazol	1982		Convulsant therapy	Withdrawn for inability to produce effective convulsive therapy, and for causing seizures

(Continued)

TABLE 1.1 *(Cont'd)*

Drug	Year	Country	Indication	Reason
Nomifensine	1981–1986	France, Germany, Spain, United Kingdom, United States, others	Depression	Hemolytic anemia, hepatotoxicity, serious hypersensitive reactions
Amobarbital	1980	Norway	Hypnotic	Self-poisoning
Cyclobarbital	1980	Norway	Insomnia	Self-poisoning
Pentobarbital	1980	Norway	Sedative hypnotic	Self-poisoning
Ticrynafen (Tienilic acid)	1980	Germany, France, United Kingdom, United States, others	Hypertension	Liver toxicity and death
Alclofenac	1979	United Kingdom	Pain, fever	Vasculitis, rash
Methapyrilene	1979	Germany, United Kingdom, United States	Allergy insomnia	Animal carcinogenicity
Pyrovalerone	1979	France	Lethargy	Abuse
Buformin	1978	Germany	Diabetes	Metabolic toxicity
Phenformin and Buformin	1977	France, Germany, United States	Diabetes	Severe lactic acidosis
Azaribine	1976	United States	Psoriasis	Thromboembolism
Oxeladin	1976	Canada, United Kingdom, United States (1976)	Antitussive	Carcinogenicity
Pifoxime (=Pixifenide)	1976	France	Anti-inflammatory	Neuropsychiatric reaction
Dipyrone(Metamizole)	1975	United Kingdom, United States, others	Analgesic, antispasmodic, and antipyretic	Agranulocytosis, anaphylactic reactions
Diethylstilbestrol	1970s		Gonorrheal vaginitis, atrophic vaginitis, menopausal symptoms, and postpartum lactation suppression	Risk of teratogenicity
Mebanazine	1975	United Kingdom	Antidepressant	Hepatotoxicity, drug interaction

Drug	Year	Canada	Category	Reason
Phenacetin	1975		Analgesic	An ingredient in "A.P.C." tablet; withdrawn because of risk of cancer and kidney disease [24] Germany Denmark, United Kingdom, United States, others Reason: nephropathy
Nialamide	1974	United Kingdom, United States	Antidepressant	Hepatotoxicity, drug interaction
Clioquinol	1973	France, Germany, United Kingdom, United States	Antifungal, antiprotozoal	Neurotoxicity
Dimazol (Diamthazole)	1972	France, United States	Antifungal	Neuropsychiatric reaction
Diacetoxydiphenolisatin	1971	Australia	Laxative	Hepatotoxicity
Triacetyldiphenolisatin	1971	Australia	Laxative	Hepatotoxicity
Amoproxan	1970	France	Antiarrhythmic, antianginal	Dermatologic and ophthalmic toxicity
Chlormadinone (Chlormenadione)	1970	United Kingdom, United States	Contraceptive	Animal carcinogenicity
Dihydrostreptomycin	1970	United States	Anti-infective	Neuropsychiatric reaction
Fenclozic acid	1970	United Kingdom, United States	Anti-inflammatory	Jaundice, elevated hepatic enzymes
Anagestone acetate	1969	Germany	Contraceptive	Animal carcinogenicity
Chlorphentermine	1969	Germany	Appetite suppressant	Cardiovascular toxicity
Cloforex	1969	Germany	Appetite suppressant	Cardiovascular toxicity
Ibufenac	1968	United Kingdom	Anti-inflammatory	Hepatotoxicity, jaundice
Bithionol	1967	United States	Antihelminthic	Dermatologic toxicity
Phenoxypropazine	1966	United Kingdom	Antidepressant	Hepatotoxicity, drug interaction
Metofoline	1965	United States	Analgesic	Unspecific experimental toxicity
Pronethalol	1965	United Kingdom	Hypertension	Animal carcinogenicity
Xenazoic acid	1965	France	Antiviral	Hepatotoxicity
Benziodarone	1964	France, United Kingdom	Hypertension	Jaundice
Butamben (Efocaine) (Butoforme)	1964	United States	Local anesthetic	Dermatologic toxicity; psychiatric reactions

(Continued)

TABLE 1.1 (*Cont'd*)

Drug	Year	Country	Indication	Reason
Dithiazanine iodide	1964	France, United States	Antihelminthic	Cardiovascular and metabolic reaction
Iodinated casein strophanthin	1964	United States	Appetite suppressant	Metabolic reaction
Iproniazid	1964	Canada rest of world	Antidepressant	Interactions with food products containing tyrosine
Bunamiodyl	1963	Canada, United Kingdom, United States	Contrast agent	Nephropathy
Ethyl carbamate	1963	Canada, United Kingdom, United States	Antineoplastic	Carcinogenicity
Triparanol	1962	France, United States	Cardiovascular disease	Cataracts, alopecia, ichthyosis
Thalidomide	1961	Germany	Sedative hypnotic antisickness	Withdrawn because of risk of teratogenicity; returned to market for use in leprosy and multiple myeloma under FDA orphan drug rules
Thenalidine	1960	Canada, United Kingdom, United States	Antipruritic	Neutropenia
Lysergic acid diethylamide (LSD)	1950s–1960s		Psychedelic therapy	Marketed as a psychiatric drug; withdrawn after it became widely used recreationally
Oxyphenisatin (Phenisatin)		Australia, France, Germany, United Kingdom, United States	Laxative	Hepatotoxicity
Secobarbital		France, Norway, others	Anesthetic, anticonvulsant, anxiolytic, sedative, and hypnotic	Self-poisoning

Adapted from http://en.wikipedia.org/wiki/List_of_withdrawn_drugs

In recent years, there have been several high-profile examples including Bayer's Baycol, GlaxoSmithKline's Avandia, and Merck's Vioxx Baycol (cerivastatin), a cholesterol-lowering statin that was approved in 1997 and voluntarily withdrawn by Bayer in 2001 after 31 deaths were reported due to severe rhabdomyolysis in patients taking the drug [39]. Vioxx, the cyclooxygenase 2 inhibitor, also approved by the FDA in 1999, was withdrawn by Merck in 2004, after concerns of increased risk of heart attack and stroke after long-term use. It was widely used to treat rheumatoid arthritis and other diseases involving chronic pain, and it was estimated it had been used in over 50 million patients before withdrawal but also associated with between 88,000 and 140,000 cases of severe heart disease [40].

As the safety–efficacy balance has become increasingly driven in the direction of safety, these cases will probably increase. Perhaps what is most worrying about some of these instances is that they have been driven not by overt observation of toxicity in controlled trials but by retrospective meta-analysis of trials outside of the formal regulatory process. In some cases, this had led to rounds of challenge and counterchallenge between regulatory agencies and their critics, as exemplified by the controversy surrounding GSK's diabetes drug Avandia. Rosiglitazone is the active ingredient in Avandia, and it is a PPAR agonist used in monotherapy and combination treatment for diabetes. Avandia could control a patient's blood sugar levels for longer than traditional oral antidiabetic drugs and therefore was an important option to help patients control their sugar levels. It was approved in United States in 1999 and achieved peak sales of over $2.5B per year by 2006. However, its label was amended in 2001, as concerns surfaced over hepatic risks and its cardiovascular safety when used in combination therapy with insulin compared to insulin therapy alone [41]. In 2001, along with the label change, the FDA also requested GSK embark on the 6-year "RECORD" study comparing cardiovascular outcomes of Avandia to other commonly prescribed antidiabetic drugs.

Meanwhile, in 2004, in a settlement after a lawsuit concerning the withholding of negative clinical trial data on another of GSK's drugs, Paxil, the company agreed to publish all its clinical trial data on its own website. However, this step caused GSK more controversy over Avandia. A meta-analysis published in *New England Journal of Medicine* (NEJM) in 2007 across 42 clinical trials published on GSK's own website linked the drug's use to an increased risk of heart attacks [42]. To counter the NEJM article, GSK published an interim analysis of the RECORD study and showed the cardiovascular risk of Avandia was not significantly different from the control groups in the key outcomes of hospitalization of death through cardiovascular outcomes. The FDA issued a prescriber safety update in 2007 [43] highlighting the contradictory evidence and later that year amended the box label for Avandia. The label stated:

"A meta-analysis of 42 clinical studies (mean duration 6 months; 14,237 total patients), most of which compared Avandia to placebo, showed Avandia to be associated with an increased risk of myocardial ischemic events such as angina or myocardial infarction. Three other studies (mean duration 41 months; 14,067 patients), comparing Avandia to some other approved oral antidiabetic agents or placebo, have not confirmed or excluded this risk. In their entirety, the available data on the risk of myocardial ischemia are inconclusive".

There were calls for the drug to be withdrawn, but in 2010, the FDA did not agree and it remained on sale in the United States, although in September of that year the EMA suspended marketing authorization of rosiglitazone-containing medications. In February 2011, the FDA recommended label changes that imposed severe restrictions on its use, limited prescribing to patients already taking rosiglitazone, to patients whose blood sugar could not be controlled with other antidiabetic medicines and who, after consulting with their healthcare professional, did not wish to use pioglitazone-containing medicines. The FDA also required a Risk Evaluation and Mitigation Strategy, which would severely limit the availability of rosiglitazone-containing medicines [44]. The FDA also called on GSK to convene a panel of independent scientists to readjudicate the results of the RECORD study, which compared the cardiovascular safety of AVANDIA to standard type II diabetes drugs. In 2010, from their peak in 2006, sales had crashed. In 2011, sales were just $205M, and as a final blow, its patent expired in 2012.

In 2013, the independent re-review of the results of the 2009 RECORD study confirmed the original conclusions and failed to show an increased risk of heart attack linked to the drug over standard of care diabetes drugs, and the FDA withdrew all restrictions on the use of Avandia [45]. This came perhaps somewhat late for GSK and many patients who might have benefitted from its use. The Avandia story certainly provides a cautionary tale, but it is common for high-profile drugs to be targets for controversy, and the debate over safety and efficacy can rage throughout the drug's patent life.

1.7 FIRST IN CLASS, BEST IN CLASS, AND THE ROLE OF THE PAYER

The pharmaceutical industry has been very successful in delivering valuable drugs that have changed the course of medical treatment. The introduction of antibiotics; cardiovascular drugs; steroids, both topical and inhaled; statins; anti-TNF biologics; antiulcer drugs; histamine antagonists; analgesics; antidepressants; immunosuppressants; and even contraceptives, to make an incomplete list, has changed the course of patients' lives. As a striking example, between 1995 and 1997, deaths from HIV/AIDS fell from 16.2/100,000 of the population to 6/100,000 of the population due to the widespread introduction of indinavir to HIV combination therapy, and with further drug introductions, by 2011, deaths had fallen even further to 2/100,000 of the population [46–48].

But drug companies only have the life of the intellectual property, and its regulatory market exclusivity, to reap a return on investment. After the expiry of these, generic competition reduces the ability of the innovating company to continue to make a financial return. A survey found that only 2/10 drugs discovered between 1990 and 1994 had lifetime sales that exceeded the average cost of development [49].

A natural next step for companies with a franchise in a particular disease area, or seeking a place within the market dominated by competitor companies, was to develop follow-on compounds, which address identified weaknesses in the earlier compounds. Pfizer's amlodipine became the best-in-class calcium channel antagonist and largely took the market from the earlier compounds such as felodipine. Likewise, AstraZeneca's proton pump inhibitor Nexium became a replacement for its own Losec, and GSK's histamine antagonist ranitidine became the best-selling follow-on to SKB's cimetidine, which in their respective heydays were both the world's biggest selling drugs. The oral neuraminidase inhibitor oseltamivir for influenza became a success at the expense of GSK's first-in-class inhaled zanamivir. While in the past there were many followers, nowadays, the follow-on drugs apparently cannot be economically further followed.

The market is rather satisfied with what it has, and hence, the hurdles to show differentiated profile have become significantly higher. In the last 5 years, it could be argued that the impact of payers' decisions (or probable decisions) have had an even bigger impact on drug development projects than regulatory concerns. The effect of national advisory bodies in those countries with government health systems (e.g., National Institute for Clinical Excellence (NICE) in the United Kingdom) on strategic thinking in project teams is now driving the introduction of "real-world evidence" early in project planning. The introduction of hard cutoffs on price might render certain therapeutic areas simply commercially nonviable in the countries where it is applied. Is the quality-adjusted life year (QALY) even index-linked to inflation by those users?

If this trend continues, many projects in exciting areas of emerging biology will probably be strangled at birth on the basis of commercial analysis. It has been argued that pricing agreements in Europe might transfer the full burden of development costs to those markets where higher prices can be obtained [50]. A move to a situation where the first-in-class drug takes the vast majority of the available market might not be desirable for patients since accumulated experience shows different patients may do better on different drugs within a class. In fact, doctors have been operating personalized healthcare for many years by matching superficially similar drugs to patients based on a patient-by-patient assessment of efficacy versus side effects. Probably, the best known example is in control of hypertension [51]. Can we afford the same range of drugs to work with in the future in other disease areas?

While some companies are still trying to innovate in these tight spaces, because of the success of the industry, the opportunity for innovation in follow-on compounds appears diminished (unless a niche for the new compound can be found through a personalized healthcare approach). Small biotechnology companies may thrive in this space, but for multinational pharmaceutical companies, the likely returns may be too small, with the need for blockbuster drugs expected to earn >1$B/year to sustain multinational profitability.

Even new targets in old areas are difficult territory, as AstraZeneca has found with the phase III failure of it's first-in-class SYK inhibitor fostamatinib for rheumatoid arthritis. Pfizer succeeded with it's also first-in-class JAK inhibitor tofacitinib, but not in Europe [52], at least so far. A number of CRTh2 antagonists from different companies have failed to show meaningful differentiated efficacy in asthma compared to inhaled steroids and β_2 agonists often with FEV1 as a clinical endpoint. GSK's recent FLAP inhibitor, GSK2190915 also failed to demonstrate meaningful clinical differentiation from the now generic cysteine leukotriene receptor 1 antagonist montelukast, even though montelukast is only partly effective in mild asthmatics. Commercial pressures are so high that even whole therapy areas have been sources of attrition as we shall see later.

As follow-on compounds are no longer rich picking grounds for blockbusters, and even new mechanism modulators in old diseases are challenging, the hunt is on for new targets in new diseases, where medical need is high. These are areas where both regulators and payers would welcome new innovation. But the focus on new targets that failed to translate into clinical efficacy has been a major source of attrition in modern portfolios. Indeed, in a review of sources of attrition by the management consultancy firm McKinsey's, novel mechanisms were twice as likely to suffer attrition in clinical development than known mechanisms [53]. Selecting which biological mechanisms we choose, in most disease areas, remains the primary challenge.

1.8 PORTFOLIO ATTRITION

While we struggle with our understanding of attrition due to biology and chemistry, we should not fail to mention human decision making as a major source of attrition in drug discovery and development pipelines. Projects can be stopped on the whims of new management or a management change of heart over the projected future value of a drug target, family of drug targets, or even disease areas. Even whole company portfolios can be at stake.

R&D is seen as an expense on the bottom line with little value being ascribed to an early portfolio. In business, it can be financially more attractive to acquire a company with its on-brand products and late-stage development opportunities than to develop your own. Even very large companies can be takeover targets to capitalize on the quirky tax regimens across global economies. Mergers, acquisitions, takeovers, and the closure of R&D pipelines can be financially viable propositions.

In 2014, Allergan announced it was to cut 1500 jobs, in a preemptive measure to cut costs as it attempted to defend itself from a $53 billion hostile bid from Valeant pharmaceuticals. It was a clash of corporate ideologies. Allergan with arguably a more traditional belief in sustainable growth driven through R&D innovation, versus Valeant and it's acquisition-based growth model and a focus on strong financial discipline. Throughout 2014, it was a battle fought in the boardroom, the courtroom, and in the press-room. However a final showdown was avoided, when a more welcome suitor emerged for Allergan. Activis, a company with corporate values apparently more closely aligned to Allergan's own, came forward with an acceptable $66 billion offer.

For US company AbbVie, the takeover of the Irish pharmaceutical company Shire for $53 billion ticked all the boxes. Shire provided AbbVie with products, late-stage development candidates, and a new tax domicile in Ireland with a lower rate of corporation tax from 22 to 13%. But following the international furor over the profit versus patients battle between Pfizer and AstraZeneca, the AbbVie–Shire deal was reputedly scuppered by changes in the tax regulations in the United States, which made the "tax inversion" much less attractive for US-based companies seeking to avoid US taxation on worldwide profits.

Neuroscience R&D has been hit hard in recent years, as pharmaceutical companies come under pressure to deliver value from their R&D investments, and the unpredictability and cost of clinical trials that seem particularly apparent in neuroscience. In 2010, GSK ended neuroscience R&D in England and Italy, and Novartis closed neuroscience R&D in Basel in 2011. Likewise, AstraZeneca closed its neuroscience R&D units in Wilmington, United States; Montreal, Canada; and Södertälje, Sweden, and replaced these with a small virtual neuroscience R&D group, following a largely opportunistic and completely outsourced R&D model.

Anti-infectives are another disease area that has been receiving lower investment from major pharmaceutical companies. Anti-infective research has its own challenges, and over the past 30 years, only two new classes of antibiotics have been introduced (in 2000 the oxazolidinones and in 2003 the lipopeptides) and only three antibiotic NMEs in this decade [54]. The requirement to kill rapidly growing (and mutating) bacteria requires high plasma levels relative to other drugs and to be safe at these higher levels [55]. Unfortunately, anti-infectives when effective are given for only a short time to cure their disease or maybe are reserved only for second-line therapy. They can have a limited market life due to emerging resistance, all limiting the commercial opportunity relative to chronic therapies in other disease areas [56]. Many pharmaceutical companies have withdrawn from anti-infective R&D, while paradoxically the need for new antibiotics has never been higher. Governments are waking up to the fact that the supply of new antibiotics has dried up, as recognized in a recent World Health Organization (WHO) report [57]. Perhaps indicating a change of view, or at least a gap in the market, is the recent re-entrance of companies like Roche and Merck into antibacterial R&D. Since 2013 Roche has acquired rights to a number of new antibiotic development programs while in January 2015 Merck spent an estimated $9.5B in it's aquisition of anti-infectives specialist Cubust Pharmaceuticals.

Lagging behind the opposition can cause the termination of otherwise interesting projects: there is a strong current perception that first in class is dominant. That would be supported to date in the case of DPP-IV inhibitors, where Januvia has maintained leadership over later market entrants in the class. However, there are clear examples of the reverse, especially when the fast follower can benefit from experience acquired by the leader: two examples could be Tarceva and Iressa in oncology, or more strikingly, Sovaldi and Incivek in hepatitis virus C therapy. Maybe Aesop's tortoise can indeed beat the hare.

While pharmaceutical companies may avoid certain disease areas where the risk/financial reward balance appears unfavorable, new models of drug discovery are emerging. Charity-funded R&D is now becoming a major player. The Cystic Fibrosis Foundation provided $75million dollars for Vertex to develop their CFTR channel modulator Kalydeco [58]. In December 2013, the Dementia Consortium launched a £3M fund to bolster dementia drug discovery. In 2009, members of the Association of Medical Research Charities funded $1.1B of research in the United Kingdom alone. In 2012, the Bill and Melinda Gates Foundation alone made grants of $892M for global health projects. To stimulate R&D into new anti-infectives, and following a European Parliament resolution to establish an European Union-wide plan to combat antimicrobial resistance, the Innovative Medicines Initiative in partnership with five pharmaceutical companies has launched a $280M program to spur new anti-infective R&D [59] and will fund a phase III program for GSK's peptide formylase inhibitor for community-acquired pneumonia. However, one wonders whether this really is a solution to the problem. While the discovery and development of Vertex's Kalydeco is a triumph for CF treatment and an undoubted success for charity–industry partnerships, patient groups who raised and donated the $75M toward the costs of discovery of Kalydeco are now possibly consider they required to pay again, albeit more likely through their medical insurances, this time in excess of $311,000 per year per patient to receive it [60]. It is no wonder with examples such as this that even organizations as eminent as the WHO are questioning the value of relying on a commercial pharmaceutical industry, to meet the needs of the world's sick. They are looking to open innovation and a fully funded "idea to market R&D" model, in the interests of world health. As an example, the Indian government, recognizing the healthcare needs of its growing population, has embraced and is investing in open innovation. India would rather consider fully funding pharmaceutical R&D without industrial property protection and allow generic pharmaceutical manufacturing companies to sell the discovered drug with market competition to restrict pricing as a more economical healthcare model than the one currently operating with pharmaceutical companies. With an investment of $35 million so far committed, it is leading a global open innovation initiative called Open Source Drug Discovery, with the vision to provide affordable healthcare to the developing world [61].

Although it feels like the major pharmaceutical company model of drug discovery is broken, and the pharmaceutical industry is in decline, global pharmaceutical companies still spent an estimated $135B in 2011 of R&D. So it is probably fairer to say that it is chang-ing. Large and significant grants are being made by the government and charity sectors, and they are liable to increase. New models of pharmaceutical R&D are being explored, such as open innovation. Taken together, these ventures can only be seen as a good thing.

1.9 "AVOIDING" ATTRITION

If new drugs at new targets are proving too tough a challenge, pharmaceutical companies seek other opportunities to bring drugs to the market that meet unmet patient need, at lower overall risk. These opportunities include new formulations, new drug combina-tions, new indications for existing drugs, and even new drug modalities, among others.

1.9.1 Drug Combinations and New Formulations

New formulations have always been a source of innovation, intellectual property, and therefore profits. Many diseases require polypharmacology, and as patient compliance to any one drug is already a major source of efficacy variability, polypharmacology increases the problem. Thus, fixed-dose combinations for oral topical or inhaled formulations have been a major interest and major commercial and clinical success. The combination of amoxicillin and the β-lactamase inhibitor clavulanic acid has been a longtime success for the treatment of penicillin-resistant bacteria. The fixed-dose combinations of a β_2 ago-nist and a steroid, such as Symbicort (budesonide and formoterol) and Advair (fluticasone and salmeterol), are world's leading therapies for asthma and COPD. Gilead in combination with various pharmaceutical partners has a portfolio of fixed-dose combina-tions for HIV/AIDS treatment including Stribild, approved in 2012, a quad combination of elvitegravir/cobicistat/emtricitabine/tenofovir; Complera, a triple combination of rilpivirine + emtricitabine + tenofovir approved in 2011; as well as Atripla, a triple combination of efavirenz + emtricitabine + tenofovir approved in 2006, and Truvada, a double combination of emtricitabine + tenofovir approved in 2004.

The regulators, patent authorities, and payers all wish to see that combinations should show significant advantages over dosing the drugs individually, as the monotherapies could be used in combination, often at lower cost, as they may now be generic. The FDA [62], EMA [63], and WHO [64] have all issued guidance on fixed-dose combinations and indicate the likely situations where fixed-dose combinations are more and less likely to be approved.

The regulatory approval itself provides drugs with market protection through a period of data exclusivity. In the United States, the Drug Price Competition and Patent Term Restoration Act of 1984 (Hatch-Waxman amendments) provided a more streamlined pathway for generic drugs to be brought to market, whereby some or all of the efficacy and safety data relied upon for approval were not conducted by the applicant or for which the applicant did not have a right of reference. But to still provide incentives for pharma-ceutical innovation, the Act also authorized a period of NCE data exclusivity, preventing generic drug applications citing the original data for a period of 5 years. In Europe, the data exclusivity can be 10 years, and this protection can exceed the protection provided by the patent. But in the United States, where one part of the fixed dose combination has been previously registered, data exclusivity was only 3 years. After petitions from companies such as Gilead and Bayer, the FDA has recently amended this to 5 years for future drug applications [65].

But the path to success with drug combinations is not an easy one. In 2012, the FDA rejected Merck's application for their lipid-lowering combination product Liptruzet, a combination of ezetimibe with atorvastatin (a generic version of Pfizer's Lipitor), indicating a requirement for more data. This drug was seen as a replacement for Merck's existing drug combination of ezetimibe with the generic statin, simvastatin. This was almost the end of the road for this particular combination. The FDA had previously rejected the application in 2009, but it was third time lucky for Merck, as in May 2013 the FDA finally approved the combination.

Intellectual property challenges to drug combinations are also causing problems for the industry. Patent challenges have led to the revocation of some high-profile combination patents, opening the door to generic competition. In 2004, the United Kingdom combination patent for GSK's Advair, a fixed-dose combination of salmeterol and fluticasone proprionate, was revoked on the basis of obviousness over GSK's existing combination of salbutamol and beclomethasone dipropionate sold in a metered dose inhaler device before the relevant priority date, after challenge from four generic companies. In 2007, the European Patent Office revoked AstraZeneca's combination patent for Symbicort, its fixed-dose combination of formoterol and budesonide.

In 2013, the UK patents court revoked the patent for GSK's malaria combination Malarone, a 5:2 mixture of atovaquone and proguanil, due to the presence of prior art found in a presentation and an abstract to a lecture [66]. In June 2013, the US federal court handed down a decision in Novo Nordisk A/S versus Caraco Pharmaceutical Laboratories Ltd. revoking a patent covering Novo's diabetes treatment involving a combination of drugs metformin and repaglinide as an obvious combination of known diabetes treatments [67]. The case highlights the difficulties in establishing nonobviousness when claim elements are individually known even where there is evidence of synergistic benefits of the combination. Novo has petitioned for a rehearing, and this was granted, partly on the recognition of the industry-wide implications of the decision.

The Indian patent authority in particular is currently taking a strict approach on patents for new formulations, salts, and other improvements. It has recently revoked a patent granted to GSK for a new salt for its dual tyrosine kinase inhibitor lapatinib, concluding that improved efficacy should be interpreted strictly on a therapeutic basis, taking no account of pharmaceutical properties such as flow, stability, or hygroscopicity [68]. Drug combinations are clearly a complicated area for pharmaceutical companies.

1.9.2 Biologics versus Small Molecules

Biologic drugs have been a major clinical and commercial success. Fifteen years ago, there were only small molecules in the world's top 10 selling drugs list. In 2012, the top 3 selling drugs were all biologics targeting TNF-alpha, that is, Humira, Remicade, and Embrel, with combined sales of $25.3 billion. Pharmaceutical companies have been rapidly trying to buy up or strike deals with the biologics companies. Biologic drugs are attractive due to the high levels of efficacy and specificity and have achieved rapid development times with lower levels of attrition than with small-molecule drugs. A cross-industry analysis found that between 2006 and 2010 25% of large molecules in phase II reached the market compared to only 10% for small molecules the 13 contributing pharmaceutical companies [69]. An analysis by Tufts University covering top 50 pharmaceutical companies for drugs entering development from 1999–2004 up to 2009 found similar results and found the clinical success rate for large molecules was 32% compared to 13% for small molecules [70]. Biologic drugs have also been less susceptible to generic competition. Apart from the difficulty in the production of biologic drugs, which in itself gives a drug protection from competition, a process for biologic generic registration was only clarified in 2010 in the United States with the Biosimilars Act. The act itself provides a market exclusivity advantage to the innovator biologic of 12 years relative to a small molecule, which is granted only 5 years. Generic biologics have to wait longer. Even then the first generic to market is itself granted between 12 and 42 months of exclusivity, relative to only 180 days for small-molecule generic drug. The costs of biologic drug production are also high, and with the high efficacy, biologic drugs have been able to command high prices. It has been seen as a potential advantage for small-molecule drugs to undercut biologics on price due to the lower production costs. But this may also change. There is a global production overcapacity for biologic drugs, and some estimates suggest that the production cost is only 5% of the total price of the biologic drug [71]. It is therefore likely that any production cost differential could disappear, meaning small-molecule drugs have to meet biologics head-on on efficacy differentiation, safety, and compliance. It all sounds like bad news for small-molecule drugs. But small molecules also have advantages over biologics, for example, they can target both intracellular and extracellular mechanisms and have wider options for administration. They also can demonstrate a pleiotropic effect through targeting more than one pharmacology, or by modulating divergent signaling pathways, which has been shown to be a key contributor to the efficacy of a number of small-molecule drugs [72]. There is room and opportunity for both small- and large-molecule therapeutics.

1.9.3   Small-Molecule Compound Quality

From Hansch [73] to Lipinski [74], we've been aware of the controlling influence of bulk physical properties on the key pharmacokinetic properties of drug molecules. It is not too surprising that lipophilicity, molecular size, charge type, and hydrogen bond donor counts in particular have important controlling influences on the pharmacokinetic and toxicological profile of drugs. Lipophilic compounds tend to be less water soluble, distribute into fat, are more highly metabolized, and, as many receptors tend to have hydrophobic active sites, may be more promiscuous with respect to off-target pharmacology [75]. Database studies indicate that more lipophilic compounds also have a higher chance of attrition in clinical development [76]. Molecular size can be understood to have a direct impact on permeability, as molecular diffusion is dependent upon molecular size as described by the Stokes–Einstein equation. Where the lipophilicity scale is described by n-octanol-water partitioning, hydrogen bond donor counts need to be considered separately, because n-octanol likely overestimates the partitioning ability of donors into a hydrophobic phase [77]. Charge type is apparently an independent contributor to promiscuity, as a number of database analyses have shown bases are more promiscuous than similarly lipophilic acids or neutrals, that is, are more frequent hitters in broad selectivity screens. Certain chemical motifs are also disfavored in drugs, most often due to their chemical or metabolic reactivity with the potential to lead to *in vivo* covalent adducts or highly reactive genotoxic intermediates. Many companies maintain lists of "ugly" functionality to eliminate compounds containing these features from screening collections and discourage their inclusion in designed drug molecules [78, 79].

Other descriptors have also been variously implicated as playing a role, including sp2/sp3 count and numbers of aromatic rings. The statistical validity of some of the claims has recently been questioned [80] as have the overall conclusions [81]. Different companies place different degrees of emphasis on these compound quality indicators, which may suggest a lack of acceptance or a different "organizational culture" over how compound quality optimization can lead to development success [82]. There is a continuing debate over compound quality considerations, which are variously viewed as either focusing innovation in areas of property space more likely to give success or overly limiting the opportunity for chemical innovation [83]. Whichever camp you are in, it is certain that there is an opportunity for innovation at the edges of "drug-like space." Of the drugs that are Lipinski violations, many are cyclic/macrocyclic drugs, or drugs known to be subject to active transport. Large molecules have advantages in affinity, and this may be particularly important where the drug is required to antagonize a protein–protein interaction. Protein–protein molecular recognition interfaces often yield high binding energies, as the interacting proteins may be present at low concentrations in cells. It therefore may require a high-affinity small-molecule ligand to be able to compete. But the protein–protein interaction interface is often spread over a large surface area, adding to the difficulty for small-molecule low molecular weight antagonists. Lipinski-compliant small molecules just don't have the affinity capacity required, but the larger and more complex macrocycles and linear or cyclic peptides do, and hence, these are an exciting new area for drug discovery innovation. A number of biotech companies offer screening and drug discovery services based on libraries of cyclic and linear peptides of macrocycles. But while nature and evolution, with a little help from formulation science, has helped molecules such as cyclosporine and FK-506 to become successful oral drugs, the edges of Lipinski space carry with them the attendant problems of solubility and permeability as might be expected, and these will need some pharmaceutical ingenuity to overcome them. But with some ingenuity and determination progress can be made. Roche and Abbvie's BCL-2 inhibitor Venetoclax may be classified as an "ugly" molecule by compound quality measures (Mwt = 839.4, ACDlogP = 10.9), it has recently been granted FDA breakthrough status after successful phase II results in chronic lymphocytic leukaemia patients with a 17p gene deletion.

1.10   GOOD ATTRITION VERSUS BAD ATTRITION

It is desirable to stop projects that are likely to fail as early as possible and instead invest in projects that have a better chance to succeed. Certainly, finding out at the end of a phase IIb clinical study that the biological hypothesis that linked a target to disease is not valid is not good attrition. We want to be masters of attrition and not the victims of it. Attrition through the informed scientific decisions we make is good attrition. The earlier we can make that good scientific stop/go decision for a project, the better. Pharmaceutical companies have been developing guidelines, such as AstraZeneca's 5 Rs [84, 85], which are used to ask questions of R&D projects throughout their lifetime, to ensure projects are on track to delivering medicines that meet patient's needs and that payers will pay for (5 Rs = right target, right patients, right safety, right tissue/exposure, right commercial). AstraZeneca's 5 "rights" were based on a detailed retrospective analysis of the successes and failures of 142 small-molecule R&D projects from 2005 to 2011, and with only a 15% success rate in phase II compared with an industry average in that period of 33%, there was a justifiable reason to ask searching questions. The outcome was a framework for five determinants of technical success in projects and portfolio quality, along with a sixth determinant, which was a culture for truth-seeking and rigorous decision making based on these determinants. The framework encourages those involved in the projects to question themselves against the 5 Rs at every stage of project progression.

Right target
- Strong link between target and disease
- Differentiating efficacy from existing therapies
- Available and predictive biomarkers
- Right tissue exposure
- Adequate bioavailability and tissue exposure
- Human pharmacokinetics/pharmacodynamics (PD) prediction
- PD biomarkers
- Drug–drug interaction
- Right safety
- Clear assessment of safety risks
- Clear understanding of risk/benefit
- Availability of predictive biomarkers

Right patients
- Scientific evidence in lead indication
- Risk/benefit stratification of patient population
- Personalized healthcare strategy including diagnostic/biomarkers
- Right commercial
- Differentiated value proposition versus future standard of care
- Priority geographies
- Market access/payer/provider focus
- Personalized healthcare strategy including diagnostic/biomarkers

AstraZeneca (and other companies) hope the aspiration to reduce attrition in R&D, and particularly late-phase attrition will be met.

1.11   SUMMARY

For the beleaguered drug project team battling these challenges, what hope can be offered? One major disadvantage of the decreasing size of major pharmaceutical company model is the decreasing opportunity for R&D scientists to gain experience over many projects. Drug discovery is still an empirical science; we continue to learn by our successes and failures. The experience if properly applied can have a major impact in reducing attrition as previous mistakes are avoided. As the industry becomes more and more fragmented, the chances of mistakes being repeated, or inappropriate R&D strategies being followed, increase.

Attrition can be summarized in three stages. At the initiation of drug research, there is strategic attrition in deciding to exit or not pursue certain therapeutic areas on the basis of low likelihood of technical success or failure to envisage a return on investment. We have discussed the withdrawals from CNS and antibiotic research, but other areas such as critical care, stroke, and septic shock are areas of major unmet need that are now less attractive. Clearly, regulators and payers have a role to play here in changing the landscape. In the middle stage of drug discovery and development, the major form of attrition is technical, for example, unpredicted toxicity, lack of target engagement, poor translation of effect, and patent competition. This phase is at the same time the most challenging and the stage where our understanding is improving most rapidly. Finally, there is late-stage attrition. Short of outright therapeutic or safety failure, attrition is often the result of a combination of technical complexity and commercial uncertainty. As the patent clock is ticking, reducing the potential time to make a return on investment, research companies can be unwilling to risk further resources in new trials to investigate effects seen in the first clinical evaluations. This creates a "one strike and you are out" approach, which seems potentially wasteful. Again, this is an area where a new framework for commercial viability is needed. The breakthrough status designation from the FDA is a promising step in this direction.

This book can suggest many tools and strategies to maximize the chance of success. It is an opportunity to distill that empirical understanding into something of real value for the way we will do drug discovery over the next few years. Great progress has been made in the last 20 years in the basic science of producing both small- and large-molecule drugs. As discussed in detail previously, the drive to understanding the translational link from target to disease is still a major difficulty, but here the impact of genetics of disease (and likely soon the epigenetic basis of disease) will make a rapidly accelerating contribution. The understanding of the major players in drug metabolism and drug transporters now allows reasonable prediction (and hence the opportunity to design out) drug–drug interactions or nonlinear kinetics. The benefits of investments in clinical pharmacology in phase II for better dose ranging are now being seen in many development programs. Investments in personalized healthcare, genetics of disease, patient segmentation, and biomarkers are helping to direct the right drug to the patients who would benefit most. Lastly, chemists have progressed in their understanding of what compound quality means. It is now up to the reader to implement these ideas in selecting the target and through the chemical structures you design, synthesize, and test.

Perhaps ultimately, there will need to be (further) discussion about the benefit of medicines to society, in an era of aging populations and rising expectations for healthcare. The whole commercial basis of drug development might need to be revaluated, as it has been in the past, with patent term extensions being increased to encourage even more thorough investigations of drugs in phase III. But bolder steps could be taken. We can see the commercial model failing in areas of still high unmet medical need, and pharmaceutical companies turn to withdraw from unprofitable or high-risk areas such as anti-infectives, third world diseases, and neuroscience. It is reassuring to see charities and governments stepping in and becoming themselves new and growing players in the global R&D. This is a fascinating time to be in the industry, and we look forward to monitoring the progress of alternative funding models, such as open innovation. The current system is far from perfect, and any attempt to provide additional ways of meeting areas of unmet medical need is to be welcomed the goose that laid the golden eggs.

REFERENCES

1 Paul, S.M., Mytelka, D.S.; Dunwiddie, C.T., Persinger, C., Munos, B.H., Lindborg, S.R., Schacht, A.L. (2010). How to improve R&D productivity: the pharmaceutical industry's grand challenge. *Nat. Rev. Drug Discov.* 9, 203–214.

2 Pharmaceutical Manufacturers of America Industry Profile 2015. http://www.phrma.org/sites/default/files/pdf/2015-phrma_profile_membership_results.pdf (accessed September 8, 2015).

3 http://www.fda.gov/AboutFDA/WhatWeDo/History/ProductRegulation/Summary ofNDAApprovalsReceipts1938tothepresent/default.htm (accessed July 16, 2015).

4 http://www.forbes.com/sites/matthewherper/2012/02/10/the-truly-staggering-cost-of-inventing-new-drugs/ (accessed July 16, 2015).

5 Prentis, R.A., Lis, Y., Walker, S.R. (1988). Pharmaceutical innovation by the seven UK-owned pharmaceutical companies (1964–1985). *Br. J. Clin. Pharmacol.* 25, 387–396.

6 Kola, I., Landis, J. (2004). Can the pharmaceutical industry reduce attrition rates? *Nat. Rev. Drug Discov.* 3, 711–716.

7 Cong, L., Ran, F. A., Cox, D., Lin, S., Barretto, R., Habib, N., Hsu, P.D., Wu, X., Jiang, W., Marraffini, L.A., Zhang, F. (2013). Multiplex genome engineering using CRISPR/Cas systems. *Science* 339, 819–823.

8 Fennell, M., Xiang, Q., Hwang, A., Chen, C., Huang, C.-H., Chen, C.-C., Pelossof, R., Garippa, R. J. (2014). Impact of RNA-guided technologies for target identification and deconvolution. *J. Biomol. Screen.* 18, 1327–1337.

9 Hartung, T. (2013). Look back in anger—what clinical studies tell us about preclinical work. *ALTEX 30*, 275–291.

10 van der Worp, H.B., Howells, D.W., Sena, E.S., Porritt, M.J., Rewell, S., O'Collins, V., Macleod, M.R. (2010). Can animal models of disease reliably inform human studies? *PLoS Medi. 7*, e1000245.

11 Nellesen, D., Person. A., Yee. K., Chawla, A. (2009). *Personalized medicine: trends in clinical studies based on National Registry Data.* Poster presented at ISPOR, 11. Orlando, FL, May 2009. Available at http://www.analysisgroup.com/personalized_medicine_trends (accessed July 16, 2015).

12 Andersen, D.H. (1938). Cystic fibrosis of the pancreas and its relation to celiac disease: a clinical and pathological study. *Am. J. Dis. Child 56*, 344–399.

13 Riordan, J.R., Rommens, J.M., Kerem, B., Alon, N., Rozmahel, R., Grzelczak, Z., Zielenski, J., Lok, S., Plavsic, N., Chou, J.L. (1989). Identification of the cystic fibrosis gene: cloning and characterization of complementary DNA. *Science 245*, 1066–1073.

14 http://www.fda.gov/NewsEvents/Newsroom/PressAnnouncements/ucm269856.htm (accessed July 16, 2015).

15 http://www.astrazeneca.com/Media/Press-releases/Article/20142609--iressa-receives-chmp-positive-opinion-to-include-blood-based-diagnostic-testing-in-european-label (accessed July 16, 2015).

16 http://www.fda.gov/downloads/RegulatoryInformation/Guidances/UCM126957.pdf (accessed July 16, 2015).

17 Corren, J., Lemanske, R.F. Jr., Hanania, N. A., Korenblat, P. E., Parsey, M.V., Arron, J.R., Harris, J.M., Scheerens, H., Wu, L.C., Su, Z., Mosesova, S., Eisner, M.D., Bohen, S.P., Matthews, J.G. (2011). Lebrikizumab treatment in adults with asthma. *N. Eng. J. Med.* 365, 1088–1098.

18 Morgan, P.; Van Der Graaf, P.H., Arrowsmith, J., Feltner, D. E., Drummond, K. S., Wegner, C. D., Street, S.D.A. (2012). Can the flow of medicines be improved? *Fundamental pharmacokinetic and pharmacological principles toward improving Phase II survival. Drug Discov. Today 17*, 419–424.

19 http://www.reuters.com/article/2012/10/18/us-abbott-trial-kidneydisease-idUSBRE89 H0RB20121018 (accessed July 16, 2015).

20 http://www.sarossaplc.com/archive/reports/ar2011.pdf (accessed July 16, 2015).

21 https://investor.lilly.com/releasedetail.cfm?ReleaseID=793403 (accessed July 16, 2015).

22 https://investor.lilly.com/releasedetail.cfm?ReleaseID=763858 (accessed July 16, 2015).

23 http://www.astrazeneca.com/Media/Press-releases/Article/20130504-astrazeneca-announces-topline-results-from-phase-iii-o (accessed July 16, 2015).

24 http://ir.prosensa.eu/releasedetail.cfm?ReleaseID=791929 (accessed July 16, 2015).

25 http://www.roche.com/media/store/releases/med-cor-2013-07-10.htm (accessed July 16, 2015).

26 Arrowsmith, J., Miller, P. (2013). Trial watch: Phase II and phase III attrition rates 2011–2012. *Nat. Rev. Drug Discov. 12*, 569.

27 http://www.fda.gov/downloads/Training/ClinicalInvestigatorTrainingCourse/UCM283378.pdf (accessed July 16, 2015).

28 http://www.fda.gov/NewsEvents/Newsroom/PressAnnouncements/ucm327152.htm (accessed July 16, 2015).

29 http://www.ema.europa.eu/docs/en_GB/document_library/Summary_of_opinion_-_Initial_authorisation/human/002542/WC500146629.pdf (accessed July 16, 2015).

30 Holford, N., Ma, S.C., Ploeger. B.A. (2010). Clinical trial simulation: a review. *Clin. Pharmacol. Therapeut. 88*, 166–182.

31 http://www.takeda.com/news/2013/20131227_6117.html (accessed July 16, 2016).

32 http://www.reatapharma.com/investors-media/news/news-timeline/archive/company-statement-termination-of-the-beacon-trial.aspx (accessed July 16, 2015).

33 http://bms.newshq.businesswire.com/press-release/financial-news/bristol-myers-squibb-discontinues-development-bms-986094-investigationa (accessed July 16, 2015).

34 Redfern, W.S., Carlsson, L., Davis, A.S., Lynch, W.G., MacKenzie, I., Palethorpe, S., Siegl, P.K.S., Strang, I., Sullivan, A.T., Wallis, R., Camm, A.J., Hammond, T.G. (2003). Relationships between preclinical cardiac electrophysiology, clinical QT interval prolongation and torsade de pointes for a broad range of drugs: evidence for a provisional safety margin in drug development. *Cardiovasc. Res. 58*, 32–45.

35 Woosley, R.L. (2013). One hundred years of drug regulation: where do we go from here? *Annu. Rev. Pharmacol. Toxicol. 53*, 255–273.

36 Peltzman, S. (1973). An evaluation of consumer protection legislation: the 1962 drug amendments. *J. Polit. Econ. 81*(5), 1049–1091. Reprinted in Chicago Studies in Political Economy, edited by George J Stigler, 303–348. Chicago, University of Chicago Press, 1988. Thomas, L.G. (1990). Regulation and firm size: FDA impacts on innovation. *Rand J. Econ. 21*(4), 497–517.

37 http://www.fda.gov/NewsEvents/Newsroom/PressAnnouncements/ucm280536.htm (accessed July 16, 2015).

38 http://www.fda.gov/downloads/drugsGuidanceComplianceRegulatoyInformation/Guidance/UCM071590.pdf (accessed July 16, 2015).

39 Charatam, F. (2001). Bayer decides to withdraw cholesterol lowering drug. *Brit. Med. J., 323*, 359.

40 Zwillich, T. (2005). How Vioxx is changing US drug regulation. *Lancet 366*, 1763–1764.

41 http://www.fda.gov/downloads/Drugs/GuidanceComplianceRegulatoryInformation/EnforcementActivitiesbyFDA/WarningLettersandNoticeofViolationLetterstoPharmaceutical Companies/UCM166426.pdf (accessed July 16, 2015).

42 Nissen, S.E., Wolski, K. (2007). Effect of rosiglitazone on the risk of myocardial infarction and death from cardiovascular causes. *N. Engl. J. Med. 356*, 2457–2471.

43 http://www.fda.gov/newsevents/newsroom/pressannouncements/2007/ucm108917.htm (accessed July 16, 2015).

44 http://www.fda.gov/Drugs/DrugSafety/ucm255005.htm (accessed Julyu 16, 2015).

45 http://www.fda.gov/Drugs/DrugSafety/ucm376389.htm (accessed July 16, 2015).

46 http://www.phrma.org/sites/default/files/pdf/ChartPack_4%200_FINAL_2014MAR25.pdf (accessed July 16, 2015).

47 Samji, H., Cescon, A., Hogg, R.S., Modur, S.P., Althoff, K.N., Buchacz, K., Burchell, A.N., Cohen, M., Gebo, K.A., Gill, M.J., Justice, A., Kirk, G., Klein, M.B., Korthuis, P.T., Martin, J., Napravnik, S., Rourke, S.B., Sterling, T.R., Silverberg, M.J., Deeks, S., Jacobson, L.P., Bosch, R.J., Kitahata, M.M., Goedert, J.J., Moore, R., Gange, S.J. (2013). Closing the gap: increases in life expectancy among treated HIV-positive individuals in the United States and Canada. *PLoS One 8*, e81355.

48 HHS, CDC, NCHS. "Health, United States, 2003 with Chartbook on Trends in the Health of Americans." Hyattsville, MD: HHS, 2003 HHS, CDC, NCHS. "Health, United States, 2009 With Chartbook on Medical Technology." Hyattsville, MD: HHS, 2010; 2007 data from J. Xu, K.D. Kochanek, and B. Tejada-Vera. "Deaths: Preliminary Data for 2007." *Natl. Vital Stat. Rep.* (2009); *58*(1): 5. Hyattsville, MD: NCHS. Available at www.cdc.gov/nchs/data/nvsr/nvsr58/nvsr58_01.pdf (accessed December 2009); 2009 data from K.D. Kochanek et al. Deaths: final data for 2009. *Natl. Vital Stat. Rep.* (2011), *60*(3), 41. Hyattsville, MD: NCHS. Available at www.cdc.gov/nchs/data/nvsr/nvsr60/nvsr60_03.pdf (accessed December 2012); 2011 data from D.L. Hoyert and J. Xu. Deaths: preliminary data for 2011. *Natl. Vital Stat. Rep.* (2012), *61*(6), 38. Hyattsville, MD: NCHS. Available at www.cdc.gov/nchs/data/nvsr/nvsr61/nvsr61_06.pdf (accessed December 2012).

49 Vernon, J.A., Golec J.H., DiMasi, J.A. (2010). Drug development costs when financial risk is measured using the Fama-French three-factor model. *Health Econ. 19*, 1002–1005.

50 http://www.ita.doc.gov/td/chemicals/drugpricingstudy.pdf (accessed July 16, 2015).

51 Donnelly, R., Meredith, P. A., Elliott, H.L. (1991). The description and prediction of antihypertensive response: and individualised approach. *Br. J. Clin. Pharmacol. 31,* 627–634.

52 http://www.ajmc.com/publications/ebiid/2014/april2014ebid/despite-slow-start-ra-specialists-see-place-for-tofacitinib-in-the-formulary/2 (accessed July 16, 2015).

53 http://www.google.com/url?sa=t&rct=j&q=&esrc=s&frm=1&source=web&cd=1&cad=rja& uact=8&ved=0CCEQFjAA&url=http%3A%2F%2Fwww.mckinsey.com%2F~%2Fmedia%2 Fmckinsey%2Fdotcom%2Fclient_service%2Fpharma%2520and%2520medical%2520 products%2Fpdfs%2Frdemergingmarkets.ashx&ei=2TGOU5nTLaikyAPbooCoAQ&usg= AFQjCNFgyNE39G5t7WdelDRMNsxL1TMIlg&bvm=bv.68191837,d.bGQ (accessed July 16, 2015).

54 Kinch, M.S., Patridge, E., Plummer, M., Hoyer D. (2014). An analysis of FDA-approved drugs for infectious disease: antibacterial agents. *Drug Disc. Today 19*, 1283–1287.

55 http://www.gsk.com/content/dam/gsk/globals/documents/pdf/Policies/GSK-antibacterial-randd.pdf (accessed July 16, 2015).

56 Leung, E., Weil, D.E., Raviglione, M., Nakatani, H. (2011). The WHO package to combat antimicrobial resistance. *Bull. World Health Organ. 89*, 88–89.

57 http://whqlibdoc.who.int/publications/2012/9789241503181_eng.pdf (accessed July 16, 2015).

58 Ekins, S., Waller, C.L., Bradley., M.P., Clark, A.M., Williams, A.J. (2013). Four disruptive strategies for removing drug discovery bottlenecks. *Drug Discov. Today. 18*, 265–271.

59 Editorial (2012). *Nat. Rev. Drug Discov. 11*, 507.

60 O'Sullivan, B.P., Orenstein, D.M., Milla, C.E., (2013). Pricing for orphan drugs: will the market bear what society cannot? *JAMA 310*, 1343.

61 http://www.osdd.net/about-us/faq-s (accessed July 16, 2015).

62 http://www.fda.gov/regulatoryinformation/guidances/ucm125278.htm (accessed July 16, 2015).

63 http://www.ema.europa.eu/docs/en_GB/document_library/Scientific_guideline/2009/09/WC500003689.pdf (accessed July 16, 2015).

64 http://apps.who.int/medicinedocs/documents/s19979en/s19979en.pdf (accessed July 16, 2015).

65 http://www.fda.gov/downloads/Drugs/GuidanceComplianceRegulatoryInformation/Guidances/UCM386685.pdf (accessed July 16, 2015).

66 http://www.bailii.org/cgi-bin/markup.cgi?doc=/ew/cases/EWHC/Patents/2013/148.html&query=glenmark+and+wellcome&method=boolean (accessed July 16, 2015).

67 http://e-foia.uspto.gov/Foia/RetrievePdf?system=FCA&flNm=11-1223_2 (accessed July 16, 2015).

68 http://www.worldipreview.com/news/section-3-d-strikes-again-in-gsk-india-patent-defeat (accessed July 16, 2015).

69 http://www.genengnews.com/keywordsandtools/print/3/26751/ (accessed July 16, 2015).

70 DiMasi, J.A., Feldman, L., Seckler A., Wilson, A. (2010). Trends in risks associated with new drug development: success rates for investigational drugs. *Nat. Clin. Pharm. Therap.* 87, 272–277.

71 Kelley, B. (2009). Industrialization of mAb production technology—the bioprocessing industry at a crossroads. *mAbs 1*, 443–452.

72 Morphy, J.R., Harris, C. (2012). *Designing multi-target drugs.* Cambridge: RSC Publications

73 Fujita, T., Iwasa, J., Hansch, C. (1964). A new substituent constant p, derived from partition coefficients. *J. Am. Chem. Soc. 86*, 5175–5180.

74 Lipinski, C.A., Lombardo, F., Dominy, B.W., Feeney, P.J. (2001). Experimental and computational approaches to estimate solubility and permeability in drug discovery and development settings. *Adv. Drug Deliv. Rev. 46*, 3–26.

75 Leeson, P.D., Springthorpe, B. (2007), The influence of drug-like concepts on decision-making in medicinal chemistry, *Nat. Rev. Drug Discov. 6*, 881–890.

76 Wenlock, M.C., Austin, R.P., Barton, P., Davis, A,M., Leeson, P.D. (2003). A Comparison of physicochemical property profiles of development and marketed oral drugs. *J. Med. Chem. 46*, 1250–1256.

77 Leeson, P.D., Davis, A.M. (2004). Time-related differences in the physical property profiles of oral drugs. *J. Med. Chem. 47*, 6338–6348.

78 Cumming, J.G., Davis, A.M., Muresan, S., Haeberlein, M., Chen, H. (2013). Chemical Predictive modeling to improve compound quality. *Nat. Rev. Drug Discov. 12*, 948–962.

79 Bakken, G.A., Bell, A.S., Boehm, M., Everett, J.R., Gonzales, R., Hepworth, D., Klug-McLeod, J.L, Lanfear, J., Loesel, J., Mathias, J., Wood T.P. (2012). shaping a screening file for maximal lead discovery efficiency and effectiveness: elimination of molecular redundancy. *J. Chem. Inf. Model. 52*, 2937–2949.

80 Kenny, P.W., Montanaro, C.A. (2013). Inflation of correlation in the pursuit of drug-likeness. *J. Comput. Aided Mol. Des., 27*, 1.

81 Muthas, D., Boyer, S., Hasselgren, C. (2013) A critical assessment of modeling safety-related drug attrition *MedChemComm 4*, 1058–1065.

82 Leeson, P.D., St. Gallay, S. A. (2011). The influence of the 'organizational factor' on compound quality in drug discovery. *Nat. Rev. Drug Discov. 10*, 749–765.

83 Baell, J., Congreve, M., Leeson, P., Abad-Zapatero, C., (2013). Ask the experts: past, present and future of the rule of five. *Future Medicinal Chemistry 5*, 745.

84 http://www.astrazeneca.com/Partnering/our-partnering-process (accessed July 16, 2015).

85 Cook, D., Brown, D., Alexander, R., March, R., Morgan, P., Satterthwaite, G., Pangalos, M. (2014). Lessons learned from the fate of AstraZeneca's drug pipeline a five dimensional framework. *Nat. Rev. Drug Discov. 13*, 419–431.

2

COMPOUND ATTRITION AT THE PRECLINICAL PHASE

CORNELIS E.C.A. HOP

Department of Drug Metabolism & Pharmacokinetics, Genentech, South San Francisco, CA, USA

2.1 INTRODUCTION: ATTRITION IN DRUG DISCOVERY AND DEVELOPMENT

Publications about attrition in the pharmaceutical and biotechnology industry have become extremely popular in the scientific literature [1–3]—most of them with dire warnings about the unsustainable future of current practices in the industry. The most frequently presented graph is shown in Figure 2.1 and highlights the gradually increasing research and development (R&D) expenses versus the stagnating or declining number of drugs approved by the FDA. Consequently, the fully burdened cost per new molecular entity is probably somewhere around $1.5B. There is some glimmer of hope though because the number of approvals by the FDA seems to be going up again after years of steady decline with 35 drugs approved in 2012 and 34 drug approvals in 2013. This is at least in part fuelled by FDA initiatives such as breakthrough status, fast track, priority review, and accelerated approval—all implemented to help address unmet medical needs, in particular in oncology and other therapeutic areas without effective treatments, and for drugs to reach patients quicker than via traditional approval routes. For example, ibrutinib went from initiation of phase I studies to FDA approval in slightly less than 5 years in part due to breakthrough designation. Although greatly beneficial for both companies and patients, these processes to accelerate approval have been accompanied with larger numbers of postmarketing commitments and postmarketing requirements. It is also clear that not all companies are equal in their level of success [4]. Table 2.1 shows the R&D expenses from 2000 to 2012 divided by the number of approvals in the same period [5]. Although this analysis obviously has its limitations, such as not including the commercial

Attrition in the Pharmaceutical Industry: Reasons, Implications, and Pathways Forward,
First Edition. Edited by Alexander Alex, C. John Harris and Dennis A. Smith.
© 2016 John Wiley & Sons, Inc. Published 2016 by John Wiley & Sons, Inc.

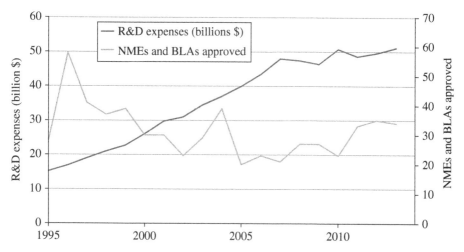

FIGURE 2.1 R&D expenses of PhRMA member companies (billion $) and number of new molecular entities (NMEs) and biologics license applications (BLAs) approved by the FDA.

TABLE 2.1   The Number of Drugs Approved From 2003 to 2012 and the R&D Expenses Per Approved Drug for 20 Major Pharmaceutical Companies [5]

	Company	Number of new drugs approved	R&D expenses per approved drug in $M
1	Abbott	1	13,183
2	Sanofi	6	10,128
3	AstraZeneca	4	9,561
4	Hoffmann-La Roche	8	8,866
5	Pfizer	10	7,779
6	Wyeth	3	7,567
7	Eli Lilly	4	6,678
8	Bayer	5	6,624
9	Schering-Plough	3	6,282
10	Novartis	10	6,073
11	Takeda	4	6,033
12	Merck & Co	9	5,459
13	GlaxoSmithKline	11	5,236
14	J & J	13	5,202
15	Novo Nordisk	2	4,625
16	UCB	1	4,325
17	Yamanouchi	1	4,321
18	Fujisawa	1	4,286
19	Amgen	5	4,270
20	Astellas	3	4,194

success of launched drugs, it does illustrate the ability of companies to launch drugs successfully—or not—in a sustained fashion. One reason for the difference across companies could be the difference in therapeutic area focus; it is known that the attrition differs across therapeutic areas with higher attrition for cardiovascular and neurology indications and lower attrition in infectious diseases and attrition is lower for biologics than for small molecules [6]. A more sophisticated analysis was published online recently [7], and it emphasized the relationship between corporate growth and R&D productivity. The analysis was based on the cumulative R&D expenditure from 2003 to 2012, including costs for mergers and acquisitions and a 7% cost of capital, and the risk-adjusted net present values of marketed products launched in the last 5 years and pipeline products. This analysis also identified several other drivers of corporate success, and these include a manageable R&D organization (between 500 and 700 people per site) and budget (between $1.4 and $2.7 billion), a high degree of focus on specific therapeutic areas, a focus on the United States, higher value concentration in their R&D portfolio, and strong internal R&D capabilities complemented with earlier (pre-phase III) in-licensing activities and unique R&D models. In large part, it highlights the importance of sufficient critical mass and talent while keeping room for individual initiative, creativity, and innovation. Nevertheless, attrition is large even for otherwise successful organizations. The target space is finite and the value of preclinical models is limited. Indeed, one of the key areas of concern in the pharmaceutical industry is the very high attrition in phase II (>60% is quite common and it is trending upward), and the main reason is lack of efficacy or insufficient efficacy compared to standard of care [2, 3, 8]. However, in some cases, the lack of efficacy may be due to insufficient exposure, and it is not inconceivable that efficacy may have been achieved if absorption had been better and the clearance lower. The latter is particularly worrisome because the clinical validity of the target may not have been properly assessed and, therefore, the decision to nominate a backup compound or not cannot be made reliably. Indeed, 12 out of 44 phase II programs failed at Pfizer because no downstream pharmacological effect was observed in human and confidence that sufficient exposure was achieved to test the mechanism was low [9]. This analysis highlights the importance of incorporation of pharmacodynamic (PD) biomarkers in the clinical development plan. This should include target biomarkers, pathway biomarkers, and disease biomarkers. The availability of these biomarkers leads to a greater understanding of the extent of target modulation by the drug in question and the downstream consequences for disease modulation. Even though disease modulation can never be guaranteed for novel targets, the availability of multiple PD biomarkers should be a prerequisite and will ultimately lead to a greater understanding of disease biology and (down the road) lead to reduced attrition.

Obviously, the goal should be to reduce attrition, but attrition is inevitable and not necessarily a bad thing. As mentioned earlier, despite preclinical "evidence," not all targets will result in disease modulation in humans. In other cases, target modulation may result in unacceptable adverse events either in preclinical studies or in the clinic, which prevents achieving clinical success. The two key factors to consider in this regard are as follows:

- Take a disciplined approach when identifying a target to pursue and gather as much information as reasonably possible to make a "weight of evidence" argument and to derisk the target. This evidence should include relevant information regarding efficacy and safety related to the target such as phenotypic information from humans with single nucleotide polymorphs in that particular target (e.g., humans with SNPs in CCR5 are not or less prone to HIV infection), rigorous understanding of downstream changes associated with target modulation in cells and in preclinical models, knockout (KO) animals to assess safety of target inhibition, and so on.

TABLE 2.2 Stringent Five "R" Selection Criteria Presented by AstraZeneca in 2011 [10, 11]

Right target engagement	Link between target and disease
	Predictive biomarkers
Right tissue exposure	Bioavailability and tissue exposure
	Human PK/PD prediction
Right safety	Differentiating safety
	Reactive metabolites
Right patients	Scientific evidence in lead indication
	Stratification of patient population
Right commercial	Differentiated value proposition
	Embedded payer perspective

- Identify potential liabilities early and pursue specific preclinical and clinical studies early to derisk those liabilities. Taking attrition early at least saves money that can be spent wisely elsewhere. Unfortunately, there are many (mostly proprietary) examples where liabilities were identified early, but compounds still proceeded to phase III or sometimes registration failed for foreseeable reasons.

All of this highlights the need for a disciplined approach in R&D without being rigid. As information becomes available, projects should be reviewed against specific success criteria (see later), but the criteria applied to one target or disease area may be very different from those applied elsewhere. Thus, attrition should be viewed as a way to identify superior targets early. Recently, AstraZeneca performed a rigorous in-house analysis of success and failure in development and identified the five "R" criteria: the right target, the right exposure, the right safety, the right patient, and the right commercial opportunity [10, 11]. The five "R"s provide a clear and customizable framework when starting a project in drug discovery and to keep it on track in drug development (see Table 2.2).

Determining drug absorption, distribution, metabolism, and excretion (ADME) properties of drug candidates early on in the drug discovery process is also critical for achieving drugs with optimal properties and, in particular, an acceptable dose and dosing regimen. The attrition due to ADME has declined; in the 1980s, inappropriate ADME characteristics were the leading cause of attrition in drug development and about 40% of drugs failed for drug metabolism and pharmacokinetic (DMPK) reasons, and this number was reduced to about 10% by the year 2000 [12]. The latter was confirmed by another analysis of drug development project terminations from 1992 until 2002 with ADME being cited as the reason for attrition in 14, 17, and 4% in phase I, II, and III, respectively [13]. The main reason for this trend is the availability of suitable biological reagents, such as microsomes, and LC–MS/MS equipment to allow incorporation of ADME parameters early on in drug discovery. Nevertheless, it is important not to overinterpret the reduced attrition due to ADME reasons and assume that ADME issues are no longer of concern. As mentioned earlier, efficacy and toxicity have to be seen in light of drug exposure. Indeed, the lack of efficacy may be due to an inability to achieve sufficient exposure. In addition, some of the attrition due to toxicity may be mediated by (reactive) metabolites, and therefore, a greater understanding of the metabolic fate of drug candidates could influence survival in drug development. Toxicity due to inhibition of transporters, such as the bile salt export pump, has been observed as well. Thus, there are still ample opportunities for ADME sciences to have an impact on attrition. Indeed, several components of AstraZeneca's five "R"s have a direct bearing on ADME properties. In particular, pharmacokinetic (PK)/PD modeling has become a very popular and critical activity in both drug discovery and development. Using preclinical data, PK/PD modeling can illustrate the (quantitative) relationship between exposure and the impact on proximal and more distant biomarkers, and this knowledge can be applied to the determination of the anticipated human dose as well as design of clinical studies. In some cases, the preclinical models may be reliable surrogates for clinical efficacy, but in other cases, the translatability is tenuous at best. For example, the value of preclinical models for depression, schizophrenia, and Alzheimer's disease is very limited. In light of this, it would be best to include techniques such as imaging in clinical development to assess receptor occupancy early. Although receptor occupancy does not guarantee clinical efficacy, it at least confirms target engagement. Once clinical data is available, the PK/PD model can be revisited to firm up the relationships and reflect on translatability of the preclinical models. All of this highlights the rigor that should be associated with understanding the target and disease.

Subsequent sections will address past, current, and future practices in the pharmaceutical and biotechnology industry and their impact on attrition. Note that this chapter will include more small-molecule examples than large-molecule examples, but many of the same principles apply. However, there are also distinct differences such as the optimization process and, consequently, the preclinical attrition. For small molecules, the optimization process frequently necessitates synthesis of thousands of novel compounds after a high-throughput screening (HTS). Dozens of properties are determined for each synthesized compound such as biochemical and cellular potency, selectivity, *in vitro* and *in vivo* ADME properties, *in vitro* and *in vivo* safety properties, and so on. This multiparameter optimization process results in dramatic but desirable and structured attrition to identify the one compound or few compounds that have the best balance of properties and chance of success. For large molecules, such as antibodies, the number of constructs made is very limited (usually around a dozen), and their properties are usually fairly similar. Antibody-drug conjugates (ADCs) are conceptually somewhere in between small molecules and antibodies. The optimization process of ADCs involves many parameters—the choice of antibody, the site and type of conjugation, the linker type, the conjugated small-molecule drug, the number of conjugated small molecules, and so on—and each could have a significant impact on the properties of the ADCs. Thus, ADCs require significant and time-consuming preclinical optimization.

2.2   TARGET IDENTIFICATION, HTS, AND LEAD OPTIMIZATION

Currently, drug discovery usually begins with identification of the target of interest based on compelling preclinical and human genetic data. For example, an inactivating mutation in the Nav1.7 gene makes homozygous people insensitive to pain, and mutations in the LRRK2 gene are prevalent in Parkinson's disease patients. The target classes include peptide and aminergic GPCRs, metalloproteases, serine/cysteine/aspartyl proteases, hydrolases, kinases, phosphodiesterases, ion channels, nuclear hormone receptors, and so on. Some types of targets have historically proven to be harder to drugs than others. Nevertheless, each target usually has unique challenges. For example, because of sequence homology among kinases, selectivity is a key parameter to assess early on in drug discovery for kinase targets. Some degree of off-target activity may be acceptable and could even contribute to efficacy in some cases. For example, neratinib is an inhibitor of both ERBB2 and EGFR, and it is assumed that both contribute to the clinically observed efficacy. However, off-target activity could also be the driver for toxicity. For targets that involve the central nervous system (CNS), penetration of the blood–brain barrier and avoiding efflux can sometimes be a significant challenge. An analysis by AstraZeneca [14] has shown that the probability of a project transitioning from an HTS to a successful lead identification effort varies significantly across target classes. The attrition was assigned to chemistry issues (no tractable hits were found) or lack of target validation. Surprisingly, chemistry failures were more pronounced than target validation failures, and the success rate was lower with GPCRs than, for example, ion channels and nuclear hormone receptors. Increased attrition with certain types of targets may also reflect the limited organizational experience with those types of targets.

It is imperative to define a target candidate profile (TCP) early on that addresses desirable and/or essential properties regarding potency, selectivity, ADME, PD and efficacy, diagnostics and PD biomarkers, safety, pharmaceutical sciences, and so on. This profile lists the properties of a potential candidate—usually with numerical cutoffs. At an early stage, the TCP is not that well defined and it can be refined as the project proceeds. Moreover, it is important to keep in mind that some compromises in the profile of the candidate may be necessary or inevitable (provided all essential properties have been met) and, therefore, it is important not to apply the TCP criteria in a very rigid fashion.

Most companies have in-house capabilities for biochemical assay development and HTS. A chemical library may have been compiled over many years of in-house synthetic chemistry supplemented by acquisition of targeted libraries. Most large pharmaceutical companies have in-house libraries comprised of one to five million compounds. The latter may look impressive, but the overall chemical space has been estimated to exceed 10^{60} [13]. Moreover, the chemical diversity among the one to five million compounds in the library may be relatively limited because the compounds are derived from a finite number of internal projects or derived from simple and similar synthetic scaffolds to ease library synthesis [15]. Assuming a library of one million compounds and a not uncommon hit rate of 0.1%, the HTS will result in 1000 hits. The hits are rarely very potent (i.e., in the desirable single-digit nanomolar range), and most hits are in the 0.5–10 μM range. First, known promiscuous compounds should be removed [16]. Second, scaffolds or singletons need to be identified for chemical expansion. If a few compounds or scaffolds stand out in potency, this process is easy. However, identification of promising leads to pursue can be rather subjective. Lajiness *et al.* [17] performed a blinded analysis of small libraries of compounds, including a subset of 250 compounds previously rejected by "a very experienced senior medicinal chemist," and compared the compounds selected and rejected by a group of 13 medicinal chemists. Not surprisingly, there were distinct differences among the 13 medicinal chemists, and only one of the 250 compounds was rejected by all medicinal chemists. Thus, there is merit to the identification of promising leads or scaffolds by a group of experienced medicinal chemists instead of one individual medicinal chemist to avoid unfounded bias. Another aspect that may influence medicinal chemists is ease of synthesis. Although that is probably unavoidable and may even be advantageous because it may provide a quicker

TABLE 2.3   Examples of Drugs Discovered By Phenotypic Screening and Target-based Screening Between 1998 and 2009 [18]

Drugs discovered by phenotypic screening	Drugs discovered by target-based screening
Aripiprazole	Aprepitant
Caspofungin	Bortezomib
Daptomycin	Gefitinib
Ezetimibe	Imatinib
Fulvestrant	Maraviroc
Linezolid	Raltegravir
Sirolimus	Sitagliptin
Zonisamide	Sunitinib

route to compounds with the right balance of properties, it could also result in losing out on promising leads or scaffolds.

Sometimes, it appears that drug discovery is rather cyclical in the sense that approaches that have fallen out of favor a while ago are dusted off and reposed as the way to go. The discussion over phenotypic screening versus target-based screening falls in that category [18]. Originally, most of the screening involved phenotypic screening in cells or in some cases *in vivo* with obvious throughput limitations. This was driven in large part because the actual target of interest may have been unknown or the interest was driven by impact on a downstream PD endpoint somewhat independent of the actual mechanism behind the change in the PD. Phenotypic screening can either involve screening of a "random" compound library or it could be a more compound-specific library based on significant prior knowledge of compound properties (e.g., a subset with specific structural features). The latter approach overcomes the capacity issues for phenotypic screening. A nice example of a drug discovered via phenotypic screening is the cholesterol-lowering agent ezetimibe. The drug was launched in 2002 showing significant LDL lowering mainly driven by an unknown mechanism in the intestine. Ezetimibe was discovered with a phenotypic screen, and detailed *in vivo* preclinical data were published in 1995 [19]. Ezetimibe was later—after the launch—discovered to bind to a critical mediator of cholesterol absorption, the Niemann-Pick C1-Like 1 protein, on the epithelial cells of the gastrointestinal tract as well as in hepatocytes [20]. With advances in molecular biology, the pharmaceutical and biotechnology industries started to take a more target-based approach looking for modulators of specific targets. A few drugs discovered via phenotypic and target-based screening are listed in Table 2.3. An analysis showed that more first-in-class drugs have been discovered via phenotypic screening than target-based screening between 1999 and 2008: 37% versus 23%, respectively (the remainder are modified natural substances, 7%, and biologics, 33%) [18]. The opposite is true for follower drugs with 51% discovered via target-based screening and 18% via phenotypic screening. However, a recent publication suggested that the percentage of first-in-class drugs (launched from 1999 until 2013) discovered via phenotypic screens is much lower: 7% via phenotypic screens and an additional 23% via chemocentric approaches in which a compound with known pharmacology (but unknown target) served as the starting point [21]. Nevertheless, it appears that phenotypic screening has received renewed attention and is being pursued by several companies in specific therapeutic areas where there are easy to measure PD endpoints: antibiotics, antivirals, drugs for metabolic disorders, drugs that interfere with tumor metabolism, and so on. Phenotypic screening is clearly a tool with value and it should be pursued where appropriate. However, after hits are obtained, an effort should be made to identify the target in question, and this can be quite time-consuming. After the target has been identified, larger libraries can be screened to facilitate further optimization of the chemical matter. It is possible that the PD change observed in the phenotypic screen is caused by modulation of multiple targets. If that is the case, the translation of the phenotypic screen to a target-based screen is problematic, if not impossible. Interestingly, there are some people who argue that the industry should shift from a single to a multitarget paradigm [22]. *In this context, instead of pursuing highly selective compounds for unique targets, that is, 'single keys for specific locks,' the goal is shifting towards identifying ideal 'master keys' that selectively operate a set of 'multiple locks' to gain access to a clinical benefit that is usually associated with a complex biological process. Of course, the 'master key' should not open any lock (antitargets) to avoid adverse effects. In this scenario, the challenge is to also identify the set of targets that are associated with a desired clinical effect* [22].

Target-based screening could also involve screening of a library of fragments up to a molecular weight (MW) of around 250 Da. The technical approach is very similar to screening the complete library, and the fragment library is nowadays frequently appended as a subset. However, the potency of the fragments is limited with IC50 values usually exceeding 10 μM. Translation of fragment hits into lead molecules can be facilitated by structural biology and determination of cocrystals of the fragment and the target in question to identify the site of interaction. Once the site of interaction is known, it is easier to add and/or replace structural features to improve the potency significantly. Several drugs are in development that result from a fragment-based approach, but none of them have been approved yet [23].

In case the target of interest is precedented with promising chemical matter published in the literature or patents, the lead identification and optimization strategy could be very different and truncated. The medicinal chemists could bypass HTS and resort to a "participation strategy" whereby relatively small changes are made to compounds published in patents or the literature while making sure that they are sufficiently differentiated to assure freedom to operate and proprietary intellectual property space. Overall, this is an easy way to accelerate a project, but the chemists do run the risk that other companies may take the same approach and their chemical matter may be covered in patents not yet published. Whether a "participation strategy" is taken or novel chemical matter is pursued through HTS, it is always prudent to synthesize a few competitor compounds (if available) for benchmarking purposes and to compare them against the TCP. If it proves hard to show superiority over the competitor compounds, it may be time to reassign resources to another project.

The next step in the drug discovery process is establishing a high-quality and efficient screening cascade—an example of a screening cascade for a CNS target is shown in Figure 2.2. The screening cascade should involve measurement of biochemical and cellular potency, appropriate off-target screens, *in vitro* and *in vivo* ADME properties, preclinical PD and efficacy studies, and *in vitro* and *in vivo* safety evaluation but will look different from target to target. The capacity of various assays differs and, therefore, appropriate quantitative filters have to be inserted to manage the throughput of assays further downstream in the cascade. In addition, it is of great importance to optimize the

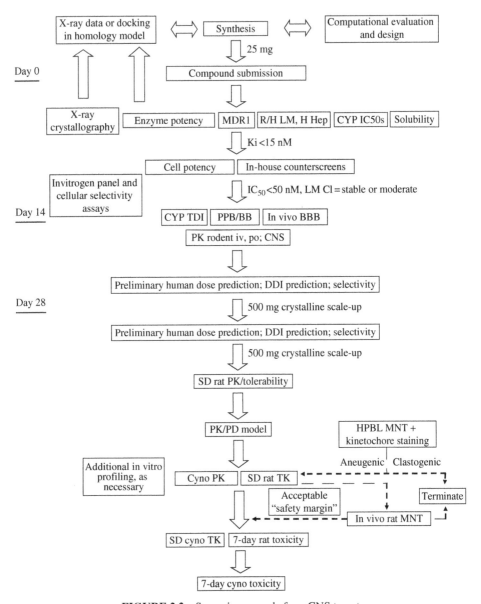

FIGURE 2.2   Screening cascade for a CNS target.

handoff between different assays and departments to assure timely generation of data to influence design of the next generation of compounds and efficient triage. If turnaround time is rather slow, the medicinal chemists are forced to design and synthesize new compounds without incorporating these data. Nevertheless, for certain assays—in particular *in vivo* assays—it may take several days or weeks to obtain data. Therefore, *in silico* models should be incorporated in screening cascades because they can provide preliminary data in an instant.

2.3   RESURGENCE OF COVALENT INHIBITORS

Covalent inhibitors have been around for a long time, and quite a few drugs bind covalently to their target of interest. Examples include aspirin, penicillin, omeprazole, clopidogrel, oseltamivir, telaprevir, and carfilzomib. However, in most cases, generation of these covalent inhibitors was not done by design, but established in hindsight after the drug was already introduced on the market. There has been a resurgence of covalent inhibitors lately, and several covalent inhibitors, such as ibrutinib (BTK inhibitor) and afatinib (EGFR inhibitor), were designed as such [24, 25].

The first step would be identification of the specific amino acid (usually cysteine, serine, lysine, or histidine) modified by the small-molecule drug and the mechanism (reversible vs. irreversible covalent binding). For example, aspirin inhibits cyclooxygenase 1 and 2 enzymes via acetylation of serine residues (serine-530 for COX-1 and serine-516 for COX-2) in close proximity to the aspirin binding site. The Bruton tyrosine kinase (BTK) inhibitor ibrutinib covalently binds via Michael addition to the cysteine-481 residue, and this is illustrated in Figure 2.3. The potential advantages of covalent inhibitors are as follows:

- High potency because the small-molecule inhibitor renders the protein permanently inactive resulting in degradation of the protein.
- Extended duration of action beyond the PK half-life of drug. The dosing regimen is determined by the resynthesis rate of the protein. If the resynthesis is rather short (<1 h), the advantages of a covalent inhibitor versus a competitive inhibitor are drastically reduced.
- Covalent inhibitors may still be active in mutated forms of the protein. However, if the specific residue that is responsible for the covalent interaction with the small molecule is mutated, the inhibitor will be virtually inactive.

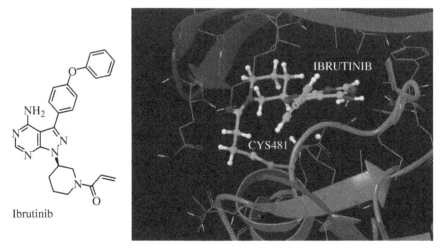

Ibrutinib

FIGURE 2.3   Structure of ibrutinib and proposed binding mode of ibrutinib in a homology model of BTK. The highlighted residue in the center is the cysteine-481 residue covalently bound to ibrutinib.

Because of the high potency and the extended duration of action, the demands for superior PK properties are less and it can reduce the dose and/or dose frequency for efficacy. However, oncology drugs are frequently administered at the maximum tolerated dose, and therefore, the dose of a covalent inhibitor in oncology may still be high; for example, the daily dose for ibrutinib is 560 mg.

The potential risks associated with covalent inhibitors cannot be ignored. It is possible that the inhibitor also covalently inhibits structurally related proteins, which could result in a specific biological response, potentially adverse but not necessarily. Alternatively, the inhibitor could nonselectively modify multiple proteins that could result in immune-mediated idiosyncratic drug toxicity. The latter response is rare, unpredictable, without a clear dose–response relationship, and there are usually no preclinical models available to inform about the clinical risk. The author encountered one situation where a covalent inhibitor displayed a whole-body rash in the dose-escalation phase, which was a sufficient reason to abandon the compound. The overall risk can be reduced by creating a selective covalent inhibitor. *In a nutshell, the therapeutic applicability or the success of irreversible binding kinase inhibitors is dependent on whether or not the covalent bond can be confined solely to the protein kinase of interest* [25]. Alternatively, the risk can be reduced by keeping the dose relatively low (preferably less than 50 mg) or by creating a reversible covalent inhibitor, which is presumed to have a reduced risk for an idiosyncratic adverse response. Nevertheless, most organizations are quite hesitant to nominate a covalent inhibitor for more benign indications and/or indications associated with prolonged drug use. It is too early to tell if covalent inhibitors have higher or lower attrition rates than competitive inhibitors.

Thus, the screening cascade to identify a covalent inhibitor will look quite different with increased emphasis on selectivity, and it may also be advantageous to insert assays that assess the general risk of forming reactive species (reactive metabolite trapping with glutathione, cyanide, or semicarbazide, time-dependent inhibition of cytochromes P450 (CYPs), etc.).

2.4 *IN SILICO* MODELS TO ENHANCE LEAD OPTIMIZATION

In silico models have matured tremendously over the last decade, and there are now many models available for various properties. The most mature models are those for physico-chemical properties, such as $\log P$, $\log D$, pK_a, and topological polar surface area (TPSA), and they are routinely used in drug discovery. *In silico* models for ADME parameters have become more popular as well. Over the last 10 years, the capacity of *in vitro* ADME screens has increased rapidly due to assay automation and LC–MS/MS for compound detection. Many companies are now able to screen in excess of 500 compounds per week through a standardized suite of *in vitro* ADME assays such as metabolic stability in microsomes and hepatocytes, reversible and time-dependent CYP inhibition, permeability in MDCK or LLC-PK1 cells, efflux assays in cells overexpressing MDR1, plasma protein binding, and so on. A collateral benefit from higher-throughput *in vitro* ADME screening is that there is now a wealth of data available that can be used to build sophisticated *in silico* ADME models that can enhance the drug discovery process. Some *in silico* models are available from vendors. However, a significant fraction of the commercial models are of limited value because the number of compounds to train the model and the corresponding chemical diversity are limited and the data were usually obtained from the literature using different experimental protocols. Hence, many pharmaceutical companies have built various *in vitro* ADME models based on their own extensive in-house dataset. The models incorporate a range of chemical descriptors and/or the structures of the compounds in the training set. Different statistical methods (regression methods: partial least squares; Bayesian methods; supervised learning methods: decision trees, support vector machine; neural networks) are used to build the models. It is of critical importance to validate the models with a separate dataset.

The output of the model can either be categorical or numerical. For numerical models, the correlation between measured and calculated data or the average fold error can be used to assess the value of the model. For categorical model, it is important to look at the percentage false positive or false negative compounds. Besides the output, the confidence in the prediction should be defined. The confidence in the prediction is usually derived from (i) the structural similarity of the compound under investigation with those in the training set and (ii) the number of structurally similar compounds (nearest neighbors) in the training set. How to use a model is illustrated in Figure 2.4. The model predicts the probability of a compound being metabolically stable in human liver microsomes: <0.2 suggests that it is quite likely that the compound is unstable and >0.8 suggests that it is quite likely that the compound is stable. Of the 163 compounds that had a probability value of <0.2, only one compound was actually measured to be stable (false negatives). On the other hand, of the 530 compounds that had a probability value of >0.8, only 27 compounds were measured to be unstable (false positives). Thus, the number of false negatives and false positives is quite low, and hence, a minimum probability value can be defined prior to synthesis to enrich the compound set in meta-bolically stable compounds without running the risk that promising compounds are missed. The minimum probability will be project dependent because the model may not be as predictive for different chemical space.

The *in silico* ADME models can be of great value, but they obviously have their limitations as well, and ultimately their use should be viewed in light of the implica-tions of the *in silico* data on the project. In other words, *in silico* data should be used in

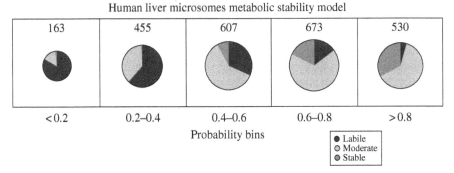

FIGURE 2.4 Performance of a human liver microsomes metabolic stability model. The bins pro-vide the predicted probability of the compound being stable: <0.2 suggests that it is quite likely that the compound is unstable and >0.8 suggests that it is quite likely that the compound is stable. The color provides the measured metabolic stability in microsomes: gray, CLhep <6.2 ml min^{-1} kg^{-1}; light gray, CLhep between 6.2 ml min^{-1} kg^{-1} and 14.5 ml min^{-1} kg^{-1}; dark gray, CLhep >14.5 ml min^{-1} kg^{-1}.

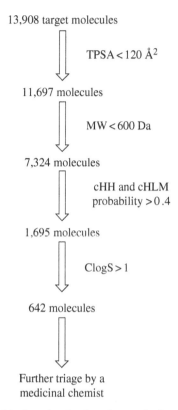

FIGURE 2.5   Triaging of virtual molecules based on calculated physicochemical properties (topological polar surface area and molecular weight) and *in silico–in vitro* ADME models (metabolic stability in human liver microsomes and human hepatocytes) and calculated solubility.

a "fit-for-purpose" fashion. Despite significant process, the models aren't always that good at making accurate predictions for a particular property, and this limits their use in the later stages of drug discovery where differentiation among compounds with relatively similar properties is a necessity. The value of the *in silico* ADME models is most pronounced in the early stages of drug discovery. Prior to synthesis of compounds, a range of physicochemical or ADME properties can be calculated to improve the odds of synthesizing compounds with desirable properties. An example is presented in Figure 2.5. A virtual library of compounds was designed and it was run through a series of successive filters for different properties: TPSA, MW, calculated human hepatocyte and microsomal stability, and calculated solubility [26]. This reduced 13,908 potential compounds to be synthesized to 642 compounds; the 642 compounds were subsequently reviewed by a medicinal chemist and narrowed down further based on synthetic feasibility and potency SAR information. Overall, this is a nice example of informed and desirable attrition. Thus, these *in silico* models are generally not meant to replace experimental data—or at least not at this stage. The main goal should be to increase the probability of identifying compounds with improved properties, such as metabolic stability or permeability. This effect is illustrated by an analysis of in-house data at Genentech showing that the number of compounds with good microsomal stability

(CL_{hep} in human liver microsomes $< 6.2\,ml\,(min\,kg)^{-1}$ increased from less than 10% of all new compounds synthesized to about 30% after implementation of the *in silico* metabolic stability model in 2010, but without dramatic changes in the project portfolio. *In silico* ADME models can also be used as a filter prior to experimental determination of the property in question if the capacity of the assay is limited. For example, if a compound is predicted to be metabolically unstable and if the confidence in the prediction is high, the compound can be set aside and more promising compounds can be submitted instead. Finally, *in silico* models provide a quick means to get a (preliminary) view of the ADME SAR, which can then be compared with the potency SAR. Ideally, multiple *in silico* models, including potency models, should be used in a parallel fashion to triage large virtual libraries. Unfortunately, progress with building reliable *in silico* models for potency—pharmacophore and docking models—appears to lag behind the advances made with *in silico* ADME models. Moreover, industrial medicinal chemists seem hesitant to use the models where they do exist.

As mentioned before, some models are commercially available and models such as MetaSite and StarDrop have proven to be relatively reliable. Although there are some fundamental differences in the computational algorithms for MetaSite and StarDrop, both are capable of predicting the site of metabolism by certain CYP enzymes. Both incorporate the intrinsic reactivity of potential sites for oxidative metabolism and the proximity of the site of metabolism to the active oxyheme species in the CYP enzymes. It was shown that both MetaSite and StarDrop are capable of predicting the right site of metabolism about 60–90% of the time for CYP2C9, CYP2D6, and CYP3A4 for a substantial set of literature and in-house compounds [27]. This could provide valuable structural information for lead optimization to guide design away from metabolic liabilities. However, neither software package predicts the rate of metabolism.

2.5 STRUCTURE-BASED AND PROPERTY-BASED COMPOUND DESIGN IN LEAD OPTIMIZATION

There are two common approaches in drug discovery to guide design of compounds with superior potency:

- Structure-based design
- Property-based design

Structural biology can play a critical role in the identification of appropriate chemical matter to pursue and/or how to make appropriate changes to improve potency or reduce off-target activity. The first step is obtaining the crystal structure of the target itself. Crystallization may be problematic and time-consuming; frequently several constructs have to be generated to obtain one that provides crystal structures with a good resolution (usually better than 3 Å). Next, the structure of the target with the small molecule embedded in the active site should be obtained, and this can be achieved by either cocrystallization or soaking. The crystal structure may identify the point(s) of interaction and the type(s) of interaction between the small molecule and the amino acid residues in the protein and that information can be used in lead optimization to make appropriate structural changes to improve potency. For example, a side chain may be added to reach further into

(a)

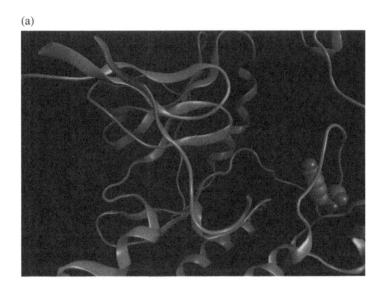

(b)

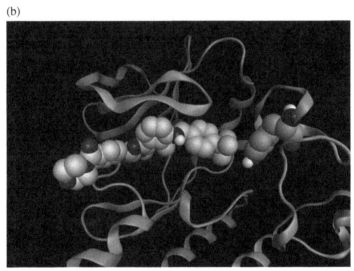

FIGURE 2.6 (a) Crystal structure of human apo-BTK. (b) Crystal structure of CGI1746 bound to human BTK.

the active site and establish a critical contact with the protein. A visual example is shown in Figure 2.6. Figure 2.6a shows the structure of BTK, and Figure 2.6b shows the structure of CGI1746 bound in the active site of BTK. It can be seen that a part of the active site moved out of the way to accommodate the inhibitor and create a point of interaction with a tyrosine residue, which stabilizes an inactive conformation. Thus, structure-based design can be an effective way to create or enhance specific points of interaction, and this can lead to improvement of the binding mode.

Another approach in lead optimization is property-based design. The first thorough analysis of the impact of physicochemical parameters was performed by Christopher

Lipinski at Pfizer in the late 1990s [28], and he introduced the famous "rule of 5." He argued that poor absorption is more likely if:

- $\log P > 5$
- MW > 500 Da
- Number of hydrogen bond donors (HBD) > 5
- Number of hydrogen bond acceptors (HBA) > 10

Fortunately, all these parameters can be calculated easily and incorporated in compound design prior to synthesis. Subsequently, many manuscripts have been published about the importance of physicochemical parameters. Different cutoffs have been proposed or other parameters (such as TPSA, number of sp³ carbon atoms, number of aromatic atoms, number of rotatable bonds) have been included, but overall the messages were quite similar. Even though there are quite a few successful drugs outside the traditional "rule of 5" space (see later for details), the odds of clinical success are generally lower. Thus, it is better to stay within "rule of 5" space unless there is a compelling structural reason to deviate. For example, if the goal is to disrupt a protein–protein interaction with a much larger surface of interaction and a large and shallow binding pocket, it may be necessary to synthesize larger lipophilic molecules with a larger number of HBD and/or HBA.

Several authors have shown that the MW and lipophilicity of approved drugs have been gradually increasing (see Table 2.4 based on data in Refs. 29 and 30). This may have been an unfortunate side effect of HTS and subsequent lead optimization with an intense push to identify more potent compounds; it has been shown that potency increases with MW and $\log P$ in a statistically significant fashion [31]. Consequently, compounds may have been optimized to increase lipophilic interactions resulting in mainly entropically driven gains in potency. However, this inevitably results in less desirable properties elsewhere (e.g., poor solubility, higher clearance, and/or higher plasma protein binding resulting in lower unbound drug levels). Indeed, an analysis by the author of a large corporate database showed that—relatively speaking—a higher percentage of compounds with "rule of 5" violations attrite for PK reasons than compounds that are "rule of 5" compliant. In addition, an

TABLE 2.4 Average Physicochemical Properties of oral Drugs in Phase I and Marketed oral Drugs

	Phase I oral drugs	Marketed oral drugs	Oral drugs launched pre-1983	Oral drugs launched 1983–2002
Molecular weight (Da)	423	337	331	377
$c\log P$	2.6	2.5	2.3	2.5
$c\log D_{7.4}$	1.3	1.0		
Hydrogen bond donors	2.5	2.1	1.8	1.8
Hydrogen bond acceptors	6.4	4.9	3.0	3.7
Rotatable bonds	7.8	5.9	5.0	6.4
Rings			2.6	2.9

Data from Wenlock *et al.* [29] and Leeson and Davis [30].

analysis of a database of oral drugs and the ChEMBL database of "hits" indicated that the mean MW of drugs and "hits" are 333 and 430 Da, respectively, and the mean log$P$ values are 2.5 and 3.5, respectively [31]. It has also been shown that the percentage of sp^3 carbon atoms in approved drugs is higher than that in phase I drugs [32]. These data suggest that "rule of 5" violations are associated with higher attrition in drug development and, therefore, physicochemical properties should be considered and incorporated in drug design.

It is also important to consider the organizational culture of a pharmaceutical company in this context; some companies rigorously stick to "rule of 5" compliance or similar criteria, whereas other companies may be more lenient, and either attitude can be inappropriate depending on the circumstances. A detailed analysis by Leeson and St-Gallay [33] showed that the log$P$ and MW values of compounds in patents published by various companies vary widely and that differences are maintained if only shared targets are considered. There is almost one log unit difference in the mean $c$log$P$ between Takeda, Lilly, and Wyeth on the one hand (mean $c$log$P$ between 4.4 and 4.5) and Pfizer, Vertex, and AstraZeneca on the other hand (mean $c$log$P$ between 3.7 and 3.5). Similarly, there is a distinct difference in the mean MW between Novartis, Boehringer Ingelheim, and Schering-Plough on the one hand (mean MW between 475 and 485 Da) and Lilly, Pfizer, and Vertex on the other hand (mean MW between 440 and 425 Da). There are also remarkable differences across companies in other physicochemical properties such as the number of HBD and HBA, the number of sp^3 carbon atoms, the number of aromatic carbon atoms, and the number of rotatable bonds. A nice illustration is represented by the CCR5 antagonists published by AstraZeneca, GlaxoSmithKline, Merck, and Pfizer showing a 2 log unit difference in mean $c$log$P$ and 100 Da difference in mean MW between leads from these four companies [34].

Finally, different target classes and/or therapeutic areas are inevitably associated with differences in physicochemical properties [33, 35, 36]. For example, protease inhibitors tend to be much less lipophilic than antagonists of nuclear hormone receptors. In addition, drugs that need to penetrate the blood–brain barrier tend to have a lower MW (<400 Da) and much lower TPSA (<90 Å^2). Thus, each target may have a unique physicochemical sweet spot that needs to be incorporated in design and/or filtering of virtual libraries. The physicochemical constraints associated with the target in question may also be a source of attrition in the sense that it may be hard to find the right balance between potency and appropriate DMPK and pharmaceutical properties. Indeed, some targets are considered undruggable for that particular reason. This topic will be discussed further in the next section.

2.5.1   Risks Associated with Operating in Nondrug-Like Space

Historically, some targets or target classes are considered undruggable because it is (nearly) impossible to find the right balance between various properties such as potency, selectivity, ADME properties, and pharmaceutical properties. However, more companies are moving beyond the traditional "rule of 5" compliant space to go after biologically compelling targets. One example is navitoclax, ABT-263 (see Fig. 2.7). Navitoclax is an inhibitor of Bcl-2 and Bcl-xl, and the preclinical data suggest that inhibitors of these targets are attractive for the treatment of hematological malignancies such as chronic lymphocytic leukemia and non-Hodgkin lymphoma [37]. Briefly, cancer cells frequently overexpress the prosurvival Bcl-2 protein, and Bcl-2 prevents apoptosis by sequestering proapoptotic counterparts such as Bax and Bak. By inhibiting Bcl-2, this interaction is disrupted and Bax and Bak oligomerize leading ultimately to caspase activation and

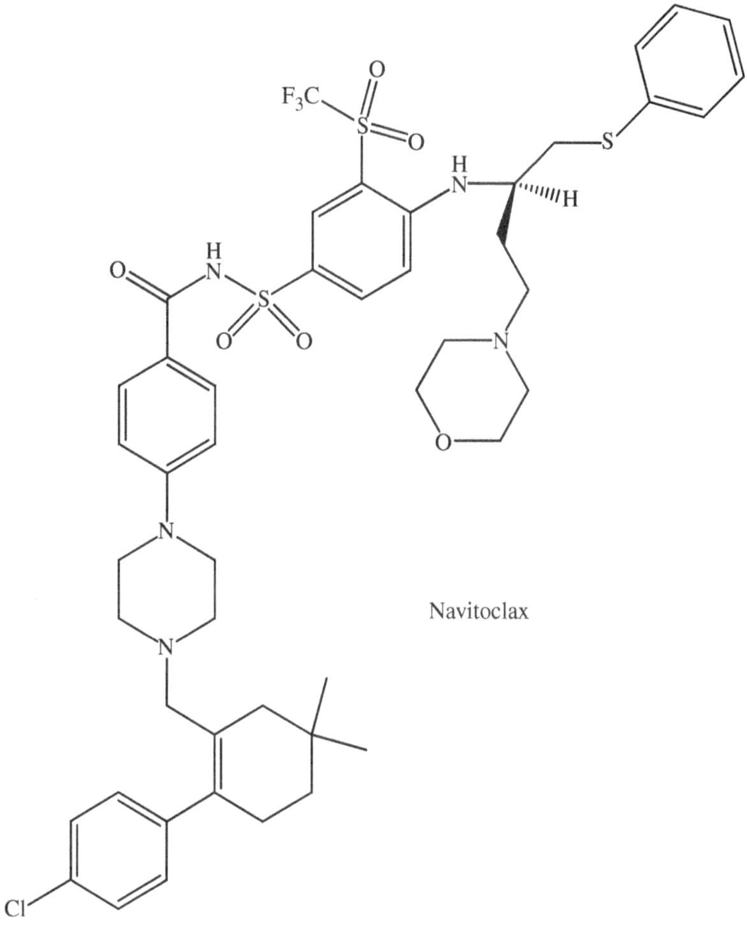

FIGURE 2.7 Structure of navitoclax.

apoptosis. Navitoclax has a high MW (974 Da), is extremely lipophilic ($clogP$ = 9.7), and has a moderately high TPSA (128 Å^2) [38]. Consequently, the aqueous solubility is very low (<0.0002 mg ml^{-1} at pH = 7.4), but it is soluble in lipophilic vehicles such as soybean oil (9.17 mg ml^{-1}). The physicochemical properties clearly do not conform with the "rule of 5," and therefore, the bioavailability was predicted to be quite low after oral dosing. Nevertheless, the plasma exposure in humans is quite reasonable with a C_{max} of about 4–5 μg ml^{-1} following a 250 mg of navitoclax, and single-agent clinical efficacy was observed with a dramatic drop in lymphocyte count [39]. Subsequently, it was shown that lymphatic absorption may play a role in the better than expected PK [38]. Indeed, the bioavailability was 56.5% in intact dogs and dropped to 21.7% in thoracic lymph duct-cannulated dogs. The lymphatic absorption not only improved absorption, but it also avoided the hepatic first-pass effect. Development of the clinical formulation was quite challenging though. Unfortunately, exposure in the clinic was limited by thrombocytopenia, which was mediated by Bcl-xl [39]. Subsequently, a more selective Bcl-2 inhibitor that did not inhibit Bcl-xl was identified, and it achieved increased efficacy due to its platelet-sparing profile [40]. Thus, clinical success can be achieved in chemical space beyond the "rule of 5" and should be pursued if the biological rationale is compelling. However, the undesirable physicochemical properties may create challenges in development such as identification of a suitable formulation. A good example is telaprevir (MW is 680 Da and the log$P$ is 4.0), which was launched in 2011 for the treatment of hepatitis C virus infection [41]. The bioavailability in preclinical and clinical studies with the initial formulation was very low (≤2%) and variable. Indeed, scientists from Vertex joked that marble was probably more soluble than telaprevir. Extensive formulation development was required, and it was on the critical path toward approval, but a stable amorphous dispersion was identified with good bioavailability, 20–40%. Despite these heroic and laudable efforts, the dosing regimen is not ideal: 750 mg three times a day. Thus, the pursuit of compounds outside "rule of 5" space has a direct impact on the development path, and the quote by Teague [42] is broadly applicable.

> Where there is a choice in a therapeutic area between a lipophilic agent that violates the rules and one with a better balance of properties, senior management would still be best advised to back the latter. Imperfect agents might simply validate a target and encourage competitors to enter with a drug having the improved properties. [42]

Natural products deserve special mention and have become more popular [43, 44] as part of diversity-oriented synthesis and biology-oriented synthesis strategies. Despite pronounced "rule of 5" violations, many natural products and in particular macrocyclic natural products such as cyclosporine, erythromycin, and tacrolimus display acceptable PK properties including bioavailability. The reasons are still a topic of discussion and debate but may involve (i) specific uptake transporters to enhance absorption and/or distribution and (ii) conformational flexibility and internal hydrogen bond formation [45]. In a liquid environment, a macrocyclic natural product may have improved solubility because of hydrogen bond formation with water, whereas in a lipid membrane, internal hydrogen bonds are formed that "shield" polarity and enhance the permeability. Cyclosporine is a natural product obtained from a fungus found in Norway and is a cyclic peptide comprised of 11 amino acids, including 1 D-amino acid. The physicochemical properties of cyclosporine (MW is 1202 Da; clog$P$ is 4.0; TPSA is 279 Å^2) suggest poor absorption. However, a detailed nuclear magnetic resonance (NMR) and gas-phase computational analysis [45] has shown that cyclosporine forms four internal hydrogen bonds, which reduces the number of HBD from 5 to 1. Consequently, the effective polar surface area is much lower than that predicted by the TPSA (279 Å^2) and leads to a more compact conformation with a lower molecular volume. These factors could explain the remarkably good oral bioavailability of cyclosporine in humans, 30% [45]. Hence, natural product and macrocycle research has encountered a revival, in particular for antibiotics and antivirals. However, dialing in the properties described earlier by design still remains a formidable task.

2.6   ATTRITION DUE TO ADME REASONS

The main reason that attrition for ADME-related reasons has been reduced significantly in drug development is that various *in vitro* and *in vivo* ADME assays have been incorporated in drug discovery. ADME properties are optimized in parallel with potency, selectivity, and certain toxicity-related parameters by incorporating these assays in screening cascades (see preceding text). Consequently, the ADME assays can be a significant source of attrition prior to nomination, which is a good thing because it leads to compounds with superior ADME properties that have a better chance of achieving an efficacious and safe exposure in humans at a reasonable dose and without major drug–drug interactions. The following *in vitro* ADME assays are routinely employed in drug discovery:

- Permeability in Caco-2 or Madin-Darby canine kidney (MDCK) cells or the parallel artificial membrane permeability assay (PAMPA)
- Efflux in MDCK or LLC-PK cells overexpressing transporters such as P-glycoprotein (P-gp), breast cancer resistance protein (BCRP), organic anion transporter polypeptides (OATPs), organic anion transporters (OATs), and organic cation transporters (OCTs)
- Metabolic stability in microsomes, S9, cytosol, hepatocytes, or recombinant CYP enzymes, including reaction phenotyping and metabolite identification
- Reversible CYP inhibition and mechanism-based or time-dependent CYP inhibition
- CYP induction
- Plasma protein binding or drug binding in other matrices such as microsomes or brain
- Blood to plasma partitioning

It is beyond the scope of this review to provide specifics for these assays. The format of most assays is the same independent of the target or the therapeutic area. However, the timing of the assays and the risk associated with an adverse finding does depend on the target and the therapeutic area. For example, the desirable ADME profile in oncology or for a drug that has a shorter duration of administration (e.g., antibiotics) may be more lenient than that of a drug that will be administered for a long time to relative healthy patients (e.g., cholesterol-lowering drugs or weight loss drugs). An inhaled drug obviously has a different set of desirable attributes than an oral drug; poor metabolic stability is usually desirable for an inhaled drug to reduce systemic exposure. Finally, the timing of these studies is situation dependent. For example, if the permeability in the chemical series of interest is high, the need for transporter studies is less and can be reduced to occasional spot checking. Thus, the positioning of these *in vitro* ADME assays in the screening cascade varies from project to project, and the same applies to the numerical cutoffs to proceed to the next level in the screening cascade (see Fig. 2.2 for an example).

To meet project demands, most major pharmaceutical companies have made major investments to increase the capacity and throughput of these *in vitro* ADME assays. A capacity of 500 compounds per week for some of these assays (e.g., metabolic stability in microsomes, competitive CYP inhibition, and passive permeability in cells) is not uncommon with a turnaround time of 1 week or less. This has been achieved due to the extensive use of (i) automation, (ii) 96-well and, preferably, 384-well plates, and (iii) analysis by LC–MS/MS [46–48]. In most cases, the assays operate in a tiered fashion with most compounds going through more routine assays (such as metabolic stability in microsomes and reversible CYP inhibition) and the outcome of these assays (as well as assays such as biochemical or cellular potency) influencing what is going through more capacity limited assays. The sheer quantity of data generated using standardized assays has greatly facilitated building reliable *in silico* ADME models (see preceding text). Although these data can and should be used to influence design in drug discovery, they are not rigorous enough for regulatory submissions and more customized and validated assays are required for that purpose.

In vivo PK studies are commonly performed in drug discovery, usually in rats or mice, but dog and/or cynomolgus or rhesus monkey studies can provide value in the later stages of drug discovery. The purpose of these studies is as follows:

- Assessment of the exposure in preclinical species to inform PD or efficacy studies. Sometimes, it may be necessary to perform a PK study in parallel to PD or efficacy studies to avoid impact of blood withdrawal on the PD or efficacy endpoint. These studies can provide preclinical proof of concept (POC) and increase the confidence in the target. However, it is understood that the translatability of preclinical models varies greatly. For example, the value of preclinical models for depression and schizophrenia is questionable, whereas preclinical models for antibiotics are reasonably reliable.

- Assessment of the exposure in preclinical species to aid in the interpretation of toxicology studies. The exposure observed in toxicology studies can be compared to that required to achieve a preclinical PD response or preclinical efficacy or that anticipated in humans for efficacy.

- Determination of PK parameters in preclinical species to aid in the identification of compounds with appropriate human PK. The use of cassette or "n-in-one" dosing [49], sample pooling [50], and LC–MS/MS has increased the capacity of *in vivo* studies greatly. The preclinical PK data can be used to allow allometric scaling to obtain the human clearance or volume of distribution and to get a measure of the fraction absorbed. Alternatively, preclinical PK data, in particular the clearance, can be used to assess the extent of *in vitro–in vivo* correlation in preclinical species to build confidence in the use of *in vitro* data for human clearance prediction.

Other information that can be obtained from *in vivo* ADME studies is more detailed knowledge about tissue distribution as well as metabolism and excretion pathways. Preclinical tissue distribution is desired by most regulatory agencies to enhance their understanding of tissue distribution and exposure as it relates to preclinical efficacy and in particular toxicity. Tissue distribution can be done by excising tissues and measuring drug and/or metabolite levels using routine LC–MS/MS assays. A more comprehensive view of tissue distribution can be obtained by quantitative whole-body autoradiography (QWBA), but this requires radiolabeled material, which is usually not available in drug discovery. Moreover, QWBA cannot distinguish the parent compound from metabolites. Alternatively, matrix-assisted laser desorption/ionization (MALDI) imaging can be used [51]. Tissue or whole-body slices are obtained in the same fashion as for QWBA, but the slices are subsequently coated with a finely dispersed matrix solution (e.g., 2,5-dihydroxybenzoic acid in methanol/water). The analytes are extracted out of the tissue by the solvent. The matrix absorbs the laser lights, which results in evaporation and ionization of the matrix followed by proton transfer and ionization of the analytes. Next, the laser is

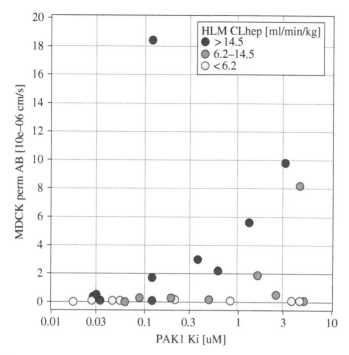

FIGURE 2.8 Potency, MDCK permeability, and metabolic stability in human liver microsomes of various PAK1 inhibitors.

scanned across the tissue and an image of the distribution of the parent drug and/or its metabolites is obtained. In most cases, detection is in the MS/MS mode to prevent interference by the abundant matrix ions. Recently, MALDI imaging was used to examine the distribution of PI3K inhibitors in the brain for the treatment of glioblastomas [52]. Mouse xenograft models with different glioblastoma cell lines were used. It was shown that GDC-0941, a compound that is a substrate of P-gp, does not penetrate the healthy part of the brain and the GS2 tumors, but does penetrate the more "leaky" U87 tumors. In contrast, GNE-317, a compound that isn't a P-gp substrate, was detected across the healthy brain and both tumor models. This highlighted the importance of blood–brain barrier penetration and the absence of P-gp efflux for successful treatments or glioblastomas. Although significant progress has been made, the two main disadvantages of MALDI imaging are limited sensitivity and lengthy data acquisition. Overall, tissue distribution studies can be quite informative but rarely affect "go/no go" decisions.

In general, it is still remarkably hard to optimize ADME properties in parallel with other properties such as potency, selectivity, and various toxicity endpoints, and it can be a rather iterative and time-consuming process. An example is presented in Figure 2.8 for a number of PAK1 inhibitors showing the PAK1 Ki value versus the MDCK permeability and color coded according to the metabolic stability in human liver microsomes. It can be seen that there are no PAK1 inhibitors with Ki < 100 nM with good permeability in this series. Moreover, the most permeable compounds display poor metabolic stability. Combined with selectivity issues, this was sufficient reason to look elsewhere for lead matter. In some cases, optimization is made harder by a corporate culture that may put more emphasis on potency—not uncommon in organizations with dominant medicinal chemistry leaders. Indeed, Hann [53] referred to *"our perceived need for potency, bordering on an addiction."* First, potency may already have been optimized to a certain extent by the time ADME is incorporated seriously and the consequence of that approach is that dialing in ADME properties may be much harder because of the narrow chemical space provided by the potency SAR. Second, the best optimum may well be a slightly less potent compound with a higher fraction absorbed and/or lower clearance. The latter highlights a general feature that compromises and trade-offs are probably inevitable in drug discovery as long as they don't come at the expenses of the critical criteria defined in the TCP.

2.6.1 Metabolism, Bioactivation, and Attrition

Metabolite identification and, in particular, formation of reactive metabolites deserve special consideration. *In vitro* metabolite identification is a quick means to identify metabolic soft spots. *In vivo* studies with either nonradiolabeled material or radiolabeled material can provide valuable information about the metabolic pathways as well as excretion. Sometimes, the exact site of metabolism may be unambiguous (e.g., for simple N-dealkylations), but for common pathways, such as aliphatic and aromatic hydroxylation, it is usually impossible to obtain the exact structure of the metabolite by LC–MS/MS. Nevertheless, the information can be valuable to guide design because it narrows down the part of the molecule to be further optimized. The optimization can be in the form of blocking the site of metabolism (which reduces V_{max} of the catalytic cycle) or involve disruption of binding to the metabolizing enzyme (which increases the K_M of the catalytic cycle). In addition, metabolite identification and quantitation play an important role to ensure that there is sufficient exposure in preclinical toxicology studies to derisk human safety, that is, metabolites in safety testing (MIST) criteria.

The formation of reactive metabolites can be of particular concern because it hints at intrinsic reactivity of the parent molecule or a metabolite, which could result in immune-mediated toxicity and attrition. The link between bioactivation and toxicity is mostly based on circumstantial evidence. Nevertheless, bioactivation is of concern because the toxicity could be idiosyncratic in nature and, consequently, not be picked up until phase III trials are ongoing or until after approval. Thus, the risk may be relatively low, but the impact could be very significant. Many companies have established assays to assess the risk in drug discovery [54, 55]. Although the reactive intermediates cannot be detected, it is possible to trap them with reagents such as glutathione (abundant in the liver and an effective nucleophile by virtue of its cysteine sulfhydryl group), potassium cyanide (a "hard" nucleophile for detection of iminium species), and methoxylamine and semicarbazide (to detect aldehydes). Routine LC–MS/MS assays are available to identify these adducts.

Some labs have taken this a step further and are examining covalent binding to cellular components (usually proteins). Although covalent binding is of concern, the study requires radiolabeled material and that is usually not available in discovery. In addition, it does not really lend itself to iterative studies to establish SAR. Nakayama *et al.* [56] examined the correlations between covalent binding, glutathione adduct formation, and time-dependent CYP inhibition using a dataset of 42 structurally diverse compounds. They found no statistically significant correlation between the formation rate of glutathione adduct formation and the extent of covalent binding. For example, pioglitazone and ritonavir showed extensive covalent binding (353 and 253 pmol mg^{-1} protein, well above the 50 pmol mg^{-1} protein threshold suggested by Evans *et al.* [54]), but glutathione adducts were not detected. Not all reactive metabolites lend themselves to trapping with glutathione, and it is conceivable that adducts would have been observed with other trapping reagents. In addition, the authors also did not observe a statistically significant relationship between the enzyme inactivation rate (as assessed by the time-dependent CYP inhibition) and the extent of covalent binding. To obtain a more comprehensive picture of bioactivation, the authors calculated the intrinsic clearance for forming reactive metabolites by summing the intrinsic clearance value associated with (i) the formation of glutathione adducts and (ii) the enzyme inactivation in the time-dependent CYP inhibition assay. Fortunately, there was a statistically significant correlation between the intrinsic clearance for forming reactive metabolites as defined earlier and the extent of covalent binding, and there were no obvious false negatives (i.e., compounds that were clean in the time-dependent CYP assay and/or glutathione trapping assay but showed more than 50 pmol mg^{-1} protein in the covalent binding assay). Thus, glutathione adduct formation and time-dependent CYP inhibition can be examined in drug discovery as a surrogate for covalent binding. However, there is no clear relationship between the extent of covalent binding and hepatotoxicity [57]. Indeed, atorvastatin displayed extensive covalent binding (352 pmol mg^{-1} protein), but it is remarkably safe in the clinic and was the best-selling drug in the world for many years in a row. A simpler but potentially more reliable way to assess the risk for idiosyncratic adverse drug reactions (IADRs) is based on the clinical dose. It has been well established that the vast majority of drugs withdrawn from the market due to IADRs or having a black box warning due to IADRs have a dose in excess of 100 mg [58]. Thus, the level of concern about formation of reactive metabolites should be viewed in light of the following parameters:

- Clinical dose and exposure
- Duration of therapy (short vs. long duration)
- Therapeutic area (life threatening or not)
- Target population (young vs. old, relatively healthy vs. seriously ill individuals)
- First in disease/class or not
- Lead compound versus backup
- Species differences (bioactivation may be unique to one species)

2.6.2 PK/PD Modeling in Drug Discovery to Reduce Attrition

The availability of extensive preclinical exposure and PD or efficacy data allows PK/PD/ efficacy modeling. This will enhance our understanding—in a quantitative fashion—of the relationship between the exposure and the observed effect and provide mechanistic insights about pathway modulation [59, 60]. Next, this relationship can be combined with the predicted human PK parameters to predict the human efficacious dose. The importance of this approach was also recognized by AstraZeneca by including it in their 5 "R" criteria for project success [11]. A detailed analysis of preclinical tumor xenograft data and clinical data in oncology patients has shown the value of these types of analyses as well [61]. There was no correlation between the percent tumor growth inhibition at the maximum tolerated dose in preclinical mouse models and the clinical response from single-agent trials. However, if the preclinical exposure/response relationship from mouse xenograft studies was combined with the observed human PK, there was a clear correlation between the simulated human tumor growth inhibition and the clinical response. Finally, this retrospective analysis also helped set a benchmark of at least 60% simulated human tumor growth inhibition to have confidence in clinical viability and that value is now routinely used at Genentech to assess clinical viability. Note that in a discovery setting, the mouse xenograft data need to be combined with the *predicted* human PK. The latter increases the uncertainty, but it is still a good way to identify the best compound to nominate, to predict the human efficacious dose, and to increase the confidence in clinical viability. Overall, continuous assessment of preclinical predictions with clinical results allows for calibration of preclinical models and increases the confidence to translate across species; in other words, continuously "learn and confirm." PK/PD/efficacy modeling has matured significantly over the last decade, and it is now an essential endeavor in drug discovery. However, it is still important to keep in mind that the preclinical models don't always translate. For example, a similar in-house analysis of rodent arthritis models, such as rat and mouse collagen-induced arthritis and adjuvant-induced arthritis models, showed that they are not very predictive for the clinical response. The exposure/response relationships appear to be quantitatively different in rodent models of arthritis compared to human. Nevertheless, there is still value to these studies if signs of activity are observed in multiple preclinical models and/or the activity stacks up favorable against other agents in the same class. Thus, PK/PD/efficacy modeling can't prevent attrition of compounds nominated for development, but it can increase the confidence in the candidate and the target and—combined with a clinical PD biomarker—it can provide reassurance that sufficient exposure was achieved in the clinic to test the mechanism. Finally, it will enhance our understanding of biological pathways leading to systems pharmacology [62].

2.6.3 Human PK Prediction Uncertainties

Prediction of human PK has been examined extensively, and methods are available to predict the human PK. The key parameters to predict are the fraction absorbed, the clearance, and the volume of distribution. The parameter that is most prone to uncertainty is the human clearance. It is beyond the scope of this chapter to describe the prediction methods in detail here. In general, the prediction methods for human clearance fall in two categories:

- Methods based on human *in vitro* matrices such as microsomes and hepatocytes. The anticipated human clearance is obtained after incorporation of appropriate scaling factors, that is, *in vitro–in vivo* extrapolation. One area that is still somewhat controversial is the incorporation of corrections for the free fraction in plasma and/or the *in vitro* matrix. Scientific principles suggest that methods based on unbound levels make most sense. However, a review of the literature indicates that some obtain better predictions after incorporation of the free fractions and others get better results using total drug levels [63, 64].

- Methods based on allometric scaling of the clearance in preclinical species. Various allometry methods are available such as regular allometry, allometry with correction factors such as maximum life span and brain weight, free allometry, and the rat fraction unbound corrected intercept method [64, 65]. It is worth emphasizing that allometry is of little value if different clearance mechanisms are operational across species. Out of convenience, single-species scaling methods are applied by some as well with the most popular being those based on the rat or monkey [66, 67]. Finally, the use of humanized rodents, mainly mice, is gathering momentum, but this topic is beyond the scope of this review.

The use of integrated physiologically based pharmacokinetic (PBPK) modeling has become more popular in the last 10 years [68, 69]. It is essentially based on the use of *in vitro* metabolic stability data, but it incorporates many other parameters that describe the molecule as well as the system (i.e., human physiology) to predict the human PK. PBPK modeling is also quite useful for prediction of drug–drug interaction.

Despite considerable progress, there are occasions where—even under the best of circumstances—there is significant uncertainty about the human PK. In those situations, a phase 0 strategy may be warranted. A dose of 100 µg (or 1/100th of the antici-pated pharmacologically active dose, but not exceeding 100 µg) is administered following single-dose safety testing in a rodent at a 100-fold margin. A microdose study in humans can either involve nonradiolabeled material accompanied by a very sensitive LC–MS/MS method or the use of radiolabeled material and accelerator mass spectrom-etry for analysis of the trace of radioactivity. Dose linearity of the human PK from a microdose to a therapeutically relevant dose is invariably a concern, but data from two comprehensive trails—the EUMAPP and the CREAM trials—combined with other lit-erature data have shown that 21 out of 26 drugs displayed essentially linear PK (the exposure associated with the microdose extrapolated to the pharmacological dose was within twofold of the exposure measured for the pharmacological dose) [70]. Where nonlinearity is observed, it is associated with oral dosing; IV dosing was shown to scale extremely well. One of the exceptions for scaling from a microdose is propafe-none because it exhibits dose-dependent first-pass metabolism mediated by CYP2D6. Interestingly, fexofenadine displayed linear oral and IV PK despite being a P-gp and OATP substrate. Thus, microdosing is a viable strategy unless the drug in question has saturable clearance and/or distribution or dissolution problems at the pharmacological dose, but not at the microdose. Despite its promise, microdosing is not used extensively in the industry. The main reason is that it is only cost-effective if the compound displays poor human PK with a microdose. If the microdose PK data of a compound look good in humans, additional toxicology studies are needed in preparation for a conventional IND and much time is lost in the process. Thus, ideally, multiple compounds are taken into a phase 0 trial to "pick the winner"; the latter approach would be quite costly using traditional INDs.

A disadvantage of the microdosing strategy is that you only obtain PK data and no pharmacologically relevant information is obtained. Genentech encountered great uncertainty about the anticipated human PK for GDC-0834 [71] but wanted to get PK data as well as PD data out of the first clinical studies, which ruled microdosing out. The uncertainty about the human PK was derived from several conflicting pieces of data about the anticipated human clearance. First, the clearance in preclinical species was reasonable, except mouse and monkey (see Table 2.5), and various forms of allometry suggested a moderate human clearance. Moreover, there was a reasonable *in vitro–in vivo* correlation in preclinical species, except rat (see Table 2.5). However,

TABLE 2.5 Predicted Clearance in Microsomes with and without NADPH and Hepatocytes for Human, Rat, Mouse, Dog, and Monkey and Observed Clearance in Rat, Mouse, Dog, and Monkey for the BTK inhibitor GDC-0834 [71]

	Predicted CLhep ($ml\,min^{-1}\,kg^{-1}$)				
	Human	Rat	Mouse	Dog	Monkey
Liver microsomes with NADPH	19	39	79	28	36
Liver microsomes without NADPH	18	7.8	25	6.8	2.4
Hepatocytes	19	25	63	19	32
In vivo clearance		4.7	43	14	33

GDC-0834 was unstable in human hepatocytes and human liver microsomes, even in the absence of NADPH, and the enzyme involved was unknown. The non-NADPH-dependent pathway was present in other species but appeared to be of relatively minor importance. The biological rationale for a BTK inhibitor was compelling, and GDC-0834 had many other favorable attributes. Therefore, there was a strong organizational desire to pursue this compound further. However, the uncertainty about the human clearance and the possibility for high clearance in humans—voiced clearly by the department of DMPK—ruled out a conventional approach. Thus, the decision was made to perform single-dose rat and dog toxicology studies to enable a single ascending dose in humans to address the PK and PD and allow for more rapid continuation of the trial if the compound was successful. This approach greatly reduced the CMC demands and allowed us to go in humans quickly. The preclinical toxicity data were remarkably clean, and the single ascending dose study in human was initiated. Unfortunately, it was clear within one cohort that the human clearance was extremely high. The GDC-0834 levels were negligible, but a hydrolysis metabolite was observed. In hindsight, this was an excellent example of the value of human *in vitro* data. However, we have not yet been able to unambiguously identify the enzyme responsible for the hydrolysis despite considerable effort. Despite the clinical failure, the team got an answer quickly and made appropriate chemical changes to avoid this liability in future candidates. Other excellent examples about dealing with uncertainty in human PK predictions are presented in Ref. [72].

2.7 ATTRITION DUE TO TOXICITY REASONS

Preclinical safety is another source of considerable attrition in drug discovery and development, and it is trending upward [12]. The attrition due to preclinical safety findings is most pronounced in preclinical development, and it can be as high as 70% [3]. Therefore, many companies have incorporated safety considerations in drug discovery. The main factors to consider are as follows:

1. Chemical space that leads to more or less toxicity findings
2. *In vitro* assays to obtain a mechanistic understanding of toxicity findings and to enable SAR to avoid it in the future
3. Empirical *in vivo* toxicology studies in rodents, dogs, or monkeys

Several companies have shown that safety-related attrition of compounds is influenced by various physicochemical properties and structural features. These properties can be calculated and incorporated in design to reduce, but obviously not eliminate, the risk of safety-related attrition. For example:

- Price *et al.* [73] showed that the likelihood for preclinical toxicity is about sixfold greater for compounds with a TPSA less than 75 Å^2 and a $clogP$ more than 3 versus those with a TPSA more than 75 Å^2 and a $clogP$ less than 3. The authors suggested that this could at least in part be due to increased promiscuity and off-target activity, which resonates with the analysis by Gleeson *et al.* [31] showing that the number of off-target hits increases with increasing $logP$ and MW. Alternatively, worse PK properties are more likely if the TPSA is less than 75 Å^2 and the $clogP$ is more than 3, which would result in a higher dose and a dose of more than 100 mg is a known risk factor for IADRs [58] and drug-induced liver injury (see later).

- Chen *et al.* [74] provided data to show that attrition due to drug-induced liver injury in humans is more likely for compounds with a $logP \geq 3$ and a daily dose of ≥ 100 mg. This was subsequently coined the "rule of 2." Note that the positive predictive value was good with more than 80% of the compounds that violated the "rule of 2" showing signs of hepatoxicity in humans. However, the negative predictive value, in other words not violating the "rule of 2" and not showing signs of hepatoxicity in humans, is much more limited; about 40% of the compounds that do not violate the "rule of 2" do not show signs of hepatoxicity in humans. In a drug discovery setting, the anticipated efficacious dose or dose range should be used.

- Certain structural features in molecules can give rise to bioactivation—structural alerts—and that can result in idiosyncratic toxicity [58, 75]. Therefore, most companies have a list of features to avoid and public web-based tools, such as ToxAlerts [75], are available. Albeit valuable, the structural alerts should not be applied too rigidly, because the overall properties of the molecule will determine if bioactivation will happen and, if so, if it will result in toxicity in preclinical and clinical studies.

Another example of the impact of physicochemical parameters on toxicity endpoints is provided by the preclinical toxicology studies performed for two Met inhibitors [76]. The authors examined two closely related Met inhibitors, GEN-203 and GEN-890, but the second compound is considerably less basic due to an extra fluorine atom on a piperidine ring to reduce the basicity of the nitrogen atom (pK_a values of 7.45 and 5.93, respectively). The basicity of GEN-203 resulted in a high volume of distribution and high tissue levels—most likely due to accumulation in acidic organelles such as endosomes and the Golgi apparatus. These high tissue levels were rationalized as the reason for the severe bone marrow and liver toxicity in mice of this compound. In contrast, GEN-890 is less basic and has a lower volume of distribution, and the bone marrow and liver concentrations are about 1000-fold lower even though the clearance of GEN-890 was only twofold greater than that of GEN-203. In contrast to GEN-203, no detectable bone marrow and liver toxicities were observed for GEN-890 up to doses of 600 mg/kg. The volume of distribution was also observed to be correlated with a higher likelihood for toxicity by Sutherland *et al.* [77]. Thus, physicochemical parameters can be optimized to reduce the likelihood of toxicity.

Most companies use an extensive battery of *in vitro* toxicity assays to identify liabilities such as promiscuity, intrinsic cytotoxicity, cardiovascular risks, bioactivation, and genotoxicity in drug discovery [78]. In addition, it is beneficial to counter screen for off-target inhibition or activation of proteins closely related to the target of interest. This is of particular importance for kinase inhibitors because they have closely related ATP binding pockets; the so-called Gini coefficient is applied to indicate the level of selectivity with a coefficient close to 1 indicating that the compound is very selective compound and a coefficient close to 0 implying that the compound is very nonselective. Complete selectivity is generally unachievable. Therefore, if a compound is nonselective, the physiological role of the off-target should be investigated to determine if it is a liability or not. Phenotypic cellular screens, such as hepatocytes to gauge intrinsic cytotoxicity and cardiomyocytes to investigate cardiovascular risk, are also becoming more popular [79]. Finally, various genotoxicity assays should be incorporated in screening cascades: Ames or mini-Ames, *in vitro* micronucleus, and so on. Some of these *in vitro* toxicity assays can be performed at higher throughput and incorporated upstream in screening cascades; the assays can be accompanied by specific cutoffs to triage compounds. The cutoffs can be numerical (to quantify a specific margin) or absolute such as for several genotoxicity assays. A positive *in vitro* finding in some of these assays may be followed by an *in vivo* study to assess the liability further. Finally, once a significant quantity of data is available for a specific *in vitro* toxicity assay, *in silico* models can be built to allow incorporation earlier on in the screening cascade.

In vivo toxicology studies are a regulatory requirement and have been incorporated in drug discovery as well to reduce attrition. Companies run studies with a truncated design—usually just rodents, up to 7 days of dosing, no histopathology or histopathology of a limited number of tissues—in drug discovery followed by more comprehensive studies. The concordance between preclinical *in vivo* findings and human findings continues to be debated. The most cited work is the comprehensive study by Olson *et al.* [80] that suggests that there is 71% concordance between preclinical findings in rodents and nonrodents and findings in humans. Nonrodents alone are 63% predictive and rodents alone are 43% predictive. Albeit valuable, the analysis has its limitations because (i) the dose–response relations in the preclinical species may well be very different from that in humans and (ii) compounds with significant toxicity findings in preclinical species frequently do not proceed to clinical testing and, therefore, it is not known if the preclinical findings for those compounds would translate to humans. Currently, the International Consortium for Innovation and Quality in Pharmaceutical Development is performing a similar analysis with a more recent set of compounds. No matter what the translatability is, they are currently a regulatory requirement. However, doing multiple-dose toxicology studies is time-consuming and costly, and scale-up of the compound can be challenging. Therefore, Wager *et al.* [81] performed an analysis to identify parameters that are indicative for the likely outcome of these toxicology studies. They found that the most predictive parameter was the predicted efficacious concentration in humans with a $C_{eff} \leq 250$ nM (total drug) and $C_{eff} \leq 40$ nM (free drug) more suggestive of an acceptable toxicology profile in exploratory rodent and nonrodent studies (see Table 2.6 for detailed statistics). Nevertheless, a low predicted efficacious concentration in humans did not guarantee success with 39% of the compounds with $C_{eff} < 250$ nM total drug still failing the exploratory toxicology studies. It should come as no surprise that animal models such as zebrafish are explored as alternative models to identify liabilities early [82].

TABLE 2.6   Percentage of Compounds Passing or Failing Exploratory Toxicology Studies in Rodents and Nonrodents As a Function of Total and Free Predicted Efficacious Concentration in Humans [81]

	C_{eff} total drug		C_{eff} free drug	
	<250 nM	≥250 nM	<40 nM	≥40 nM
Passed exploratory toxicology study	61%	5%	51%	24%
Failed exploratory toxicology study	39%	95%	49%	76%

2.8   CORPORATE CULTURE AND NONSCIENTIFIC REASONS FOR ATTRITION

A major source of attrition highlighted in the article by Kola and Landis [12] as well as a recent analysis by Tufts University [83] is "commercial reasons." The specific reasons can be quite diverse but may involve lack of differentiation against a competitor (either in a preclinical or clinical setting), insufficient value proposition, a strategic change in the direction of the company accompanied by stopping certain projects or exiting certain therapeutic areas, and so on. Proper use of AstraZeneca's 5 Rs principles [11] could avoid some of this because it injects an essential level of foresight in a project. Early benchmarking against competitor molecules—when available—is also a critical activity. If there is no obvious differentiation, one would have to rely on attrition of the competitor compound to increase the chance of commercial success. Obviously, the latter is a tenuous strategy at best. Many companies have switched attention to oncology and large molecules in the hope of a more steady revenue stream. The generics threat for large molecules is less with the path forward for biosimilars still ill-defined in the United States. However, a recent analysis by Tufts University suggested that the attrition of small molecules in oncology is actually slightly less than that of biologics with the phase I to NDA/BLA approved survival being 14.3 and 11.5%, respectively [84].

One endemic problem in the pharmaceutical industry is the frequent reorganizations and mergers. Subsequently, organizations lose focus and/or champions of a specific promising target are no longer there and good opportunities are missed. At best, reorganizations and mergers lengthen the discovery and development process. Another unfortunate source of attrition is lack of persistence. Many drug discovery projects encountered significant hurdles in drug discovery and development, and through the dogged persistence of a small group of individuals and organizational supports, the hurdles were overcome and projects succeeded. For example, a project at Genentech encountered a specific kind of toxicity in rats, but not dogs. At first, it was perceived as a clinical showstopper. However, subsequent studies with KO rats showed that it was on targets in rats. Moreover, it was pronounced in Sprague-Dawley rats, but much less so, if not absent, in other rat strains. Finally, there are humans with a rare mutation in the target gene that do not display the kind of toxicity observed in rats. Thus, the organization managed to convince the regulatory agency that Sprague-Dawley rats are uniquely sensitive and that these findings are unlikely to translate to humans. The regulatory agency approved the IND, and clinical trials have started with appropriate safety precautions built in. However, you have to walk a very fine line, because many projects also attrite for reasons that were already apparent earlier on, but not fully appreciated. Nevertheless, persistence is valuable to *unambiguously* show whether there is or isn't a path forward. It is also important to realize when a compound has the desired attributes and nominate it for development instead of continue to optimize it with diminishing returns. AstraZeneca showed that the candidate was frequently synthesized relatively early in a series of compounds; on average, the candidate was the 19th compound in a series of 147 compounds [85].

Aside from the scientific approaches to reduce attrition—described in the previous sections—novel strategies are proposed. One common theme is public–private collaboration involving academia, government-funded research institutes, privately funded research institutes, patient advocacy groups and industry, as well as collaboration across companies in precompetitive space. Organizations such as the European Innovative Medicines Initiative, the Accelerating Medicines Partnership, the International Consortium for Innovation and Quality in Pharmaceutical Development, and the National Biomarker Development Alliance have been formed in the last 5 years to facilitate this process [86, 87]. The goals are diverse and range from pursuit of new targets and better understanding of disease biology, identification of common biomarkers, removal of research bottlenecks, advancing nonproject/target-related science to improved interaction with regulatory agencies. The changing landscape of the drug discovery and development compelled NIH director Francis S Collins to comment that *We have moved beyond the idea of the competitive nature of targets in drug discovery* [87]. These collaborations look promising, but they have specific hurdles related to intellectual property and the ability to share and compare data that need to be overcome. An alternative may be open innovation and crowd sourcing.

2.9 SUMMARY

In summary, attrition is prevalent in drug discovery and can be ascribed to reasons such as lack of confidence in the target, suboptimal DMPK properties, toxicity (off-target activity and/or *in vivo* toxicity in preclinical species), or sometimes poor pharmaceutical properties (e.g., poor solubility). Moreover, attrition is inevitable in drug discovery. Attrition should be viewed as a prioritization process, and the goal should be to incorporate appropriate studies and decision points to eliminate the "losers" early. A disciplined approach is necessary when identifying a target and one should gather as much information as reasonably possible to make a "weight of evidence" argument and to derisk the target. The validity of the target should be explored with both biological (e.g., RNAi) and chemical tools, and once lead optimization is initiated, the SAR of all key parameters (potency, selectivity, permeability, clearance, etc.) should be explored and the physicochemical "sweet spot" that provides the best balance of properties should be identified. A TCP and detailed screening cascade are essential to keep teams on track. However, it is also important to be flexible and to be prepared to compromise and take appropriate risks.

REFERENCES

1 Munos B. (2009). Lessons from 60 years of pharmaceutical innovation. *Nature Reviews Drug Discovery, 8*, 959–968.
2 Paul S.M., Mytelka D.S., Dunwiddie C.T., Persinger C.C., Munos B.H., Lindborg S.R., Schacht A.L. (2010). How to improve R&D productivity: the pharmaceutical industry's grand challenge. *Nature Reviews Drug Discovery, 9*, 203–214.

3 Pammolli F., Magazzini L., Riccaboni M. (2011). The productivity crisis in pharmaceutical R&D. *Nature Reviews Drug Discovery, 10*, 428–438.

4 Tollman P., Morieux Y., Murphy J.K., Schulze U. (2011). Identifying R&D outliers. *Nature Reviews Drug Discovery, 10*, 653–654.

5 Herper M. How much does pharmaceutical innovation cost? A look at 100 companies. http://www.forbes.com/sites/matthewherper/2013/08/11/the-cost-of-inventing-a-new-drug-98-companies-ranked/ (accessed June 25, 2015).

6 Hay M., Thomas D.W., Craighead J.L., Economides C., Rosenthal J. (2014). Clinical development success rates for investigational drugs. *Nature Biotechnology, 32*, 40–51.

7 Thunecke M., Scholefield G., Meyer C. Value-Based Pharma R&D Productivity: Is There A Sweet Spot? http://www.pharmamedtechbi.com/publications/in-vivo/32/6/valuebased-pharma-rampd-productivity-is-there-a-sweet-spot (accessed June 25, 2015).

8 Arrowsmith J. (2011). Phase II failures: 2008–2010. *Nature Reviews Drug Discovery, 10*, 328.

9 Morgan P., Van Der Graaf P.H., Arrowsmith J., Feltner D.E., Drummond K.S., Wegner C.D., Street S.D.A. (2012). Can the flow of medicines be improved? Fundamental pharmacokinetic and pharmacological principles towards improving phase II survival. *Drug Discovery Today, 17*, 419–424.

10 AstraZeneca investor relations general presentation November 2011. http://www.astrazeneca.com/cs/Satellite?blobcol=urldata&blobheader=application%2Fpdf&blobheadername1=Content-Disposition&blobheadername2=MDT-Type&blobheadervalue1=inline%3B+filename%3DInvestor-relations-general-presentation.pdf&blobheadervalue2=abinary%3B+charset%3DUTF-8&blobkey=id&blobtable=MungoBlobs&blobwhere=1285630934426&ssbinary=true (accessed June 25, 2015).

11 Cook D., Brown D., Alexander R., March R., Morgan P., Satterthwaite G., Pangalos M.N. (2014). Lessons learned from the fate of AstraZeneca's drug pipeline: a five-dimensional framework. *Nature Reviews Drug Discovery, 13*, 419–431.

12 Kola I., Landis J. (2004). Can the pharmaceutical industry reduce attrition rates? *Nature Reviews Drug Discovery, 3*, 711–715.

13 Schuster D., Laggner C., Langer T. (2005). Why drugs fail—a study on side effects in new chemical entities. *Current Pharmaceutical Design, 11*, 3545–3559.

14 Barker A., Kettle J.G., Nowak T., Pease J.E. (2013). Expanding medicinal chemistry space. *Drug Discovery Today, 18*, 298–304.

15 Dandapani S., Marcaurelle L.A. (2010). Accessing new chemical space for 'undruggable' targets. *Nature Chemical Biology, 6*, 861–863.

16 Bruns R.F., Watson I.A. (2012). Rules for identifying potentially reactive or promiscuous compounds. *Journal of Medicinal Chemistry, 55*, 9763–9772.

17 Lajiness M.S., Maggiora G.M., Shanmugasundaram V. (2004). Assessment of the consistency of medicinal chemists in reviewing sets of compounds. *Journal of Medicinal Chemistry, 47*, 4891–4896.

18 Swinney D.C., Anthony J. (2011). How were new medicines discovered? *Nature Reviews Drug Discovery, 10*, 507–519.

19 Salisbury B.G., Davis H.R., Burrier R.E., Burnett D.A., Boykow G., Caplen M.A., Clemmons A.L., Compton D.S., Hoos L.M., McGregor D.G., Schnitzer-Polokoff R., Smith A.A., Weig B.C., Zilli D.L., Clader J.W., Sybertz E.J. (1995). Hypocholesterolemic activity of a novel inhibitor of cholesterol absorption, SCH 48461. *Atherosclerosis, 115*, 45–63.

20 Altmann S.W., Davis Jr. H.R., Zhu L.-J., Yao X., Hoos L.M., Tetzoff G., Iyer S.P.N., Maguire M., Golovko A., Zeng M., Wang L., Murgolo N., Graziano M.P. (2004). Niemann-Pick C1 like 1 protein is critical for intestinal cholesterol absorption. *Science, 303*, 1201–1204.

21 Eder J., Sedrani R., Wiesmann C. (2014). The discovery of first-in-class drugs: origins and evolution. *Nature Reviews Drug Discovery, 13*, 577–587.

22 Medina-Franco J.L., Giulianotti M.A., Welmaker G.S., Houghten R.A. (2013). Shifting from the single to the multitarget paradigm in drug discovery. *Drug Discovery Today, 18*, 495–501.

23 De Kloe G.E., Bailey D., Leurs R., De Esch I.J.P. (2009). Transforming fragments into candidates: small becomes big in medicinal chemistry. *Drug Discovery Today, 14*, 630–646.

24 Singh J., Petter R.C., Baillie T.A., Whitty A. (2011). The resurgence of covalent drugs. *Nature Reviews Drug Discovery, 10*, 307–317.

25 Barf T., Kaptein A. (2012). Irreversible protein kinase inhibitors: balancing the benefits and risks. *Journal of Medicinal Chemistry, 55*, 6243–6262.

26 Ortwine D.F., Aliagas I. (2013). Physicochemical and DMPK *in silico* models: facilitating their use by medicinal chemists. *Molecular Pharmaceutics, 10*, 1153–1161.

27 Shin Y., Le H., Khojasteh C., Hop C.E.C.A. (2011). Comparison of metabolic soft spot predictions of CYP3A4, CYP2C9 and CYP2D6 substrates using MetaSite and StarDrop. *Combinatorial Chemistry and High Throughput Screening, 14*, 811–823.

28 Lipinski C.A., Lombardo F., Dominy B.W., Feeney P.J. (1997). Experimental and computational approaches to estimate solubility and permeability in drug discovery and development settings. *Advanced Drug Delivery Reviews, 23*, 3–25.

29 Wenlock M.C., Austin R.P., Barton P., Davis A.M., Leeson P.D. (2003). A comparison of physicochemical property profiles of development and marketed oral drugs. *Journal of Medicinal Chemistry, 46*, 1250–1256.

30 Leeson P.D., Davis A.M. (2004). Time-related differences in the physical property profiles of oral drugs. *Journal of Medicinal Chemistry, 47*, 6338–6348.

31 Gleeson M.P., Hersey A., Montanari D., Overington J. (2011). Probing the links between in vitro potency, ADMET and physicochemical parameters. *Nature Reviews Drug Discovery, 10*, 197–208.

32 Lovering F., Bikker J., Humblet C. (2009). Escape from flatland: increasing saturation as an approach to improving clinical success. *Journal of Medicinal Chemistry, 52*, 6752–6756.

33 Leeson P.D., St-Galley S.A. (2011). The influence of the 'organizational factor' on compound quality in drug discovery. *Nature Reviews Drug Discovery, 10*, 749–765.

34 Leeson P.D., Springthorpe B. (2007). The influence of drug-like concepts on decision-making in medicinal chemistry. *Nature Reviews Drug Discovery, 6*, 881–890.

35 Morphy R. (2006). The influence of target family and functional activity on physicochemical properties of pre-clinical compounds. *Journal of Medicinal Chemistry, 49*, 2969–2978.

36 Paolini G.V., Shapland R.H.B., Van Hoorn W.P., Mason J.S., Hopkins A.L. (2006). Global mapping of pharmacological space. *Nature Biotechnology, 24*, 805–815.

37 Tse C., Shoemaker A.R., Adickes J., Anderson M.G., Chen J., Jin S., Johnson E.F., Marsh K.C., Mitten M.J., Nimmer P., Roberts L., Tahir S.K., Xiao Y., Yang X., Zhang H., Fesik S., Rosenberg S.H., Elmore S.W. (2008). ABT-263: a potent and orally bioavailable Bcl-2 family inhibitor. *Cancer Research, 68*, 3421–3428.

38 Choo E.C., Boggs J., Zhu C., Lubach J.W., Catron N.D., Jenkins G., Souers A.J., Voorman R. (2014). The role of lymphatic transport on the systemic bioavailability of the Bcl-2 protein family inhibitors navitoclax (ABT-263) and ABT-199. *Drug Metabolism and Disposition, 42*, 207–212.

39 Roberts A.W., Seymour J.F., Brown J.R., Wierda W.G., Kipps T.J., Khaw S.L., Carney D.A., He S.Z., Huang D.C.S., Xiong H., Cui Y., Busman T.A., McKeegan E.M., Krivoshik A.P., Enschede S.H., Humerickhouse R. (2011). Substantial susceptibility of chronic lymphocytic

leukemia to BcCL2 inhibition: results of a phase I study of navitoclax in patients with relapsed or refractory disease. *Journal of Clinical Oncology*, *30*, 488–496.

40 Souers A.J., Leverson J.D., Boghaert E.R., Ackler S.L., Catron N.D., Chen J., Dayton B.D., Ding H., Enschede S.H., Fairbrother W.J., Huang D.C.S., Hymowitz S.G., Jin S., Khaw S.L., Kovar P.J., Lam L.T., Lee J., Maecker H.L., Marsh K.C., Mason K.D., Mitten M.J., Nimmer P.M., Oleksijew A., Park C.H., Park C.-M., Phillips D.C., Roberts A.W., Sampath D., Seymour J.F., Smith M.L., Sullivan G.M., Tahir S.K., Tse C., Wendt M.D., Xiao Y., Xue J.C., Zhang H., Humerickhouse R.A., Rosenberg S.H., Elmore S.W. (2013). ABT-199, a potent and selective BCL-2 inhibitor, achieves antitumor activity while sparing platelets. *Nature Medicine*, *19*, 202–208.

41 Kwong A.D., Kauffman R.S., Hurter P., Mueller P. (2011). Discovery and development of telaprevir: an NS3-4A protease inhibitor for treating genotype 1 chronic hepatitis C virus. *Nature Biotechnology*, *29*, 993–1003.

42 Teague S.J. (2011). Learning lessons from drugs that have recently entered the market. *Drug Discovery Today*, *16*, 398–411.

43 Newman D.J., Cragg G.M. (2012). Natural products as sources of new drugs over the 30 years from 1981 to 2010. *Journal of Natural Products*, *75*, 311–335.

44 Giordanetto F., Kihlberg J. (2014). Macrocyclic drugs and clinical candidates: what can medicinal chemists learn from their properties? *Journal of Medicinal Chemistry*, *57*, 278–295.

45 Alex A., Millan D.S., Perez M., Wakenhut F., Whitlock G.A. (2011). Intramolecular hydrogen bonding to improve membrane permeability and absorption in beyond rule of five chemical space. *Medicinal Chemistry Communications*, *2*, 669–674.

46 Janiszewski J.S., Liston T.E., Cole M.J. (2008). Perspectives on bioanalytical mass spectrometry and automation in drug discovery. *Current Drug Metabolism*, *9*, 986–994.

47 Hop C.E.C.A., Cole M.J., Davidson R.E., Duignan D.B., Federico J., Janiszewski J.S., Jenkins K., Krueger S., Lebowitz R., Liston T.E., Mitchell W., Snyder M., Steyn S.J., Soglia J.R., Troutman M.D., Umland J., West M., Whalen K.M., Zelesky V., Zhao S.X. (2008). High throughput ADME screening: Practical considerations, impact on the portfolio and enabler of *in silico* ADME models. *Current Drug Metabolism*, *9*, 847–853.

48 Halladay, J.S., Delarosa E.M., Tran D., Wang L., Wong S., Khojasteh S.C. (2011). High-Throughput, 384-well, LC-MS/MS CYP inhibition assay using automation, cassette analysis technique and streamlined data analysis. *Drug Metabolism Letters*, *5*, 220–230.

49 Berman J., Halm K., Adison K., Shaffer J. (1997). Simultaneous pharmacokinetic screening of a mixture of compounds in the dog using API LC/MS/MS analysis for increased throughput. *Journal of Medicinal Chemistry*, *40*, 827–829.

50 Korfmacher W.A., Cox K.A., Ng K.J., Veals J., Hsieh Y., Wainhaus S., Broske L., Prelusky D., Nomeir A., White R.E. (2001). Cassette-accelerated rapid rat screen: a systematic procedure for the dosing and liquid chromatography/atmospheric pressure ionization tandem mass spectrometric analysis of new chemical entities as part of new drug discovery. *Rapid Communications in Mass Spectrometry*, *15*, 335–340.

51 Khatib-Shahidi S., Andersson M., Herman J.L., Gillespie T.A., Caprioli R.M. (2006). Direct molecular analysis of whole-body animal tissue sections by imaging MALDI mass spectrometry. *Analytical Chemistry*, *78*, 6448–6456.

52 Salphati L., Shahidi-Latham S., Quiason C., Barck K., Nishimura M., Alicke B., Pang J., Carano R.A., Olivero A.G., Phillips H.S. (2014). Distribution of the phosphatidylinositol 3-kinase inhibitors pictilisib (GDC-0941) and GNE-317 in U87 and GS2 intracranial glioblastoma models – assessment by matrix-assisted laser desorption ionization imaging. *Drug Metabolism and Disposition*, *42*, 1110–1116.

53   Hann, M.M. (2011). Molecular obesity, potency and other addictions in drug discovery. *Medicinal Chemistry Communications*, 2, 349–355.

54   Evans D.C., Watt A.P., Nicoll-Griffith D.A., Baillie T.A. (2004). Drug-protein adducts: an industry perspective on minimizing the potential for drug bioactivation in drug discovery and development. *Chemical Research in Toxicology*, 17, 3–16.

55   Hop C.E.C.A., Kalgutkar A.S., Soglia J.R. Importance of early assessment of bioactivation in drug discovery. *Annual Reports in Medicinal Chemistry*, Vol. 41, ed. by A.M. Doherty, Elsevier, Amsterdam, 2006, 369–381.

56   Nakayama S., Takakusa H., Watanabe A., Miyaji Y., Suzuki W., Sugiyama D., Shiosakai K., Honda K., Okudaira N., Izumi T., Okazaki O. (2011). Combination of GSH trapping and time-dependent inhibition assays as a predictive method of drugs generating highly reactive metabolites. *Drug Metabolism and Disposition*, 39, 1247–1254.

57   Park B.K., Boobis A., Clarke S., Goldring C.E.P., Jones D., Kenna J.G., Lambert C., Laverty H.G., Naisbitt D.J., Nelson S., Nicoll-Griffith D.A., Obach R.S., Routledge P., Smith D.A., Tweedie D.J., Vermeulen N., Williams D.P., Wilson I.D., Baillie T.A. (2011). Managing the challenge of chemically reactive metabolites in drug development. *Nature Reviews Drug Discovery*, 10, 292–306.

58   Stepan A.F., Walker D.P., Bauman J., Price D.A., Baillie T.A., Kalgutkar A.S., Aleo M.D. (2011). Structure alert/reactive metabolite concept as applied in medicinal chemistry to mitigate the risk for idiosyncratic drug toxicity: a perspective based on the critical examination of trends in the top 200 drugs marketed in the United States. *Chemical Research in Toxicology*, 24, 1345–1410.

59   Gibbs J.P. (2010). Prediction of exposure-response relationships to support first-in-human study design. *AAPS Journal*, 12, 750–758.

60   Gabrielsson J., Fjellström O., Ulander J., Rowley M., Van Der Graaf P.H. (2011). Pharmacodynamic-pharmacokinetic integration as a guide to medicinal chemistry. *Current Topics in Medicinal Chemistry*, 11, 404–418.

61   Wong H., Choo E.F., Alicke B., Ding X., La H., McNamara E., Theil F.-P., Tibbits J., Friedman L.S., Hop C.E.C.A., Gould S.E.. (2012). Anticancer activity of targeted and cytotoxic agents in murine subcutaneous tumor models correlates with clinical response. *Clinical Cancer Research*, 18, 3846–3855.

62   Van Der Graaf P.H., Benson N. (2011). Systems pharmacology: bridging systems biology and pharmacokinetic-pharmacodynamic (PKPD) in drug discovery and development. *Pharmaceutical Research*, 28, 1460–1464.

63   Obach R.S., Baxter J.G., Liston T.E., Silber B.M., Jones B.C., MacIntyre F., Rance D.J., Wastall P. (1997). The prediction of human pharmacokinetic parameters from preclinical and in vitro metabolism data. *Journal of Pharmacology and Experimental Therapeutics*, 283, 46–58.

64   Ring B.J., Chien J.Y., Adkinson K.K., Jones H.M., Rowland M., Jones R.D., Yates J.W.T., Ku M.S., Gibson C.R., He H., Vuppugalla R., Marathe P., Fischer V., Dutta S., Björnsson T., Lavé T., Poulin P. (2011). PhRMA CPDCD initiative on predictive models of human Pharmacokinetics, part 3: comparative assessment of prediction methods of human clearance. *Journal of Pharmaceutical Sciences*, 100, 4090–4110.

65   Lombardo F., Waters N.J., Argikar U.A., Dennehy M.K., Zhan J., Gunduz M., Harriman S.P., Berellini G., Rajlic I.L., Obach R.S. (2013). Comprehensive assessment of human pharmacokinetic prediction based on in vivo animal pharmacokinetic data, part 2: clearance. *Journal of Clinical Pharmacology*, 53, 178–191.

66   Ward K.W., Smith B.R. (2004). A comprehensive quantitative and qualitative evaluation of extrapolation of intravenous pharmacokinetic parameters from rat, dog, and monkey to humans. I Clearance. *Drug Metabolism and Disposition*, 32, 603–611.

67 Hosea N.A., Collard W.T., Cole S., Maurer T.S., Fang R.X., Jones H., Kakar S.M., Nakai Y., Smith B.J., Webster R., Beaumont K. (2009). Prediction of human pharmacokinetics from preclinical information: comparative accuracy of quantitative prediction approaches. *Journal of Clinical Pharmacology*, 49, 513–533.

68 De Buck S.S., Sinha V.K., Fenu L.A., Nijsen M.J., Mackie C.E., Gilissen R.A.H.J. (2007). Prediction of human pharmacokinetics using physiologically based modelling: a retrospective analysis of 26 clinically tested drugs. *Drug metabolism and Disposition*, 35, 1766–1780.

69 Jones H.M., Gardner I.B., Watson K.J. (2009). Modelling and PBPK simulation in drug discovery. *AAPS Journal*, 11, 155-165.

70 Lappin G. (2010). Microdosing: current and the future. *Bioanalysis*, 2, 509–517.

71 Liu L., Halladay J.S., Shin Y., Wong S., Corragio M., La H., Baumgardner M., Le H., Gopaul S., Boggs J., Kuebler P., Davis Jr. J.C., Liao X.C., Lubach J.W., Deese A., Sowell C.G., Currie K.S., Young W.B., Khojasteh S.C., Hop C.E.C.A., Wong H. (2011). Significant species differences in amide hydrolysis of GDC-0834, a novel and potent and selective Bruton's tyrosine kinase inhibitor. *Drug Metabolism and Disposition*, 39, 1840–1849.

72 Harrison A., Gardner I., Hay T., Dickens M., Beaumont K., Phipps A., Purkins L., Allan G., Christian R., Duckworth J., Gurrell I., Kempshall S., Savage M., Seymour M., Simpson M., Taylor L., Turnpenny P. (2012). Case studies addressing human pharmacokinetic uncertainty using a combination of pharmacokinetic simulation and alternative first in human paradigms. *Xenobiotica*, 42, 57–74.

73 Price D.A., Blagg J., Jones L., Greene N., Wager T. (2009). Physicochemical drug properties associated with in vivo toxicological outcomes: a review. *Expert Opinion in Drug Metabolism and Toxicology*, 5, 921–931.

74 Chen M., Borlak J., Tong W. (2013). High lipophilicity and high daily dose of oral medications are associated with significant risk for drug-induced liver injury. *Hepatology*, 58, 388–396.

75 Sushko I., Salmina E., Potemkin V.A., Poda G., Tetko I.V. (2012). ToxAlerts: a web server of structural alerts for toxic chemicals and compounds with potential adverse reactions. *Journal of Chemical Information and Modeling*, 52, 2310–2316.

76 Diaz D., Ford K.A., Hartley D.P., Harstad E.B., Cain G.R., Achilles-Poon K., Nguyen T., Peng J., Zheng Z., Merchant M., Sutherlin D.P., Gaudino J.J., Kaus R., Lewin-Koh S.C., Choo E.F., Liederer B.M., Dambach D.M. (2013). Pharmacokinetic drivers of toxicity of basic molecules: strategy to lower pKa results in decreased tissue exposure and toxicity for a small molecule Met inhibitor. *Toxicology and Applied Pharmacology*, 266, 86–94.

77 Sutherland J.J., Raymond J.W., Stevens J.L., Baker T.K., Watson D.E. (2012). Relating molecular properties and in vitro assay results to in vivo drug disposition and toxicity outcomes. *Journal of Medicinal Chemistry*, 55, 6455–6466.

78 Hornberg J.J., Laursen M., Brenden N., Persson M., Thougard A.V., Toft D.B., Mow T. (2014). Exploratory toxicology as an integrated part of drug discovery. Part II: screening strategies. *Drug Discovery Today*, 19, 1137–1144.

79 Kleinstreuer N.C., Yang J., Berg E.L., Knudsen T.B., Richard A.M., Martin M.T., Reif D.M., Judson R.S., Plokoff M., Dix D.J., Kavlock R.J., Houck K.A. (2014). Phenotypic screening of the ToxCast chemical library to classify toxic and therapeutic mechanisms. *Nature Biotechnology*, 32, 583–591.

80 Olson H., Betton G., Robinson D., Thomas K., Monro A., Kolaja G., Lilly P., Sanders P., Sipes G., Bracken W., Dorato M., Van Den K., Smith P., Berger B., Heller A. (2000). Concordance of the toxicity of pharmaceuticals in humans and in animals. *Regulatory Toxicology and Pharmacology*, 32, 56–67.

81 Wager T.T., Kormos B., Brady J.T., Will Y., Aleo M., Stedman D., Kuhn M., Chandrasekaran R. (2013). Improving the odds of success in drug discovery: choosing the best compounds for in vivo toxicology studies. *Journal of Medicinal Chemistry*, 56, 9771–9779.

82 Nadanaciva S., Aleo M.D., Strock C.J., Stedman D.B., Wang H., Will Y. (2013). Toxicity assessments of nonsteroidal anti-inflammatory drugs in isolated mitochondria, rat hepatocytes, and zebrafish show good concordance across chemical classes. *Toxicology and Applied Pharmacology*, *272*, 272–280.

83 Causes of clinical failures vary widely by therapeutic class, phase of study. *Tufts CSDD Impact Report*, 2013, *15* (5), 1–4.

84 DiMasi J.A., Reichert J.M., Feldman L., Malins A. (2013). Clinical approval success rates for investigational cancer drugs. *Clinical Pharmacology and Therapeutics*, *94*, 329–335.

85 Cheshire D.R. (2011). How well do medicinal chemists learn from experience? *Drug Discovery Today*, *16*, 817–821.

86 Khanna I. (2012). Drug discovery in pharmaceutical industry: productivity challenges and trends. *Drug Discovery Today*, *17*, 1088–1102.

87 The rise and rise of research consortia. *Chemical & Engineering News*, 2014, *92* (30), 17–19.

3

ATTRITION IN PHASE I

DENNIS A. SMITH[1] AND THOMAS A. BAILLIE[2]

[1] Department of Chemistry, University of Cape Town, Cape Town, South Africa; The Maltings, Walmer, Kent, UK

[2] School of Pharmacy, University of Washington, Seattle, WA, USA

3.1 INTRODUCTION

The term "phase I study" has a dual meaning. In the first case, it describes the initial introduction of a new candidate drug into humans. In most cases, these studies are performed in healthy volunteers, but in some indications such as oncology, the studies may be conducted in patients. In the initial study, a single dose (often as a simple solution or suspension) is administered. The vast majority of such studies are randomized and placebo controlled. The starting dose is determined by consideration of preclinical toxicology and pharmacology studies and is selected to give no effects. Dose escalation is performed on a weekly basis (usually by a doubling of the previously administered dose) until a well-tolerated dose is exceeded. The previous dose is then termed the maximum tolerated dose (MTD). In addition to an understanding of drug safety after a single dose, the study also allows the first examination of the pharmacokinetics (PK) of the drug allowing half-life and dose linearity to be determined in a provisional manner.

Assuming successful completion of the single-dose study, the phase I clinical work is extended to a multiple-dose study. The design may incorporate some of the knowledge gained from the single-dose study, such as PK, in the dosage regimen. For instance, nonlinearities in plasma concentration and administered dose may mean a smaller dose range examined in later studies. The multiple-dose phase I study will usually consist of three dose levels spanning the expected therapeutic range, with the top dose level close to the MTD. The objectives are again safety and tolerability. PK profiles now allow examination of the effects of multiple dosing, such as accumulation, autoinduction, and so on. These studies are restricted to 10–14 days. Toxicities such as hepatotoxicity may first become apparent in these studies where a high incidence occurs.

Attrition in the Pharmaceutical Industry: Reasons, Implications, and Pathways Forward, First Edition. Edited by Alexander Alex, C. John Harris and Dennis A. Smith.
© 2016 John Wiley & Sons, Inc. Published 2016 by John Wiley & Sons, Inc.

In the second case, the term phase I study refers to all volunteer studies. These occur throughout the development and lifetime of a drug to study specific aspects of clinical pharmacology. Such studies include the following:

Formulation bioavailability. These studies are conducted whenever significant changes in formulation of the drug occur. This will often be the case where the initial solution or suspension is changed to a capsule or tablet for phase II patient studies and repeated as the formulation is optimized and commercialized.

Radiolabeled metabolism and excretion (ADME, mass balance). This study is usually performed with a single dose of radiolabeled drug with the purpose of identifying circulating and excreted metabolites and determining overall mass balance.

Specific populations. Certain populations can pose additional risk of toxicities due to the possibility of changes in PK produced by accompanying disease. These include studies in renal insufficiency and hepatic impairment. In addition, as the range of patient groups likely to be treated is widened, studies in the elderly, the young, and different sex or race may be performed.

Drug interactions. The possibility of drug interactions, particularly those that alter the PK of a drug rather than the pharmacodynamics (PD), is explored in volunteer studies. The interactants are normally selected from intensive preclinical work examining the enzymology involved in the clearance of the drug. These studies often revert to the worst-case interactant (e.g., for a drug cleared by CYP3A4, the very potent inhibitor itraconazole, or previously ketoconazole, will be dosed to steady state alongside the candidate drug).

Pharmacogenomics of clearance pathways. There is considerable variation in the expression of the enzymes and transporters involved in drug clearance. This can be accentuated where an enzyme or transporter shows genetic polymorphism. With simple genomic tests, it is possible to select volunteers to examine the degree of variation in PK that this causes. Such studies can be incorporated into others, even the initial phase I study or the radiolabeled study.

Special safety. Where a biomarker exists, studies can incorporate this measurement. Again, this may occur in all or selected studies or be the subject of an individual study. An example of such a measurement is QTc, and further investigation if the drug causes lengthening of the signal.

3.2 ATTRITION IN PHASE I STUDIES AND PAUCITY OF PUBLISHED INFORMATION

Statistics collected over different time frames indicate that first-in-human (FIH) studies have a 60–70% success rate in moving to the next phase II stage. Although phase I studies are subject to regulatory approval and are listed on databases, information on their outcomes is hard to obtain. In a review of phase I studies, Decullier *et al.* [1] concluded that only 17% of phase I studies were published in contrast to 43% of phase II to IV studies. Confidentiality was cited as the main reason, but it is likely that in addition those FIH studies, which led to compound attrition, would be of low priority in terms of publication policy to a commercial organization. It is understandable that FIH studies take place in a competitive arena and companies may also be progressing backup compounds and feel reluctant to impart information that could be of high value to other researchers. The considerable time taken to resolve these aspects means considerable delay and difficulty in publication so that few full papers emerge (several years possibly after the study). This view of lack of publication and lack of information was also stated by Sibile *et al.* in a review of adverse events in phase I studies [2].

3.3 DRUG ATTRITION IN NOT FIH PHASE I STUDIES

It is important to separate FIH studies from other volunteer studies conducted to examine specific aspects of clinical pharmacology. The nature of many of these clinical pharmacology studies is that they were introduced after particular drug risks were identified. Thus, some of the drugs described in postmarketing attrition had risk factors clarified by clinical pharmacology studies or are direct triggers for the need for such studies. This is exemplified by benoxaprofen (see Chapter 5). This nonsteroidal "profen" anti-inflammatory was introduced with a clinical dose of 600 mg with no subdivision or dosing advice for special patient populations. This "one-size-fits-all" dose was seen as commercially advantageous in a very competitive therapeutic area with many similar drugs. Detailed PK studies in different patient populations were only conducted postlaunch when the side effects of the drug became apparent. The plasma elimination half-life of the drug was in the range of 19–26 h in younger volunteers and patients commensurate with its commercially attractive once-a-day dosage regimen. Studies in elderly patients revealed a very different PK profile with half-lives extending out to 150 h, which coincided with reduced creatinine clearance in these subjects. Side effects such as phototoxicity and hepatotoxicity were associated with the elderly and, by inference, with this PK profile leading to pronounced drug accumulation. Studying the elderly and renal impairment in phase I studies would have indicated the need for dosage adjustments to be made and specified. It is possible that with the correct dosages for elderly or renally impaired patients, the drug would have had a safety profile comparable with other NSAIDs when marketed. Alternatively, if this now meant that the drug was no longer deemed commercially attractive, then the huge costs of phase III and drug launch would have been saved.

Another drug with significant risks, which would routinely be identified in preclinical and phase I studies today, is terfenadine, also detailed in Chapter 5. Terfenadine was marketed as a nonsedating antihistamine, properties that the drug (a lipophilic amine) owes to its very rapid and extensive conversion to a zwitterionic metabolite fexofenadine. The nonsedation differentiated terfenadine from the first-generation sedating antihistamines. Terfenadine interacted with a number of secondary pharmacology receptors, one of which was the I_{Kr} channel. Secondary pharmacology was not observed under normal conditions because the concentrations of terfenadine were subpharmacological due to the drug's very rapid metabolism. Fexofenadine, the active principal, is highly selective for the H_1 receptor with no activity against I_{Kr}. At the time of launch, little was known about the enzymology of the conversion of terfenadine to fexofenadine or indeed the role of the I_{Kr} channel. Following observation of deaths of patients on terfenadine due to cardiotoxicity, subsequent studies dissected the risk factors. Conversion of terfenadine to fexofenadine was mediated by CYP3A4, and co-administration with potent inhibitors such as ketoconazole or erythromycin led to concentrations of terfenadine present in patients and volunteers that exceeded the affinity for the I_{Kr} channel. In some individuals, these concentrations caused prolongation of QT, and this effect can clearly be associated with the fatal cardiac arrhythmias. Concentration–response relationships were not straightforward, possibly because terfenadine's polypharmacology also included affinity for the Ca^{++} channel. Again, the risks of terfenadine would today have been clearly identified in preclinical and phase I studies by ion channel screening, enzymology of clearance (metabolism), and *in vitro* and *in vivo* studies. For instance, terfenadine's metabolism by CYP3A4 would have led to a clinical phase I drug interaction study using a perturbant such as ketoconazole. The dramatic increase in parent terfenadine concentrations would have been a cause for concern. With the benefit of hindsight, the predictability of terfenadine's toxicity would have meant that the molecule would never advance from preclinical testing.

3.4 ATTRITION IN FIH STUDIES DUE TO PK

These studies examine preliminary PK and toleration of the drug, so it is not surprising that "failure" in terms of these characteristics is a major reason for discontinuation. Failure in PK will normally be low bioavailability and/or too short a half-life. Low bioavailability may be caused by poor absorption or rapid gut and/or liver metabolism. A judgment can be made on the early data that the PK make it impossible to achieve a reasonable dosage regimen in terms of dose size or frequency of administration. Toleration of the new drug has to be considered in the light of the expected efficacy dose and the clinical indication. Failure to influence a biomarker that is seen as a surrogate for pharmacology or efficacy is the other major reason.

Prior to 2000, the predictive tools for human PK and pharmacokinetic/pharmacodynamic modeling (PK/PD) derived from preclinical data were limited. Analysis suggested that PK were a major reason for discontinuation [3].

The availability of human reagents such as hepatocytes or liver microsomes was severely limited. Drug candidates were often selected based on pharmacological screening data alone or with minimal PK in the test animal. Advancement to man was an act of faith based on loose allometric principles: in general, dose size and duration of action (half-life) would be more favorable in human than laboratory species. While many drugs were discovered by these principles, it is not too surprising that when tested in human, many failed.

An example of PK attrition is the cardiovascular agent carbazeran, which was discovered principally using the dog as the preclinical species. PK studies were also conducted in this species prior to human studies. The drug was well absorbed in the dog with a bioavailability of 68% and biotransformation mainly by O-demethylation. In man [4], bioavailability was too low to be measurable and, in contrast to dog, carbazeran was almost completely cleared via 4-hydroxylation of the phthalazine moiety (Fig. 3.1). Subsequent further *in vitro* work indicated that carbazeran was a substrate for liver aldehyde oxidase in addition to cytochrome P450 (CYP). While the aldehyde oxidase pathway was minor in dog with low rates of metabolism, in human the rate of metabolism by aldehyde oxidase was extremely high [5].

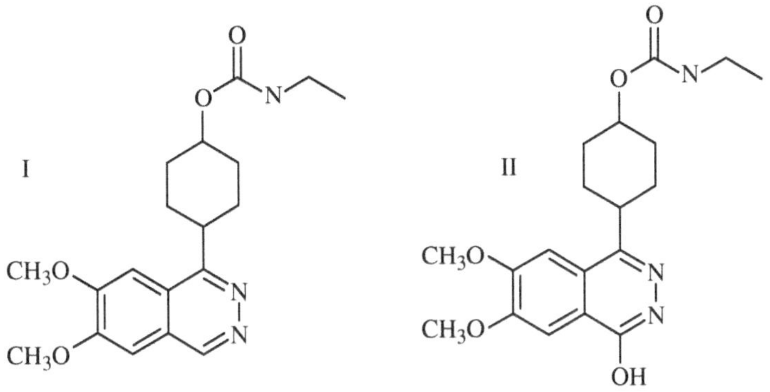

FIGURE 3.1   Structures of carbazeran (I) and its hydroxylated metabolite (II) formed by rapid metabolism in human by aldehyde oxidase.

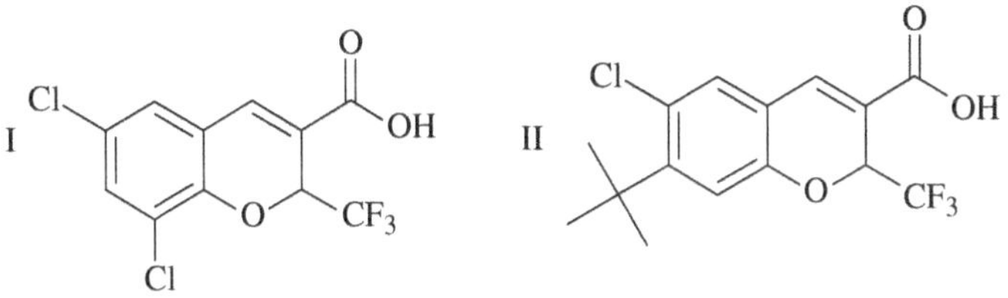

FIGURE 3.2   Structures of COX-2 selective anti-inflammatories SD8381 (I) that had an unacceptably long half-life in phase I studies and a subsequent compound SC75416 (II) with a half-life commensurate with daily dosing.

Rapid metabolism, high clearance, and short half-lives are typical examples of "poor PK." The opposite can also be seen as a problem. Compounds have been halted in phase I studies due to too long a half-life and the worry of finding an acceptable dosage regimen in chronic conditions. SD8381 (Fig. 3.2) is a COX-2 selective anti-inflammatory, which, when administered in phase I studies, had a half-life of 360 h. More metabolically labile compounds [6] were sought and SC75416 was discovered, which, when tested in humans, had a half-life of 34 h.

These examples represent fairly clear-cut decision-making studies. Later examinations of attrition indicate that commercial reasons have become increasingly dominant. In a survey of the 2008–2009 cohort of phase I studies and project discontinuation, commercial reasons were the single most important factor. Stating it as a single factor is probably false as poor PK and resultant high dose leading to commercially unfavorable cost of goods issues could be classed in either PK or commercial reasons for project termination. The case of benoxaprofen, earlier, could have been a commercial decision after appropriate phase I studies demonstrated that the need for several dose size adjustments for different patient groups was not commercially viable in a very crowded market (nonselective COX inhibitors).

Commercial reasons may also cover backup compounds. These may be progressed as far as phase I and then held (and later terminated), while the frontrunner moves forward through development. The converse of this is also true in that the backup has properties that are an improvement over those of the prototype. In the preclinical phase leading to

TABLE 3.1   Considerations on Continuing or Terminating a Follow-on Medicine Based on a Phase I First-in-man Study

Same primary and secondary pharmacology	Different secondary pharmacology
Pharmacokinetics and better dosage regimen	Pharmacokinetics and better dosage regimen. Lower side effect burden
Response to new therapy	Response to new therapy
Uptake of therapy by clinical community	Uptake of therapy by clinical community
Cost of development	Cost of development

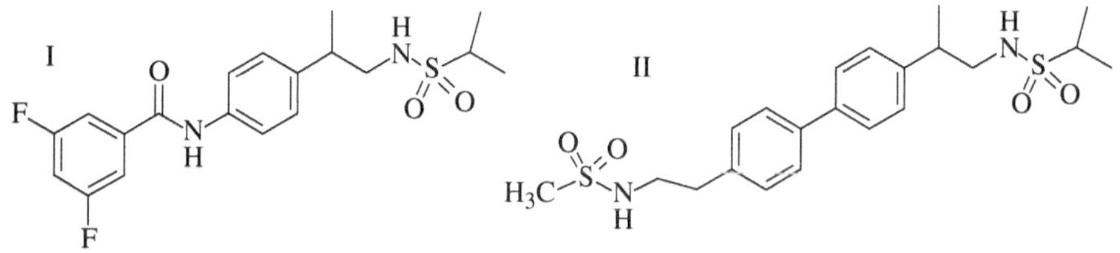

FIGURE 3.3   Structures of LY450108 (I) and LY451395 (II) that were parallel tracked into FIH studies.

the phase I study or the phase I study itself, these properties become problematic slowing development. Switching to the backup will be recorded as attrition in phase I of the prototype.

A large proportion of drugs registered each year, particularly from 1970 to the early 2000s, were follow-on compounds, looking to exploit a recently introduced or discovered new mode of action. If companies began competitor activity around the end of phase II for a medicine with a new mode of action, it is likely that phase I studies would be conducted for competitor compounds postlaunch of the prototype. A large amount would then be known about this new medicine. Various considerations around the decision to terminate or continue a follow-on medicine are given in Table 3.1.

PK advantages can be observed in a phase I study, but their impact on PD advantages can only be reliably extrapolated, from a phase I study, for drugs that act by a rapid onset–offset, reversible pharmacological mechanism. Similarly, advantages in selectivity can only be probed where the secondary pharmacology is observed as a side effect in volunteers.

Parallel development tracks have been utilized for pairs of molecules into phase I FIH studies. LY450108 and LY451395, both AMPA receptor potentiators, were evaluated in phase I studies in parallel [7]. Assessment in volunteers of both compounds included plasma and cerebrospinal fluid concentration evaluation. LY451395 was advanced into phase II studies for Alzheimer's disease (Fig. 3.3).

Clearly, the usual outcome for such a strategy is that the compound deemed to be inferior will be held and probably be terminated at phase I.

3.4.1   Attrition due to Pharmacogenetic Factors

In addition to poor intrinsic PK behavior (low systemic exposure following oral administration, inappropriate half-life in humans, etc.), a related source of attrition in phase I has been unexpectedly high intersubject variability in PK due to genetic

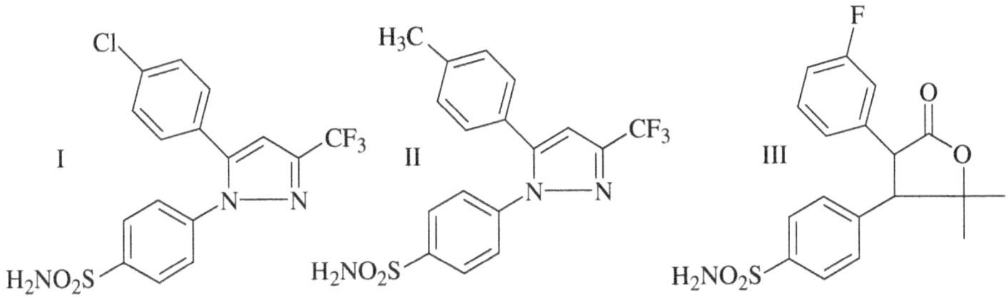

FIGURE 3.4 Structures of COX-2 inhibitors SC-236 (I), SC-58635 (celecoxib) (II), and DFU (III).

polymorphisms in key drug-metabolizing enzymes, notably the CYP system. The field of COX-2 inhibitors provides examples of such genetic polymorphisms, in one case leading to a warning in the label of an approved product and in another case to outright failure during early development.

As described earlier in this chapter, when animal PK studies predict a long half-life of a new chemical entity in humans, medicinal chemistry efforts typically focus on the introduction of a metabolic "soft spot" with the specific goal of increasing clearance and thereby shortening elimination half-life. This was the case in the discovery of celecoxib, the first marketed COX-2 selective inhibitor that was developed by Searle in the late 1990s. When SC-236 (Fig. 3.4), a promising early lead compound from the 1,5-diarylpyrazole series of COX-2 inhibitors, was found to have a half-life of 117 h upon oral administration to rats, a number of analogues were prepared in which the chlorine substituent at the 4-position of the phenyl ring was replaced with a variety of functional groups known to be susceptible to metabolic attack [8]. This strategy led to the synthesis of the 4-methyl compound (SC-58635; Fig. 3.4) that retained high potency and selectivity against the COX-2 enzyme and was metabolized extensively to the corresponding (inactive) 4-hydroxymethyl and carboxylic acid derivatives. The half-life of this compound was much shorter at 3.5 h, yet it exhibited good oral bioavailability and had a high volume of distribution and excellent safety characteristics in preclinical studies. Based on this dataset, SC-58635 entered phase I clinical trials, which demonstrated that the compound possessed a favorable PK profile in humans with a half-life of 12 h. Ultimately, SC-58635 became the anti-inflammatory drug celecoxib (Celebrex®), which was approved by the FDA in 1998.

The potential role of genetics in the metabolism of Celebrex did not emerge until 2001, when it was reported that the clearance of the drug in humans was mediated largely by CYP2C9, an enzyme that exhibits genetic polymorphism in the human population [9]. In a clinical study designed to assess the consequences of this phenomenon in terms of intersubject variability in PK, a >2-fold reduced oral clearance of the drug was observed in homozygous carriers of CYP2C9*3 compared to carriers of the wild-type CYP2C9*1 [10]. Consequently, it was predicted that approximately 0.5% of Caucasians with the homozygous CYP2C9*3/*3 genotype would have significantly increased systemic exposure to celecoxib. Based on this analysis and the associated potential safety implications, celecoxib became one of the first therapeutic agents for which the manufacturer's drug information recommended caution when administering celecoxib to poor metabolizers of CYP2C9.

A more dramatic example of the role of a CYP polymorphism in influencing the fate of a COX-2 inhibitor, and one that led to the demise of a promising drug candidate in phase I, came from the discovery program at Merck. In an effort to identify a highly selective "second-generation" COX-2 inhibitor, structurally distinct from celecoxib and Merck's own first entry into this market, rofecoxib (Vioxx®), a substituted furanone derivative termed DFU, emerged as an attractive candidate and was selected for clinical evaluation [11]. Phase I studies with DFU immediately revealed a striking variability in the compound's PK, with half-lives that ranged in individual subjects from ~12 h to >72 h. It was then discovered that DFU was metabolized exclusively by CYP2C19, a polymorphic CYP enzyme, and when subjects who had received DFU were genotyped retrospectively for their CYP2C19 status, an excellent correlation emerged between their genotype and the corresponding DFU elimination half-life. Since it was deemed impracticable to develop different dosing regimens of DFU based upon a patient's CYP2C19 phenotype (extensive vs. poor metabolizer), development of the compound was terminated. It should be noted that, at the time these studies were performed, CYP phenotyping was not a routine component of preclinical drug development activities, as it is today, and compounds such as DFU whose clearance is highly dependent upon a single polymorphic CYP would not be progressed to phase I trials in the current environment.

3.5 ATTENUATION OF PK FAILURE

While failure due to PK is still a problem, a large focus has been placed on preclinical systems to predict compound absorption, distribution, and clearance. In addition, microdose or phase 0 studies in human subjects have been adopted where preclinical experiments give conflicting answers in the prediction of PK. These studies are conducted at a single low dose well below that required for pharmacological effect. Plasma or blood concentrations are measured either using conventional technology with a highly sensitive assay or using accelerator mass spectrometry (AMS) (after appropriate separation of the analyte) and dosing a ^{14}C derivative of the compound of interest.

3.5.1   Preclinical Methods (In Vivo)

Allometry scales PK parameters to the body weight of the animal. To maintain a temperature of 37°C, the metabolic rate of an organism is related to its body surface area. Thus, factors important in PK such as organ blood flow show an allometric relationship [12–14]. Allometric relationships provide an equation of the general form: PK parameter = $A.BW^a$ where BW is the body weight of the animal.

In defining drug duration (half-life or mean residence time), the key parameters for prediction are volume of distribution and clearance. For volume of distribution, an allometric relationship is not surprising as this value will be dependent upon the relative affinity for tissue compared to plasma, and as the makeup of tissues is similar across species, the ratio will remain relatively constant, particularly when species-specific differences in plasma protein binding are corrected for. Retrospective analyses of development compounds have shown that allometry, factoring in species differences in plasma protein binding, results in 69% of predictions being within twofold of that observed in the clinic compared to 64% when plasma protein binding differences were not included [15].

Where the clearance process is similar across species (such as glomerular filtration), the allometric relationship is strong since the clearance process depends on organ blood flow. For drugs cleared by metabolism, some elements of allometry are still important. The organs of elimination represent a smaller proportion of body weight as the overall size of the mammal increases. These findings, which will directly affect the total animal content of enzymes such as CYP450, probably explain why allometry is a reasonable method for the prediction of metabolically cleared drugs.

To improve the predictability, the basic equation involving BW^a has been modified [12, 13, 16] when calculating clearance (CL) to correct for animal brain weight (BrW) and maximum life span potential (MLP). These are implemented when the exponential term exceeds that expected from simple scaling for terms such as blood flow. The "rule of exponents" states that if the exponent from the simple allometric equation is less than 0.50 or lies between 0.55 and 0.70, then simple allometry is applied. After simple scaling, if the exponent is between 0.70 and 1.0, then the CL × MLP is applied. Finally, if the simple allometry exponent is greater than 1.0, then the CL × BrW is used. MLP can be calculated according to the following equation: MLP = $185.4 \times (BrW)^{0.636} \times (BW)^{-0.225}$ [17].

An example of how these modified equations can improve prediction is provided by ST-246, a new drug for smallpox. The drug was evaluated in mice, rabbits, monkeys, and dogs and later compared to humans. A species-specific difference was observed for CL/F (oral clearance as F, fraction absorbed is unknown) with an increase across species from mouse to dog. Modified allometry was used, and the human MLP-corrected CL/F, terminal half-life, and extraction ratios were in close proximity to the final clinical values [18].

Despite the above example of ST-246, when considering clearance, an allometric relationship is less conserved since the mechanisms of clearance, and expression patterns of the enzymes and transporters involved, usually vary between species.

3.5.2 Preclinical Methods (*In Vitro*)

Because of the known species differences in metabolism, estimations of human intrinsic clearance can also be obtained from *in vitro* human liver microsomal and hepatocyte assays. Human hepatocytes make up 70–80% of the human liver and contain the enzymes responsible for the metabolism of drugs. Human liver microsomes are formed from centrifugation of whole liver homogenate, resulting in a concentrated membrane fraction containing CYP450s, flavin-monooxygenases, carboxylesterases, epoxide hydrolases, UDP-glucuronosyltransferases, and other drug-metabolizing enzymes. Enzymes such as aldehyde oxidase and sulfotransferase are not present being soluble. Aldehyde oxidase, the enzyme responsible for carbazeran metabolism, is also a soluble enzyme. Typically, soluble fraction or low-speed supernatant (S9) are used to screen for these enzymes. Hepatocytes provide a universal screen, but rates of overall metabolism for CYP450 reactions may be lower than microsomes due to the enriched nature of this preparation. The rate of clearance in such systems can be scaled knowing liver content to a prediction of human intrinsic clearance [19]. Human intrinsic clearance can be converted into human hepatic extraction, blood clearance due to hepatic metabolism, and so on using equations such as the well-stirred model. Considerable confusion and debate surround the choice of the best scaling factors in terms of yield of microsomes or hepatocytes from human liver. Importantly, the preparations do not represent the architecture of the liver, so exact numerical scaling is probably flawed. Consensus suggests that individual laboratories are best served by the use of empirical scaling factors based on standard test compounds [20]. Transporter-mediated clearance through hepatic uptake is a clearance method that does not have a routine approach or methodology. Sandwich-cultured hepatocytes, which allow for the development of intact bile canaliculi and the ability to measure hepatic uptake and biliary clearance, are being used to predict hepatobiliary clearance in a similar manner to hepatocytes and metabolic clearance, but uncertainty to the accuracy of the predictions remains. Expression of transporter proteins in these systems is not constant, with a decrease of BSEP/Bsep protein and an increase of BCRP/Bcrp protein in both rat and human hepatocytes over time. In contrast, Mrp2 in rat hepatocytes is significantly diminished, while MRP2 constantly increased in human hepatocytes during the cell culture process. Quantification of these proteins and measured rates of known substrates show that absolute protein amount is a key determinant for hepatobiliary clearance. This suggests that scaling factors can be calculated, which will provide for future extrapolation of biliary secretion data from the *in vitro* to the *in vivo* situation [21].

Culture technology is continuously improving and *in vitro* systems are beginning to replicate the architecture of the liver. For instance, HepatoPac® is a micropatterned hepatocyte/fibroblast coculture system that can be used for continuous incubations of up to 7 days or more. This allows compounds of low intrinsic clearance and turnover to be studied. An evaluation [22] of this system has been performed, and for 17 commercially available compounds with low to intermediate clearance (<12 ml/min/kg), hepatic clearance was accurately predicted for 10 compounds (59%; predicted clearance within twofold of observed human *in vivo* clearance values). Further evaluation looked at three slowly metabolized compounds (alprazolam, meloxicam, and tolbutamide). The metabolism rate of alprazolam and tolbutamide was approximately twofold greater using HepatoPac compared with suspended hepatocytes. Furthermore, HepatoPac, but not suspended hepatocytes, demonstrated significant turnover of meloxicam. It is likely that these types of coculture procedures will bring more precision to human PK predictions.

Commercial physiologically based pharmacokinetic (PBPK) models, such as GastroPlus™, Simcyp®, and PKSim®, are now available to aid in the prediction of human clearance from *in vitro* data. Simcyp also predicts the potential range of clearance values in different populations as well as the nonexistent "average" individual.

3.5.3   Phase 0, Microdose Studies in Humans

These studies are performed in a small number of human subjects at a subpharmacological dose. They may also be termed exploratory IND studies in the United States. The dose is typically 1% of the predicted pharmacological dose with a maximum dose of 0.1 mg. They can be performed by any route of administration and need support from only a small selective toxicology program. Typically, that is a single-dose extended observation toxicology study in a single species with necroscopy at 2 and 14 days. Data from these studies are broadly similar to conventional dose phase I studies, especially in terms of drug duration (half-life) [23, 24]. Very sensitive analytical technology is needed, which can be provided by positron emission tomography (PET), liquid chromatography–mass spectrometry/mass spectrometry (LC–MS/MS), and AMS. Because of the short half-life of ^{11}C, PET does not allow capture of extensive PK data. AMS provides exquisite sensitivity but needs the drug to be ^{14}C labeled. The adoption of these methods has not been widespread as a means of obtaining early human clinical data, probably due to the added cost and time delay in performing the study compared to the conventional route. Phase 0 studies only add value to an overall drug discovery portfolio in identifying compounds that have unacceptable PK in humans. An individual program is unlikely to recommend this route unless it has a number of compounds with similar properties to explore and no other means of prioritizing them. Interestingly, such a situation would mean that phase 0 studies would have high attrition rates. For instance, one publication describes analytical methods and a clinical study [25] to investigate five HIV integrase fusion inhibitors in human microdose studies. The compounds were administered at a dose of 50 µg by the oral and intravenous routes, and the sensitivity allowed plasma concentrations down to 2 pg/ml to be determined. Peak concentrations after oral administration ranged from 0.6 to 1.1 ng/ml. PK of the compounds in terms of duration (half-life) were similar with raltegravir (already further in development) showing the best bioavailability. Raltegravir's inclusion allowed comparison with conventional phase I data and a conclusion that the results were linear between the two studies. This one study gives rise to several debates. In a similar situation, would this study's results effectively and neatly terminate a backup program? Clearly, a compound with better duration than the lead may have been identified and triggered further development. Does the ability to relatively cheaply profile compounds in humans mean that more compounds will be profiled and lead to expensive phase 0 programs? Janssen is one pharma company that has used microdosing in candidate selection in this manner [26]. In an unspecified program, the lead compound showed a suboptimal human PK profile (rapid clearance and short half-life). Two backups (a metabolite and a structural analogue) showed lower *in vitro* clearance. A microdose study using LC–MS/MS was run to compare the three compounds. Eighteen healthy volunteers were randomized to receive one of the three compounds as a single oral microdose of 100 µg with PK sampling over a 72 h period. Both metabolite and structural analogues showed an improved human PK profile over the lead, with the longest half-life and highest AUC demonstrated by the structural analogue.

The lack of concordance in preclinical PK data from several species represents a dilemma in advancing a compound into phase I. Traditionally, researchers may favor data from higher species and perhaps believe that the cynomolgus or rhesus monkey represents the gold standard. Microdose studies can be used to test an otherwise promising candidate. An example of this is the EP_1 antagonist TGSK269984A, which was administered to two groups of healthy human volunteers by the intravenous and oral routes at a dose of 100 µg [27]. Following the IV microdose, the terminal elimination half-life was 8.2 h and the absolute oral bioavailability was estimated to be 95%. These data were inconsistent with predictions of human PK based on allometric scaling of *in vivo* PK data from three preclinical species (rat, dog, and monkey). Only the rat, which had been the species originally tested and provoked interest in the compound, scaled correctly. Monkey and dog PK data would have suggested the drug would have an unacceptably short half-life and not suitable for clinical evaluation.

There is a concern that when more detailed preclinical work is performed in toxicology to support the follow-on conventional dose-escalation phase I study, problems will be encountered that severely inhibit further progression of the compound. Phase 0 studies are supported by single-dose toxicology studies in one species, normally the rat, with limited pathology, whereas conventional FIH studies are supported by 14-day (or longer) studies in two species, one a nonrodent, with full pathology in both species. Since preclinical toxicity is a major cause of attrition, the concern is supported by existing statistics. Around 30% of attrition is ascribed to this factor, usually leading to termination of the compound at an early stage prior to gaining substantial, or indeed any, human experience.

3.5.4 Responding to Unfavorable PK Characteristics

In many cases, potential drug candidates are deselected at the preclinical stage based on a PK profile in laboratory animals that is taken to be predictive of inadequate PK in humans for the intended therapeutic indication. Where the problem lies in rapid metabolic clearance, knowledge of the primary site(s) of biotransformation provides the basis for structural redesign to block, or at least minimize, the responsible metabolic pathway. When this can be accomplished without unduly compromising pharmacological activity, an otherwise unattractive candidate can be "salvaged" and progressed to clinical evaluation. A novel variation on this approach is the use of specific deuterium labeling to retard oxidation at a metabolic "soft spot" through the introduction of a kinetic deuterium isotope effect. In favorable cases, the resulting decrease in metabolic clearance can result in a substantial improvement in the PK profile, and since deuterium labeling has minimal effect on the interaction of a drug candidate with its biological target, pharmacological activity is preserved [28]. To date, however, no deuterated compound has been approved as a human medicine.

An interesting example of the use of specific deuterium labeling to potentially salvage an otherwise promising compound with a poor preclinical PK profile is found in CTP-354, a subtype-selective $GABA_A$ modulator currently being developed by Concert Pharmaceuticals as a nonsedating treatment for spasticity in both patients with spinal cord injury and multiple sclerosis [29]. The corresponding unlabeled compound, L-838417, was evaluated preclinically at Merck Research Laboratories in the 1990s as part of a program aimed at the discovery of novel anxiolytics with reduced sedation and ataxia. Although this compound was found to exhibit a particularly attractive pharmacological profile, with partial agonism at the alpha-2 and alpha-3 $GABA_A$ receptor and antagonism at alpha-1 [30] rapid *in vitro* metabolism and corresponding high *in vivo* clearance in rats and dogs, meant that the compound was not advanced to human studies. However, interest remained high in L-838417, albeit as a pharmacological tool, and a retrospective analysis of its metabolism indicated that two CYP-mediated pathways were responsible for its high clearance in animals, namely, O-dealkylation of the methyltriazole moiety and hydroxylation at the *t*-butyl group. An assessment of the effects of selective deuteration at each of the two vulnerable sites demonstrated that the [*t*-butyl-$^2H_9$] variant of L-838417 (CTP-354) exhibited a three- to fivefold increase in exposure upon oral dosing to rats and dogs that was attributed to a deuterium isotope effect on first-pass metabolism, whereas labeling at the methylene bridge to the methyltriazole moiety has a lesser effect on the PK profile [29]. Structures of CTP-354 and L-838417 are shown in Figure 3.5.

On the basis of the above animal findings, and on the results of *in vitro* studies that indicated that CTP-354 exhibited greater metabolic stability than L-838417, the deuterated compound was evaluated in a phase I study. When administered orally to healthy human volunteers, CTP-354 had excellent PK characteristics, with good systemic exposure in terms of expected therapeutic concentrations and a terminal elimination

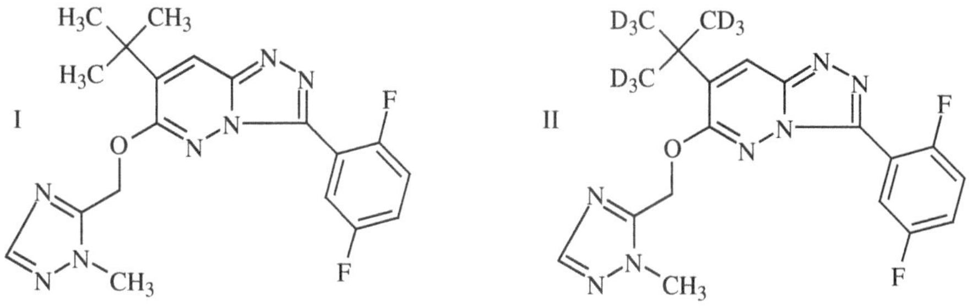

FIGURE 3.5 Structures of GABA$_A$ modulator L-838417 (I) and its deuterated analogue CTP-354 (II) that possesses superior PK properties.

half-life (18–22 h) suitable for once-a-day dosing. That therapeutic concentrations were achieved was evidenced by high and sustained occupancy of the GABA$_A$ receptor following both single and repeat doses [31]. Since L-838417 was never dosed to humans, it has not been possible to compare CTP-354 with L-838417 in man. However, this example highlights the potential utility of deuterium substitution in reducing the metabolic clearance of a new chemical entity to the point where the risk of failure due to inadequate human PK was considered sufficiently low to permit its entry into clinical development.

3.6 PHASE I ONCOLOGY STUDIES

Oncology phase I studies may be performed in patients. The traditional studies, mainly examining cytotoxic agents, used late-stage cancer patients and a variety of dose-escalation strategies to reach a toxicity rate of 33% or less. Having established an unacceptable level of toxicity, further patients entering the trial were treated at the next lower dose level(s) (MTD) in order to fully evaluate the adverse event profile of the candidate drug. If acceptable safety was obtained, larger phase II studies would be conducted to determine efficacy. These studies had an ethical context, which does not apply so strictly to volunteers, that a minimal number of patients were exposed to probably subtherapeutic doses of the candidate drug. Like volunteer studies, the ethics and safety requirements mean that a minimum number of patients are exposed to the more severe dose-limiting toxicities.

The rationally designed targeted drugs (such as kinase inhibitors) have imposed new constraints on the design of these studies. These drugs are not cytotoxic and are targeted at specific proteins present uniquely or upregulated in tumors. Patient populations are more precisely defined by the tumor type and target expression. Recruitment into studies can therefore be slow and patient selection will involve detailed work-ups. Because of relative selectivity and therefore safety of targeted therapies, the study objectives shift from the definition of an MTD to the evaluation of a recommended phase II dose. This dose may be defined by indices such as target occupancy, which may be estimated by unbound free drug concentrations based on preclinical work. Studies may include imaging or collection of fresh tumor tissue for surrogate marker analyses to help define the optimum dose regimen for efficacy [32].

In contrast to other therapeutic areas, side effects and toxicities may be more severe as evidenced by the commonly accepted grades of effects observed outlined below. These grades are illustrated with examples of neurological side effects.

1. Mild; with no or mild symptoms; no interventions required (mild somnolence, agitation)
2. Moderate; minimal intervention indicated; some limitation of activities (moderate somnolence, agitation)
3. Severe but not life-threatening; hospitalization required; limitation of patient's ability to care for himself/herself (severe somnolence, agitation, confusion, disorientation, hallucination)
4. Life-threatening; urgent intervention required (coma, seizures, toxic psychosis)
5. Death related to adverse event

These side effect profiles become less relevant for targeted therapies where the profiles are largely confined to the first two grades.

Because of the nature of therapies, phase I trials are today often designed to generate preliminary biomarker evidence of target engagement and to identify subgroups of responsive patients. To help in this aim, phase 0 studies are being conducted (differing from other phase 0 microdose studies referred to earlier). These small FIH studies to test an agent's ability to affect a molecular target are being pioneered prior to initiation of phase I testing. These studies use several dose escalations to approach the relevant pharmacologically active dose and utilize imaging or a biomarker to probe target occupancy.

Overall attrition rates of oncology drugs in phase I do not differ from other therapeutic areas with 60–70% success [33], suggesting that the differences in trial design are not a major factor. Differences in attrition rates in phase II and III, with oncology having a lower rate of success, could stem in part from favorable interpretation of early efficacy signals in phase I studies. Observations on tumor regression, for instance, may not translate into significant outcomes in larger and longer clinical trials.

The term phase I in oncology can be extended into comparatively large efficacy and toleration studies. One such study was pivotal in the development of crizotinib (Fig. 3.6). It arose from a low rate of efficacy in the original clinical studies. In preclinical studies, crizotinib showed high activity against c-MET-positive and ALK-positive cell lines. An FIH, phase I, dose-escalation trial evaluated crizotinib as an oral single agent in 37

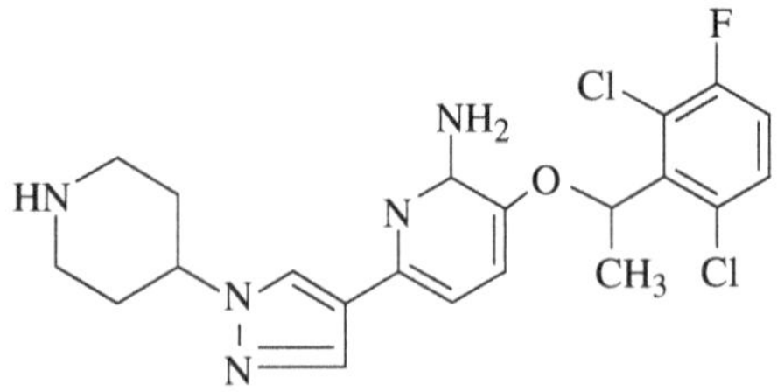

FIGURE 3.6 Structure of crizotinib, a c-MET/ALK inhibitor that showed a strong clinical effect in a subset of patients in phase I testing.

patients with advanced solid tumors, identifying 250 mg twice daily as the recommended dose. There were two patients with NSCLC harboring EML4-ALK rearrangement who showed dramatic improvement in their symptoms when treated with crizotinib during the dose-escalation phase. Because of previous clinical data, a large and prolonged study was conducted in patients with ALK-positive NSCLC [34]. In this study, patients with ALK-positive stage III or IV NSCLC received oral crizotinib 250 mg twice daily in 28-day cycles. Sixty percent of patients had an objective response, including three complete responses. Time to response was around 8 weeks with a duration of 50 weeks. The most common adverse events were visual effects, nausea, diarrhea, constipation, vomiting, and peripheral edema. A small incidence of more severe adverse events was recorded including neutropenia, raised alanine aminotransferase, hypophosphatemia, and lymphopenia. These observations on side effects should be contrasted to other therapeutic areas. In addition, crizotinib was identified as a moderate clinical inhibitor of CYP3A4, a finding that may have precluded development in other therapeutic areas.

Crizotinib has been approved following more detailed studies. The success of crizotinib indicates how vital selection of patients is and also how previous commercial guidelines such as disease incidence, patient numbers, and so on are no longer reliable in prioritizing and selecting drugs for further development.

3.7 TOLERATION AND ATTRITION IN PHASE I STUDIES

Although toleration is a major reason for stopping dose escalation, the toleration of many drugs in phase I studies is surprisingly high [2]. An examination of side effects in phase I studies for 23 different drugs showed 110 different adverse events, but the frequency was only around 13% on drug and 8% on placebo. Most common side effects were headache, diarrhea, and dyspepsia. Ninety-seven percent of side effects were minor. The 3% of severe side effects included loss of consciousness, atrial fibrillation, hyperthyroidism, and bicytopenia. Some of these side effects, although severe, are not limited to drug treatment and may occur in placebo groups. Despite these figures, data from individual companies in some cases indicates toleration or safety as a major reason for phase I attrition. AstraZeneca states a 60% success rate in phase I. The main reason for failure (62%) is due to safety [35].

The importance of placebo results and historic databases is underlined by further studies. A retrospective review [36] of pooled transaminase data collected from the placebo arms of 13 phase I trials in 93 healthy volunteers hospitalized for 14 days has been conducted. A surprisingly high proportion (20.4%) of the subjects had at least one alanine transaminase (ALT) value above the upper limit of the normal range (ULN) and 7.5% had at least one value twice ULN. The authors conclude that laboratory safety results of phase I trials should be interpreted with caution to avoid premature discontinuation of development. Although anecdotal, the authors' experience has been that, on several occasions, the findings of questionable raised liver function tests (LFTs) in phase I studies (questionable due to rapidity of onset, no dose relationship, low dose, etc.) have been used to rationalize the substitution of a superior backup (potency, selectivity, etc.) into development. Thus, these compounds would be recorded as attrition due to toxicity rather than replacement of the candidate.

3.7.1 Improving the Hepatic Toleration of Compounds

Recognizing that safety issues in general and liver toxicity in particular represent a major source of attrition in early clinical trials, several pharmaceutical companies have implemented preclinical screening protocols designed to identify risk factors for hepatotoxicity in drug candidates prior to their nomination for development. For example, AstraZeneca established an *in vitro* approach that integrates cellular effects of new chemical entities with their propensity to undergo metabolism to chemically reactive, potentially toxic intermediates in an effort to minimize the risk of idiosyncratic adverse reactions [37]. In this paradigm, compounds are tested in a panel of five *in vitro* assays that quantify (i) toxicity to THLE cells (SV40 T-antigen-immortalized human liver epithelial cells) that lack CYP activity, (ii) toxicity to a THLE cell line that selectively expresses CYP3A4, (iii) cytotoxicity in HepG2 cells in glucose and galactose media (indicative of mitochondrial injury), (iv) inhibition of the human bile salt export pump (BSEP), and (v) inhibition of the rat multidrug resistance associated protein 2 (Mrp2). In addition, estimates are made of the human "covalent binding body burden," a measure of projected whole-body exposure to chemically reactive metabolites of the compounds of interest that alkylate cellular proteins. The aggregate results from these assays then are used to score compounds for their respective risk of causing liver toxicity *in vivo*, and only those drug candidates that meet certain threshold criteria are advanced to formal development status. GlaxoSmithKline also has published details of a preclinical screening strategy designed to reduce the potential of drug candidates to cause clinical hepatotoxicity, in which an emphasis is placed on identifying metabolism-dependent inhibition of CYP enzymes and reactive metabolite formation as assessed by glutathione adduct formation and covalent binding to protein in human hepatocytes [38]. It is to be hoped that strategies of these types will reduce the incidence of attrition in phase I studies and beyond due to adverse liver effects.

3.7.2 Rare Severe Toxicity in Phase I Studies

Many phase I studies that fail to reach acceptable endpoints are not reported. Thus, an MTD that is close to the predicted efficacious dose may lead to cessation of a compound or program, but the institutional learnings are unlikely to be widely shared. Although this chapter is mainly aimed at small-molecule drugs, a widely publicized failure in a phase I study due to severe toxicity was TGN1412 [39], a novel superagonist anti-CD28 monoclonal antibody that directly stimulates T-cells. The antibody was administered to healthy young male volunteers in the first phase I clinical trial. The dose was a small fraction of the tolerated dose in monkeys (0.1 mg/kg) or 1/500th of the no-observed-adverse-effect dose in nonhuman primates (50 mg/kg). At the time, the dose was deemed to be an appropriate starting dose for the phase I trial. However, the dose proved to yield pharmacologically active concentrations in humans, probably toward the top of the dose–response curve. All volunteers had a systemic inflammatory response characterized by a rapid induction of proinflammatory cytokines and accompanied by headache, myalgias, nausea, diarrhea, erythema, vasodilatation, and hypotension. Twelve to sixteen hours after drug administration, they became critically ill, with pulmonary, renal, and lung injury, together with intravascular coagulation, followed by severe depletion of lymphocytes and monocytes. Prolonged cardiovascular shock and acute respiratory distress syndrome developed in two patients, who required intensive organ support for 8 and 16 days. The clinical study design was clearly misled by the animal data. Subsequent work has shown [40] that it is likely that activation of CD4+ effector memory T-cells by TGN1412 was responsible for the cytokine storm. Lack of CD28 expression on the CD4+ effector memory T-cells of species used for preclinical safety testing of TGN1412 would explain the absence of severe reactions in the preclinical safety assessment.

This study had a significant impact on the development of other antibodies, particularly agonists, in ensuring that the starting dose was set below a size that could cause any pharmacological effects. Phase I studies with antibodies now have to define the minimum anticipated biological effect level (MABEL) for selection of the initial dose [41]. This is not just based on a safety study, as was the case for TGN1412. The selection of dose is now based on specifically designed preclinical pharmacology and toxicology studies, mechanistic *ex vivo/in vitro* investigations with human and animal cells, and PK/PD modeling. Moreover, a thorough understanding of the biology of the target and the relative binding and pharmacological activity of the mAb in animals and humans is required. Essentially, the same practice has become established for small-molecule drugs.

3.8 TARGET OCCUPANCY AND GO/NO-GO DECISIONS TO PHASE II START

As outlined earlier, phase I studies rely heavily on plasma or blood concentrations of drug as a biomarker for effects extrapolated to and from other studies including preclinical. This is reliable in many cases, especially where the target does not reside in an organ with pronounced barriers (such as the central nervous system (CNS)) or the drug is not a reversible inhibitor or agonist. To improve the decision-making power of phase I studies, a measure of target occupancy as an additional biomarker is very valuable. Pfizer and AstraZeneca have analyzed their failures and successes and concluded that the candidates that had an increased likelihood of phase II survival were those for which there was an integrated understanding of three key principles: exposure at the site of action, target occupancy, and expression of pharmacological activity in the target organ [35, 42]. Often, these issues can be addressed in phase I studies as shown below, but it should be noted that pharma has a history of not doing this. Conducting these studies could increase phase I attrition (although better preclinical work may help to attenuate this) but substantially reduce more expensive late-stage attrition.

Target occupancy can be determined by various direct techniques or via a downstream signal, which may or may not relate to efficacy. Examples of these approaches are illustrated with the first two showing how positive and negative results influence the programs and then expanding into technology and PK/PD separations:

1. Terminate compound: "For this CNS compound with a narrow safety margin from animal data, no predictive safety and tolerability were indentified and, therefore, biomarkers reflecting brain activities (direct or indirect) were needed. Aepodia recommended that the Sponsor introduce CSF (cerebrospinal fluid) collections in the first multiple ascending-dose study (MAD) of this compound with a new mechanism of action (MOA). The results of the clinical study confirmed that the compound was crossing the blood brain barrier. However, no effect was detected on

CSF biomarkers specific to this MOA. Consequently, compound development has been stopped ("Quick Kill") before launching a costly Phase 2 program in target patient populations" [43].

2. Proceed into further studies: "Omeros Corporation announced positive results from its Phase 1 clinical program evaluating OMS824, the lead compound from its phosphodiesterase 10 (PDE10) program. This study measured the extent to which OMS824 binds to PDE10 in the basal ganglia, a region of the brain that has been linked to a wide range of diseases that affect cognition. The results show that the selected dose of OMS824 achieved approximately 50% occupancy of PDE10 without triggering the extrapyramidal symptoms (loss of muscle control, e.g., muscle rigidity, tremors, or involuntary muscle movements) reported as side effects with other PDE10 inhibitors that achieved similar or significantly lower occupancy levels" [44].

In further studies, it is often not practical to routinely use such biomarkers. In their design, therefore, it is important to establish a relationship between receptor occupancy and plasma concentrations (the universal biomarker). This is illustrated by a PET study with GSK1144814, a potent, insurmountable antagonist at human NK_1 and NK_3 receptors. In this study, NK_1 receptor occupancy data were measured in parallel with the first-time-in-human study. [^{11}C]GR-205171, a selective NK_1 receptor PET ligand, was used to estimate the time-plasma concentration–occupancy relationship. A direct relationship was seen between the GSK1144814 plasma concentration and its occupancy of the brain NK_1 receptor, indicating target occupancy can be accurately calculated from the measured plasma concentration of the compound [45]. Without establishing this relationship, much of the development program will have major uncertainty.

Where compounds have a slow offset from the receptor, including covalent interaction, the relationship between plasma concentration and target occupancy can be complex. Simple observation of plasma concentration and its decline may grossly underestimate the duration of activity. The occupancy of the target is dependent on the circulating concentrations and the rate of onset and offset. In the case of covalent interactions, the rate of offset is dependent on the synthesis of new target and disappearance of the complexed one. In these situations, some measure of the complex is highly valuable to guide future dosing in subsequent studies. Bruton's tyrosine kinase (Btk) is required for activation of B lymphocytes through the B-cell receptor (BCR) signaling network and is strongly implicated in rheumatoid arthritis. AVL-292 (Fig. 3.7) covalently binds to the Cys481 of Btk and irreversibly inhibits the kinase.

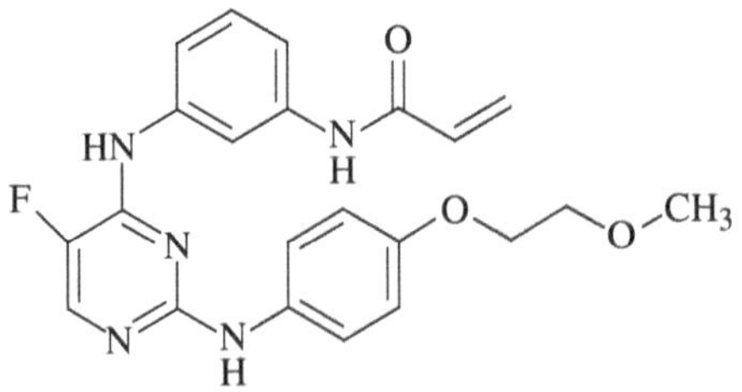

FIGURE 3.7   Structure of the covalent Btk inhibitor AVL-292 that exhibits a complex relationship between plasma concentration and receptor occupancy.

TABLE 3.2 Receptor Occupancy Required for Efficacy for a Range of Drug Targets

Target	MOA	Occupancy
GPCR	Antagonists	60–80% (3×)
	Agonists (high efficacy)	2–30% (0.02–0.4×)
	Agonists (low efficacy)	60–95% (2–10×)
CCR5	Antagonists, anti-infectives	>99% (>10×)
Ligand-gated ion channels	Antagonists	65–95% (3–10×)
	Agonists	5–80% (0.04–4×)
Transporters	Inhibitors	60–85% (3×)
Enzymes (e.g., ACE, renin, PDE5, various kinases)	Inhibitors	80–99% (4–10×)

The figures in brackets approximate how much the unbound free drug concentration would need to exceed the Kd or Ki of the drug against its target [47–50].

An ELISA assay was developed to quantitatively determine the degree of AVL-292-Btk occupancy so that both enzyme occupancy and plasma concentrations (using conventional parent compound assays) could be followed. In healthy human subjects, AVL-292 was found to be safe and well tolerated following oral administration at dose levels ranging from 0.5 to 7.0 mg/kg. All subjects that received an oral dose of 1.0 mg/kg of AVL-292 demonstrated >80% Btk occupancy. Five of the six subjects administered AVL-292 at 2.0 mg/kg achieved complete Btk occupancy. The covalent mechanism of binding resulted in a Btk occupancy that was sustained through 24 h, even though plasma concentrations of AVL-292 had declined substantially by that time [46].

It is important to place receptor occupancy and target occupancy into context. The type of target dictates how much occupancy is required. Table 3.2 lists a number of drug targets and the occupancy normally associated with efficacy at that target. These values show considerable variation, for instance, a pure agonist of a GPCR requires a much lower occupancy than an antagonist for efficacy. Ideally, early studies should include a biomarker and compare the readout to receptor occupancy. Preclinical studies may also give valuable insight into the levels of occupancy required.

An example of the need for a biomarker (or other readout) in addition to receptor occupancy is provided by maraviroc. Maraviroc is a CCR5 antagonist (nonaminergic GPCR). Its binding interferes with HIV entry into cells. Viral gp120 binds to CCR5 or CXCR4 coreceptor and enters CD4 T-cells by fusion. Early clinical studies looked at both HIV viral load and its reduction and receptor occupancy and developed PK/PD models. Receptor occupancy had a Kd value of 0.17 nM in contrast to the antiviral IC_{50} of 16 nM. High occupancy was achieved at low doses (3 mg b.i.d), but was not antivirally effective. Mechanistic modeling indicated that only 1.2% of free activated receptors were utilized to elicit 50% of the maximum infection rate. To be antivirally effective, very high receptor occupancy was required, leading to daily dosage regimens of 150 mg or above [51, 52].

Getting proof of target engagement is highly valuable, but it should be used with caution to accelerate drug development, even when the target occupancy is measured by a relevant biomarker. Vatalanib (a VEGF inhibitor) was shown to reduce tumor blood supply during phase I studies, using dynamic contrast-enhanced magnetic resonance imaging. The compound was accelerated into phase III studies in combination with FOLFOX in colorectal cancer. These studies showed negligible extra benefit and increased side effects. They can be questioned on the combination of agents and the dosage regimen in terms of frequency of vatalanib as reasons for failure, and undoubtedly a more cautious structured approach may have achieved a successful outcome or lead to termination in a less costly smaller phase II program [53].

3.9   CONCLUSIONS

Phase I studies have a reasonable success rate that preclinical nomination of compounds now includes extensive predictions for PK. Understanding of target engagement and occupancy adds further dimensions to phase I studies in being decision making in terms of compound progression. A success rate of 60–70% is perhaps a good criterion to aim for, provided that the failures are compounds that do not meet the exact criteria in terms of PK and activity, or at least access to the target. As shown by the pharma retrospective portfolio analyses, a high proportion of candidates advanced into phase II and even phase III do not test the mechanism [26, 31]. The place of phase 0 studies is still open to debate. Evidence suggests that these studies are being undertaken more frequently but probably represent only around 5% of FIH studies. An attractive addition to human studies is the ability to give a concomitant IV microdose of ^{14}C-labeled material to accompany a conventional oral dose. Such studies give absolute bioavailability data in the same individuals and could be conducted as part of phase I or at any subsequent time.

REFERENCES

1   Deullier, E., Chan, A.-W., Chapuis, F.,(2009). Inadequate dissemination of phase 1 trials: a retrospective cohort study. *PLoS Med., 6*, 202–208.

2   Sibile, M., Deigat, N., Janin, A., Kirkesseli, S., Durand, D. (1998). Adverse events in phase-1 studies: a report in 1015 healthy volunteers. *Eur. J. Clin. Pharmacol., 54*, 13–20.

3   Prentis, R. A., Lis Y., Walker S.R. (1988). Pharmaceutical innovation by the seven UK-owned pharmaceutical companies (1964-1985). *Br. J. Clin. Pharmacol., 25*, 387–396.

4   Kaye, B., Offerman, J.L., Reid, J.L., Elliott, H.L., Hillis, W.S. (1984). A species difference in the presystemic metabolism of carbazeran in dog and man. *Xenobiotica, 14*, 935–945.

5   Kaye, B., Rance, D., Waring, L. (1985). Oxidative metabolism of carbazeran in vitro by liver cytosol of baboon and man. *Xenobiotica, 15*, 237–242.

6   Wang, J.L., Limburg, D., Graneto, M.G., Springer, J., Rogier, J., Hamper, B., Liao, S., Pawlitz, J.L., Krumball, R.G., Maziasz, T., Talley, J.J., Kiefer, J. R., Carter, J. (2010). The novel benzopyran class of selective cyclooxygenase-2 inhibitors. Part 2: the second clinical candidate having a shorter and favourable human half-life. *Biorg. Med. Chem. Lett., 20*, 7159–7163.

7   Jhee, S.S., Chappell, A.S., Zarotsky, V., Moran, S.V., Rosenthal, M., Kim, E., Chalon, S., Toublanc, N., Brandt, J., Coutant, D.E., Ereshefsky, L. (2006). Multiple dose plasma pharmacokinetic and safety study of LY450108 and LY 451395 (AMPA receptor potentiators) and their concentration in cerebrospinal fluid in healthy human subjects. *J. Clin. Pharmacol., 46*, 424–432.

8   Penning, T.D., Talley, J.T., Bertenshaw, S.R., Carter, J.S., Collins, P.W., Docter, S., Graneto, M.J., Lee, L.F., Malecha, J.W., Miyashiro, J.M., Rogers, R.S., Rogier, D.J., Yu, S.S., Anderson, G.D., Burton, E.G., Cogburn, J.N., Gregory, S.A., Koboldt, C.M., Perkins, W.E., Seibert, K.,

Veenhuizen, A.W., Zhang, Y.Y., Isakson, P.C. (1997). Synthesis and biological evaluation of the 1,5-diarylpyrazole class of cyclooxygenase-2 inhibitors: identification of 4-[5-(4-methylphenyl)-3-(trifluoromethyl)-1-H-pyrazol-1-yl]benzenesulfonamide (SC-58635, Celecoxib). *J. Med. Chem.*, *40*, 1347–1365.

9 Tang, C., Shou, M., Rushmore, T.H., Mei, Q., Sandhu, P., Woolf, E.J., Rose, M.J., Gelmann, A., Greenberg, H.E., De Lepeleire, I., Van Hecken, A., De Schepper, P.J., Ebel, D.L., Schwartz, J.I., Rodrigues, A.D. (2001). In-vitro metabolism of celecoxib, a cyclooxygenase-2 inhibitor, by allelic variant forms of human liver microsomal cytochrome P450 2C9: correlation with CYP2C9 genotype and in-vivo pharmacokinetics. *Pharmacogenetics*, *11*, 223–235.

10 Kirchheiner, J., Störmer, E., Meisel, C., Steinbach, N., Roots, I., Brockmöller, J. (2003). Influence of CYP2C9 genetic polymorphisms on pharmacokinetics of celecoxib and its metabolites. *Pharmacogenetics*, *13*, 473–480.

11 Chauret, N., Yergey, J.A., Brideau, C., Friesen, R.W., Mancini, J., Riendeau, D., Silva, J., Styhler, A., Trimble, L.A., Nicoll-Griffith, D.A. (2001). In vitro metabolism considerations, including activity testing of metabolites, in the discovery and selection of the inhibitor etoricoxib. *Bioorg. Med. Chem. Lett.*, *11*, 1059–1062.

12 Boxenbaum, H. (1982). Interspecies scaling, allometry, physiological time, and the ground plan of pharmacokinetics. *J. Pharmacokin. Biopharm.*, *10*, 201–227.

13 Boxenbaum, H., D'Souza, R. (1988). Physiological models, allometry, neoteny, space time and pharmacokinetics in Pharmacokinetics (eds A. Pecile and A. Rescigno), *NATO ASI Ser.*, *Ser. A, 145*, 191–214.

14 Huang, Q., Riviere, J.E. (2014). The application of allometric scaling principles to predict pharmacokinetic parameters across species. *Exp. Opin. Drug Met Toxicol.*, *10*, 1241–1253.

15 Obach, R.S., Baxter, J.G., Liston, T.E., Silber, M., Jones, B.C., MacIntyre, F., Rance, D.J., Wastall, P. (1997). The prediction of human pharmacokinetic parameters from preclinical and in vitro metabolism data. *J. Pharm. Exp. Ther.*, *283*, 48–58.

16 Mahmood, I., Balian, J.D. (1996). Interspecies scaling: predicting clearance of drugs in humans. Three different approaches. *Xenobiotica*, *26*, 887–95.

17 Sacher, G.A. (1959). Relation of life span to brain weight and body weight in mammals. In: editors. *The Lifespan of Animals* (eds. Wolstenholme G.E.W. and O' Connor, M., London: Little, Brown and Company. pp. 115–133.

18 Amantana, A., Chen, Y., Tyavanagimatt, S. R., Jones, K. F., Chinsangaram, J., Bolken, T. C., Leeds, J. M., Hruby, D. E. (2013). Pharmacokinetics and Interspecies Allometric Scaling of ST-246, an oral antiviral therapeutic for treatment of orthopoxvirus infection. *PLoS One*, *8*, 1–10.

19 Smith. R., Jones, R.D., Ballard, P.G., Griffiths. H.H. (2008). Determination of microsome and hepatocyte scaling factors for in vitro/in vivo extrapolation in the rat and dog. *Xenobiotica*, *38*, 1386–1398.

20 Ito, K., Houston, J. B. (2005). Prediction of human drug clearance from in vitro and preclinical data using physiologically based and empirical approaches. *Pharm Res.*, *22*, 103–112.

21 Li, N., Bi, Y.A., Duignan, D.B., Lai, Y. (2009). Quantitative expression profile of hepatobiliary transporters in sandwich cultured rat and human hepatocytes. *Mol Pharm. 6*, 1180–1189.

22 Chan, T. S., Yu, H., Moore, A., Khetani, S., Tweedie, D. (2013). Meeting the challenge of predicting hepatic clearance of compounds slowly metabolized by cytochrome P450 using a novel hepatocyte model, HepatoPac®. *Drug Met. Disp. 41*, 2024–2032.

23 Lappin, G., Garner, R.C. (2008). The utility of microdosing over the past 5 years. *Exp. Opin. Drug Met. Toxicol.*, *4*, 1499–1506.

24 Rowland, M. (2012). Microdosing: A critical assessment of human data. *J. Pharm. Sci.*, *101*, 4067–4074.

25 Sun, L., Hankun, L., Wilson, K., Breidinger, S., Rizk, M.L., Wenning, L., Woolf, E. J. (2012). Ultrasensitive liquid chromatography-tandem mass spectrometric methodologies for quantification of five HIV-1 integrase inhibitors in plasma for a microdose clinical trial. *Anal. Chem., 84*, 8614–8621.

26 Verhaeghe, T. (2014). LC-MS/MS as an enabler for a broader application of microdose studies in drug development; the Janssen experience. EBF Meeting Barcelona, http://bcn2014. europeanbioanalysisforum.eu/site/ebf_bcn2014/assets/slides/pdf/Tom_Verhahe.pdf (accessed on June 28, 2015).

27 Ostenfeld, T., Beaumont, C., Bullman, J., Beaumont, M., Jeffrey, P. (2012) Human microdose evaluation of the novel EP1 receptor antagonist GSK269984A. *Br J Clin Pharmacol. 74*(6):1033–44.

28 Baillie, T.A. (1981). The use of stable isotopes in pharmacological research. *Pharmacol. Rev. 33*, 81–132.

29 Liu, J. F., Harbeson, S., Uttamsingh, V., Morales, A.J., Nguyen, S., Bridson, G., Cheng, C., Asianian, A., Wu, L. (2012). CTP-354: A novel deuterated subtype-selective GABA(A) modulator for treatment of neuropathic pain, spasticity and anxiety disorders. Presented at the American College of Neuropsychopharmacology Meeting, December 2–6, 2012. http://www. concertpharma.com (accessed June 28, 2015).

30 McKernan, R.M., Rosahl, T.W., Reynolds, D.S., Cur, C., Wafford, K.A., Atack, J.R., Farrar, S., Myers, J., Cook, G., Ferris, P.,Garrett, L., Bristow, L., Marshall, G., Macaulay, A., Brown, N., Howell, O., Moore, K.W., Carling, R. W., Street, L.J., Castro, J.L., Ragan, C.I., Dawson, G.R., Whiting, P.J. (2000). Sedative but not anxiolytic properties of benzodiazepines are mediated by the GABA$_A$ receptor α$_1$ subtype. *Nat. Neurosci. 3*, 587–592.

31 Braman, V., Liu, J.F., Harbeson, S., Uttamsingh, V., Bridson, G., Wu, L., Shipley, J. E. (2014). Preliminary clinical outcomes for CTP-354, a novel subtype-selective GABA(A) modulator. Presented at the American Neurological Association Annual Meeting, October 12–14, 2014. http://www.concertpharma.com (accessed June 28, 2015).

32 Salzberg, M. (2012). First-in-human phase 1 studies in oncology: the new challenge for investigative studies. *Rambam Maimonides Med. J., 3*, e0007.

33 Kola, I., Landis, J. (2004). Can the pharmaceutical industry reduce attrition rates. *Nat. Rev. Drug Disc. 3*, 712–715.

34 Camidge, D.R., Bang, Y.J., Kwak, E. L., Iafrate, A.J., Varella-Garcia, M., Fox, S.B., Riely, G.J., Solomon, B., Ou, S.H., Kim, D.W., Salgia, R., Fidias, P., Engelman, J. A., Gandhi, L., Jänne, P.A., Costa, D.B., Shapiro, G.I., Lorusso, P., Ruffner, K., Stephenson, P., Tang, Y., Wilner, K., Clark, J. W., Shaw, A.T. (2012). Activity and safety of crizotinib in patients with ALK-positive non-small-cell lung cancer: updated results from a phase 1 study. *Lancet Oncol., 13*, 1011–1019.

35 Cook, D., Brown, D., Alexander, R., March, R., Mrgan, P., Satterthwaite, G., Pangelos, M. N. (2014). Lessons learned from the fate of AstraZeneca's drug pipeline: a five-dimensional framework Lessons learned from the fate of AstraZeneca's drug pipeline: a five-dimensional framework. *Nat. Rev. Drug Disc. 13*, 419–431.

36 Rosenzweig, P., Miget, N., Brohier, S.(1999). Transaminase elevation on placebo during phase I trials: prevalence and significance. *Br. J. Clin. Pharmacol. 48*, 19–23.

37 Thompson, R.A., Isin, E.M., Li, Y., Weidolf, L., Page, K., Wilson, I., Swallow, S., Middleton, B., Stahl, S., Foster, A.J., Dolgos, H., Weaver, R., Kenna, J.G. (2012). In vitro approach to assess the potential for risk of idiosyncratic adverse reactions caused by candidate drugs. *Chem. Res. Toxicol., 25*, 1616–1632.

38 Sakatis, M.Z., Reese, M.J., Harrell, A.W., Taylor, M.A., Baines, I.A., Chen, L., Bloomer, J.C., Yang, E.Y., Ellens, H.M., Ambroso, J.L., Lovatt, C.A., Ayrton, A.D., Clarke, S.E. (2012).

Preclinical strategy to reduce clinical hepatotoxicity using in vitro bioactivation data for >200 compounds. *Chem. Res. Toxicol., 25,* 2067–2082.

39 Suntharalingam, G., Meghan, R.P., Ward, S., Brett, S.J., Castello-Cortes, A., Brunner, M.D., Panoskaltsis, N. (2006). Cytokine storm in a phase 1 trial of the anti-CD28 monoclonal antibody TGN1412. *N. Engl. J. Med., 355,* 1018–1028.

40 D Eastwood, D., Findlay, L., Poole, S., Bird, C., Wadhwa, M., Moore, M., Burns, C., Thorpe, R., Stebbings, R. (2010). Monoclonal antibody TGN1412 trial failure explained by species differences in CD28 expression on CD4$^+$ effector memory T-cells *Br. J. Pharm., 161,* 512–526.

41 Muller, P.Y., Milton M., Lloyd, P., Sims, J., Brennan, F.R.(2009) The minimum anticipated biological effect level (MABEL) for selection of first human dose in clinical trials with monoclonal antibodies. *Curr. Opin. Biotechnol., 20,* 722–729.

42 Morgan, P., Van Der Graaf, P.H., Arrowsmith, J., Feltner, D.E., Drummond, K.S., Wegner, C. D., Street, S.D. (2012) Can the flow of medicines be improved? Fundamental pharmacokinetic and pharmacological principles toward improving Phase II survival. *Drug Discov. Today, 17,* 419–424.

43 Aepodia (2014). Cerebrospinal Fluid Collection for a Compound with a Narrow Safety Margin and a Novel Mechanism of Action http://www.aepodia.com/en/case-studies/.

44 Huntington's Disease Society of America (2013). Omeros Reports Positive Results from OMS824 Positron Emission Tomography Clinical Trial. OMS824 Achieves High Target Occupancy without Causing Movement Disorders Seen with Other PDE10 Inhibitors http://www.hdsa.org/research/news/omeros-phase-i-oms824-tril-results-positive.html.

45 Ridler, K., Gunn, R.N., Searle, G.E., Basrletta, J., Passchier, J., Dixson, L., Hallett, W.A., Gray, F.A., Burgess, C., Poggesi, I., Bullman, J.N., Ratti, E., Laruelle, M.A., Rabiner, E.A. (2014). Characterising the plasma-target occupancy relationship of the neurokinin antagonist GSK1144814 with PET. *J. Psychopharmacol., 28,* 244–253.

46 Evans, E., Tester, R., Aslanian, S., Chaturvedi, P., Karp, R., Labenski, M., Mazdiyasni, H. (2011). Translational medicine of a selective inhibitor of Btk in rheumatic diseases: pharmacology and early clinical development. *Arthritis Rheum., 63,* 1757.

47 Smith, D.A., Morgan, P., Vogel, W.M., Walker, D. (2010). The use of Cav rather than AUC in safety assessment. *Reg Toxicol Pharmacol, 57,* 70–73.

48 Grimwood, S., Hartig, P.R. (2009). Target site occupancy: emerging generalizations from clinical and preclinical studies. *Pharmacol Ther, 122,* 281–301.

49 Allikmets, K. (2007). Aliskiren—an orally active renin inhibitor. Review of pharmacology, pharmacodynamics, kinetics, and clinical potential in the treatment of hypertension. *Vascular Health Risk Manage., 3,* 809–815.

50 Cleland J.G.F., Poole-Wilson P.A. (1994). ACE inhibitors for heart failure: a question of dose. *Br. Heart J., 72,* S106–S110.

51 Rosano, M. C., Jacqumin, P., van der Ryst, E. (2008). Population pharmacokinetic / pharmacodynamic analysis of CCR5 receptor occupancy by maraviroc in healthy subjects and HIV positive patients. *Br. J. Clin Pharmacol., 65,* 86–94.

52 Jacqumin, P., Mc Fadyan, L., Wade, J. R. (2008) A receptor-theory based semi-mechanistic PD model for the CCR5 non competitive antagonist maraviroc. *Br. J. Clin. Pharmacol, 65* (Suppl. 1) 95–106.

53 Lacombe, D., Liu, Y. (2013). The future of clinical research in oncology: where are we heading to? *Chin. Clin. Oncol., 2,* 9–16.

4

COMPOUND ATTRITION IN PHASE II/III

Alexander Alex[1], C. John Harris[2], Wilma W. Keighley[3]
and Dennis A. Smith[4]

[1] Evenor Consulting Ltd, Sandwich, Kent, UK
[2] cjh Consultants, Kent, UK
[3] WK Life Sciences, Kent, UK
[4] Department of Chemistry, University of Cape Town, Cape Town, South Africa; The Maltings, Walmer, Kent, UK

4.1 INTRODUCTION

As mentioned in the introduction to this book, a strong cohort of new drug approvals by the FDA toward the end of the year increased the total to 41 for 2014, the largest number in 18 years (following 39 NCEs in 2012, which was the previous last high). However, despite these very encouraging numbers, the total number of drugs approved in the last five years still is not sufficient for a sustainable business model.

The challenges facing the pharmaceutical industry in terms of compound attrition in discovery and early clinical phases have already been outlined in the three previous chapters of this book. The issues with the level of productivity in the industry versus the high level of investment have also been discussed in much detail in the previous chapters and will not be revisited here unless it is relevant to particular case histories. The focus of this chapter will be on attrition of candidate drugs, which have already overcome many significant hurdles, including safety studies in healthy volunteers, and which have, as a result of these, been considered sufficiently safe to progress further into clinical studies in patients particularly with the goal of establishing efficacy but also to extend the safety data to significantly larger groups of individuals in clinical trials. We ask the question if more could be done to take attrition in phase II rather than phase III or beyond.

Phase II/III studies are by far the most expensive part in the development of any drug, dwarfing the investment in discovery and early clinical studies. However, the outcome of those phase II/III clinical trials, although the actual readout is not known at this stage, is already predetermined by the physicochemical and pharmacological properties of the

Attrition in the Pharmaceutical Industry: Reasons, Implications, and Pathways Forward,
First Edition. Edited by Alexander Alex, C. John Harris and Dennis A. Smith.
© 2016 John Wiley & Sons, Inc. Published 2016 by John Wiley & Sons, Inc.

compound going into the trials. This may be an uncomfortable fact, but it certainly highlights the importance of state-of-the-art drug discovery and modern compound design strategies with a clear view of the potential risks, which may arise only years down the line in clinical development. Therefore, the significance and importance of compound design as well as extensive testing in disease-relevant animal models to the eventual phase II/III success rate cannot be overstated. Despite efforts in the early stages of discovery and development to ensure that only the most promising drug candidates, with acceptable safety, are progressed into trials in patients, the proportion of failures in those later phases is still staggeringly high. So what is going wrong? Although we aim to understand and predict the outcomes of pharmacological studies, the fact of the matter is that we don't understand biological systems well enough to do so with any degree of confidence. There are good reasons why these clinical phases are called "clinical trials," implying the "trial and error" concept as the prevailing (if not the only) business model. Pharmaceutical organizations have tried to improve the success rates of clinical studies by introducing improved principles and processes: the use of more structured screening cascades; the introduction of relevant, and clinically useful, biomarkers to early stage discovery; and go/no go organizational principles. Unfortunately, successes with these attempts have been rather limited. These tactics will be discussed later in more detail (see Chapters 7 and 8).

4.2 ATTRITION RATES: HOW HAVE THEY CHANGED?

There have been a great number of publications over the last two decades on the subject of attrition and productivity in the pharmaceutical industry, outlining the issues facing the industry as a whole and in some cases pointing out potential ways forward [1–4]. The high overall attrition rate from first in man to registration of 89% is clearly an alarming finding, indicating that productivity at that level is not a sustainable business model [1]. In the late 1990s, pharmacokinetics (PK) and ADME were popularly cited as the major causes of compound attrition [5]. On closer inspection, particularly if the special case of anti-infective drugs is excluded, the data are less convincing with toxicity and lack of efficacy already a major factor in attrition (46%) [6]. Analysis of clinical data between 2004 and 2011 highlights a further shift from ADME as the main cause of attrition toward efficacy and toxicity, and therefore from the earlier stages of clinical development into phase II/III [1, 7]; however, the overall relative attrition rate in terms of percentage for both clinical safety and efficacy has remained virtually unchanged between 1990 and 2000 [1]. An analysis in 2009 of attrition in the oncology field showed that the clinical attrition rate was 82% for all anticancer agents, with the highest attrition rate at the phase II/III transition, which is significantly lower than for other disease areas. This may indicate that if there is a clear mode of action or a specific target that can be relied upon to be efficacious, attrition risk appears to be lower. Interestingly, the attrition rate was only 53% for kinase inhibitors designed against molecular targets, and kinase inhibitors were also more successful in the high-risk transition between phase II and phase III [8]. Attrition rate also varied across disease areas, with cardiovascular development candidates showing the lowest attrition at 80% and women's health the highest at 96%. It may not be so surprising that attrition rate varies across therapeutic areas—for example, one might expect that cardiovascular candidates would have a good translation rate because the tests for efficacy are well understood and can be applied well in animal models, whereas one would expect CNS drugs to be the ones with a higher phase II/III attrition because there are fewer relevant and reliable animal models. Potentially, this reflects the relative utility of animal models and availability of relevant biomarkers for different disease areas, where cardiovascular medicine is relatively well served, whereas women's health (if this includes libido enhancement) has relatively few reliable indicators. Overall, phase II has the highest overall failure rate by some margin of any development stage, whereas phase I and III attrition rates are quite similar, clearly indicating that lack of efficacy is often the highest hurdle in development [1].

4.3 WHY DO DRUGS FAIL IN PHASE II/III? LACK OF EFFICACY OR MARGINAL EFFICACY LEADING TO LIKELY COMMERCIAL FAILURE

A study conducted by the Bay Clinical Research and Development Services in 2009 discussed the reasons why late-phase clinical trials, in particular phase III, have been failing with regard to the development of stroke and cancer drugs [9]. Some major reasons given in this report for the high failure rate include misleading or inaccurate information recorded in phase II trials and ineffective programs, which apparently do not provide enough information for moving on to phase III trials or designing an adequate subsequent experiment. In addition, some phase II trials suffer because of shortened timelines before approval in order to compress the total trial times before submission, which is often due to pressures to get drugs to market as quickly as possible, and it seems that drug development regularly focuses on the phase II stage for time reductions. In a series of very strong statements, Retziaos posits that many phase II trials are improperly executed or provide incomplete data [10]. Taken at face value, this suggests that properly conducted phase II might incur even higher attrition rates since current practice allows progression to phase III where it should not. Additionally, the smaller size of patient populations in earlier trials may lead to safety signals being missed. As a potential way forward, Retziaos suggests that a more adequate collection of information at phase II to allow decision to progress or not, although more costly in time and money in phase II, could significantly reduce overall developmental costs. Solutions to these late-phase failures could include developing a clear view of what needs to be achieved in terms of readout in phase III when designing the phase II trials, and analyzing previous development and historical data of the test compounds, as well as choosing a relevant endpoint for both the medication tested and the clinical patient population [5].

In 2012, Gan et al. reviewed 235 recently published phase III randomized clinical trials (RCTs) [11]. They report that 62% of the trials did not achieve results with statistical significance. Trying to explain the high failure rate, they note the actual magnitude of benefit achieved in a clinical trial is nearly always less than what was predicted at the time the trial was designed. In their conclusion, they state that "investigators consistently use overly optimistic assumptions of treatment benefits when designing RCTs to evaluate treatments for cancer, likely contributing to the high percentage of negative RCTs. Attempts to reduce the frequency of negative RCTs should focus on better ways of specifying the expected benefit. More frequent use of interim analyses, adaptive designs, and better patient selection are possible ways of mitigating this problem." The data of Gan et al. should serve as an indication of the risk of clinical trials losing their focus and of being undersized. In contrast, in a 2012 publication, Amiri-Kordestani et al. claim that conducting larger trials and doing more interim analyses or using adaptive trial designs are not the solutions to the high attrition in phase II and III trials [12]. They further state that "we do not need more marginal results that are then pronounced 'new treatment paradigms' or a 'new standard of therapy.' What we need instead are meaningful goals and better drugs—much better drugs aimed at targets that are really important" [12]. Strong statements indeed, proposing that change is needed on a scale that goes far beyond just the design of clinical trials but also how expectations are set and how success is measured in phase II trials. In particular, this leads to the view that even statistically significant effects are not of value if they do not offer meaningful improvements to the patient's well-being and quality of life.

A study of clinical trials in rheumatoid arthritis (RA) published in 2012 [13] assessed the risk of drug failure by clinical testing in patients with moderate to severe RA. Therapies for RA were investigated by reviewing phase I to phase III studies conducted from December 1998 to March 2011, and clinical trial success rates were calculated and compared to industry standards. Trial failures were analyzed in detail and were classified as either commercial or clinical failures [13]. A search conducted by the authors in clinicaltrials.gov and approved drugs for RA yielded a total of 69 drugs that met the study criteria. The overall success rate was found to be 16%, which is apparently equivalent to the industry standard of 16%. In addition, it was noted that for each phase, the frequency of clinical failures exceeded commercial failures. Although clinical studies equally consisted of investigations of small molecules and biological agents, biologics exhibited a higher success rate overall. The conclusion of the authors is that clinical trial risk in RA with the 84% failure rate is at par with industry performance and phase II success rate seems to be highly predictive of phase III success [13] in contrast with, for example, the attrition rate for kinase inhibitors for cancer therapy. Several examples are illustrated for efficacy failure. Pomaglumetad methionil (LY-2140023) is an example where early preclinical and clinical data raised expectations, and despite some further disappointing clinical data, a phase III study was conducted. The balance between mixed results in phase II studies and further development is a very difficult decision, and some of the factors are stated in the example. Dimebon represents another example. In this case, no preclinical rationale or data were supportive, yet phase II study conducted in Russia provided some of the most impressive results in the disease (Alzheimer's) yet seen. Phase III was initiated without further phase II trials and ended in the absence of any clinical efficacy.

Further case examples highlight three antibody drugs aimed at the treatment of Alzheimer's disease: bapineuzumab (4), gantenerumab (5), and solanezumab (6). These antibodies all were aimed at a similar target: all Alzheimer patients have extracellular deposits of amyloid composed primarily of the $A\beta$ peptide. Based largely on the study of genetic mutations that can cause Alzheimer's disease, the amyloid plaques have been increasingly viewed as a critical step in the pathogenesis of the disease leading to dementia. The antibody drugs were designed to lower $A\beta$ and did produce beneficial effects in mouse models although how these actually work is not known [14]. Proposed mechanisms are that the antibodies act to catalytically modify the $A\beta$ conformation to one that is less prone to fibril formation; the antibodies opsonize the $A\beta$ deposits, leading to degradation and removal of the material by phagocytes or other activated immune components; the antibodies act as a sink to pull $A\beta$ into the circulation, thereby reducing its content in the brain through modifying the equilibrium between these compartments.

All these mechanisms could apply in different degrees with these drugs. Just considering the sink hypothesis, it is easy to see how dosing schedule would be absolutely critical in optimizing the therapy so that free antibody in terms of binding capacity should be available throughout the treatment. For the sink hypothesis, antibodies would need to be high capacity and highly selective (see later). One of the antibody drugs has had the mechanism partly clarified [15]. Gantenerumab was shown to induce cellular phagocytosis of human amyloid-β deposits in brain slices when cocultured with primary human macrophages. When administered to transgenic mice, gantenerumab showed sustained binding to cerebral amyloid-β and reduced small amyloid-β plaques by recruiting microglia and prevented new plaque formation. Unlike other Aβ antibodies, gantenerumab did not alter plasma Aβ suggesting undisturbed systemic clearance of soluble Aβa and that the drug did not work according to the sink hypothesis. Gantenerumab, therefore, preferentially interacts with aggregated Aβ in the brain and lowers amyloid-β by eliciting effector cell-mediated clearance. The epitope specificity of anti-Aß antibodies may dictate efficacy in the clinic. Those directed at the N-terminal amino acids of the Aß peptide have been found effective in mouse models as have those binding to the central domains of the Aß peptide, but efficacy in mouse may not translate to man. In addition, selectivity is critical. If an antibody was not selective, it would rapidly become saturated with nontarget proteins or bind to the wrong or multiple on- and off-target proteins. In either case, much higher doses would be needed and the possibility of off-target side effects is raised. A 2014 study [16] showed that antibody drugs bapineuzumab, solanezumab, and crenezumab were able to bind Aβ in mouse tissue. However, significant differences were observed in human brain tissue. Bapineuzumab was able to bind a variety of N-terminally truncated Aβ species; in contrast, the binding by solanezumab and crenezumab was minimal. None of the antibodies were able to detect any Aβ species in human blood. Surprisingly, both solanezumab and crenezumab have extensive cross-reactivity with non-Aβ-related proteins. Bapineuzumab demonstrated target engagement with brain Aβ, consistent with published clinical data. Solanezumab and crenezumab did not, most likely as a result of a lack of specificity due to cross-reactivity with other proteins containing epitope overlap. Since Alzheimer's is a disease of the brain and drugs have to cross the blood–brain barrier (BBB), it is surprising to find statements about studies published in 2002 regarding low brain penetration (approximately 0.1% of injected antibodies crossing the BBB) being advance in 2014 for the low clinical success rate of several anti-Aβ therapies [17]. If this is the case, such factors need to be related to PK/PD incorporating dose, amount of target, and affinity of antibody in simulations to see if therapy is feasible or to select appropriate dose schedules. It is not possible to ascertain which of these issues were researched or discussed by the various sponsor companies for these expensive trials. Some of the drugs seem to have had more extensive investigation than others. In each case, the drugs appear to have entered phase III trials prematurely, suggesting commercial and organizational pressures to be the "winner." The commitment to the studies is probably reinforced by the knowledge that others are doing exactly the same. This helps increase the pressure to be first and also reassures the companies that others are thinking the same (must be right). Multiple failures can produce a backlash deeming the area difficult, lacking good targets, and so on. It is of concern that in spite of these mechanism and PK/PD issues, a number of antibody drugs have been advanced into phase III studies and not too surprisingly failed due to efficacy. A similar situation exists for small-molecule agents, and we detail examples for both types of agent in our case studies.

4.4 TOXICITY

Compound safety remains one of the major causes of attrition; therefore, more accurate prediction of ADMET and adverse reactions could vastly improve productivity [18, 19]. The prediction of adverse reactions and side effects related to polypharmacology received a lot of attention in the early 2000s when the idea of mapping and understanding polypharmacology through compound profiling against a large number of (often pharmacologically unrelated) targets was developed to support the prediction of adverse reactions [20–22]. In addition, large-scale analysis of data for adverse reactions has provided some insights into links between adverse reactions and chemical structure. These retrospective analyses do not aim to understand the mechanistic cause of adverse reactions but rather provide a tool for compound prioritization and for managing attrition risk [23]. Experimental wide ligand profiling and chemoinformatic analysis of adverse reactions have become standard tools in many large pharmaceutical companies and have certainly impacted in some specific areas (e.g., hERG); however, the general and overall impact is less easy to assess. The difficulties with predicting toxicity and polypharmacology risk based on molecular structure and properties have been highlighted by two studies of toxicity risk in a set of drug candidates within two large organizations. In the first study, a group at Pfizer looked at 245 preclinical compounds and their toxicity and observed a strong relationship between two physical chemistry measures (total polar surface area and calculated logP) and the relative promiscuity and polypharmacology of the compounds (clogP < 3, TPSA < 75 Å^2 gives higher likelihood of safety) [24]. However, a later publication by a group at AstraZeneca could not find similar trends in their analysis of 150 drug candidates and drugs approved between 2009 and 2011 [25]. The authors also point out that, for example, four of the ten top selling drugs in the United States in 2010 had a clogP greater than 3 and a polar surface area of less than 75 Å^2 (Plavix, Abilify, Singulair, and Cymbalta) and that a large number of drugs approved between 2009 and 2011 would never have reached patients if such guidelines had been applied at an early stage. These studies clearly highlight the difficulties with predicting safety and attrition based on compound structure and molecular properties alone and also indicate that the solution to the problem most likely lies in improved experimental safety studies.

In addition to undesirable polypharmacology, genotoxicity and carcinogenicity are important safety considerations in compound development. An extensive analysis of genotoxicity and carcinogenicity data for 472 marketed drugs highlighted that assessment and prediction of these risk factors based on current standard assays is also extremely difficult [26]. It is certainly debatable whether toxicological outcomes could ever be predicted with confidence based on, for example, the structure or physicochemical properties of a given small molecule, much less for biological agents. The use of statistical analysis and machine learning may open up some opportunities in this area, but the sparseness of toxicological data is very likely to limit the ability to build reliable models and hence the confidence placed in these predictions. Pooling of toxicological data from within the industry to give access to larger numbers of outcomes on which to build models would surely be of general benefit, and such efforts could be seen as of great precompetitive value. Candidate drugs can fail at any stage, but those with the potential to be undetected in earlier clinical and preclinical studies are the low incidence/high severity toxicity often termed idiosyncratic. Such toxicities may be evident only in large phase III studies or after the drug is launched (see Chapter 5) and are typically seen in 1/1000 patients or rarer.

Hepatotoxicity, severe skin effects, and blood dyscrasias are typical examples. In many cases, reactive metabolites of the drug have been implicated (see Chapter 5). An example of this is ximelagatran, which we detail in the case examples. The other form of toxicity seen in phase III involves disadvantages to the outcome measure by the drug, possibly due to the actual mechanism. Torcetrapib, again detailed in our case examples, exemplifies a mechanism of toxicity (increased incidence of cardiovascular events), which may relate to the primary or secondary pharmacology.

4.5 ORGANIZATIONAL CULTURE

In a recent very insightful review, Cook et al. summarized the lessons learned from AstraZeneca's drug pipeline [27] based on the development of a five-dimensional framework used to analyze their 2005–2010 pipeline. This framework has been named the five "R"s: the right target, the right patient, the right tissue, the right safety, and the right commercial potential. They also mention a sixth factor, the right culture, which is crucial in encouraging effective decision making based on the technical determinants outlined in the five "R"s. It is interesting to note that organizational culture didn't seem to quite make it into the top tier of "R"s and that the proposed approach to improve the effectiveness of a pharmaceutical organization is entirely down to "technical" improvements and, effectively, more and better science. This may be considered an underestimation of the importance and impact of organizational culture, including a willingness to value "fail early," on productivity and organizational effectiveness. This important aspect will be discussed in more detail in some of the later chapters.

The aim of this chapter is not to extract or rewrite some of the excellent reviews published on the subject, but rather to analyze some of the specific case histories in order to learn from worked examples and to identify some common features that could lead to valuable conclusions that would provide a way forward.

4.6 CASE STUDIES FOR PHASE II/III ATTRITION

4.6.1 Torcetrapib

Torcetrapib was a cholesteryl ester transfer protein inhibitor (CETP inhibitor) intended to increase HDL and to lower LDL resulting in more favorable ratios. The drug was intended to be used in combination with a statin, particularly Lipitor, thereby maintaining or even increasing Pfizer's share of the cholesterol-lowering drug market. Development of the drug began in 1990 with clinical trials starting in 1999, but submission to the FDA wasn't until 2006. The ILLUMINATE study was intended to determine the clinical outcome of raising HDL by treatment with the CETP inhibitor torcetrapib in combination with atorvastatin [28]. It had enrolled around 15,000 patients at high coronary heart disease (CHD) risk who were randomized to treatment with torcetrapib plus atorvastatin versus atorvastatin alone. As reported in the *Wall Street Journal* (December 4, 2006), the study was terminated abruptly and unexpectedly after just over a year of treatment, because of an excess of deaths in the torcetrapib/atorvastatin versus atorvastatin groups (82 vs. 51, respectively). Increases in rates of heart failure, angina, and revascularization procedures were also observed [28]. The question raised by Tall et al. is whether this was the awakening from a dream of a highly effective way to raise HDL or whether it was simply an issue of unintended pharmacological effects of this particular CETP inhibitor. As there is no clear answer yet, it is of strong interest to follow the development of other CETP inhibitors and to reflect on the potential underlying reasons for this particular failure [28]. Torcetrapib was the first inhibitor of CETP activity to enter extensive evaluation in humans and was associated with increases in blood pressure in several clinical trials. These blood pressure increases appear to be a compound-specific, off-target effect and not related to CETP inhibition [29]. They relate to effects on the renin–angiotensin–aldosterone system [30]. The compound specificity is shown by comparative preclinical studies comparing torcetrapib with dalcetrapib, which showed not only elevated aldosterone but changes in genes related to the renin–angiotensin–aldosterone system [31]. The phase III ILLUMINATE was halted after only one year due to increased CVD events and all-cause mortality compared with atorvastatin, the balance of adverse effects clearly outweighing any potential benefit of torcetrapib treatment in patients [29]. Further insights into the feasibility of CETP inhibitors as therapeutic agents came from the clinical trials of dalcetrapib, our next case example.

The most important question from the program in terms of future research is why the compound entered phase III with identified cardiovascular risks and why this was disregarded as a significant risk through the discovery and development program. Identification of possible mechanisms seems to have been researched after phase III and also by competitor companies. Perhaps the commercial imperative to provide a lifeline for the Lipitor franchise was of such an influence that the decision to proceed was taken regardless. It is possible that identification of the blood pressure changes or their significance occurred relatively late in the program and any backups lacking this profile were very far behind torcetrapib, again raising the pressure to enter phase III with torcetrapib. A final scenario is that all backup compounds showed the cardiovascular side effects, meaning that research would have to reenter a discovery program putting Pfizer (the leader) years behind other companies. No simple answer emerges except the maxim "to test a medical hypothesis with a drug the researchers must use a highly selective compound at the correct dose."

4.6.2 Dalcetrapib

Roche's dalcetrapib, a CETP inhibitor structurally unrelated to torcetrapib, already had five phase II trials started by 2010. There was no evidence of the raised blood pressure seen with torcetrapib, and no clinically relevant differences versus placebo in adverse events, laboratory parameters including aldosterone, electrocardiograms, and other vital signs. Dalcetrapib had no measurable, clinically relevant effect on lymph node size [29]. Phase III trials commenced against this background of a seemingly clean safety profile (JTT-705/RO4607381) [32]. The dal-OUTCOMES phase III trial passed its first interim review in July 2011, but development of dalcetrapib was halted on May 7, 2012, due to a lack of efficacy [33] in terms of cardiovascular endpoints despite a 30% increase in HDL-C. Subsequent genomic analysis of patients showed that a single SNP in the ADCY9 gene (which codes for adenyl cyclase 9) was associated with cardiovascular events and dalcetrapib response. Patients with genotype AA showed a 39% reduction in the composite cardiovascular endpoint with dalcetrapib compared to placebo, while patients with genotype GG showed a 27% increase in events with dalcetrapib versus placebo [34].

These findings help explain lack of overall response and potentially signal the end of CETP inhibitors as sole or add-on therapy in the commercially attractive "one-size-fits-all approach" in the area of lipid-lowering drugs dominated by the statins but suggest that some patients could benefit from CETP inhibitors. Careful genomic profiling would allow patients who would benefit to be separated from patients who would have increased risk on the medicine. The commercial implications of this requirement in a market with many generic drugs of high efficacy would seem to be a major stumbling block. The CETP episode is typical of programs spanning the blockbuster era, target identification by genomics without a full understanding of all the ramifications, and the beginning of personalized medicine. In the light of the adverse outcome of ILLUMINATE and lack of positive results from OUTCOMES, other approaches to increasing HDL, macrophage cholesterol efflux, and reverse cholesterol transport will likely receive more attention, including approaches to increasing apoA-1 by infusion or synthesis, and induction of a variety of genes involved in macrophage cholesterol efflux, transport, and excretion via LXR activation [28].

4.6.3 Onartuzumab

Onartuzumab, a cancer drug candidate being developed by Genentech/Roche, was a targeted therapy attacking the protein MET, which can be found at high levels in patients with non-small-cell lung cancer (NSCLC) [35]. Increased hepatocyte growth factor/MET signaling is associated with poor prognosis and acquired resistance to epidermal growth factor receptor (EGFR)-targeted drugs in patients with NSCLC. Erlotinib, an EGFR inhibitor, is currently the standard treatment for patients whose NSCLC tumors have EGFR mutations. Although erlotinib improves survival in this group of patients, many of them eventually develop resistance to erlotinib and stop responding. MET, when expressed at high levels, is known to be involved in the development of resistance to erlotinib. To address the erlotinib resistance problem, onartuzumab was tested in patients with EGFR-mutant NSCLC. Results from a phase II clinical trial comparing treatment with a combination of onartuzumab and erlotinib versus erlotinib alone were published in 2013 [35]. The clinical study investigated whether dual inhibition of MET/EGFR results in clinical benefit in patients with NSCLC. Patients with recurrent NSCLC were randomly assigned at a ratio of one to one to receive onartuzumab plus erlotinib or placebo plus erlotinib; crossover was allowed at progression. Tumor tissue was acquired to assess MET status by immunohistochemistry (IHC), and coprimary endpoints were progression-free survival (PFS) in the intent-to-treat (ITT) and MET-positive (MET IHC diagnostic positive) populations; additional endpoints included overall survival (OS), objective response rate, and safety. There was no improvement in PFS or OS in the ITT population. MET-positive patients treated with erlotinib plus onartuzumab showed improvement in both PFS and OS. Conversely, clinical outcomes were worse in MET-negative patients treated with onartuzumab plus erlotinib. MET-positive control patient groups had worse outcomes versus MET-negative control patient groups. Incidence of peripheral edema was increased in onartuzumab-treated patients. The positive results for the combination treatment in MET-positive patients in contrast with the worse outcomes observed in MET-negative patients highlight the importance of diagnostic testing in drug development and selection of an appropriate patient population for treatment. The onartuzumab/erlotinib combo did not show an OS advantage for the patients across all groups; however, further analysis by the authors indicated that when MET levels were taken into account, the onartuzumab/erlotinib combo showed a clear advantage in patients with MET-positive tumors. The MET-positive patients who received both drugs had an OS time of 12.6 months compared to 3.8 months for MET-positive patients who received erlotinib only. The authors of the study concluded that "these results combined with the worse outcomes observed in MET-negative patients treated with Onartuzumab highlight the importance of diagnostic testing in drug development" [35]. With this lesson learned, the phase III trial enrolled only patients with MET-positive tumors and, based on the results of the phase II trial, expectations were high. However, the phase III study was stopped since there was no evidence that the addition of onartuzumab to erlotinib has any positive effect [36]. One of the learnings from this drug is the importance of selecting the correct patient population or subpopulation, although, as in this case, this is not sufficient to guarantee a successful outcome.

4.6.4 Bapineuzumab

Bapineuzumab, a monoclonal antibody to target Alzheimer's disease, had been developed jointly by Pfizer with Johnson & Johnson and the Irish pharmaceutical company Elan. The phase III trials read out in 2012 and the program was officially stopped shortly afterward. The drug addressed the amyloid-beta hypothesis, as the accumulation of a naturally occurring protein, beta amyloid, is believed to lead to nerve death and cognitive decline. A recent study found that a rare gene mutation that reduces beta amyloid production appears to protect people from developing the disease [37]. However, in recent years, a number of drug candidates including bapineuzumab, which target beta amyloid, have failed in trials to improve memory or cognitive functioning in Alzheimer's patients while effectively reducing accumulation of beta amyloid. In the first of four phase III trials, bapineuzumab failed to outperform a placebo in moderating symptoms of mild-to-moderate Alzheimer's disease [38]. Although biomarker results in the carrier study suggest that bapineuzumab may modify $A\beta$ accumulation and a downstream biomarker (phospho-tau), neither trial showed a benefit of bapineuzumab with respect to clinical outcomes. Two recent phase III clinical trials with solanezumab, another anti-$A\beta$ monoclonal antibody, also did not show benefit with respect to the primary clinical outcomes in patients with mild-to-moderate Alzheimer's disease [39]. It is believed that amyloid accumulation probably starts many years before the onset of the first symptoms of the disease, and starting antiamyloid treatment only after dementia develops may be too late to affect the clinical course of the disease [39].

4.6.5 Gantenerumab

In December 2014, Swiss pharmaceutical company Roche announced the closure of a phase III clinical trial (SCarletRoAD) for the experimental drug gantenerumab [40] (codeveloped with MorphoSys), a monoclonal antibody therapy. It appears that the drug has failed to help prevent Alzheimer's disease [41] after initially promising results [42]. Gantenerumab was designed to be administered early in the disease's progression before severe symptoms occur (prodromal Alzheimer's disease patients). The SCarletRoAD study had enrolled 799 presymptomatic patients with so-called prodromal Alzheimer's disease. Two additional clinical studies are ongoing [43], the second Roche-sponsored phase II trial of gantenerumab in mild Alzheimer's disease patients (Marguerite RoAD) [44] in approximately 1,000 patients with mild Alzheimer's disease and a third trial, run by the Dominantly Inherited Alzheimer Network (DIAN), to assess the safety, tolerability, and biomarker efficacy of gantenerumab in individuals who have a genetic predisposition for Alzheimer's disease [37]. The outcome of the Marguerite RoAD study may provide insights into the most appropriate patient population to benefit from gantenerumab, but the ScarletRoAD failure also highlights the importance of the development and use of relevant biomarkers to back up any proof-of-concept study.

4.6.6 Solanezumab

Solanezumab is a further example of an Alzheimer's drug not achieving the desired clinical outcome. In 2012, Eli Lilly reported that solanezumab had failed both primary endpoints in a large phase III study for Alzheimer's disease and the findings of the study were published in 2014 [31]. Solanezumab, a humanized monoclonal antibody that binds amyloid, failed to improve cognition or functional ability. The trials showed no significant improvement in the primary outcomes.

4.6.7 Pomaglumetad Methionil (LY-2140023)

Just days after the solanezumab clinical trial failure in 2012, Lilly stopped its late-stage schizophrenia program for pomaglumetad methionil, also known as mGlu2/3 or LY-2140023. LY-2140023 was a methionine amide prodrug of the selective metabotropic glutamate receptor (mGluR)2/3 agonist LY-404039 [45]. Justification for mGluR2 as a target for schizophrenia includes activity in mGluR2−/− and mGluR3−/− mice and mGluR2 effects in the prefrontal cortex, striatum, hippocampal formation, and the thalamus. The antipsychotic activity of LY-2140023 was supported by preclinical behavioral screens. Initial clinical trials were successful with a randomized, placebo-controlled, phase II clinical trial in patients with schizophrenia. However, in a second four-week trial, treatment with neither the mGluR2/3 agonist nor olanzapine was more effective than placebo. A greater-than-average placebo response was blamed for the lack of effect of both drugs [46]. A third study in which the drug was administered for seven weeks to patients with an acute exacerbation of schizophrenia showed no effect on the primary endpoint (PANSS score) with LY-2140023 in contrast to risperidone.

The phase III study [47] tested whether combination treatment with pomaglumetad methionil (LY-2140023 monohydrate), a metabotropic glutamate receptor 2/3 agonist with a second-generation antipsychotic, demonstrated significantly greater reduction of negative symptoms in patients with schizophrenia compared with treatment with the antipsychotic agent alone. Patients treated with LY-2140023 and standard of care (SOC) treatment failed to demonstrate a statistically significant improvement over patients treated with placebo and SOC during the study. Changes in secondary efficacy measures were not significantly different between groups at the endpoint. With the exception of vomiting that was greater in the LY-2140023 group, there were no statistically significant differences in safety and tolerability measures. This study found no benefit of adjunctive LY-2140023 versus placebo for negative symptoms in patients with schizophrenia receiving treatment with SOC. LY-2140023 was generally well tolerated in these patients [47].

4.6.8 Dimebon (Latrepirdine)

Latrepirdine (also known as dimebolin and sold as Dimebon) is an antihistamine drug that has been used clinically in Russia since 1983 [48]. Research was conducted in both Russia and western nations into potential applications as a neuroprotective agent to combat Alzheimer's disease and, possibly, also as a nootropic [49]. In 2008, positive data from a Russian study and 10 years of sales experience to underscore its safety encouraged the further investigation of Dimebon jointly by Pfizer and Medivation. In March 2010, the results of a clinical trial phase III were released, indicating that Dimebon failed in the pivotal CONNECTION trial of patients with mild-to-moderate disease. The failure of the remaining Pfizer and Medivation phase III trial for latrepirdine in Alzheimer's disease in 2012 effectively ended development for this indication [50].

The very promising Russian data is in marked contrast to the results of phase II, and this has led to several areas of comment and criticism. Most comments question the validity of the Russian data, which were unprecedented in terms of outcome especially the duration of effect. Also, the Dimebon formulation used in the Russian study appears different from that of the multinational phase III study, the former having a bitter taste and a numbing effect on the tongue, which is highly significant if it led to unblinding [51]. This unblinding would not occur in the phase III study where the tablet was film coated. Unexplained differences such as the high incidence of dry mouth in the phase II study that was not observed in the phase III study and other side effect differences have further raised the question of validity of the phase II study. Scathing criticism has also been made of the lack of rationale for the development of the drug in this indication: Dimebon being described as a "drug with no plausible mechanism that emerged from an incomprehensible series of screens." Other observers have cited the lack of detailed PK [51]. What little preclinical pharmacology data appears to have emerged following phase II studies and gives little indication of a mechanism. In addition to its antihistamine activity, Dimebon in $vitro$ was shown to act as an inhibitor of NMDA receptors (IC50 = 10 μM) and voltage-gated calcium channels (IC50 = 50 μM), which could provide protective effects in Alzheimer's. However, these high concentrations suggest that such effects are extremely unlikely to be clinically relevant. Any effects could be only reliably attributed to potent inhibition of H1 receptors, which is not recognized as a likely mechanism in Alzheimer's disease [52].

It is difficult with hindsight not to question the decision to go to phase III with a drug of unknown mechanism in a disease such as Alzheimer's. The phase II results could have been questioned in the absence of a preclinical rationale.

4.6.9 BMS-986094

Bristol-Myers Squibb (BMS) acquired BMS-986094, a nucleotide polymerase inhibitor drug candidate for the treatment of hepatitis C, through its $2.5 billion purchase of Inhibitex Inc. in January 2012. Just 7 months after the purchase, the development was discontinued after one patient died and others were hospitalized while taking the drug in a clinical study. The estimated loss to BMS is around $1.8 billion [53]. The mechanism of this toxicity is not yet known. Patients showed lowered left ejection fractions and some required hospitalization for suspected cardiotoxicity. Pathological analysis of cardiac tissue revealed severe myocyte damage with elongated myofibrils without gross necrosis

[54]. There is no evidence that this toxicity is due to the pharmacological class. It would appear that the development of this drug may have been an extreme case of unexpected findings [54].

4.6.10   TC-5214 (S-Mecamylamine)

Mecamylamine is a nonselective and noncompetitive antagonist of nicotinic acetylcholine (NAC) receptors that was introduced in the 1950s as an antihypertensive drug. The S-enantiomer had been developed by Targacept and AstraZeneca under the name of TC-5214, and its potential as a treatment of depression had been highlighted in 2009 [55]. The rationale for this mode of action is that depressive states involve hypercholinergic tone. S-Mecamylamine demonstrated positive effects in a number of animal models of depression and anxiety (forced swim test in rats, behavioral despair test in mice, social interaction in rats, light/dark chamber in rats). Although TC-5214 shows modest selectivity among NNR subtypes, the antidepressant and anxiolytic effects seen in these studies are believed to be due to antagonism of the alpha4beta2 neural NACs. In 2009, Targacept conducted a double-blind, placebo-controlled phase IIb trial in India, for the indication of major depression, with TC-5214, as an adjunct treatment with citalopram. The results were very positive and demonstrated superiority to placebo in both primary and secondary outcomes. Development of the drug continued with patients recruited from western centers including the United States. Four late-stage studies of TC-5214, also known as Inversine, by AstraZeneca partnered with Targacept, were all discontinued. According to the AstraZeneca press release from December 2009 [56], the second of four phase III efficacy and tolerability studies did not meet its primary endpoint. AstraZeneca continued with the development of the two remaining fixed-dose phase III RENAISSANCE efficacy and tolerability studies and one long-term safety study. A potential new drug application filing in the United States was at the time planned for the second half of 2012, with an EU marketing authorization application targeted for 2015 [56]. However, development was stopped by AstraZeneca in 2012 [57]. Some comparison can be made with Dimebon, above, concerning the scrutiny and ready acceptance of the results from the phase IIb study.

4.6.11   Olaparib

The story of olaparib (AZD-2281), developed by AstraZeneca, runs contrary to a majority of our case studies due to the fact that after initial failures in clinical trials and halting of development, further trials were conducted and the drug eventually made it to market. Initially a phase III failure, olaparib was finally approved by the FDA in December 2014 under the trade name Lynparza. Olaparib is a chemotherapeutic agent, developed by KuDOS Pharmaceuticals and later by AstraZeneca, which is an inhibitor of poly ADP ribose polymerase (PARP), an enzyme involved in DNA repair. It acts against cancers in people with hereditary BRCA1 or BRCA2 mutations, which include many ovarian, breast, and prostate cancers. In December 2011, following interim analysis of a phase II study, AstraZeneca announced that its investigational compound olaparib would not progress into phase III development for the maintenance treatment of serous ovarian cancer [43]. However, data from a phase II study, comparing the efficacy and tolerability of olaparib and cediranib in combination to olaparib alone, for the treatment of women with recurrent ovarian cancer, showed that the combination of olaparib and cediranib nearly doubled the PFS benefit compared to olaparib alone [58]. The decision to discontinue development of the drug was then reversed in 2013 [59], with AstraZeneca posting a new phase III trial of olaparib for patients with BRCA-mutated ovarian cancer in April 2013 [60]. On December 19, 2014, the FDA approved olaparib as monotherapy for patients with germline BRCA-mutated (gBRCAm) advanced ovarian cancer who have been treated with three or more prior lines of chemotherapy [61]. What initially looked like a failed drug candidate eventually made it to market, perhaps persistence and the philosophy of trying again even after failure sometimes leads to success.

4.6.12   Tenidap

Tenidap is an anti-inflammatory agent, which Pfizer hoped would be a successor to the non-selective COX inhibitor Feldene (a nonsteroidal anti-inflammatory agent: NSAID). The drug had a number of actions including cyclooxygenase inhibition with 5-lipoxygenase inhibition and modulation of cytokines such as interleukins (1 and 6) and tumor necrosis factor [62], inhibition of xanthine oxidase [63], inhibition of Ca^{2+} influx through voltage-sensitive calcium channels [64], and action as an opener of the inwardly rectifying K^+ channel hKir2.3 [65]. Pfizer believed the compound had disease-modifying properties in rheumatoid and osteoarthritis although the pharmacology was difficult to fully reproduce in external studies [66]. Early clinical trials were extremely positive [5] in comparison to the established drug diclofenac. Tenidap, in contrast to diclofenac, was associated with significant, rapid, and sustained reductions in C-reactive protein and serum amyloid A levels and with a significant reduction in plasma interleukin-6. Tenidap was associated with an equal incidence of elevated transaminases but a higher incidence of mild (≥500 mg/24 h < 1500 mg/24 h) nonprogressive proteinuria of proximal tubular origin compared with diclofenac [67]. The proteinuria, which was reversible, was a common clinical finding and was probably related to inhibition of albumin reabsorption. The drug was approved in a number of countries including the Netherlands and Spain but not launched. Longer-term studies looked initially very encouraging and reviews were favorable to the drug, supporting its disease-modifying properties [68]. The drug was submitted but was rejected at the proposed dose of 120 mg by the FDA in May 1996 [69]. Of principal concern was an apparent loss of bone mineral density over long-term usage, which may relate to its role in osteoclast stimulation and its inhibition of various cytokines. The safety profiles, including the renal effects, which had a high incidence (13%) (proteinuria), and hepatotoxicity, which now had some examples of severe abnormalities, were also regarded unfavorably. Pfizer issued a press release shortly afterward saying that despite tenidap's unique characteristics, the company had decided not to pursue commercialization of tenidap 120 mg for RA. An FDA Advisory Committee further recommended that the agency not approve the drug for either RA or osteoarthritis (OA) based on a risk/benefit assessment [51].

Several points are worth stating, given the various decisions that were taken on tenidap. The COX-2 enzyme had been discovered in 1988 with the promise of a new generation of safe anti-inflammatory agents and many pharma companies had active discovery and development programs in place. Celebrex (celecoxib) the first of these agents was approved in 1999, indicating how the timescales of COX-2 and tenidap overlapped. In addition to COX-2 inhibition, tenidap was a potent inhibitor of COX-1 and like other nonselective COX inhibitors demonstrated a high level of gastrointestinal side effects that were expected to be significantly attenuated by the newer generation of COX-2 selective agents.

For a drug with multiple effects like tenidap, selection of an optimum dose is problematic. Was 120 mg the optimum dose or are more dose selection studies required to find doses which have better risk/benefit? Clearly, it would be very difficult to design and discover such an agent with such a combination of properties at the right potency for each to ensure the optimum effects. Having discovered such an agent dose selection studies would be complex and require multiple biomarkers.

Finally, long-term outcomes are always uncertain, as with the problem of bone density changes. Would other dose levels have shown similar effect? Based on the available data, it appears that tenidap was a drug development program that was simply at the wrong time. New developments were around the corner (which may have been highly significant for internal decision making), and the FDA was increasingly vigilant on drug safety in defining risk benefit.

4.6.13   NNC0109-0012 (RA)

The story of NNC0109-0012 highlights the impact of clinical attrition on pharmaceutical companies and their employees. In September 2014, Novo Nordisk announced a decision to discontinue all its research and development (R&D) activities within inflammatory disorders [70]. This decision followed a review of Novo Nordisk's strategic position in the therapeutic area after the recently announced discontinuation of the company's most advanced compound (NNC0109-0012), a biologic that has been in clinical trials in phase II as an anti-IL-20 candidate for the treatment of RA [71]. At the time of writing, no clinical outcomes were published, so it is not clear why the study was terminated. The decision to exit inflammatory R&D directly impacted approximately 400 employees [52].

4.6.14   Omapatrilat

The kallikrein–kinin pathway and the natriuretic peptides (atrial natriuretic peptide (ANP)) have important effects on the cardiovascular system, and the concept of dual inhibitors that simultaneously block angiotensin II production and increase ANP became a focus of many companies as the next stage of cardiovascular therapy following the introduction of ACE inhibitors. The attraction of this dual approach was that both targets were zinc-dependent enzymes (ACE and NEP) with significant structural homology. Omapatrilat (discovered by BMS) was a thiol-containing ACE/NEP inhibitor that achieved the target of balanced dual inhibition (ACE IC_{50}=5 nmol, NEP IC_{50}=8 nmol) and exerted prolonged antihypertensive effects in addition to cardiorenal protection. The results were superior to ACE inhibition alone [72, 73], and early development showed that these preclinical results extended to the clinic. The FDA reviewed the original application for omapatrilat in 2000. The most serious adverse event, angioedema, occurred in 0.5% of patients and was life threatening in four (requiring emergency procedures). Since this incidence was deemed higher than that for ACE inhibitors, BMS was asked to undertake further studies. BMS decided to use 10 mg rather than 20 mg as a starting dose [65, 66].

The phase III OCTAVE study investigated the effect of omapatrilat (10–80 mg) versus enalapril (5–40 mg) on cardiovascular outcomes in 5770 patients with congestive heart failure. Omapatrilat was superior to enalapril in antihypertensive effects but demonstrated no reduction in cardiovascular deaths (although later analysis suggested a small reduction). Unfortunately, the benefits were outweighed by a significant increase in angioedema

(2.17% vs. 0.68%), a side effect that was worse in Afro-American patients (5.54% vs. 1.62%). Increasing the concern was the finding that the cases of angioedema seen with omapatrilat were more likely to be severe, with the risk for angioedema requiring hospitalization 9.5 times higher than enalapril [65, 66]. This serious side effect has been attributed largely to the rapid increase in bradykinin and its metabolite BK-8 by simultaneously blocking both ACE and NEP, the major degrading pathways. The results of OCTAVE led to a nonapproval vote by the FDA committee in 2002, and a second phase III study OPERA was dropped when BMS's strategy of labeling the drug only for severe patients and ensuring detailed product labeling and patient counseling by pharmacists, together with postmarketing surveillance, did not influence the committee [65, 66]. BMS withdrew omapatrilat shortly afterward.

In the course of its development, omapatrilat moved from "super ACE inhibitor" to a failure on recognizing risk/benefit early. Since the major side effect also occurred with ACE inhibitors, were BMS sufficiently concentrating on this increased risk of raised bradykinin concentrations rather than just the benefit? Perhaps most significantly, the initial analysis of OCTAVE did not give the beneficial outcomes on cardiovascular risk expected. Quite simply overall benefit was no better than for ACE inhibition, but risk was substantially increased. It is probable that only phase III studies would show this and this is the cost of drug development. The results also highlight the increased risk in developing drugs in an area already reasonably well served by a variety of agents from different pharmacological classes even if there is a huge commercial potential.

4.6.15 Ximelagatran

Ximelagatran was an oral thrombin inhibitor, which had the potential to provide a replacement for warfarin, developed by AstraZeneca. It was an unusual dual prodrug that provided low bioavailability of the nonabsorbable melagatran [74] and was approved in a number of European and South American countries. Clinical studies showed that ximelagatran was noninferior to warfarin in stroke prevention, and as acute and extended therapy of deep vein thrombosis (DVT), and superior to warfarin for the prevention of venous thromboembolism. Hepatic abnormalities led to the FDA rejecting an initial application for approval in 2004. The implications of the effects on the liver became progressively more serious, and further development was discontinued in 2006.

Preclinical safety testing had revealed no indication of hepatotoxic potential [75]. The initial phase III clinical trials, looking at periods of treatment of 12 days or less, showed no hepatotoxicity. Raised liver function tests (LFTs) were first noticed in long-term (>35 days) use of ximelagatran (8% of patients with >3x upper limit of normal (ULN) plasma ALT). Elevated bilirubin levels were similar to comparator groups, but individuals with ALT and bilirubin increases were five times more frequent in the treated group [68]. These changes should be judged against the current "red flag," adopted as warning of further more severe hepatotoxicity, "Hy's law," which has three components:

1. The drug causes frequent threefold or greater elevations above the ULN of ALT or AST than the control agent.
2. Among subjects showing such elevations, some cases show elevation of serum total bilirubin to ≥2xULN.
3. No other reason can be found to explain the changes such as viral hepatitis, con meds, or preexisting medical conditions.

The withdrawal of ximelagatran and termination of the ximelagatran development program was caused by further data from a 35-day study, which showed severe hepatic injury could develop after cessation of dosing (rendering liver function monitoring highly problematic as a risk strategy). In further laboratory studies with patient samples, ALT increases were associated with major histocompatibility complex (MHC) alleles DRB1*07 and DQA1*02 suggesting a possible immunogenic pathogenesis, which helps explain the absence of findings in preclinical and possibly short-term clinical studies. Ximelagatran illustrates well the risky business that is drug discovery and development. The drug had no particular structural alerts, although the prodrug is highly unusual. Moreover, the drug was used at comparatively low doses (36 mg b.d.), and toxicity was only observed in late-stage trials.

4.7   SUMMARY AND CONCLUSIONS

From the case histories reviewed, it appears that phase II/III attrition is quite evenly spread right across small molecules, antibodies, and biologics but rather more prevalent in some disease areas than others, for example, Alzheimer's disease. This is not surprising since compound-specific issues relating to PK or toleration will most likely appear in the earlier stages of development, that is, phase I, while attrition due to lack of efficacy or toxicity and adverse effect findings will often become apparent only in larger patient populations. Where issues relate to a particular mechanism of action (mechanism associated toxicity, or lack of efficacy of mechanism), then failure of the entire cohorts of compounds from different companies are also not unexpected. Without clear evidence of mechanism associated toxicity, companies are unlikely to drop a compound in development based on the failure of a competitor's candidate, particularly if that compound is considered to be ahead in the race to market. In fact, failures of competitor compounds will most likely be seen as an advantage to the business by reducing competition in the market and launching a first-in-class drug. Continued absence of clinically relevant biomarkers in many trials also results in failure of efficacy for one agent to provide no insight into the utility or not of the mechanism/target since there is no data to prove that the relevant target was adequately addressed.

Clearly, phase II and phase III failures are a major problem for the pharmaceutical industry, and while some publications put the blame for many of the failures firmly at the door of the pharmaceutical companies, it should be acknowledged also that drug discovery is inherently a high-risk business [5]. It can be proposed that some compounds have been advanced to phase II/III when they should have suffered earlier rejection, but as the example of olaparib shows, success can come after a line of failures. If a company does not advance promising compounds into phase II and phase III, there will be no chance to succeed and get drugs onto the market. It should also be remembered that safety is not an absolute measure and that special consideration ought to be given where drugs are aimed at ailments with unmet medical need such as multiple sclerosis and certain cancers, for example, vandetanib, fingolimod, and dabigatran, were all approved despite concerns regarding serious toxicity and difficulty in achieving safe dosage [76]. While the decision making at stage gates may be questioned, it should be acknowledged that a certain element of luck is involved in developing a drug and getting it to market. Many people, particularly those not involved in the industry, may think that clinical outcomes are somehow more predictable than they actually are. So, perhaps the strategy to try and try again is a valid option, if patient safety is not compromised as a consequence. Failing due to lack of efficacy, although painful for all involved and particularly for patients waiting for new medicines, is, and will probably remain, an all too common feature in drug development, and given our growing but still somewhat limited understanding of complex biological systems, this should not come as a huge surprise. Clearly, attention is needed in terms of the right endpoints for clinical trials, the right patient groups, and the relevant biomarkers and advances in these areas are happening via development and use of biomarkers in early discovery selection that translate well for clinical use. Those advances alone will not eliminate clinical trial failures but rather will increase the success rate, and even a small decrease of attrition in development (from upper to lower 80's percent) will result in several more drugs being approved and marketed each year, which would be extremely beneficial for the industry but more importantly for patients.

REFERENCES

1 Kola, I., Landis, J. (2004). Can the pharmaceutical industry reduce attrition rates? *Nat. Rev. Drug Disc., 3*, 711–715.

2 Schmid, E.F., Smith, D.A. (2005). Is declining innovation in the pharmaceutical industry a myth? *Drug Discov. Today, 15*, 1031–1039.

3 Malik, N.N. (2009). Key issues in the pharmaceutical industry: consequences on R&D. *Expert Opin. Drug Disc., 4*, 15–19.

4 Paul, S. M., Mytelka, D. S., Dunwiddie, C. T., Persinger, C. C., Munos, B. H., Lindbord, S. R., Schacht, A. L. (2010). How to improve R&D productivity: the pharmaceutical industry's grand challenge. *Nat. Rev. Drug Disc., 9*, 203–214.

5 Kennedy, T. (1997). Managing the drug discovery/development interface. *Drug Discov. Today, 2*, 436–444.

6 Kubinyi, H. (2003). Drug research: myths, hype and reality. *Nat. Rev. Drug Discov., 2*, 665–668.

7 Arrowsmith, J. (2011). Trial watch: phase II failures: 2008-2010. *Nat. Rev. Drug Discov., 10*, 328–329.

8 Walker, I., Newell, H. (2009). Do molecularly targeted agents in oncology have reduced attrition rates? *Nat. Rev. Drug Discov., 8*, 15–16.

9 http://adrclinresearch.com/Issues_in_Clinical_Research_links/Why%20Pivotal%20Clinical%20Trials%20Fail%20-%20Part%201_v12L_a.pdf (accessed July 15, 2015).

10 Retziaos, A. D. (2009). Why Phase 3 Clinical Trials Fail. Report by Bay Clinical R&D Services.

11 Gan, H. K., You, B., Pond, G. R., Chen, E. X. (2012). Assumptions of expected benefits in randomized phase III trials evaluating systemic treatments for cancer. *J. Natl. Cancer Inst., 104*, 590–598.

12 Amiri-Kordestani, L., Fojo, T. (2012). Why do phase III clinical trials in oncology fail so often? *J. Natl. Cancer Inst., 104*, 1–2.

13 Jayasundara, K. S., Keystone, E. C., Parker, J. L. (2012). Risk of failure of a clinical drug trial in patients with moderate to severe rheumatoid arthritis. *J. Rheumatol. 39*, 1–5.

14 Morgan, D. (2003). Antibody therapy for Alzheimer's disease. *Expert Rev. Vac., 2*, 89–95.

15 Bohrmann, B., Baumann, K., Benz, J., Gerber, F., Huber, W., Knoflach, F., Messer. J., Oroszlan, K., Rauchenberge,r R., Richter, W.F., Rothe, C., Urban, M., Bardrof,f M., Winter, M., Nordstedt, C., Loetscher, H. (2012). Gantenerumab: a novel human anti-Aβ antibody demonstrates sustained cerebral amyloid-β binding and elicits cell-mediated removal of human amyloid-β. *J. Alzheimer's Dis.*, *28*, 49–69.

16 Watt, A.D., Crespi, G.A.N., Down, R.A., Acher, D.B. Gunn, A., Perez, K.A., Mc Lean, C.A., Villemagne, V.L., Parker, M.W., Barnham, K.J. (2014). Do current therapeutic anti-Aβ antibodies for Alzheimer's disease engage the target? *Acta Neuropath.*, *127*, 803–810.

17 Spencer, B., Masliah, E. (2014). Immunotherapy for Alzheimer's disease: past, present and future. *Frontiers Aging Neurosci.*, *6.* 114. http://www.ncbi.nlm.nih.gov/pmc/articles/PMC4051211/ (accessed July 15, 2015).

18 Bakker, A., Caricasole, A., Gaviraghi, G., Pollio, G., Robertson, G., Terstappen, G. C., Salerno, M., Tunici, P. (2009). How to achieve confidence in drug discovery and development: managing risk (from a reductionist to a holistic approach). *ChemMedChem*, *4*, 923–933.

19 Smith, D. A. (2011). Discovery and ADMET: where are we now. *Curr. Top. Med. Chem.*, *11*, 467–481.

20 Krejsa, C. M., Horvath, D., Rogalski, S. L., Penzotti, J. E., Mao, B., Barbosa, F., Migeon, J. C. (2003). Predicting ADME properties and side effects: the BioPrint approach. *Curr. Opin. Drug Disc. Develop.*, *6*, 470–480.

21 Fliri, A. F., Loging, W. T., Thadeio, P. F., Volkman, R. A. (2005). Biological spectra analysis: linking biological activity profiles to molecular structure. *Proc. Natl. Acad. Sci.*, *102*, 261–266.

22 Fliri, A. F., Loging, W. T., Thadeio, P. F., Volkman, R. A. (2005). Biospectra analysis: model proteosome characterizations for linking molecular structure and biological response. *J. Med. Chem.*, *48*, 6918–6925.

23 Scheiber, J., Jenkins, J. L., Sukuru, S. C. K., Bender, A., Mikhailov, D., Milik, M., Azzaoui, K., Whitebread, S., Hamon, J., Urban, L., Glick M., Davies, J. W. (2009). Mapping adverse drug reactions in chemical space. *J. Med. Chem.*, *52*, 3103–3107.

24 Hughes, J. D., Blagg, J., Price, D. A., Bailey, S., DeCrescenzo, G. A., Devraj, R. V., Ellsworth, E., Fobian, Y. M., Gibbs, M. E., Gilles, R. W., Greene, N., Huang, E., Krieger-Burke, T., Loesel, J., Wager, T., Whiteley, L., Zhang, Y. (2008). Physicochemical drug properties associated with in vivo toxicological outcomes. *Bioorg. Med. Chem. Lett.*, *18*, 4872–4875.

25 Muthas, D. Boyer, S., Hasselgren, C. (2013). A critical assessment of modelling safety-related drug attrition. *Med. Chem. Commun.*, *4*, 1058–1065.

26 Brambilla, G., Martelli, A. (2009). Update on genotoxicity and carcinogenicity testing of 472 marketed drugs. *Mutation Res.*, *681*, 209–229.

27 Cook, D., Brown, D., Alexander, R., March, R., Morgan, P., Satterthwaite, G., Pangalos, M. N. (2014). Lessons learned from the fate of AstraZeneca's drug pipeline: a five-dimensional framework. *Nat. Rev. Drug Disc.*, *13*, 419–431.

28 Forrest, M.J., Bloomfield, D., Briscoe, R. J., Brown, P.N., Cumiskey, A.M., Ehrhart, J., Hershey, J.C., Keller, W.J., Ma, X., McPherson, H.E., Messina, E., Peterson, L.B., Sharif-Rodriguez, W., Siegle, P.K.S., Sinclair, P.J., Sparrow, C.P., Stevenson, A.S., Sun, S.-Y., Tsai, C., Vargas, H., Walker, M. III, West, S.H., White, V., Woltmann, R.F. (2008). Torcetrapib-induced blood pressure elevation is independent of CETP inhibition and is accompanied by increased circulating levels of aldosterone. *Brit. J. Pharmacol.*, *154*, 1465–1473.

29 Stein, E. A., Roth, E. M., Rhyne, J. M., Burgess, T., Kallend, D., Robinson, J. G. (2010). Safety and tolerability of dalcetrapib (RO4607381/JTT-705): results from a 48-week trial. *Europ. Heart J.*, *31*, 480–488.

30 Tall, A. R., Yvan-Charvet, L., Wang, N. (2007). The failure of torcetrapib. Was it the molecule or the mechanism? *Arteriosclerosis Thrombosis Vasc. Biol., 27*, 257–260.

31 Stroes, E.S.G., Kastelein, J.J.P., Benardeau, A., Kuhlmann, O., Blum, D., Campos, L.A., Clerc, R.G., Niesor, E.J. (2009). Dalcetrapib: no off-target toxicity on blood pressure or on genes related to the renin-angiotensin-aldosterone system in rats. *Brit. J. Pharmacol., 158*, 1763–1770.

32 de Grooth, G.J., Kuivenhoven, J.A., Stalenhoef, A.F., de Graaf, J., Zwinderman, A.H., Posma, J.L., van Tol, A., Kastelein, J.J. (2002). Efficacy and safety of a novel cholesteryl ester transfer protein inhibitor, JTT-705, in humans: a randomized phase II dose-response study. *Circulation, 105*, 2159–2165.

33 Bennett, S., Kresge, N., Roche drops after halting cholesterol drug development. *Bloomberg*, May 7, 2012.

34 Tardif, J.C., Rhéaume, É., Perreault, L.P.L., Grégoire, J.C., Zada, Y.F., Asselin, G., Provost, S., Barhdadi, A., Rhainds, D., L'Allier, P.L., Ibrahim, R., Upmanyu, R., Niesor, E.J., Benghozi, R., Suchankova, G., Lahrissi-Thode, F., Guertin, M.-C., Olsson, A.G., Mongrain, I., Schwartz, G.G., Dubé, M.-P. (2015). Pharmacogenomic determinants of the cardiovascular effects of dalcetrapib. *Circ. Cardiovasc. Genet., 8*(2), 372–382; CIRCGENETICS-114.

35 Spigel, D.R., Ervin, T.J., Ramlau, R.A., Daniel, D.B., Goldschmidt, J.H. Jr., Blumenschein, G.R. Jr., Krzakowski. M.J., Robinet, G., Godbert, B., Barlesi, F., Govindan, R., Patel, T., Orlov, S.V., Wertheim, M.S., Yu, W., Zha, J., Yauch, R.L., Patel, P.H., Phan, S.C., Peterson, A.C. (2013). Randomized phase II trial of Onartuzumab in combination with erlotinib in patients with advanced non-small-cell lung cancer. *J. Clin. Oncol., 31*, 4105–4114.

36 http://www.roche.com/media/media_releases/med-cor-2014-03-03.htm (accessed July 15, 2015).

37 Jonsson, T., Atwas, J.K., Steinberg, S., Snaedal, J., Jonsson, P.V., Bjomsson, S., Stefansson, H., Sulem, P., Gudbjartsson, D., Maloney, J., Hoyte, K., Gustafson, A., Liu, Y., Lu, Y., Bhangale, T., Graham, R.R., Huttenlocher, J., Bjomsdottir, G., Andreassen, O.A., Jönsson, E.G., Palotie, A., Behrens, T.W., Magnusson, O.T., Kong, A., Thorsteinsdottir, U., Watts, R.J., Stefansson, K. (2012). A mutation in *APP* protects against Alzheimer's disease and age-related cognitive decline. *Nature, 488*, 96–99.

38 Salloway, S., Sperling, R., Fox, N.C., Blennow, K., Klunk, W., Raskind, M., Sabbagh, M., Honig, L.S., Porsteinsson, A.P., Ferris, S., Reichert, M., Ketter, N., Nejadnik, B., Guenzler, V., Miloslavsky, M., Wang, D., Lu, Y., Lull, J., Tudor, I.C., Liu, E., Grundman, M., Yuen, E., Black, R., Brashear, H.R. (2014). *N. Engl. J. Med., 370*, 322–333.

39 Doody, R.S., Thomas, R.G., Farlow, M., Iwatsubo, T., Vellas, B., Joffe, S., Kieburtz, K., Raman, R., Sun, X., Aisen, P. S., Siemers, E., Liu-Seifert, H., Mohs, R. (2014). Phase 3 trials of solanezumab for mild-to-moderate Alzheimer's disease. *N. Engl. J. Med., 370*, 311–321.

40 Delrieu, J., Ousset, P. J., Vellas, B. (2012). Gantenerumab for the treatment of Alzheimer's disease. *Expert Opin. Biol. Ther., 12*, 1077–1086.

41 http://www.reuters.com/article/2014/12/19/roche-alzheimers-idUSL6N0U30D920141219 (December 2014).

42 Panza, F., Solfrizzi, V., Imbimbo, B.P., Giannini, M., Santamato, A., Seripa, D., Logroscino, G. (2014). Efficacy and safety studies of gantenerumab in patients with Alzheimer's disease. *Expert Rev. Neurother. 14*, 973–986.

43 https://clinicaltrials.gov/ct2/show/NCT01224106 (accessed July 15, 2015).

44 http://www.margueriteroadstudy.com (accessed July 15, 2015).

45 Mezler, M., Geneste, H., Gault, L., Marek, G.J. (2010). LY-2140023, a prodrug of the group II metabotropic glutamate receptor agonist LY-404039 for the potential treatment of schizophrenia. *Curr. Opin. Investig. Drugs, 11*, 833–845.

46 Kinon, B.J., Zhang, L., Millen, B.A., Osuntokun, O.O., Williams, J.E., Kollack-Walker, S., Jackson, K., Kryzhanovskaya, L., Jarkova, N. (2011). A multicenter, inpatient, phase 2, double-blind, placebo-controlled dose-ranging study of LY2140023 monohydrate in patients with DSM-IV schizophrenia. *J. Clin. Psychopharmacol.*, *31*, 349–355.

47 Stauffer, V.L., Millen, B.A., Andersen, S., Kinon, B.J., Lagrandeur, L., Lindenmayer, J.P., Gomez, J.C. (2013). *Schizophr. Res.*, *150*, 434–441.

48 Matveeva I. A., (1983). Action of dimebon on histamine receptors. *Farmakologiia i Toksikologiia (in Russian)*, *46*, 27–29.

49 Shevtsova, E. F., Kireeva, E. G., Bachurin, S. O. (2005). Mitochondria as the target for neuro-protectors. *Vestnik Rossiiskoi Akademii Meditsinskikh Nauk (in Russian)*, *9*, 13–17.

50 Sweetlove, M. (2012) Phase III CONCERT trial of latrepirdine. Negative results. *Pharm. Med.*, *26*, 113–115.

51 Jones, R. W. (2010). Dimebon disappointment. *Alzheimer's Res. Ther.*, *2*, 25.

52 Wu, J., Li, Q., Bezprozvanny, I. (2008). Evaluation of Dimebon in cellular model of Huntington's disease. *Mol. Neurodegener.*, *3*, 15.

53 http://www.bloomberg.com/news/2012-08-23/bristol-myers-drops-hepatitis-c-drug-after-patient-death.html (August 24, 2012).

54 Ahmad, T., Yin, P., Saffitz, J., Pockros, P. J., Lalezari, J., Shiffman, M., Freilich, B., Zamparo, J., Brown, K., Dimitrova, D., Kumar, M., Manion, D., Heath-Chiozzi, M., Wolf, R., Hughes, E., Muir, A.J., Hernandez, A.F. (2014). Cardiac dysfunction associated with a nucleotide polymerase inhibitor for treatment of hepatitis C. *Hepatology*, http://onlinelibrary.wiley.com/doi/10.1002 (accessed July 15, 2015).

55 Bacher, I., Wu, B., Shytle, D. R., George, T. P. (2009). Mecamylamine—a nicotinic acetylcho-line receptor antagonist with potential for the treatment of neuropsychiatric disorders. *Expert Opin. Pharmacother.*, *10*, 2709–2721.

56 http://www.astrazeneca.com/Media/Press-releases/Article/20111220-az-updates-olaparib-TC5214-development (December 2009).

57 http://www.astrazeneca.com/Media/Press-releases/Article/20032012tc5214-failed-phase-iii-endpoint (March 20, 2012).

58 Liu, J., (2013). *50th Annual Meeting of the American Society for Clinical Oncology (ASCO), Abstract LBA5500*.

59 http://www.astrazeneca.com/Media/Press-releases/Article/astrazeneca-enrollment-patient-phaseIII-olaparib. (September 2013).

60 http://www.astrazeneca.com/Media/Press-releases/Article/20140531--astrazeneca-welcomes-positive-data-on-the-combination-olaparib-cediranib-ovarian-cancer. (May 2013).

61 http://www.fda.gov/Drugs/InformationOnDrugs/ApprovedDrugs/ucm427598.htm. (December 2014).

62 Bondeson, J., Sundler, R. (1996). Differential effects of tenidap on the zymosan-and lipopolysaccharide-induced expression of mRNA for proinflammatory cytokines in macro-phages. *Biochem. pharmacol.*, *52*, 35–42.

63 Chatham, W.W., Baggott, J.E., Loose, L.D., & Blackburn, W.D. (1995). Effects of tenidap on superoxide-generating enzymes: non-competitive inhibition of xanthine oxidase. *Biochem. Pharmacol.*, *50*, 811–814.

64 Cleveland, P.L., Millard, P.J., Showell, H.J., Fewtrell, C.M.S. (1993). Tenidap: a novel inhibitor of calcium influx in a mast cell line. *Cell Calcium*, *14*, 1–16.

65 Liu, Y., Dong Liu, Printzenhoff, D., Coghlan, M. J., Harris, R., Krafte, D. S. (2002). Tenidap, a novel anti-inflammatory agent, is an opener of the inwardly rectifying K^+ channel hKir2.3. *Eur. J. Pharmacol.*, *434*, 153–160.

66 Griswold, D.E., Hillegass, L.M., Breton, J.J., Esser, K.M., & Adams, J.L. (1992). Differentiation in vivo of classical non-steroidal antiinflammatory drugs from cytokine suppressive antiinflammatory drugs and other pharmacological classes using mouse tumour necrosis factor alpha production. *Drugs under Exp. Clin. Res., 19*, 243–248.

67 Wylie, G., Appelboom, T., Bolten, W., Breedveld, F.C., Feely, J., Leeming, M.R.G., Le Loet, X., Manthorpe, R., Marcolongo, R., Smolen, J. (1995). A comparative study of tenidap, a cytokine-modulating anti-rheumatic drug, and diclofenac in rheumatoid arthritis: a 24-week analysis of a 1-year clinical trial. *Rheumatology, 34*, 554–563.

68 Canvin, J.M.G., Madhok, R. (1996). Tenidap: not just another NSAID? *Annals Rheum. Res., 55*, 79–82.

69 http://www.fda.gov/ohrms/dockets/ac/96/minutes/3178m1.pdf (May 1996).

70 Novo Nordisk press release, 2 September 2014.

71 https://clinicaltrials.gov/ct2/show/results/NCT01636817 (accessed July 15, 2015).

72 Jandeleit-Dahm K.A.M. (2006). Dual ACE/NEP inhibitors—more than playing the ACE card. *J. Hum. Hypertension, 20*, 478–481.

73 The rise and fall of omapatrilat. http://www.medscape.com/viewarticle/443224 (accessed July 15, 2015).

74 Ho, S.-J., Brighton, T. A. (2006). Ximelagatran: direct thrombin inhibitor. *Vasc. Health Risk Manag., 2*, 49–58.

75 Keisu, M., Andersson, T.B. (2010). Drug-induced liver injury in humans: the case of ximelagatran. *Handb. Exp. Pharmacol., 196*, 407–418.

76 Moore, T.J., Furberg, C.D. (2012). The safety risks of innovation. The FDA's expedited drug development pathway. *JAMA, 308*, 869–870.

5

POSTMARKETING ATTRITION

DENNIS A. SMITH

Department of Chemistry, University of Cape Town, Cape Town, South Africa; The Maltings, Walmer, Kent, UK

5.1 INTRODUCTION

Until recently, postmarketing attrition usually reflected a change in benefit–risk that only became apparent as more patients were exposed to the drug over longer periods of time. In some cases, these data were obtained as part of planned postapproval studies (phase IV) or just in the general prescribing of the drug. Cost of drugs and the availability of generic alternatives of many leading therapeutic introductions have also dramatically affected the success of some new drug introductions. Thus, although a new drug may have perceived and even demonstrated advantages over existing therapy, it may fail to be widely prescribed.

Benefit–risk with chronic therapy drugs is probably only fully understood after large numbers (100,000 plus) have taken the drug for a considerable time. This usage over time may reveal both benefits and side effects of the therapy unobserved or not present in studies with less patients over shorter periods. A confounding factor in this is that some of the side effects may be present to a comparatively high incidence in the untreated population. In considering small molecules, we will concentrate on the United States and comment on other countries and consider the period from 1980 to date.

The analysis looks at on-target and off-target pharmacology as the reasons for drug withdrawal and further subdivides off-target pharmacology into known and unknown receptors. In this classification, the drug refers to the drug and its metabolites and does not distinguish between stable and unstable (reactive metabolites). The major drug withdrawals from the US market are shown in Table 5.1.

This chapter examines drugs withdrawn from the US market since 1980. It classifies the reasons for withdrawal as on- and off-target pharmacology. Off-target pharmacology

Attrition in the Pharmaceutical Industry: Reasons, Implications, and Pathways Forward,
First Edition. Edited by Alexander Alex, C. John Harris and Dennis A. Smith.
© 2016 John Wiley & Sons, Inc. Published 2016 by John Wiley & Sons, Inc.

TABLE 5.1 Drug Withdrawals from 1980 to Date Grouped into Three Categories of Reason for Withdrawal

On-target pharmacology				Off-target pharmacology, known receptor				Off-target pharmacology, unknown receptor			
Generic name	Year of US approval	Year of US withdrawal	Daily dose mg	Generic name	Year of US approval	Year of US withdrawal	Daily dose mg	Generic name	Year of US approval	Year of US withdrawal	Daily dose mg
Alosetron*	2000	2000	1	Astemizole	1988	1999	10	Benoxaprofen	1982	1982	600
Cerivastatin	1998	2001	0.3	Cisapride	1993	2000	40	Bromfenac	1997	1998	100
Encainide	1985	1991	150	Dexfenfluramine	1994	1997	15	Nomifensine	1984	1086	125
Flosequinan	1993	1993	100	Fenfluramine	1973	1997	15	Pemoline	1975	2005	160
Rofecoxib	1999	2004	25	Grepafloxacin	1997	1999	400	Remoxipride	1990	1993	300
(Tegaserod)	2002	2007	20	Propoxyphene	1957	2010	600	Temafloxacin	1992	1992	600
				Mibefradil	1997	1998	100	Ticrynafen	1979	1982	400
				Rapacuronium	1999	2001	100	Tolcapone	1998	1998	300
				Sibutramine	1997	2010	20	Troglitazone	1997	2000	400
				Suprofen	1986	1987	800	Trovafloxacin	1998	1999	200
				Terfenadine	1985	1997	120	Zomepirac	1980	1983	400
				Thioridazine	1978	2005	600				

*Alosetron was returned to the market under restricted conditions in 2002. Tegaserod was withdrawn due to a challenged increased incidence of cardiovascular events, and it is placed in on-target pharmacology without supporting preclinical or clinical pharmacology for other mechanisms.

is further subcategorized into known receptors and unknown receptors. Off-target pharmacology drug withdrawals have triggered changes in the way new drugs are discovered and developed, with the most impactful cases being withdrawal due to a reversible interaction of a drug or its metabolite with a promiscuous ion channel such as the rapidly activating delayed rectifier potassium current channel (I_{Kr}) in cardiac tissues (hERG K$^+$, K$_v$11.1).

5.2 ON-TARGET PHARMACOLOGY-FLAWED MECHANISM

5.2.1 Alosetron

Alosetron (Fig. 5.1) is a highly selective 5-HT$_3$ antagonist introduced for the treatment of irritable bowel syndrome (IBS). The drug gave improvement in pain and discomfort in diarrhea-predominant IBS patients [1].

5-HT$_3$ antagonists suppress the reflex activation of colonic motor function in response to the eating of food. This reflex action is a normal postprandial function. However, in diarrhea-predominant IBS, it is exaggerated, and the effect of food is to induce a painful distension of the gut and accompanying diarrhea. Constipation was the major drug-related adverse event of the drug occurring in 28% of treated patients compared to a placebo rate of around 3%. In a small number, the severity of constipation led to ischemic colitis, and the drug was withdrawn for this reason. A direct link between the reduction in colonic motor tone, constipation, and ischemic colitis implicates the on-target pharmacology. The fecal impaction due to constipation results in increased colonic intraluminal pressure, compressing the mucosal vessels and impeding mucosal circulation [2]. Cilansetron is a similar 5-HT$_3$ receptor antagonist with matching efficacy in IBS and with constipation the most frequent adverse event. Treatment with cilansetron [3] is also associated with a similar low incidence of ischemic colitis events confirming the role of on-target pharmacology.

Due to patient advocacy, the FDA approved a supplemental new drug application (sNDA) that allowed restricted marketing of alosetron to restricted patients and with clear guidelines on recognition of side effects and patient–clinician participation.

5.2.2 Cerivastatin

Cerivastatin (Fig. 5.2) was the most potent of a group of drugs inhibiting hydroxymethyl-glutaryl-coenzyme A (HMG-CoA) inhibitors or statins. This enzyme is the rate-limiting enzyme of cholesterol synthesis, and the inhibitors are the first-line therapy to reduce low-density lipoprotein.

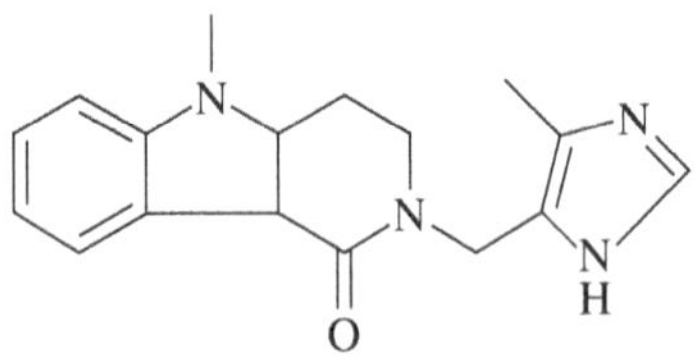

FIGURE 5.1   Structure of alosetron.

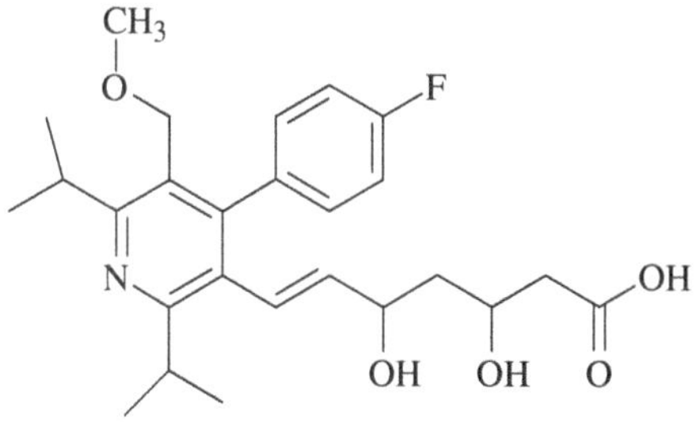

FIGURE 5.2 Structure of cerivastatin.

Muscle toxicity (myopathy) is a low-incidence adverse effect of HMG-CoA inhibitors, the extreme form of which is rhabdomyolysis [4]. Cerivastatin showed rates of rhabdomyolysis up to 10-fold greater than other HMG-CoA inhibitors. Among the likely mechanisms is the depletion of secondary metabolic intermediates (mevalonic acid and its metabolites). Mevalonic acid, whose synthesis is blocked by HMG-CoA reductase inhibitors, is important not only in cholesterol synthesis but also in the production of ubiquinone or coenzyme Q10 in the muscle cell. Ubiquinone is utilized by the electron transport chain for ATP synthesis. Thus, HMG-CoA reductase inhibitors could lead to a direct interference with ATP synthesis. Upregulation of the ubiquitin–proteasome pathway is also possible. In skeletal muscle, this pathway is responsible for the breakdown of myofibrillar proteins, including actin and myosin [5]. Cerivastatin was the most potent of the statin HMG-CoA inhibitors both preclinically and clinically [6]. The bioavailability was the highest also of any statin drug, and metabolism gave rise to circulating active metabolites. Bioavailability in this context could be viewed as a ratio of systemic burden (muscle toxicity) versus hepatic burden (liver efficacy) and may point to a key factor that differentiates this drug. Statins are known to be actively transported into the hepatocyte, and studies [7] show that the intracellular to extracellular steady-state free drug concentrations of cerivastatin (8:1) are considerably lower than a drug-like atorvastatin (18:1).

5.2.2.1 Flosequinan

5.2.2.1 Flosequinan Flosequinan is a vasodilator with mixed arterial and venous effect. The agent had mild positive chronotropic and inotropic effects [8]. Flosequinan was approved based on an improvement in exercise time and quality of life, before survival data were available. Soon after release, the high dose was withdrawn from the market due to an increase in mortality. The low dose continued on the market; however, a long-term study (PROFILE) showed all doses of flosequinan were associated with excess mortality rate and the drug was withdrawn. The drug has an active sulfone metabolite, flosequinoxan (Fig. 5.3), which also contributes to the effect and may be the active principal due to its similar intrinsic potency, higher circulating concentrations, and longer half-life (Table 5.2).

The findings with flosequinan and its metabolite may relate to its activity as a phosphodiesterase (PDE) III inhibitor [9]. Flosequinan in short to medium term showed benefit in patients with heart failure: increasing cardiac index, lowering cardiac filling pressures, and improving exercise tolerance. In these results and the subsequent increase in fatality, it shares similar properties to other PDE III inhibitors milrinone, enoximone, and vesnarinone. Although it was found that flosequinan increased circulating noradrenaline

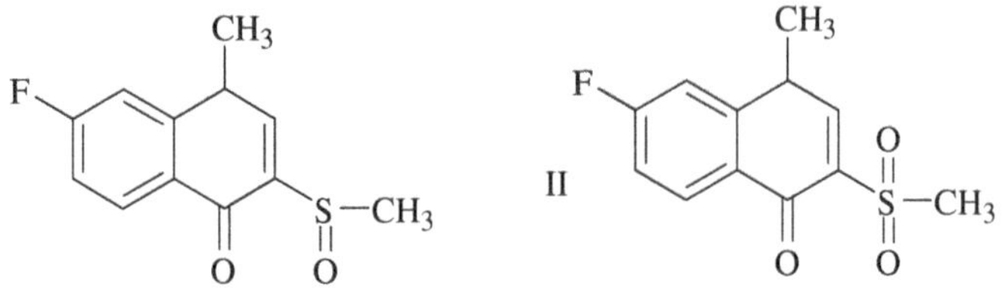

FIGURE 5.3   Flosequinan (I) and its principal metabolite flosequinoxan (II).

TABLE 5.2   *In Vitro* Potency, Steady-state Plasma Concentration, and Half-life of Flosequinan and its Principal Metabolite Flosequinoxan [11, 12]

	Inotropic activity, μM	Css, μM	$T\frac{1}{2}$, h
Flosequinan	15	10	6
Flosequinoxan	10	33	54

concentrations and heart rate and these may have been the contributing factors, PDE III inhibition is probably the main cause. A meta-analysis found that PDE III inhibitors were associated with a significant 17% increase in mortality and an increase in cardiac death, sudden death, and arrhythmias [10].

5.2.2.2 Encainide   Encainide belongs to the class 1c antiarrhythmic agents. The Vaughan Williams–Harrison classification denotes class 1c drugs as having intense (slow kinetics) blockade of Na⁺ channel excitatory current. A major series of studies (CAST) examined whether drugs of this type could prevent sudden death in high-risk patients. These patients had frequent premature ventricular extrasystoles (PVCs) and reduced ventricular function following acute myocardial infarction. The placebo group was found to have lower mortality compared to the group treated with encainide and flecainide [13]. The excess number of new ischemic events in the placebo group equalled the excess number of deaths in the treated group indicating that new ischemia was converted to sudden death by class 1c drugs. The class 1c drugs were poorly prescribed for all types of fibrillation following CAST. Encainide (Fig. 5.4) was a substrate for the polymorphic CYP2D6 and metabolized to two active metabolites (*O*-desmethyl encainide and 3-methoxy-*O*-desmethyl encainide) [14] giving it complex pharmacokinetics and pharmacodynamics.

It was voluntarily withdrawn by its manufacturers. One of the drugs, flecainide, has gradually been established as frontline therapy for atrial fibrillation (AF). AF is the most common arrhythmia in clinical practice, and its prevalence is increasing. Flecainide has a favorable safety profile in AF patients without significant left ventricular disease or coronary heart disease (i.e., excluded from the CAST studies), and the drug has an established place in restoring and maintaining sinus rhythm [15]. Flecainide is mainly cleared renally although in patients with lower renal function CYP2D6 plays a part in its clearance.

5.2.2.3 Rofecoxib   Rofecoxib belongs to the class of drugs termed COX-2 selective inhibitors (Fig. 5.5).

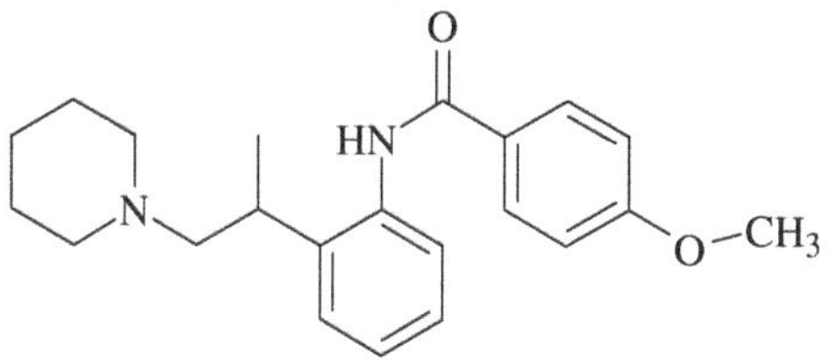

FIGURE 5.4 Structure of encainide.

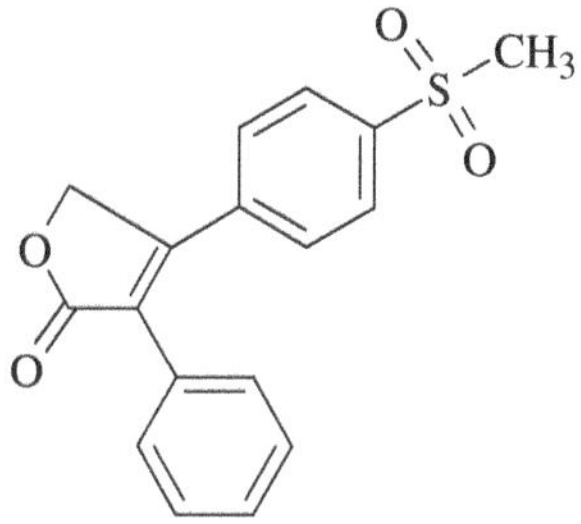

FIGURE 5.5 Structure of rofecoxib.

These were compounds with advantages over conventional nonselective cyclooxygenase (COX) inhibitors (NSAIDs) since the anti-inflammatory effects are mediated through attenuation of COX-2-derived inflammatory prostaglandin (PG), while the gastrointestinal effects such as ulceration, which is mediated by COX-1, are reduced [16]. Unfortunately, under some conditions of use, these drugs may increase the risk of thrombotic cardiovascular events [17]. The incidence of these effects has some relation to the dose and to the actual drug used. Rofecoxib at its marketed dose (50 mg) was judged superior in pain relief to celecoxib (200 mg) in clinical studies [18], which suggest that the drug was achieving a greater target occupancy than celecoxib. Many subsequent studies [19] have been conducted to clarify the cardiovascular risks of the remaining COX-2 inhibitors and the earlier nonselective compounds. A key finding has been that drugs showing <90% inhibition of COX-2 inhibition at the normal clinical doses (celecoxib, etoricoxib, ibuprofen, and meloxicam) have a relative risk of 1.18 for cardiovascular events, whereas for drugs giving >90% COX-2 inhibition, the relative risk is 1.60 (rofecoxib and diclofenac). These studies show a common risk heightened by how far the clinical dose is placing the drug on the dose–response curve. Studies in colon cancer have explored higher doses than used clinically for pain with drug such as celecoxib and have shown a clear dose–response to cardiovascular risk [20]. Pharmacological rationalization of the data can be advanced concerning suppression of endothelial vascular PGI_2 while not inhibiting platelet-derived thromboxane A_2 production resulting in an imbalance of homeostatic mechanisms [17].

5.2.3 Tegaserod

Tegaserod is a 5-HT$_4$ agonist and 5-HT$_{2B}$ receptor antagonist [21] and by stimulating gastrointestinal motility it is used for the management of IBS and chronic idiopathic constipation [22]. An FDA/company analysis indicated increased risks of heart attack or stroke.

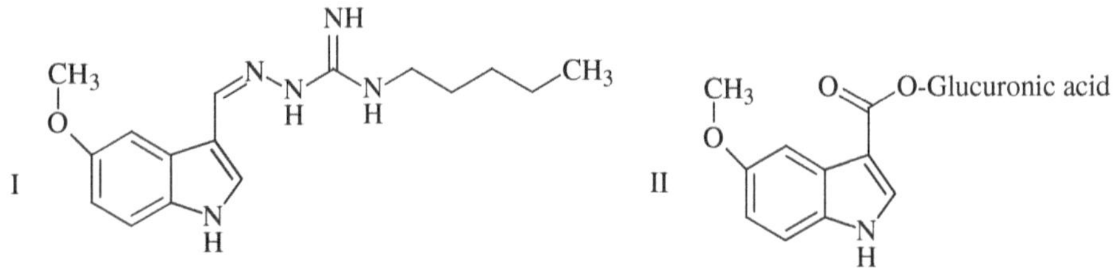

FIGURE 5.6   Structure of tegaserod (I) and its major human metabolite (II) formed by presystemic hydrolysis, oxidation, and conjugation of the resultant carboxylic acid with glucuronic acid.

TABLE 5.3   Comparison of Physicochemical Properties of Tegaserod and Cisapride

	Log$P$	pK_a	Log$D_{7.4}$	TPSA Å^2
Tegaserod	1.0	9.8	−1.4	95
Cisapride	2.9	8.2	2.1	86

A subsequent meta-analysis looking at a large historical database concluded that there was no increased risk [23]. Unlike cisapride (see later), tegaserod (which is structurally and physicochemically different; Figs. 5.6 and 5.10 and Table 5.3) and its major metabolite have considerable (over 500-fold) selectivity for the I_{Kr} channel [16, 17]. In general, potent inhibitors of the I_{Kr} are lipophilic amines. Tegaserod is 100 times less lipophilic than cisapride in terms of intrinsic lipophilicity (log$P$) and 1000 times less lipophilic when effective lipophilicity is considered (log$D$) providing a strong rationale for its much greater selectivity.

Importantly, this selectivity was also seen in clinical studies with no effect on QT interval [24]. Various studies have been conducted to examine the role of 5-HT$_4$ receptors in the effects on the vasculature and platelet aggregation. 5-HT$_4$ receptors were shown to be present in platelets. When exposed to tegaserod, small but statistically significant concentration-dependent increases in induced platelet aggregation were detected, albeit at supratherapeutic concentrations [25]. In a later study, in contrast, no effects were seen on platelet aggregation even at high concentrations [26]. Tegaserod also had no contractile activity in either porcine or human isolated coronary artery preparations and only a small and variable response in canine coronary arteries even at higher concentrations than those achieved clinically [26]. It seems likely that tegaserod may have been aligned with drugs like cisapride and been "guilty by association." Although included in the grouping of on-target effects, it may actually demonstrate the uncertainty of benefit–risk estimations and the clustering of events, which can occur in low-frequency toxicities.

Withdrawals due to On-Target Pharmacology: Impact on Drug Discovery and Development   Because of the impact of large outcomes, type of trials, or large databases, it seems natural to demand more of this data at time of registration, thus contributing hugely to the cost of drug development and in most cases denying patients quick access to new drugs. However, it is important to strike the correct balance in what is needed for market introduction. Clearly, phase IV studies are vital in this.

The balance of safety and efficacy and patient needs is illustrated by the reintroduction of alosetron. Recognizing the problems and training physicians and patients in the safe use of the medicine is still and certainly was a problem. Alosetron sets out a "contract" for its safe use and maybe is a model of how to use drugs. The future doesn't look likely to fully meet these expectations as clinical pharmacology is a subject in decline in teaching physicians.

The case of tegaserod exemplifies how hard these withdrawal decisions are and the fact that they come down to judgment. That judgment will be influenced by the need to act quickly, the context of the result (e.g., a number of drugs showing similar findings, even if the drugs are different types or even different indications). In a litigation-rich setting, early withdrawal may even favor the manufacturer if there is something really amiss.

Some of the earlier drugs described had data at registration that would be deemed inadequate today. Flosequinan, for instance, had no mechanism ascribed to its vasodilator properties compared to highly specific selective drugs such as alosetron. Indeed, the classification of flosequinan as a PDE III inhibitor does not rule out other pharmacology. Another feature of flosequinan is the active metabolite. Flosequinan itself is cleared by metabolism to the metabolite (this process is variable depending on liver disease), while flosequinan is renally cleared (variable with renal function). The enzymology of the clearance of flosequinoxan has not been defined. Development would now use sophisticated pharmacokinetics and pharmacodynamics (PK/PD) measurement of both active principals looking for covariates. This would have suggested or excluded possible pharmacokinetics (PK) reasons for a subset of patients being at risk due to, for example, high concentrations of one of the active principals. In general, early metabolism screening will identify active metabolites within the drug discovery program, and decisions can be taken early on about suitability for progression. Certainly, the sulfone metabolite (flosequinoxan) would be an expected metabolite of the sulfoxide parent (flosequinan).

Defining the dose–response is always a key part of drug development, often done badly. Proof of concept is often done at close to maximum tolerated dose, and after that, it is difficult to shy away from maximum efficacy. The COX-2 drugs illustrate that the best dose is not necessarily (maybe even not usually) the most efficacious.

Whether drug discovery development organizations can withstand trials like CAST in a major indication and prove a success in others is difficult to answer. Flecainide is a shining example of clinical research believing in a mechanism and finding the best application, eloquently expressed in the titles of two of the papers referenced: "From first class to third class: Recent upheaval in antiarrhythmic therapy – Lessons from clinical trials (1988)" and "Twenty-five years in the making: Flecainide is safe and effective for the management of atrial fibrillation (2011)."

5.3 OFF-TARGET PHARMACOLOGY, KNOWN RECEPTOR: AN ISSUE OF SELECTIVITY

5.3.1 Fenfluramine and Dexfenfluramine

The appetite suppressants fenfluramine and dexfenfluramine (single enantiomer of fenfluramine introduced later), which were both effective in the treatment of obesity, were withdrawn from the marketplace because of association with pulmonary hypertension and valvular heart disease.

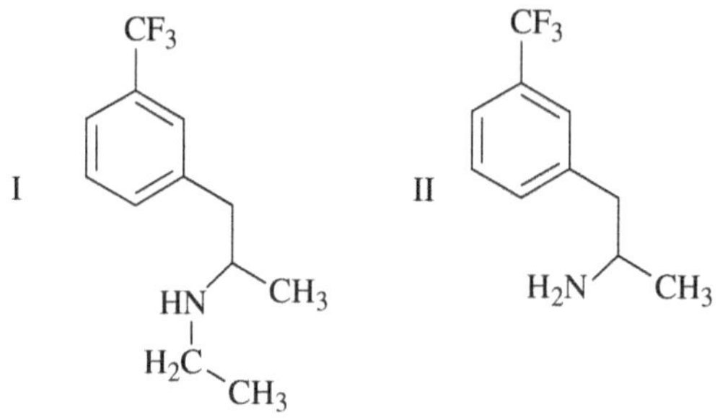

FIGURE 5.7 Structures fenfluramine (I) and its major active metabolite norfenfluramine (II).

Fenfluramine and dexfenfluramine are agonists of 5-HT receptors. A major metabolite norfenfluramine is considerably more potent than the parent and is responsible for most of the pharmacological activity (Fig. 5.7). Fenfluramine and dexfenfluramine were frequently prescribed as a combination product with phentermine (an amphetamine-like compound also licensed for weight loss). Consideration of the structure of 5-HT and fenfluramine can rationalize the increased potency of the N-deethylated metabolite, norfenfluramine, since the primary amine now mimics the functionality of the natural agonist. The drugs reduce appetite by activating 5-HT$_{2C}$ receptors. Norfenfluramine activates 5-HT$_{2C}$ receptors but is equally potent as an agonist of 5-HT$_{2B}$ receptors [27]. Activation of mitogenic 5-HT$_{2B}$ receptors on heart valves and pulmonary artery interstitial cells leads to the formation of proliferative fibromyxoid plaques that cause the deleterious changes in tissue integrity and function. Analysis of the patient data indicated that those taking the drugs were much more at risk after 4 months or more of treatment, than those taking the drugs for shorter periods. A clear dose–response in the effects could be detected, with those taking 60 mg/day or greater much more at risk of severe valvulopathy than those taking 40 mg/day or less [28].

Other drugs also implicated in valvular heart disease include ergotamine and methysergide—both drugs are also active at 5-HT$_{2B}$ receptors. Rothman et al. [29] in reviewing the valvular heart disease suggest that 5-HT$_{2B}$ receptors need to be activated to produce the adverse effect and that all clinically available medications with serotonergic activity and their active metabolites be screened for agonist activity at 5-HT$_{2B}$ receptors.

5.3.2 Rapacuronium

Rapacuronium was introduced as an ultrashort-acting nondepolarizing neuromuscular blocking agent of nicotinic receptors. The drug had an onset of approximately 90 s and a clinical duration of action of approximately 15 min. The drug was used to facilitate endotracheal intubation to assist in the controlled ventilation of unconscious patients during surgery. The chemical structure has a steroid framework similar to other neuromuscular blocking agents (Fig. 5.8). The more potent drugs such as pancuronium, decamethonium, and suxamethonium have two quaternary nitrogen atoms spaced rigidly apart by the steroid. Rapacuronium is similar to vecuronium and has one of the quaternary nitrogens replaced by a tertiary amine piperidine ring system.

The drug was withdrawn from clinical use due to a high incidence of bronchospasm, sometimes with lethal consequences. Rapacuronium was shown to be an antagonist of

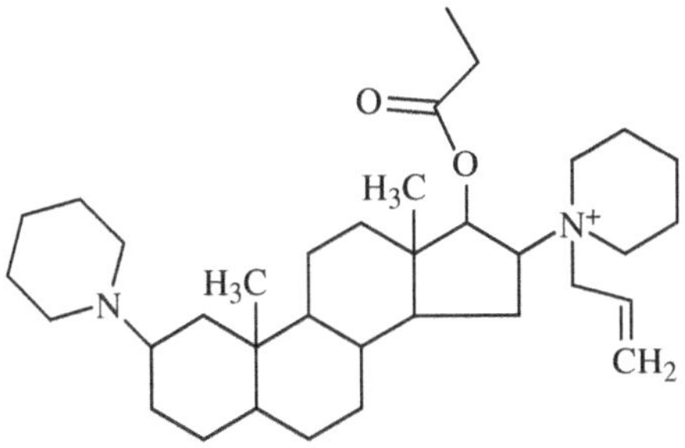

FIGURE 5.8 Structure of rapacuronium.

M2 muscarinic receptors at the concentrations achieved clinically [30, 31]. Rapacuronium displayed a 50-fold higher affinity for antagonism of the M2 versus the M3 muscarinic receptor. Preferential antagonism of presynaptic M2 receptors without antagonism of postsynaptic M3 muscarinic receptors can facilitate bronchoconstriction. Normally, the release of acetylcholine is terminated by the feedback of M2 receptors. Selective antagonism of these receptors leads to excessive release of acetylcholine, which can then act on the M3 receptors in airway smooth muscle, thereby facilitating bronchospasm.

5.3.2.1 Suprofen Suprofen was a nonselective COX inhibitor. It was withdrawn due to acute flank pain syndrome and acute renal injury. One hundred and sixty-three cases of acute flank pain syndrome were reported within the 700,000 persons who used the drug in the United States [32] giving an incidence of 1 : 3,600. Preexisting renal disease and a history of kidney stones and/or gout were corresponding risk factors. Suprofen like ticrynafen withdrawn for hepatotoxicity forms a reactive sulfoxide metabolite [33] that binds to P4502C9 and inactivates it. Both compounds cause kidney flank pain, which relates to their uricosuric properties. Acute kidney injury is a finding with all NSAIDs although of low incidence. NSAIDS inhibit prostaglandin synthesis in the kidney allowing vasoconstrictors such as angiotensin II and vasopressin to exert maximal effect. The risk of acute kidney injury varies among different NSAIDs with risk generally increasing with decrease in selectivity toward COX-2 [34]. The most selective COX inhibitors have the lowest risk. The adjusted odds ratio for various drugs for acute kidney injury is rofecoxib (0.95), celecoxib (0.96), meloxicam (1.13), etodolac (1.31), diclofenac (1.11), piroxicam (1.53), sulindac (1.61), ibuprofen (2.25), naproxen (1.72), high-dose aspirin (3.64), indomethacin (1.94), and ketorolac (2.07). Suprofen has a similar selectivity to ketorolac being around 100-fold more potent for COX-1/COX-2. The acute renal failure may therefore be a symptom of selectivity and final clinical dose. The flank pain syndrome occurs rapidly and lasts a period of several hours. Detailed clinical pharmacology studies [35] have shown that within 90 min after suprofen administration, the fractional excretion of uric acid increased from 8.8 to 35.5%. Urine became supersaturated with uric acid, while glomerular filtration rate, renal plasma flow, and the excretion of Na^+ decreased. The findings are consistent with acute uric acid nephropathy, including crystallization of uric acid in the nephron as a mechanism of suprofen-induced renal dysfunction. Consideration of structure (Fig. 5.9) could indicate that suprofen is an inhibitor of URAT1, the apical surface reuptake urate transporter, as part of its secondary pharmacology.

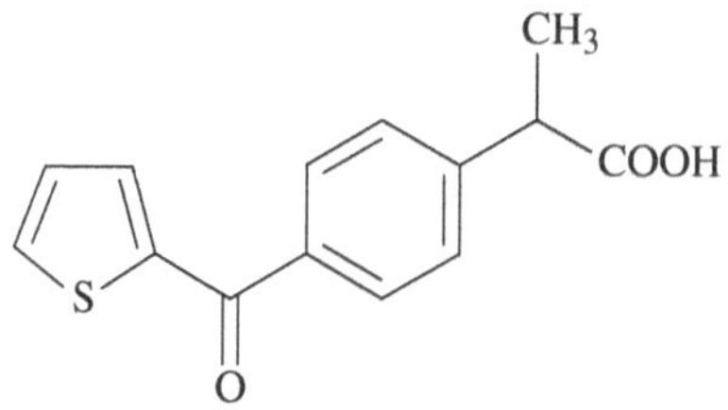

FIGURE 5.9 Structure of suprofen.

5.3.3 Astemizole, Cisapride, Grepafloxacin, and Thioridazine

These drugs represent a pharmacologically diverse group when their primary pharmacology is considered. Astemizole is a nonsedating H_1 antihistamine used to treat various allergic conditions. Cisapride is a gastrointestinal prokinetic motility agent acting by a mixed 5-HT_3 antagonist/5-HT_4 receptor agonist action, while grepafloxacin is a fluoroquinolone antibiotic. Thioridazine is an antipsychotic with typical nonselective tricycle pharmacology. All the compounds [36] inhibit the rapidly activating delayed rectifier potassium current channel (I_{Kr}) in cardiac tissues (hERG K^+, K_v11.1). Thioridazine is the most potent I_{Kr} blocker of the tricyclic drugs [37]. The slowing of repolarization by block of the I_{Kr} channel leads to QT interval prolongation and in extreme cases the triggering of potentially fatal arrhythmias (torsade de pointes, TdP). Consideration of these compounds [18] indicates that their therapeutic concentrations lie close (within 1 log unit) to their IC_{50} values against the channel; indeed, cisapride (Fig. 5.10) concentrations exceed the IC_{50}.

5.3.3.1 Propoxyphene Propoxyphene (Fig. 5.11) is an opioid analgesic that is converted to its N-demethylated metabolite norpropoxyphene by CYP3A4 [38]. Both parent and metabolite are agonists of opioid receptors, but also both metabolite and parent compound block the I_{Kr} channel [39]. In addition, they have significant blocking activity against sodium channels [40] leading to prolongation of the QRS complex. A combination of ion channel blockade is thought responsible for the drug effects causing prolonged intracardiac conduction time and ultimately heart failure [41].

5.3.3.2 Sibutramine Sibutramine is a serotonin (5-HT) reuptake inhibitor that also has activity against the noradrenaline and dopamine transporters. The drug was used to treat obesity. The satiety provided by the increase in striatal 5-HT concentrations can be compared to the 5-HT agonist activity of fenfluramine described previously. The classification therefore makes the 5-HT transporter effects the on-target pharmacology and the other transporter effects the off-target pharmacology. Like propoxyphene, the N-demethylated metabolites contribute much of the pharmacological activity due to increased half-life. Sibutramine has a half-life of 1 h and is metabolized by CYP3A4, and its primary and secondary amine metabolites have half-lives of 14 and 16 h, respectively (Fig. 5.12) [42].

By blocking noradrenaline uptake, the drug always raised a concern of hypertension as a side effect, but the effect of sibutramine appears to vary depending on the balance [42] between presynaptic and postsynaptic adrenergic receptors, which differs in the peripheral nervous system from that in the central nervous system. The postsynaptic

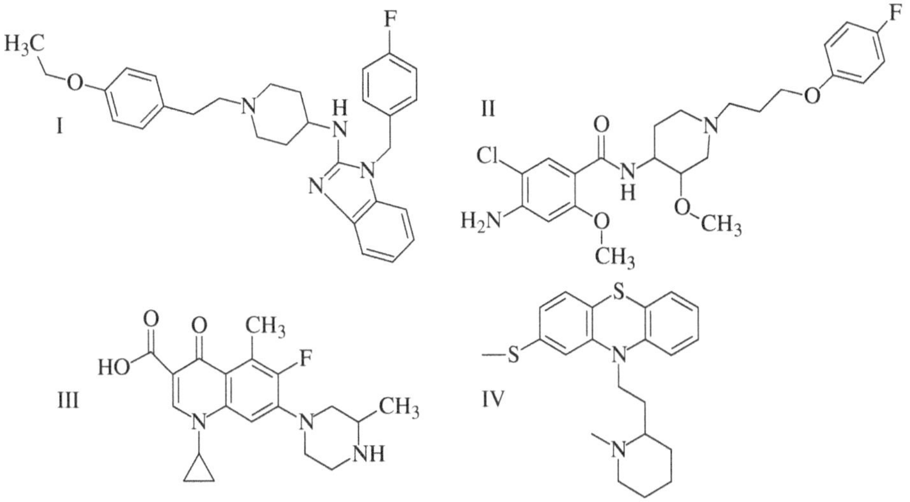

FIGURE 5.10   Structure of astemizole (I), cisapride (II), grepafloxacin (III), and thioridazine (IV).

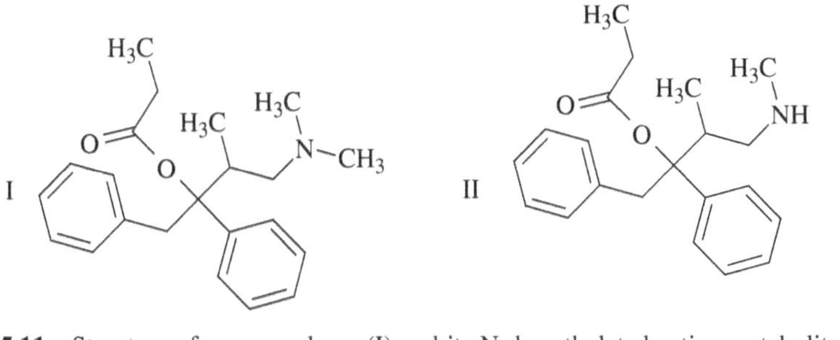

FIGURE 5.11   Structure of propoxyphene (I) and its N-demethylated active metabolite norpropoxyphene (II).

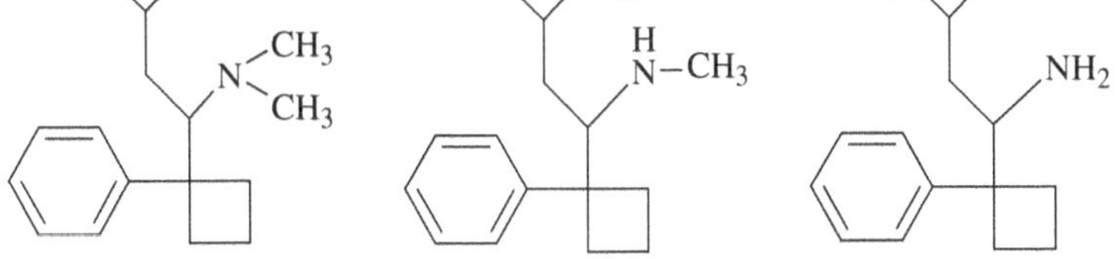

FIGURE 5.12   Structures of sibutramine and its pharmacologically active, more persistent major metabolites.

effects of adrenergic activation of $\alpha 1$ and $\beta 1$ adrenoreceptors are increases in heart rate and blood pressure, whereas the presynaptic effects of stimulation of $\alpha 2$ receptors are a decrease in heart rate and blood pressure. Thus, sibutramine will only show detrimental side effects of increased blood pressure in a subgroup of patients depending on the balance of effects. Sibutramine has also been associated with QT prolongation via effects on the I_{Krs} channel [43]. Thus, the compound has the potential to both raise blood pressure and prolong QT interval.

A key outcome of the Sibutramine Cardiovascular Outcomes Trial (SCOUT) was the finding that sibutramine increased serious adverse cardiovascular event (nonfatal heart attack and nonfatal stroke) by 16% [44].

5.3.3.3 Terfenadine Terfenadine was marketed as a nonsedating histamine, properties that the drug (a lipophilic amine) owes to its very rapid and extensive conversion to a zwitterionic metabolite fexofenadine (Fig. 5.13).

Fexofenadine has poor passive permeability and is also a substrate for P-glycoprotein leading to very low CNS penetration. Terfenadine was a potent inhibitor of the I_{Kr} channel, but the concentrations of terfenadine were normally so low that the drug did not exhibit any pharmacology, either beneficial or adverse. Fexofenadine, the active principal, is highly selective for the H_1 receptor with no activity against I_{Kr}. Conversion to fexofenadine was by CYP3A4 and coadministration with potent inhibitors such as ketoconazole or erythromycin leads to elevated concentrations of terfenadine and in some individuals prolongation of QT and in some of these fatal cardiac arrhythmias [45, 46]. Unusually for a withdrawn drug, the metabolite fexofenadine is now marketed, having reasonable (20–30%) oral bioavailability, partly due to active intestinal uptake by anion transporters (OATP1A2).

5.3.3.4 Mibefradil Mibefradil (Fig. 5.14) is a long-acting calcium channel antagonist, with particular activity against the T-type channel (transient, low voltage activated). The drug was effective in hypertension and chronic stable angina pectoris. At a clinical dose of 50–100 mg, mibefradil was a potent inhibitor of CYP3A4, which resulted in multiple clinically relevant drug interactions [47]. These included very important comedications such as simvastatin, which showed a large sevenfold elevation, sufficient to enhance potential adverse events, including rhabdomyolysis [47]. The difficulty in providing

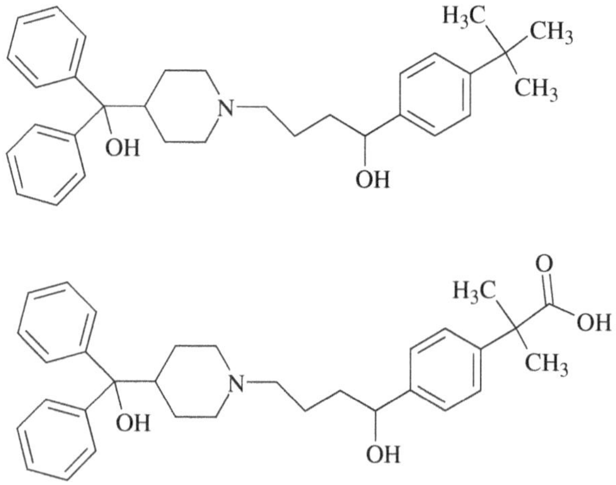

FIGURE 5.13 Structure of terfenadine (I) and its active, non-CNS-penetrating metabolite fexofenadine (II).

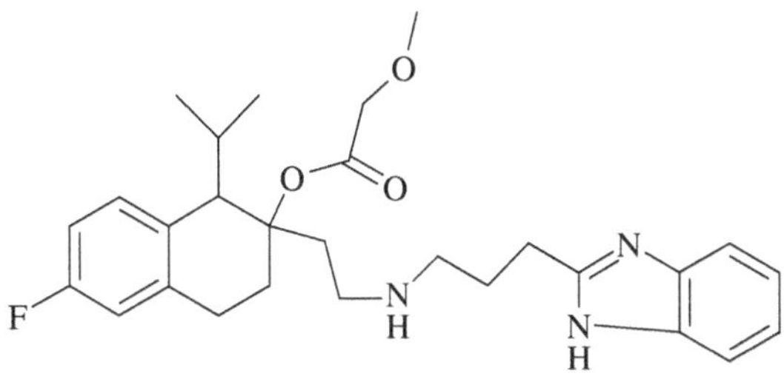

FIGURE 5.14 Structure of mibefradil.

sufficient labeling information to address the diversity and complexity of the interactions led to the voluntary withdrawal of the drug.

The mechanism of inhibition [48] is via formation of a reactive metabolite and the rate of CYP3A4 inactivation and the low partition ratio (moles of mibefradil metabolized per moles of enzyme inactivated) make mibefradil one of the most potent "mechanism-based inhibitors."

5.3.3.5 *Withdrawals due to Off-Target Pharmacology, Known Receptor: Impact on Drug Discovery and Development* The impact of lack of specificity of a drug is obvious with hindsight and can be rationalized and incorporated into screening cascades. Some of these drugs were introduced before detailed knowledge of ion channel pharmacology was available, particularly around the promiscuous nature of some of the channels such as I_{Kr}. This is particularly appropriate today, since many drug discovery programs are mounted using the "pure" protein target (either as itself or expressed in a cell line) as the primary screen. Close neighbors to the desired target as highlighted with fenfluramine/dexfenfluramine will be screened alongside the primary pharmacology. Other important screens such as I_{Kr} will also be placed high in the cascade particularly when the structure and physicochemistry of the lead chemical series make interactions with this channel likely [49]. Lipophilicity and basicity, particularly where the basic center is central to the molecule, are simple structural alert features for I_{Kr} channel activity. Screens can be configured in high-throughput format [50]. Similarly, screening for cytochrome P450 (CYP) inhibition takes place high in the cascade. Initially, most screens examined reversible inhibition, but mechanism-based inhibition, of the type seen with mibefradil, is now routinely incorporated [51]. Noticeable in the case of fenfluramine/dexfenfluramine (and earlier (primary pharmacology) in flosequinan) is the role of active metabolites in the pharmacodynamic action of the drug. Smith and Obach [52] recommend that "potential metabolites produced in human *in vitro* systems, that are not substantially different from parent drug with regard to structure and physiochemical properties should be synthesized and screened for pharmacological activity."

Broad ligand binding is also widely adopted at a later stage in drug discovery. Typical screening panels have been published [53], and such screening would clearly identify the potential problems with rapacuronium (and incidentally all of the compounds). Smith and Obach [52] suggest that drug development of any novel metabolites with circulating concentrations above 1 µm is also subject to such screening.

5.4 OFF-TARGET PHARMACOLOGY, UNKNOWN RECEPTOR: IDIOSYNCRATIC TOXICOLOGY

Benoxaprofen, bromfenac, pemoline, ticrynafen, troglitazone, tolcapone, and trovafloxacin were withdrawn from the marketplace due to hepatotoxicity. Nomifensine, remoxipride, and temafloxacin were withdrawn due to hemolytic and aplastic anemia. Zomepirac was withdrawn due to anaphylaxis. Valdecoxib was withdrawn due to serious skin toxicity and cardiac risk. In many of the cases, more than one toxicity finding also contributed to the withdrawal. For instance, benoxaprofen was a potent photosensitizer, temafloxacin was also associated with renal and liver toxicity, and ticrynafen also exhibited flank pain syndrome. In all these cases, no mechanism has been fully explained although in a number of the cases reactive metabolites are believed to be responsible for at least the initial step in the pathway that leads to toxicity.

5.4.1 Benoxaprofen

Benoxaprofen was a nonselective COX inhibitor with a long half-life (Fig. 5.15).

The clinical dose was 600 mg with no subdivision or dosing advice for special patient populations. The commonest cutaneous side effect was photosensitivity, which occurred in 86 patients (28.6%). Photosensitivity occurred in half of the patients treated in the summer. With both skin toxicity and hepatotoxicity, the elderly were most at risk [53]. The plasma elimination half-life of the drug was in the range of 19–26 h in younger volunteers and patients [54] and led to the drug being used on a once-a-day basis. Studies in elderly patients revealed a very different pharmacokinetic profile with half-lives extending out to 150 h, which coincided with reduced creatinine clearance in these subjects [55].

Although detailed studies are lacking, it is likely that renal excretion of the major acyl glucuronide metabolite was the major clearance pathway. Enterohepatic recirculation of this metabolite and subsequent hydrolysis were also likely to be important in the long duration of the drug in the body. Gastrointestinal stasis and lowered kidney function therefore probably contributed to accumulation of the drug in the elderly with the inappropriate dosage regimen.

5.4.2 Bromfenac

Bromfenac was a nonselective NSAID and was marketed for short-term relief of pain (<10 days at a time). It was often used for longer than the 10-day period, and this use resulted in a high incidence of hepatotoxicity. Flu-like symptoms associated with hepatic enzyme elevation and a case of possible drug-related hepatocellular jaundice were noted in the database of 1195 subjects reviewed by the FDA at time of approval. Following approval, rates of acute liver failure (ALF) for bromfenac were estimated to be in the

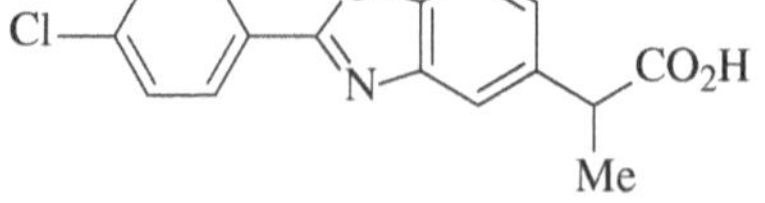

FIGURE 5.15 Structure of benoxaprofen.

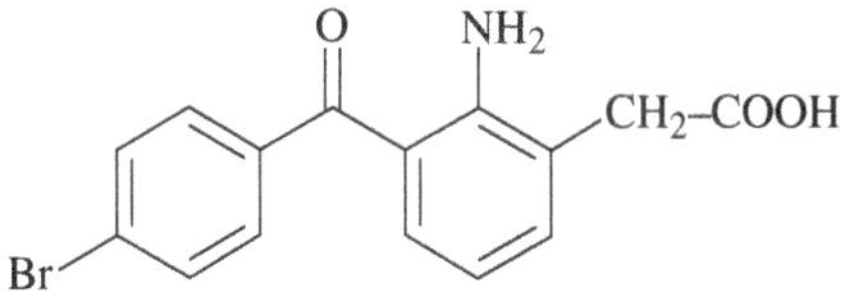

FIGURE 5.16   Structure of bromfenac.

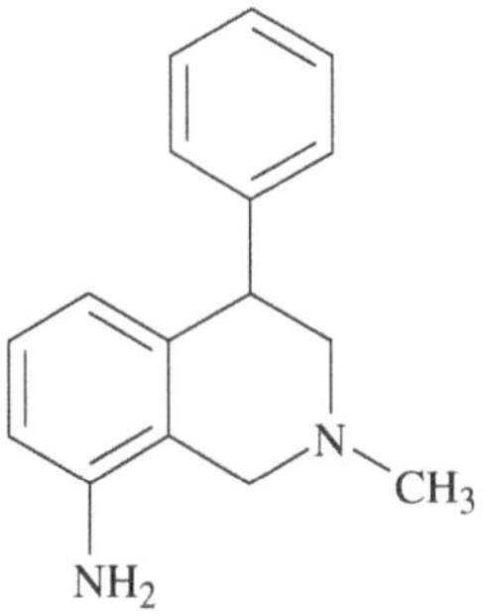

FIGURE 5.17   Structure of nomifensine.

range of 1–10,000 [56]. Bromfenac possesses arylacetic acid, aniline, and bromobenzene motifs all considered structural alerts in terms of possible reactive metabolites (Fig. 5.16).

The aniline function allows the generation of quinine-imine-type reactive metabolites which are suggested intermediates in a number of hepatotoxic drugs.

5.4.3   Nomifensine

Nomifensine is a tetrahydroisoquinoline noradrenalin and dopamine reuptake inhibitor, with relatively weak serotonin uptake inhibition [74]. It was marketed as an antidepressant. It was seen to be an improvement on imipramine and amitriptyline with similar efficacy but little or no sedation, fewer and milder anticholinergic side effects, and less likely to cause cardiotoxicity on overdosage [74]. The drug was withdrawn due to hemolytic anemia. Detailed follow-up studies showed that nomifensine can induce the production of red blood cell drug, metabolite-dependent antibody, and autoantibodies, indicating that a reactive metabolite autoimmune response triggered the toxicity [75]. Nomifensine is an aniline-containing drug (Fig. 5.17).

Other aniline precursors or aniline-containing drugs causing aplastic anemia and agranulocytosis include chloramphenicol, procainamide, aminoglutethimide, aminosalicylic acid, and dapsone [76, 77]. Multiple glutathione adducts of the oxidative products of nomifensine have been identified in human hepatocytes and liver microsomes indicating the facile formation of reactive species during metabolism [78]. In addition, an electrophilic nomifensine dihydroisoquinolinium ion metabolite can be generated by CYP450, monoamine oxidase A, human hemoglobin (supplemented by hydrogen peroxide), myeloperoxidase (supplemented by hydrogen peroxide), and whole blood. In contrast to the CYP450-generated reactive metabolites, the dihydroisoquinolinium ion was shown to react with cyanide and borohydride, but not glutathione [79].

5.4.4 Pemoline

Pemoline is a stimulant drug of the oxazoline class (Fig. 5.18). It was used as a medication to treat attention deficit hyperactivity disorder (ADHD) and narcolepsy. Pemoline hepatotoxicity ranges from asymptomatic elevations in levels of serum aminotransferases to fulminant liver failure [57–59]. Unusually, pemoline is administered at a fairly low dose, most of the cases are in children (12 years old), and no toxic metabolites have been postulated or discovered. The mechanism of pemoline-induced hepatic injury, therefore, is unclear although immunologically mediated hypersensitivity and dose-dependent hepatotoxicity reactions have been postulated.

5.4.5 Remoxipride

Remoxipride (Fig. 5.19) is an inhibitor of dopamine D_2 receptors and also has affinity for the sigma receptor [80]. It was classified as an atypical antipsychotic and used for the treatment of schizophrenia. Remoxipride was withdrawn due to a relatively high incidence of aplastic anemia (1 in 10,000 patients). Remoxipride is metabolized to phenolic metabolites (FLA 797 and FLA 908) via O-demethylation and catechol metabolites (NCQ-436) via O-demethylation and aromatic hydroxylation, which possess similar *in vitro* activity. These metabolites do not appear to be formed in sufficient amounts or penetrate the brain to contribute to the antipsychotic activity [81].

The catechol and in addition the hydroquinone (NCQ-344) (formed similarly to the catechol) are capable of forming reactive para- and orthoquinones. NCQ-344 is present in plasma and can be converted to glutathione conjugates demonstrating its reactivity [82]. These conjugates could be formed by human neutrophils illustrating that the oxidation of the metabolites to para- and orthoquinones can occur at sites other than the liver and of direct relevance to blood toxicity.

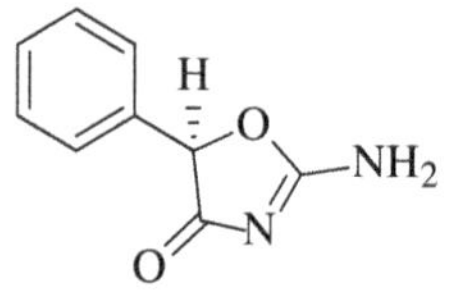

FIGURE 5.18 Structure of pemoline.

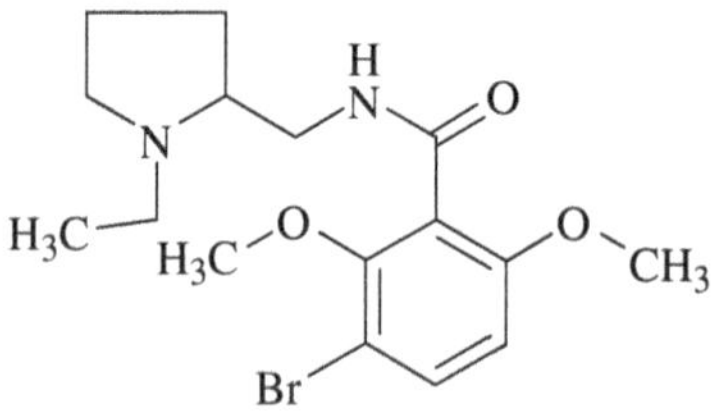

FIGURE 5.19 Structure of remoxipride.

5.4.6   Temafloxacin

Temafloxacin is a fluoroquinolone antibiotic drug (Fig. 5.20) approved to treat skin, lower respiratory tract, genital, and urinary infections, including prostate infections. Severe adverse reactions, most notably immune hemolytic anemia, developed in about 95 patients during the first four months of its use, leading to two patient deaths [83]. The time to onset was 6.4 days. Within this group, 54 cases developed renal dysfunction, while 48 showed hepatotoxicity.

Renal excretion of the drug and formation of an acyl glucuronide are major clearance pathways. Oxidative metabolism of the drug occurs on the methylpiperazine system leading to ring opening and ultimately an aniline metabolite [84]. The association of aniline-containing drugs and blood dyscrasias is well documented (see later).

5.4.7   Tienilic acid

Tienilic acid (Fig. 5.21) was a uricosuric diuretic. It produced immunoallergic hepatitis in susceptible individuals. It was withdrawn from clinical use in the United States after more than 500 cases of hepatic injury and 25 fatalities [60]. Patients with hepatitis caused by tienilic acid have circulating type 2 antibodies termed antiliver/kidney microsome (anti-LKM_2) that recognize P4502C9 but not other CYPs including CYP2C8 and CYP2C19 [61]. Tienilic acid is metabolized in the thiophene ring by P4502C9 to 5-hydroxytienilic acid.

Besides this hydroxylation, the thiophene ring also undergoes a P4502C9-catalyzed oxidation to yield a reactive thiophene-S-oxide metabolite that can bind covalently with the enzyme and inactivates it [62]. Studies have shown this alkylated form of the enzyme is transported to the cell surface by a vesicular route and triggers the formation of anti-bodies [63]. Tienilic acid is very specific in binding [61] to a protein target (CYP2C9), and if the subsequent pharmacology that linked the immune reaction, antibody formation,

FIGURE 5.20   Structure of temafloxacin.

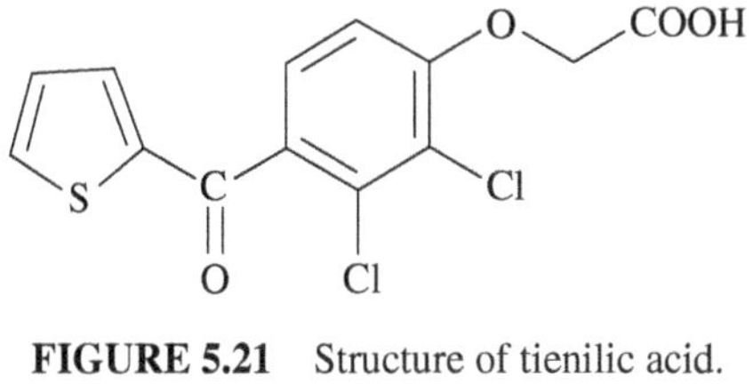

FIGURE 5.21   Structure of tienilic acid.

and ultimate toxicity were known, it could be classed as a known receptor. Unfortunately, the presence of antibodies and hepatotoxicity are circumstantial at present.

5.4.8 Troglitazone

The PPARγ agonist troglitazone was withdrawn from the market due to hepatotoxicity. ALF was seen at a comparatively high incidence of 1 in 5000 patients with ALT rises of threefold seen in over 1%. Troglitazone was an insulin sensitizer and improved insulin resistance. The acidic thiazolidinedione grouping present in this class gives the glitazones unique potency as PPARγ agonists. When troglitazone was withdrawn, the thiazolidinedione group (Fig. 5.22) was subject to detailed metabolism investigations [64].

Metabolism of this group leads to an ultimate reactive sulfonium ion formed from an initial sulfoxide, followed by a formal Pummerer rearrangement, or a C5 thiazolidinedione radical or a sulfur cation radical. Other reactive metabolites such as the quinone methide have also been identified. While a reactive metabolite may play a part in troglitazone's hepatotoxicity, it and its sulfate metabolite are also potent bile salt excretory protein (BSEP) inhibitors [65]. BSEP is an ATP-binding cassette transporter important in the secretion of bile salts into bile. Inhibition of BSEP can lead to accumulation of cytotoxic bile salts in the hepatocyte and, consequently, to hepatotoxicity. ALF was seen in approximately 1 in 5000 patients receiving troglitazone, in contrast to negligible hepatotoxicity for rosiglitazone (with the few incidents having uncertain causality) and no hepatotoxicity reported with pioglitazone.

5.4.9 Tolcapone

Tolcapone is a catechol-O-methyltransferase (COMT) inhibitor that reduces levodopa metabolism to enhance its effect in Parkinson patients (Fig. 5.23).

The drug was the first drug of this class to be marketed but was withdrawn in the European Union after deaths [66] in patients due to hepatic failure (around 1 in 20,000). In the United States, tolcapone was not withdrawn but subjected to intense liver enzyme

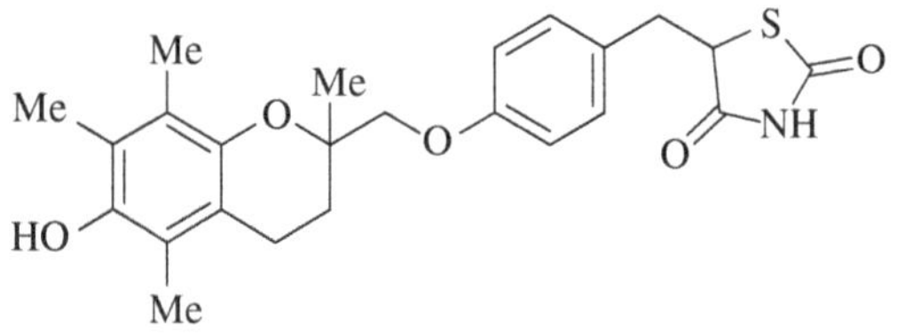

FIGURE 5.22   Structure of troglitazone.

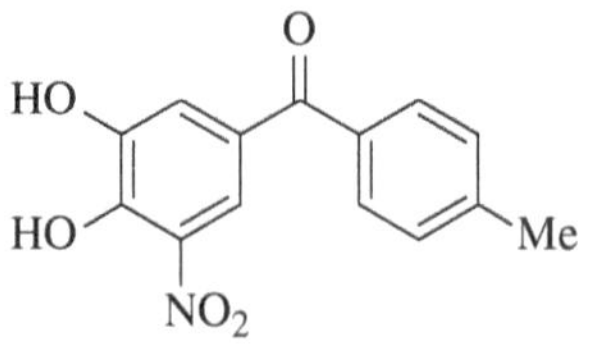

FIGURE 5.23   Structure of tolcapone.

monitoring. Tolcapone is structurally related to entacapone. Neither drug was associated with hepatotoxicity in preclinical toxicity testing. Tolcapone showed a dose–response relationship in liver function tests in clinical trials with 3X upper limit of normal (ULN) in 1% of patients at 100 mg dose and 3% in patients at 200 mg [67]. These observations led to the recommendation that periodic monitoring of liver function be performed. In contrast, clinical trials with entacapone demonstrated no increase in liver enzymes above those observed with placebo. Further, no instances of hepatic failure have been observed in postmarketing surveillance studies.

Catechol glucuronidation is the major metabolic fate of tolcapone in humans, but the aniline and N-acetyl aniline reduction products are minor metabolites [68]. The two metabolites can undergo oxidation to electrophilic quinone-imines. [69]. An association between hepatotoxicity and mutations in the uridine diphospho glucuronosyltransferase (UGT) 1A9 gene (which encodes the UGT isozyme responsible for tolcapone glucuronidation) has been observed in tolcapone-treated patients, which in poor metabolizers may raise the relative production of the electrophilic metabolites[70]. The reduced metabolites are not observed in entacapone.

5.4.10   Trovafloxacin

Trovafloxacin, like temafloxacin, is a fluoroquinolone antibiotic but with superior gram-positive antibacterial action to others in the class. It was developed as a broad-spectrum DNA gyrase and topoisomerase IV inhibitor. It was withdrawn from the market due to the risk of hepatotoxicity. Trovafloxacin had a very rapid introduction with over 3,000,000 patients per month receiving the drug within a year of launch. In an avalanche of information, trovafloxacin was "strongly associated" with 14 cases of ALF and six deaths, and further reports were highlighting a potential problem.

Trovafloxacin (Fig. 5.24) contains a cyclopropylamine moiety, which has a potential to be oxidized to reactive intermediate(s) [71], which may be significant in the mechanism.

However, reexamination of the data again gives some cause for concern in labeling trovafloxacin as hepatotoxic. Several other antimicrobials, including some fluoroquinolones [72] that are still widely available, have been associated with rare incidents of hepatotoxicity. Amoxicillin/clavulanate, for instance, has a crude incidence rate of acute liver injury of 1 in 4500. Amoxicillin alone has a rate of 1 in 25,000. Trovafloxacin has an incidence rate of 1 in 18,000 and for severe hepatotoxicity 1 in 178,000. In a study of fluoroquinolone hepatotoxicity, ciprofloxacin and moxifloxacin had the most cases, but hepatotoxicity was observed with levofloxacin and gatifloxacin. The pattern of injury was

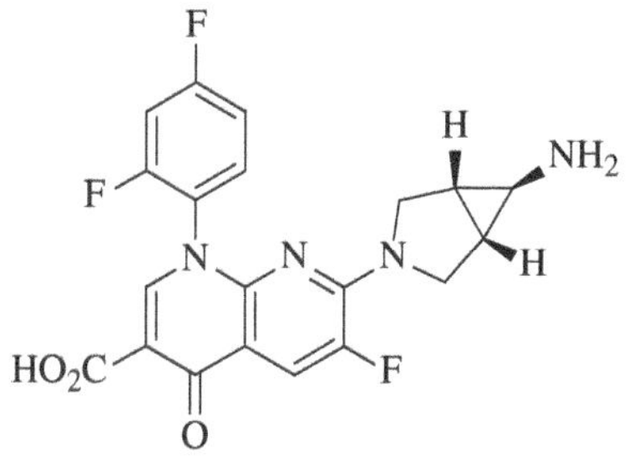

FIGURE 5.24   Structure of trovafloxacin.

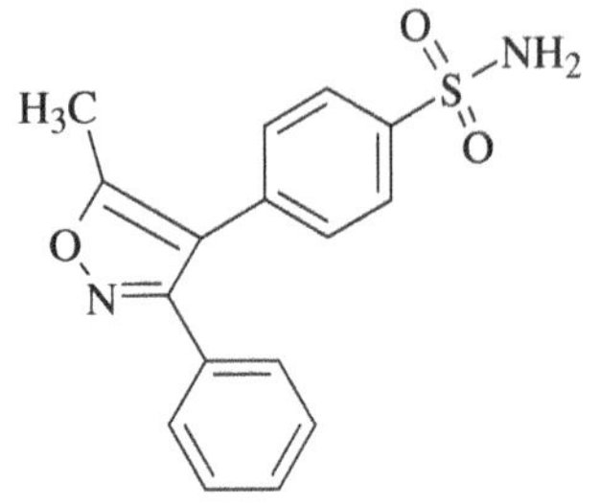

FIGURE 5.25   Structure of valdecoxib.

hepatocellular, cholestatic, or mixed with mixed cases the least severe. Acute and chronic liver failure was noted to occur. The hepatotoxicity had a rapid onset, and some patients showed immunoallergic features [73].

5.4.11   Valdecoxib

Valdecoxib was a selective COX-2 inhibitor used in the treatment of osteoarthritis, rheumatoid arthritis, and painful menstruation and menstrual symptoms. It was removed from the market due to a high incidence of serious skin toxicity, notably Stevens–Johnson syndrome [85], and concerns about possible increased risk of cardiovascular effects (see rofecoxib). The injectable water-soluble prodrug parecoxib (formed by an amide on the sulfonamide; Fig. 5.25) is still available in many countries although it was never approved in the United States.

Researchers from the FDA [86] compared valdecoxib, celecoxib, rofecoxib, and meloxicam and the incidence of Stevens–Johnson syndrome. Celecoxib was referred to as sulfonamide COX-2 inhibitors due to the presence of this function and a connection made to sulfonamide antibacterials a group of drugs widely associated with skin toxicity. Sulfonamide antibacterials cause Stevens–Johnson syndrome and toxic epidermal necrolysis, and the mechanism involved appears to be via reactive metabolites formed from the aniline nitrogen in particular the *N*-4-hydroxylamine and *N*-4-nitroso metabolites. The sulfonamide grouping in the antibacterials is not involved, so the association of the sulfonamide functionality as important in the cutaneous reactions of antibacterials and the COX-2 inhibitors is incorrect [87].

NSAIDs in general have an association with Stevens–Johnson syndrome and toxic epidermal necrolysis in particular the oxicam and butazone classes. Isoxicam was withdrawn from the market in France due to a high rate of dermal toxicity and has a methyl isoxazole group similar to valdecoxib. Metabolism studies with valdecoxib indicate hydroxylation of the methyl group and ring cleavage as routes of metabolism [88].

5.4.12   Zomepirac

Zomepirac was an acetic acid nonselective COX inhibitor (Fig. 5.26).

It was regarded as highly effective and gained a large proportion of the pain medication market. The drug however caused a high incidence of serious hypersensitivity/anaphylactic reactions in patients. Whether this more reflects the increased prescribing of the drug or a genuine chemical difference is uncertain. When examined in detail [89],

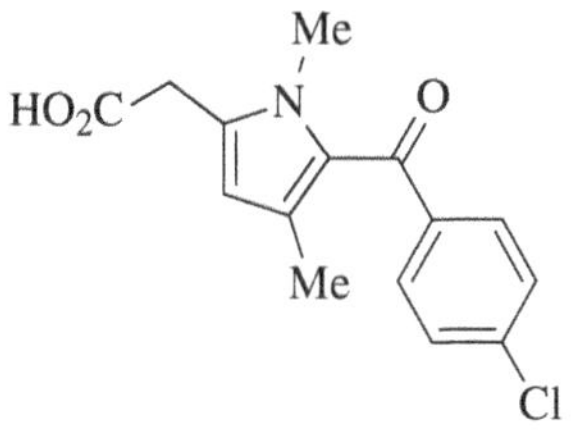

FIGURE 5.26   Structure of zomepirac.

TABLE 5.4   Incidence of Hypersensitivity Reactions for Various COX inhibitors

NSAID	Incidence per 10,000 patients	95% confidence interval
Fenoprofen	0	0–8.5
Ibuprofen	4.2	1.8–8.3
Indomethacin	5.1	1.1–15.0
Naproxen	0	0–22.8
Phenylbutazone	0	0–45.3
Sulindac	4.6	1.5–10.6
Tolmetin	0	0–18.1
Zomepirac	9.5	5.3–15.6

Note the large uncertainty in the incidence as indicated by the 95% confidence interval.

COX inhibitors were associated with an adjusted relative risk (95% confidence interval) of hypersensitivity reactions of 2.0 (1.3–2.9). The increased risk was accentuated in those with a diagnosis compatible with acute pain (3.6 [2.2–5.9]) and absent in those without such a diagnosis (1.1 [0.6–1.9]). Comparison of those exposed to zomepirac with those exposed to other COX inhibitors resulted in an age-adjusted relative risk of 2.0 (1.1–4.7). Within the drugs with data available, there is a large uncertainty of risk as indicated by the 95% confidence interval in Table 5.4. Stratification by the probable indication for COX inhibitor use again suggested that the risk may be explained by the use of the COX inhibitors for different indications.

As a class, COX inhibitors are the second most important cause of drug anaphylaxis after penicillins in hospitalized patients and the leading cause in ambulatory patients [90] indicating the problem of differentiation again.

A major route of metabolism for zomepirac is the formation of an acyl glucuronide. Zomepirac is an α-unsubstituted acetic acid, and as such, the glucuronide has a high instability and reactivity [91]. Zomepirac glucuronide readily forms positional isomers by acyl migration. In addition, it can react with lysine residues in proteins. The proposed pathway involves condensation between the aldehyde group of a rearranged acyl glucuronide and a lysine residue or an amine group (of the N terminus) leading to the formation of a glycated protein [91]. This reactivity was initially linked to the hypersensitivity reactions that lead to its withdrawal. As stated earlier, the hypersensitivity reactions occur for COX inhibitors. Many form acyl glucuronides, but compounds associated with hypersensitivity like acetaminophen, piroxicam, valdecoxib, and celecoxib do not.

Withdrawals due to Off-Target Pharmacology, Unknown Receptor: Impact on Drug Discovery and Development   In most of the cases, there is strong circumstantial evidence that the initial step in the toxicity is the formation of a reactive metabolite that is capable of covalently binding to macromolecules [92]. Certain groups or functions in molecules are particularly susceptible to metabolism to reactive intermediates. These groupings are termed "structural alerts" [93, 94]. Of particular note are aromatic amines (anilines) or precursors to this function that occur in bromfenac, temafloxacin, and nomifensine. The aniline or aniline precursor group and its activation are believed to be responsible for aplastic anemia or agranulocytosis with para-aminosalicylic acid, procainamide, chloramphenicol, aminoglutethimide, and dapsone.

Thiophenes are also easily converted to reactive species as evidenced by the mechanism-based inactivation of CYP2C9 by suprofen and ticrynafen. However, while implicated in the hepatotoxicity of ticrynafen, the withdrawal of suprofen is mainly due to reversible secondary pharmacology shared between the molecules (URAT1 inhibition).

Another feature of many of these toxicities is that the causal drugs are often administered at a high dose [95] usually of 100 mg or greater. The importance of dose size was emphasized in a meta-analysis [96] study using US and Swedish databases. Medications were divided by dose size into segments of 10 mg/day, 11–49 mg/day, and 50 mg/day and above groups. Among US prescription medicines, a statistically significant relationship was observed between the daily dose of oral medicines and reports of liver failure, liver transplantation, and death caused by drug-induced liver injury (DILI). Data from Sweden allowed closer examination and showed for all cases of DILI only 9% occurred in the 10 mg/day group rising to 14.2% in the 11–49 mg/day group. The vast majority (77%) of cases were caused by medications given at a dose of 50 mg/day or above.

Optimization of compounds in drug discovery can aim at low doses (optimal pharmacokinetics, high intrinsic potency) without obvious structural alerts that will attenuate the risk.

Many of the drugs in this classification, including those exhibiting hepatotoxicity, were withdrawn over a relatively short period postmarketing. This has lead to a more critical examination of early clinical data especially the clinical chemistry findings. The most interpretable signals are alanine transaminase (ALT) and bilirubin. Increases above background of 3X of these liver function markers are now seen as indicative of potential frank hepatotoxicity. A pragmatic rule termed Hy's law or modified Hy's law is applied when ALT >3X and bilirubin >2 or 3X ULN. Although the full predictability of these rules is not fully established, a general assumption is that incidences of patients with these combined liver function elevations will eventually lead to ALF in some patients when larger patient numbers are treated [97, 98]. Recently, decisions to not approve the thrombin inhibitor ximelagatran were based on a number of unequivocal cases of elevations in ALT and bilirubin during phase III trials [97, 98].

5.5 CONCLUSIONS

In every aspect of drug attrition, key aspects of the reasons have been taken and provided new tools, screens, and techniques for the discovery and development of new drugs. Most apparent are the attention paid to compound selectivity to avoid complications due to secondary pharmacology. Moreover, the PK characteristics of different patient groups are now closely studied. Picking the right mechanism is always a problem, but better PK/PD and dose selection should at least mean the optimum dose in terms of safety and efficacy is marketed. Dose size has now been recognized as an important factor in all toxicity, including idiosyncratic, and discovery programs recognize this in selecting the most potent compounds with appropriate PK to achieve this. A major factor in the future is the selection of patients. Genotyping of patients for particular targets will maximize efficacy and hence benefit–risk. If only appropriate patients had been treated at the correct dose with appropriate patient information, how many of the drugs described previously would have remained as marketed entities? Two examples in this chapter illustrate this: alosetron (patient information) and flecainide (patient/indication selection) where the drugs continue to be used. One thing that is fairly certain is that all the learnings mean the "market-led" approach that dictated the "one daily dose size fits all" strategy for benoxaprofen is consigned to history.

REFERENCES

1 Camilleri, M. (2000). Pharmacology and clinical experience with alosetron. *Expert Opinion of Investigational Drugs*, *9*, 147–159.

2 Beck, I.T. (2001). Possible mechanisms for ischemic colitis during alosetron therapy. *Gastroenterology*, *121*, 231–232.

3 Coremans, G. (2005). Cilansetron: A novel, high-affinity 5-HT3 receptor antagonist for irritable bowel syndrome with diarrhea predominance. *Therapy*, *2*, 559–567.

4 Jamal, S.M. (2004). Rhabdomyolysis associated with hydroxymethylglutaryl-coenzyme A reductase inhibitors. *American Heart Journal*, *147*, 956–965.

5 Chapman, M.J., Carrie A. (2005). Mechanisms of statin-induced myopathy. A role for the ubiquitin–proteasome pathway? *Arteriosclerosis, Thrombosis, and Vascular Biology*, *25*, 2441–2444.

6 Bischoff, H., Heller, A.H. (1998). Preclinical and clinical pharmacology of cerivastatin. *American Journal of Cardiology*, *82*, 18J–25J.

7 Paine, S.W., Parker, A.J.,Gardiner, P., Webborn, P.J.H.,Riley, R.J. (2008). Prediction of the pharmacokinetics of atorvastatin, cerivastatin, and indomethacin using kinetic models applied to isolated rat hepatocytes. *Drug Metabolism and Disposition*, *36*, 1365–1374.

8 Moe G.W., Rouleau J.L., Charbonneau L., Proulx G., Arnold M.O., Hall C., de Chaplan J., Barr A., Sirois P., Packer M. (2001). Neurohormonal activation in severe heart failure: Relations to patient death and the effect of treatment with flosequinan. *American Heart Journal*, *139*, 587–595.

9 Gristwood R.W., Beleta J., Bou J., Cardelus I., Fernandez A.G., Llenas J., Berga P. (1992). Studies on the cardiac actions of flosequinan in vitro. *British Journal of Pharmacology*, *105*, 985–991.

10 Varma, A., Shah, K.B., Hess, M.L. (2012). Phosphodiesterase inhibitors, congestive heart failure, and sudden death: Time for re-evaluation. *Congestive Heart Failure*, *18*, 229–233.

11 Nicholls D.P., Droogan, A., Carson, C.A., Taylor, I.C., Passmore, A.P., Johnston, G.D., Kendall, M., Dutka, D., Morris, G.K., Underwood, L.M., Hind, I.D. (1996). Pharmacokinetics of flosequinan in patients with heart failure. *Journal of Clinical Pharmacokinetics*, *50*, 289–291.

12 Falotico, R., Haertlein, B.J., Lakas-Weiss, C.S., Salata, J.J., Tobia, A.J. (1989). Positive inotropic and haemodynamic properties of flosequinan, a new vasodilator, and a sulfone metabolite. *Journal of Cardiovascular Pharmacology*, *14*, 412–418.

13 Lazzara R. (1996). From first class to third class: recent upheaval in antiarrhythmic therapy—lessons from clinical trials. *American Journal of Cardiology*, *78*, 28–33.

14 Barbey, J.T., Thompson, K.A., Echt, D.S., Woosley, R.L., Roden, D.M. (1988). Antiarrhythmic activity, electrocardiographic effects and pharmacokinetics of the encainide metabolites O-desmethyl encainide and 3-methoxy-O-desmethyl encainide in man. *Circulation*, *77*, 380–391.

15 Aliot, E., Capucci, A., Crijns, H.J., Goette, A., Tamargo, J. (2011). Twenty-five years in the making: Flecainide is safe and effective for the management of atrial fibrillation. *Europace*, *13*, 161–173.

16 Casturi, S.R., Hegde, P., Ramanujam, R. (2005). Development of COX-2 selective inhibitors-therapeutic perspectives. *Current Medical Chem-Immunology, Endocrinology, Metabolic Agents*, *5*, 241–248.

17 Malmstrong, K., Daniels, S., Kotey, P., Seidenberg, B.C. Desjardins, P.J. (1999). Comparison of rofecoxib and celecoxib, two cyclooxygenase-2 inhibitors, in postoperative dental pain: a randomized, placebo- and active-comparator-controlled clinical trial. *Clinical Therapeutics*, *21*, 1653–1663.

18 Meek, I.L., van de Laar, M., Vonkeman, H.E. (2010). Non-steroidal anti-inflammatory drugs: An overview of cardiovascular risks. *Pharmaceuticals*, *3*, 2146–2162.

19 Solomon, S.D., Pfeffer, M.A., McMurray, J.J., Fowler V., Finn, P., Levin, B., Eagle, C., Hawk, E., Lechuga, M., Nadir Arber, M.N., Wittes, J. (2006). Effect of celecoxib on cardiovascular events and blood pressure in two trials for the prevention of colorectal adenomas. *Circulation*, *114*, 1028–1035.

20 Camilleri, M. (2012). Pharmacology of the new treatments for lower gastrointestinal motility disorders and irritable bowel syndrome. *Clinical Pharmacology & Therapeutics*, *91*, 44–59.

21 Beattie, D.T., Smith, J.A., Marquess, D., Vickery, R.G., Armstrong, S.R., Pulido-Rios, T., McCullough, J.L., Sandlund, C., Richardson, C., Mai, N., Humphrey, P.P.A. (2004). The 5-HT4 receptor agonist, tegaserod, is a potent 5-HT2B receptor antagonist in vitro and in vivo. *British Journal of Pharmacology*, *143*, 549–560.

22 Loughlin, J., Quinn, S., Rivero, E., Wong, J., Huang, J., Kralstein, J., Earnest D.L., Seeger J.D. (2010). Tegaserod and the risk of cardiovascular ischemic events: an observational cohort study. *Journal of Cardiovascular and Pharmacological Therapeutics*, *15*, 151–157.

23 Drici, M.D., Ebert, S.N., Wang, W.X., Rodriguez, I., Liu, X.K., Whitfield, B.H., Woosley, R.L. (1999). Comparison of tegaserod (HTF 919) and its main human metabolite with cisapride and erythromycin on cardiac repolarization in the isolated rabbit heart. *Journal of Cardiovascular Pharmacology*, *34*, 82–88.

24 Morganroth, J., Ruegg, P.C., Dunger-Baldauf, C. (2002). Tegaserod, a 5-hydroxytryptamine type 4 receptor partial agonist, is devoid of electrocardiographic effects. *American Journal of Gastroenterology*, *97*, 2321–2327.

25 Serebruany, V.L., El Mouelhi, M., Pfannkuche, H.J., Rose, K., Marro, M., Angiolillo, D.J. (2010). Investigations on 5-HT4 receptor expression and effects of tegaserod on human platelet aggregation in vitro. *American Journal of Therapeutics*, *17*, 543–552.

26 Higgins, D.L., Ero, M.P., Loeb, M., Kersey, K., Hopkin, A., Beattie, D.T. (2012). The inability of tegaserod to affect platelet aggregation and coronary artery tone at supratherapeutic concentrations. *Naunyn Schmiedebergs Archives of Pharmacology*, *385*, 103–109.

27 Fitzgerald, L.W., Burn, T.C., Brown, B.S., Patterson, J.P., Corjay, M.H., Valentine, P.A., Sun, J.H., Link, J.M., Largent, B.L. (2000). Possible role of valvular serotonin 5-HT$_{2B}$ receptors in the cardiopathy associated with fenfluramine. *Molecular Pharmacology*, *57*, 75–78.

28 Li, R., Serdula, M.K., Williamson D.F., Bowman, B.A., Graham, D.A., Green, L. (1999). Dose effect of fenfluramine use on severity of valvular heart disease among fen-phen patients with valvulopathy. *International Journal of Obesity*, *23*, 926–928.

29 Rothman, R.B., Baumann, M.H., Savage, J.E., Rauser, L., McBride, A., Hufeisen, S.J., Roth, B.K. (2000). Evidence for possible involvement of 5-HT2B receptors in cardiac valvulopathy associated with fenfluramine and other serotonergic medications. *Circulation*, *102*, 2836–2841.

30 Jooste, E., Zhang, Y., Emala, C.W. (2005). Rapacuronium preferentially antagonizes the function of M2 versus M3 muscarinic receptors in guinea pig airway smooth muscle. *Anesthesiology*, *102*, 117–124.

31 Jooste, E., Klafter, F., Hirshman, C.A., Emala, C.W. (2003). A mechanism for rapacuronium-induced bronchospasm. *Anesthesiology*, *98*, 906–911.

32 Strom, B.L., West, S.L., Sim, E., Carson, J.L. (1989). The epidemiology of the acute flank pain syndrome from suprofen. *Clinical Pharmacology and Therapeutics*, *46*, 693–699.

33 O'Donnell, J.P., Dalvie, D.K., Kalgutkar, A.S., Obach, R.S. Mechanism-based inactivation of human recombinant P450 2C9 by the nonsteroidal anti-inflammatory drug suprofen, *Drug Metabolism and Disposition*, *31*, 1369–1377.

34 .Lafrance, J.-P., Miller, D.R. (2009). Selective and non-selective non-steroidal anti-inflammatory drugs and the risk of acute kidney injury. *Pharmacoepidemiology and Drug Safety*, *18*, 923–931.

35 Abraham, P.A., Halstenson, C.E., Opsahl, J.A., Matzke, G.R., Keane, W.F. (1988). Suprofen-induced uricosuria. A potential mechanism for acute nephropathy and flank pain. *American Journal of Nephrology*, *8*, 90–95.

36 Redfern, W.S., Carlsson, L., Davis, A.S., Lunch, W.G., McKenzie, I., Palethorpe, S., Siegl, P.K.S., Strang, I., Sullivan, A.T., Wallis, R., Camm, A.J., Hammond, T.G. (2003). Relationship between preclinical electrophysiology, clinical QT interval prolongation and torsade de pointes for a broad range of drugs: evidence for a provisional safety margin in drug development. *Cardiovascular Research*, *58*, 32–45.

37 Milnes J.T., Witchel H.J., Leaney J.L., Leishman D.J., Hancox J.C. (2006). hERG K+ channel blockade by the antipsychotic drug thioridazine: an obligatory role for the S6 helix residue F656. *Biochemical and Biophysical Research Communications*, *351*, 273–280.

38 Somogyi, A.A., Menelaou, A., Fullston, S.V. (2004). CYP3A4 mediates dextropropoxyphene N-demethylation to nordextropropoxyphene: human in vitro and in vivo studies and lack of CYP2D6 involvement. *Xenobiotica*, *34*, 875–887.

39 Ulens, C., Daenens, P., Tytgat, J. (1999). Norpropoxyphene-induced cardiotoxicity is associated with changes in ion-selectivity and gating of HERG currents. *Cardiovascular Research*, *44*, 568–578.

40 Afshari, R., Maxwell, S., Dawson, A., Bateman, D.N. (2005). ECG abnormalities in co-proxamol (paracetamol/dextropropoxyphene) poisoning. *Clinical Toxicology*, *43*, 255–259.

41 Koerner, J. DCRP consult to evaluate nonclinical cardiac safety data for propoxyphene. www.fda.gov/downloads/Drugs/DrugSafety/.../UCM234313.pdf

42 Sharma, A.M. (2005). Does pharmacologically induced weight loss improve cardiovascular outcome? Sibutramine pharmacology and the cardiovascular system. *European Heart Journal*, *7* (Supplement L), L39.

43 Harrison-Woolrych, M., Clark, D.W.J., Hill, G.R., Rees M.J., Skinner J.R. (2006). QT interval prolongation associated with sibutramine treatment. *British Journal of Clinical Pharmacology*, *61*, 464–469.

44 Torp-Pedersen, C., Caterson, I., Coutinho, W., Finer, N., Van Gaal, L., Maggioni, A., Sharma, A., Brisco, W. (2007). Cardiovascular responses to weight management and sibutramine in high-risk subjects: an analysis from the SCOUT trial. *European Heart Journal, 28*, 2915–2923.

45 Yun, C.H., Okerholm, R.A., Guengerich, F.P. (1993). Oxidation of the antihistaminic drug terfenadine in human liver microsomes. Role of cytochrome P-450 3A(4) in N-dealkylation and C-hydroxylation. *Drug Metabolism and Disposition, 21*, 403–409.

46 Honig, P.K., Wortham, D.C., Zamani, K., Conner, D.P., Mullins, J.C., Cantilena, L.R. (1993). Terfenadine-ketoconazole interaction: pharmacokinetics and electrocardiographic consequences. *Journal of the American Medical Association, 269*, 1513–1518.

47 Welker, H.A., Wiltshire, H., Bullingham, R. (1988). Clinical pharmacokinetics of mibefradil. *Clinical Pharmacokinetics, 35*, 405–423.

48 Prueksaritanont, T., Ma, B., Tang, C., Meng, Y., Assang, C., Lu, P., Reider, P.J., Lin, J.H., Baillie, T.A. (1999). Metabolic interactions between mibefradil and HMG-CoA reductase inhibitors: an in vitro investigation with human liver preparations, *British Journal of Clinical Pharmacology, 47*, 291–298.

49 Zolotoy, A.B., Plouvier, B.P., Beatch, G.B., Hayes, E.S., Wall, R.A., Walker, M.J.A. (2003). Physicochemical determinants for drug induced blockade of HERG potassium channels: effect of charge and charge shielding, *Current Medicinal Chemistry, 1*, 225–241.

50 Wible, B.A., Hawryluk, P., Ficker, E., Kurysher, Y.A., Kirsch, G., Brown, A.M. (2005). HERG-Lite: a novel comprehensive high-throughput screen for drug-induced hERG risk. *Journal of Pharmacological and Toxicological Methods, 52*, 136–145.

51 Lim, H.-K., Duczak, N., Brougham, L., Elliot, M., Patel, K., Chan, K. (2005). Automated screening with confirmation of mechanism-based inactivation of CYP3A4, CYP2C9, CYP2D6 and CYP1A2 in pooled human liver microsomes. *Drug Metabolism and Disposition, 33*, 1211–1219.

52 Smith D.A., Obach R.S. (2005). Seeing through the MIST: abundance versus percentage. Commentary on metabolites in safety testing. *Drug Metabolism and Disposition, 33*, 1409–1417.

53 Halsey, J.P., Cardoe, N. (1982). Benoxaprofen: side-effect profile in 300 patients. *British Medical Journal (Clinical Research Edition), 284*, 1365–1368.

54 Nash, J.F., Carmichael, R.H., Ridolfo, A.S., Spradlin, C.T. (1980). Pharmacokinetic studies of benoxaprofen after therapeutic doses with a review of related pharmacokinetic and metabolic studies. *Journal of Rheumatology, 6* (Supplement), 12–19.

55 Hamdy, R.C., Murnane, B., Perera, N., Woodcock, K., Koch, I.M. (1982). The pharmacokinetics of benoxaprofen in elderly subjects. *European Journal of Rheumatology and Inflammation, 5*, 69–75.

56 Goldkind, L., Laine, L. (2006). A systematic review of NSAIDs withdrawn from the market due to hepatotoxicity: lessons learned from the bromfenac experience. *Pharmacoepidemiology and Drug Safety, 15*, 213–220.

57 Marotta, P.J., Roberts, E.A. (1998). Pemoline hepatotoxicity in children. *Journal of Pediatrics, 132*, 894–897.

58 Shevell, M., Schreiber, R. (1997). Pemoline-associated hepatic failure: A critical analysis of the literature. *Pediatric Neurology, 16*, 14–16.

59 Hochman, J.A., Woodard, S.A., Mitchell, B. (1998). Exacerbation of autoimmune hepatitis: another hepatotoxic effect of pemoline therapy. *Pediatrics, 101*, 106–107.

60 Zimmerman, H.J., Lewis, J.H., Ishak, K.G., Maddrey, W.C. (1984). Ticrynafen-associated hepatic injury: analysis of 340 cases. *Hepatology, 4*, 315–323.

61 Lecoeur, S., Bonierbale, E., Challine, D., Gautier, J.C., Valadon, P., Dansette, P. M., Catinot, R., Ballet, F., Mansuy, D., Beaune, P.H. (1994). Specificity of in vitro covalent binding of tienilic acid metabolites to human liver microsomes in relationship to the type of hepatotoxicity: comparison with two directly hepatotoxic drugs. *Chemical Research in Toxicology, 7*, 434–442.

62 Dansette, P.M., Amar, C., Smith, C., Pons, C., Mansuy, D. (1990). Oxidative activation of the thiophene ring by hepatic enzymes. Hydroxylation and formation of electrophilic metabolites during metabolism of tienilic acid and its isomer by rat liver microsomes. *Biochemical Pharmacology, 39*, 911–918.

63 Robin, A.-M., Le Breton, F.-P., Bonierbale, E., Dansette, P., Ballet, F., Mansuy, D., Pessayre, D. (1996). Antigenic targets in tienilic acid hepatitis both cytochrome P450 2C11 and 2C11-tienilic acid adducts are transported to the plasma membrane of rat hepatocytes and recognized by human sera. *Journal of Clinical Investigation, 98*, 1471–1480.

64 He, K., Talaat, R.E., Pool, W.F., Reily, M.D., Reed, J.E., Bridges, A.J., Woolf, T.F. (2004). Metabolic activation of troglitazone: identification of a reactive metabolite and mechanisms involved. *Drug Metabolism and Disposition, 32*, 639–646.

65 Snow, K.L., Moseley, R.H. (2007). Effect of thiazolidinediones on bile acid transport in rat liver. *Life Sciences, 80*, 732–740.

66 Borges, N. (2003). Tolcapone-related liver dysfunction. *Drug Safety, 26*, 743–747.

67 Watkins, P. (2000). COMT inhibitors and liver toxicity. *Neurology, 55*, S51–S52.

68 Jorga, K., Fotteler, B., Heizmann, P., Gasser, R. (1999). Metabolism and excretion of tolcapone, a novel inhibitor of catechol-O-methyltransferase. *British Journal of Clinical Pharmacology, 48*, 513–520.

69 Smith, K.S., Smith, P.L., Heady, T.N., Trugman, J.M., Harman, W.D., Macdonald, T.L. (2003). In vitro metabolism of tolcapone to reactive intermediates: relevance to tolcapone liver toxicity. *Chemical Research in Toxicology, 16*, 123–128.

70 Martignoni, E., Cosentino, M., Ferrari, M., Porta, G., Mattarucchi, E., Marino, F., Lecchini, S., Nappi, G. (2005). Two patients with COMT inhibitor-induced hepatic dysfunction and UGT1A9 genetic polymorphism. *Neurology, 65*, 1820–1822.

71 Sun, Q., Zhu, R., Foss, F.W. Jr, Macdonald, T.L. (2007). Mechanisms of trovafloxacin hepatotoxicity: studies of a model cyclopropylamine-containing system. *Bioorganic Medicinal Chemistry Letters, 17*, 6682–668.

72 Moellering, R.C. (2000). Hepatotoxicity of antimicrobials: the trovafloxacin story. Program and abstracts of the 40th Interscience Conference on Antimicrobial Agents and Chemotherapy; Toronto, Ontario, Canada; September 17–20, Abstract 1880.

73 Orman, E.S., Conjeevaram, H.S., Vuppalanchi, R., Freston, J.W., Rochon, J., Kleiner, D.E., Hayashi, P.H. (2011). Clinical and histopathologic features of fluoroquinolone-induced liver injury. *Clinical Gastroenterology and Hepatology, 9*, 517–523.

74 Brogden, R.N., Heel, R.C., Speight, T.M., Avery, G.S. (1979). Nomifensine: a review of its pharmacological properties and therapeutic efficacy in depressive illness. *Drugs, 18*, 1–24.

75 Salama, A., Mueller-Eckhardt, C. (1986). Two types of nomifensine-induced immune haemolytic anaemias: drug-dependent sensitization and/or autoimmunization. *British Journal of Haematology, 64*, 613–620.

76 Uetrecht, J., Zahid, N., Rubin, R. (1988). Metabolism of procainamide to a hydroxylamine by human neutrophils and mononuclear leukocytes. *Chemical Research in Toxicology, 1*, 74–78.

77 Utrecht, J. (1989). Mechanism of hypersensitivity reactions: proposed involvement of reactive metabolites generated by activated leukocytes. *Trends in Pharmaceutical Sciences, 10*, 463–467.

78   Yu, J., Brown, D.G., Burdette, D. (2010). In vitro metabolism studies of nomifensine mono-oxygenation pathways: metabolite identification, reaction phenotyping, and bioactivation mechanism. *Drug Metabolism and Disposition*, *38*, 1767–1778.

79   Obach, R.S., Dalvie, D.K. (2006). Metabolism of nomifensine to a dihydroisoquinolinium ion metabolite by human myeloperoxidase, hemoglobin, monoamine oxidase A, and cytochrome P450 enzymes. *Drug Metabolism and Disposition*, *34*, 1310–1316.

80   Köhler, C., Hall H., Magnusson, O., Lewander, T., Gustafsson, K. (1990). Biochemical pharmacology of the atypical neuroleptic remoxipride. *Acta Psychiatrica Scandinavica Supplement*, *358*(S), 27–36.

81   Ogren, O., Lundstrom, J., Nilsson, C.B., Widman, M. (1993). Dopamine blocking activity and plasma concentrations of remoxipride and its main metabolites in the rat. *Journal of Neural Transmission*, *93*, 187–203.

82   Erve, J.C.L., Svensson M.A., von Euler-Chelpin, H., Klasson-Wehler, E. (2004). Characterization of glutathione conjugates of the remoxipride hydroquinone metabolite NCQ-344 formed in vitro and detection following oxidation by human neutrophils. *Chemical Research in Toxicology*, *17*, 564–571.

83   Blum, M.D., Graham, D.J., McCloskey, C.A. (1994). Temafloxacin syndrome: review of 95 cases. *Clinical Infectious Diseases*, *18*, 946–950.

84   Granneman, G.R., Carpentier, P., Morrison, P.J., Pernet, A.G. (1991). Pharmacokinetics of temafloxacin in humans after single oral doses. *Antimicrobial Agents and Chemotherapy*, *35*, 436–441.

85   Mockenhaupt, M. (2011). The current understanding of Stevens–Johnson syndrome and toxic epidermal necrolysis. *Expert Review of Clinical Immunology*, *7*, 803–815.

86   La Grenade, L., Lee, L., Weaver, J., Bonnel, R., Karwoski, C., Governale, L., Brinker, A. (2005). Comparison of reporting of Stevens-Johnson syndrome and toxic epidermal necrolysis in association with selective COX-2 inhibitors. *Drug Safety*, *28*, 917–924.

87   Smith, D.A., Jones, R.M. (2008). The sulfonamide group as a structural alert: a distorted story? *Current Opinion in Drug Discovery & Development*, *11*, 72–79.

88   Zhang, J.Y., Yuan, J.J., Wang, Y.F., Bible, R.H. Jr., Breau, A.P. (2003). Pharmacokinetics and metabolism of a COX-2 inhibitor, valdecoxib, in mice. *Drug Metabolism and Disposition*, *31*, 491–501.

89   Strom, B., Carson, J.L., Morse, M.L., West, S.L., Soper, K.A. (1987). The effect of indication on hypersensitivity reactions associated with zomepirac sodium and other nonsteroidal antiinflammatory drugs. *Arthritis and Rheumatism*, *30*, 1142–1148.

90   Sánchez-Borges, M., Capriles-Hulett, A., Caballero-Fonseca, F. (2004). The multiple faces of nonsteroidal antiinflammatory drug hypersensitivity. *Journal of Investigative Allergy and Immunology*, *14*, 329–334.

91   Smith, P.C., McDonagh A.F., Benet, L.Z. (1986). Irreversible binding of zomepirac to plasma protein in vitro and in vivo. *Journal of Clinical Investigation*, *77*, 934–939.

92   Li, A.P. (2002). A review of the common properties of drugs with idiosyncratic hepatotoxicity and the "multiple determinant hypothesis" for the manifestation of idiosyncratic drug toxicity. *Chemical-Biological Interactions*, *142*, 7–23.

93   Kalgutkar, A.S., Gardner, I., Obach, R.S., Shaffer, C.K., Callegari, E., Henne, K.R., Mutlib, A.E., Dalvie, D.K., Lee, J.S., Nakai, Y., O'Donnell, J.P., Boer, J., Harriman, S.P. (2005). A comprehensive listing of bioactivation pathways of organic functional groups. *Current Drug Metabolism*, *6*, 161–225.

94   Nelson, S.D. (2001). Structure toxicity relationships-how useful are they in predicting toxicities of new drugs. *Advances in Experimental Medical Biology*, *500* (Biological Reactive Intermediates VI), 33–43.

95 Utrecht, J. (2003). Screening for the potential of a drug candidate to cause idiosyncratic drug reactions. *Drug Discovery Today*, *8*, 832–837.

96 Lammert, C., Einarsson, S., Saha, C., Niklasson, A., Bjornsson, E., Chalasani, N. (2008). Relationship between daily dose of oral medications and idiosyncratic drug liver injury: search for signals. *Hepatology*, *47*, 2003–2009.

97 Lewis, J.H. (2006). "Hy's Law," the "Rezulin Rule," and other predictors of severe drug-induced hepatotoxicity: putting risk-benefit into perspective. *Pharmacoepidemiology and Drug Safety*, *15*, 221–229.

98 Kaplowitz, N., (2006). Rules and laws of drug hepatotoxicity. *Pharmacoepidemiology and Drug Safety*, *15*, 231–233.

6

INFLUENCE OF THE REGULATORY ENVIRONMENT ON ATTRITION

ROBERT T. CLAY

Highbury Regulatory Science Limited, London, UK

6.1 INTRODUCTION

Understanding the impact of regulation on the productivity of pharmaceutical research requires consideration of two dilemmas that lay at the heart of the problem. The first is described by Eichler et al. [1]: "The regulator's dilemma reflects the various needs and interests of stakeholders, all of which appear legitimate when considered in isolation, but which are difficult to reconcile with each other." The second dilemma is equally, or perhaps more, complex: the development of regulatory guidance is based on layers of experience/knowledge accumulated over at least 50 years of modern (postthalidomide) drug regulation. This second dilemma could be described as the increased regulatory burden, as each new experience, for example, QTc prolongation from mid-1990s, leads to both increased knowledge about risks associated with drugs and consequent regulatory restriction [2]. This accumulation of knowledge and the increased understanding of biological mechanisms, counterintuitively, leads to higher attrition and reducing research productivity and the developer's dilemma.

Attribution of the causes of attrition is challenging, and in the case of regulatory considerations, it may be less clear given the interplay of scientific, commercial, and social factors that different stakeholders may assign to "regulatory burden." The major focus of this discussion will be on the periapproval setting where the influence of regulatory interactions is clearer, as both developers and regulators are subject to greater public disclosure and scrutiny. This stage starts with a decision that the developer has sufficient knowledge to consider a favorable benefit–risk for a commercially viable potential product and therefore plans a marketing authorisation application (MAA) or new drug

Attrition in the Pharmaceutical Industry: Reasons, Implications, and Pathways Forward,
First Edition. Edited by Alexander Alex, C. John Harris and Dennis A. Smith.
© 2016 John Wiley & Sons, Inc. Published 2016 by John Wiley & Sons, Inc.

application (NDA). This application is required to meet the formal requirements of the relevant agency and represent a complete and valid submission. The resulting review by the regulatory authority may lead to approval of the claim made by the manufacturer; modification or imposition of restrictions on the product; or failure to gain approval. In reviewing the experience of the many submissions over several decades, it should be possible to evaluate the "regulatory causes of attrition." These factors come into play throughout the research and development (R&D) cycle, although their impact may become somewhat less clear at earlier stages since they are compounded or influenced by many layers of competitive and scientific choices made during any development program.

6.1.1 How the Regulatory Environment has Changed Over the Last Two Decades

In the early 1990s, the key criticism of the regulatory systems around the world was the time taken to make new medicines available resulting in a "drug lag." This was reflected in the development of "new" European Union approaches with the establishment of the European Medicines Evaluation Agency and the introduction of both centralized and mutual recognition giving the opportunity for the introduction of new medicines on a pan-European basis in the timescale comparable to some of the most rapid regulatory authorities, for example, the United Kingdom and Canada. Across Europe, it reduced the diversity of outcomes, although the mutual recognition still allowed either national regulators or companies to exclude individual national approvals. In the early years after introduction, the centralized approval procedure was dominated by those conditions for which the product was mandatory, for example, treatment of HIV, or products considered by their applicants to be less subject to diverse views of individual regulators. The centralized procedure has evolved to become the primary route for the approval of new molecular entities, although a decentralized procedure remains an option. The decentralized procedure allows the applicant to select a smaller number of countries, but the review occurs in parallel rather than following an initial approval in one country in the mutual recognition procedure.

In response to the debate on long approval times, the United States introduced user fees in the Prescription Drug User Fee Act of 1992 (PDUFA), thus allowing increased investment by the Food and Drug Administration (FDA) in review resources and facilitating more rapid decision making. It is notable that during this period, the introduction of fees and review targets was widespread across many national agencies around the world; the United Kingdom was one of the earliest movers where the introduction of license fees predates the United States. In the United States and Europe, these user fees contribute substantially to the costs of the system, which is consistent with the principle that the applicants pay the cost of the regulation of medicines, although it does expose regulators to potential criticism resulting from potential conflicts of interest. PDUFA V enacted in July 2012 represents the latest in a series of reauthorizations as the Food and Drug Administration Safety and Innovation Act (FDASIA) [3].

From its creation in 1990, the International Conference on Harmonisation of Technical Requirements for Registration of Pharmaceuticals for Human Use (ICH) has been the main vehicle for cooperation between regulators in the United States, EU, and Japan and through the ICH Global Cooperation Group regulators in other regions [4]. The pharmaceutical industry contributes to this process through the trade associations in the three regions. The ICH collaboration has essentially laid the foundations for a modern global regulatory system by creating common standards for submissions through development of Medical Dictionary for Regulatory Activities (MedDRA) and the Common Technical Dossier (CTD). During the 1990s, ICH activities led to the harmonization of key guidelines for safety, quality, and efficacy eliminating much of the duplication and diversity of requirements for supporting data across the concerned regions. In the subsequent decade, the focus has been on developing and updating the guidelines and improving the cooperation with other regions [5]. The model developed over the past 20 years has also provided a system for cooperation between regulators and industry through sharing expertise and data allowing the development of regulatory science. This is reflected in the many "precompetitive collaborations" supported by the Critical Path Institute (C-Path) and Innovative Medicines Initiative (IMI) exemplified by the Coalition Against Major Diseases (CAMD), which brings together academic institutions, not-for-profit organizations, government agencies (including the FDA and European Medicines Agency (EMA)), and pharmaceutical companies. CAMD focus on neurological diseases has under the auspices of the C-Path has brought together placebo-arm data from multiple sponsors to enable the creation of a database and the development of a trial simulation model [6].

Historically, many regulators have provided scientific advice during development of new products, but this advice is specific to particular products and confidential to the developer and the agency. Publicly accessible advice is usually only through guidelines, approval decisions, or in the case of FDA advisory committees held in public prior to approval. Guidelines are typically developed or updated after the approval of the first product in a new class; this creates an ongoing challenge for developers to understand evolving regulatory requirements. In response to this challenge, regulatory agencies may publish draft guidelines after they have provided scientific advice on several related topics or where it is clear there is a need for guidance for multiple sponsors requesting meetings. Manolis et al. describe a new pathway for qualification of approaches that may be applicable beyond an individual product or class [7]. This procedure uses a variant of the scientific advice procedure that results in either publication of a qualification opinion or confidential advice to the applicant to facilitate further method development. The procedure "was established as a response to the drug development bottlenecks and inefficiencies, but also to the availability of new methodologies...." This process will enable the precompetitive collaborations discussed earlier to reach agreement with regulators and minimize the resource wasted by individual developers, by not having access to information on other failures, and should enable an overall reduction in attrition rates. The FDA guidance on the review of drug development tools was published in January 2014 [8]. This guidance has a similar intent to that issued by the EMA, but the scope is somewhat narrower being limited to qualification of biomarkers, clinical outcomes, and animal models.

Substantial progress has been made through greater transparency of data and collaboration among drug developers and with regulators. However, today, the most prominent criticism of regulators arises from a perception that drug withdrawals and the occurrence of postmarketing safety events arise from poor decisions made during review. Also, the value of new medicines that do reach approval is frequently questioned by some stakeholders. A review of medicines introduced into France in 2010 illustrates these challenges, suggesting that only 4 of the 97 new drugs or indications provided a therapeutic benefit and that 19 products were introduced despite an unfavorable benefit–risk [9]. They also claim that "drug regulatory agencies can protect patients ... by refusing to grant market approval or demanding their market withdrawal." In contrast to this viewpoint, a review of the drug approval success rate in Europe for 2009 by Eichler et al. [10] considers the regulatory outcomes for a cohort of 48 new drugs in 2009. Only 29 (60%) of these new active substances received a positive CHMP opinion; the remainder received a negative opinion or were withdrawn by the applicant. These two papers may not have a great deal of overlap, although it is likely that many of the drugs receiving positive opinions in 2009 were introduced into France over the following 2 years. However, it is clear is that they represent the somewhat contradictory positions described earlier as the regulator's dilemma.

6.1.2 Past and Current Regulatory Attitude to Risk Analysis and Risk Management

Over the past 50 years, there has been a significant evolution of regulatory attitude to risk. Following the thalidomide tragedy, the focus was on the application of past experience to prevent the introduction of unsafe drugs. This is reflected in the introduction of legislation in the United States, Europe, and other countries that emphasizes the safety (in addition to quality and efficacy) of new medicines. During the next several decades, regulators such as the FDA focused on the introduction of guidances that would address these past experiences. The requirement for reproductive toxicology testing, to prevent a repeat of the thalidomide tragedy, is the first of many examples. On each occasion that the drug approval process was considered to have failed to identify an issue, additional requirements were introduced. The challenge for this approach is that it does not anticipate future problems—for example, when benoxaprofen was introduced in the 1980s, it had not been widely tested on older patients. Within months of the introduction of a potentially important new agent in the management of arthritis, there were reports of an unexpected severe adverse skin reaction in patients older than 70. The rate of unacceptable side effects in patients over 70 years was reported as 83.3%, the rate of treatment withdrawal was approximately doubled in the older cohort [11]. A major aspect of this toxicity profile was associated with phototoxicity, although Qureshi et al. [12] attribute the US withdrawal to hepatic, renal, and gastric findings. Use in this older population might have been considered off-label; nevertheless, the absence of a requirement to examine the safety of such an agent in an older population was surprising.

Eichler et al. [13] highlight a change in the public expectations regarding safety of new drugs as a result of several high-profile drug withdrawals or emergent safety findings. These withdrawals not only impact the reputation of pharmaceutical companies but also the public expectations of drug regulatory authorities. They highlight the introduction of stronger measures to manage benefit–risk assessment with the development of ICH guidelines and the implementation of EU legislation requiring the introduction of risk management plans for new medicines. The introduction of risk management plans is a substantive change in the nature of the discussion between developer and regulator by encouraging the greater openness with respect to the evolving benefit–risk information during its development and the introduction of appropriate risk management measures following approval. Prior to the development of this approach, the relevant discussions occurred during the review of the MAA or after approval, following emergent major safety issues. It seems likely that the previous approach would have resulted in delayed or denied approvals, through requests for additional information, or resulted in withdrawal of marketing authorizations due to postmarketing adverse events (see subsequent discussion on natalizumab). In contrast, the proactive application of risk management plans may reduce the likelihood of postapproval withdrawals. The development of the risk management plans along with the increased power for regulators, through EU pharmacovigilance legislation [14] and FDASIA [3], in the postmarketing setting may allow the approval of a few medicines that would not have been approved without this approach. However, it is associated with an increase in the scope and cost of postmarketing regulatory commitments, adding the overall cost of development and potential delays to approval or attrition.

Eichler et al. [13] highlight the development of research initiatives that will improve the safety assessment methodology; the introduction of better quantitative approaches will support these assessments. They consider the unintended consequences of these approaches including the increased likelihood of false safety signals and the requirement for substantial observational trials that would be required to evaluate these potential adverse effects. The pressure on regulators to assure "safe drugs" leads to higher requirements for approval and a greater likelihood of withdrawal based on the assumption that society's expectation is for greater risk aversion. They also comment that these changes are unlikely to result in an increase in trust as a result of an "upward spiral of risk awareness, in which better pharmacovigilance tools … will draw more attention to the downsides of medicines" continuing the trend for increasing regulatory requirements.

Current attitudes to risk management require a proactive approach throughout the development of new drugs and place an increased emphasis on postmarketing or continuous benefit–risk evaluation. The development of systematic approaches to benefit–risk assessment has been facilitated by the development of a small number of academic centers and the formation of independent groups including C-Path that encourage the application of the emerging discipline of "regulatory science." These offer an opportunity to bring together expertise from regulators and industry and allow pooling of information to aid improved decision making. An important aspect of these collaborations is that they increase transparency by encouraging sponsors to share data potentially improving the learning cycle. The scope of these programs has expanded substantially, and they are cofunded by public–private partnerships with the public funding from groups such as C-Path in the United States or IMI in Europe. Additional resources will be available to regulatory agencies as a result of the introduction of fees for pharmacovigilance activities in the EU [15] and user fee funding of FDA activities under "Enhancing Benefit–Risk Assessment in Regulatory Decision-Making" [3].

6.2 DISCUSSION

6.2.1 What Stops Market Approval?

The main focus of this discussion is late-stage attrition where the developer considers that they have completed an adequate program and that a favorable benefit–risk exists. As highlighted by Eichler et al. [10] for the cohort reviewed in 2009, approximately 40% applications where the applicant submitted a valid application failed to receive a positive opinion. It is likely that many of these would achieve approval based on additional information; nevertheless, it is clear that the decision of the developer to submit an MAA does not lead immediately to their desired outcome. The approval systems in the United States, EU, and Japan do allow the submission of supplementary information and negotiation regarding labeling and postmarketing commitments (PMC). However, the emphasis on speed of review has progressively limited the options to submit additional information during assessment. In the United States, submission of additional data results in a resetting of the review clock and needs the agreement of the reviewing division, increasing the likelihood of a complete response letter. The EMA actively discourages extended "clock stops" during the review in response to questions, preferring withdrawal and resubmission. A review of the causes of failure in this periapproval stage should both inform the causes of late-stage attrition and enable higher-quality decisions at previous stages of development and an overall reduction in attrition or at least the cost of attrition by taking it earlier.

In a review of outcomes for a larger cohort (January 1, 2004–December 31, 2007), Rengstrom et al. [16] examined the factors associated with success of applications. They considered a range of factors including company size, utilization of scientific advice, and therapeutic area. The approval rate in this cohort of 188 applications was 72.9% (137); the other 55 applications were either not approved or withdrawn. There was a strong positive correlation between approval and company size, ranging from 48% (26/54) success in small companies to 89% (74/83) in large pharmaceutical companies, which the authors attribute to resources and experience in drug development. However, an 11% failure rate even for large pharmaceutical companies remains a major concern given the substantial investment typically made to reach this stage of development. There was no correlation reported with regard to seeking scientific advice during development and approval. However, the authors highlight that in the cases where the application was considered to be compliant with scientific advice, the success rate was 97% (38/39) compared with 30% (6/20) applications that were considered noncompliant. The analysis of compliance is considered for three important variables in the pivotal clinical studies: primary endpoint, selection of the control group, and statistical methods. Given the relatively small proportion of the cohort seeking scientific advice, the authors suggest that a greater level of interaction with regulators during development would be beneficial, although they caution that their analysis of compliance only considered major factors relating to clinical development and did not consider other development issues related to quality or nonclinical data.

An analysis of drug approvals and failures in the United States for 2006 and 2007 [17] reported 103 approvals and 91 phase III failures during this period. During this same 2-year period, "regulatory setbacks" were reported for 65 drugs in the FDA review process: 16 of these were delays of 3 months, and authors suggest that these are mostly related to submission quality; the remaining 49 represented more serious issues resulting in either nonapprovable letters or requests for additional clinical trials. There are differences between how the EMA and FDA publish information of the success rates of applications; these reports do not consider identical cohorts, and the methods of analysis differ. However, it appears that similar patterns of failure are observed in the United States and Europe.

In a subsequent analysis of factors affecting nonapproval in Europe by Putzeist et al. [18] for a cohort of 68 applications reviewed from 2009 to 2010, the success rate was 66%. These applications were reviewed to evaluate various factors based on the CHMP assessment reports, day 120 major objections, and clinical study assessment reports. The quality of these applications was assessed with regard to three areas: development plan, clinical outcome, and clinical relevance. The development plan was classified as negative when a major objection occurred relating to one of 10 variables across the learning studies (mode of action, proof of concept, pharmacokinetics, dose finding, and safety pharmacology) or confirmatory studies (study design, choice of primary endpoint, target population, trial duration, or statistical analysis). Assessment of clinical outcome was based on statistically significant primary endpoint and the safety profile observed. Assessment of clinical relevance considered the CHMP view of medical need, the effect size for the primary endpoint, and a general consideration of clinical benefit. The clinical outcome was classified as negative when no convincing statistically significant benefit was observed and/or serious safety concerns were raised during the 210-day procedure. Overall clinical benefit was considered to be negative when none of the three aspects were considered as large, important, or compelling. For the 68 MAAs scored as positive in at least 2 of these categories, the approval rate was 34 of 36, whereas in group for which all three assessments were negative, the approval rate was 2 of 14. It is not surprising that the clinical outcome was the most important factor with 91% of those approved scoring positively. The authors concluded that the learning-phase studies were valuable in improving probability of success and thus speeding up pharmaceutical innovation. The relative risk of nonapproval was 2.3 (CI 0.6–11.6) for a positive learning phase, only; 3.4 (CI 0.6–0.2) for a positive confirmatory phase, only; and 5.3 (CI 1.2–23.6) when both were scored as negative.

In a study by Sacks et al. [19], the causes of failure in initial review of new molecular entity submissions to the FDA first submitted between 2000 and 2012 are analyzed. This study represents an important contribution based on both the size and timescale of the dataset, although it omits biologics. Biologics Licence Applications (BLA) are reviewed by the FDA through the Center for Biologics Evaluation and Research (CBER) and were not included in the analysis. There were 302 submissions identified for the analysis from 332 applications (exclusions were 23 nontherapeutic agents, and 7 were withdrawn before FDA action); 50% (151) were approved based on the initial submission, while a further 71 required at least one resubmission. First-time approval rates of drugs granted priority review was 67.9% (72/106) compared with standard review rate of 40.3% (79/196). Analysis of therapeutic use showed the highest approval rate in oncology at 72% (44/61) with the lowest observed in pulmonary/allergy at 31% (4/13). The analysis set was closed on June 30, 2013, and at that time 57.6% (87/151) of failed submissions had been resubmitted for the same indications. In the case of resubmissions, only 63.2% (55/87) were approved based on a single further review and 14.9% (13/87) on a third review. Reasons for first-cycle review failure were categorized as failure in dose selection, efficacy, safety, labeling, and chemistry, manufacturing, and controls (CMC). The latter two categories of labeling and CMC were analyzed in detail only if the first three categories were considered acceptable. In this analysis, a major cause of failure in first-cycle review was inadequate dose selection considered to be the cause in 15.9% (24) cases. Uncertainty regarding the optimal dose to maximize efficacy and minimize safety risks remains one of the most important development challenges.

The subsequent approval rates for the major categories in Figure 6.1 indicate that those applications with only safety deficiencies had the greater chance of success with 61.5% (24/39) being approved in subsequent cycles. The applications with efficacy deficiencies 31.3% (15/48) and both safety and efficacy deficiencies 31.7% (13/41) appear to have lower probabilities of success. The authors suggest that this reflects the use of labeling restrictions and risk management programs that enable the identified safety issues to be managed. It appears that regulators are less likely to accommodate concerns regarding

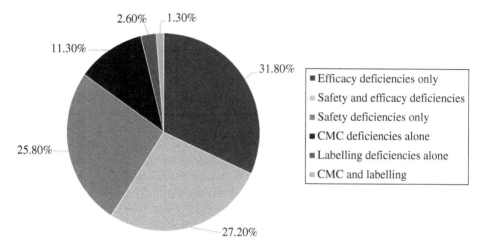

FIGURE 6.1   First-cycle review failure: new molecular entity submissions to FDA from 2000 to 2012. Data from Sacks et al. [19].

efficacy; it may also be the case that such deficiencies may take more substantial investment of resources to overcome. The reduced return on investment may lead to the applicant to abandon these programs, and it is debatable whether this should be attributed to regulatory burden.

The current estimate of the overall approval rate for these new molecular entities is 73.5%, similar to that reported earlier for positive opinions in the EU over a narrower time period [16]. The differences in regulatory procedures limit the comparison with the EU cohort where first-time approval appears higher than that observed in the United States. The EU procedures provide the applicant with greater transparency on review issues during evaluation and more opportunities to respond to issues, resulting in a higher rate of first-time approval with a similar rate in ultimate outcomes. There will be drugs in the later stage of the FDA cohort that will be approved following future submissions, so the ultimate approval rate will rise. However, it seems reasonable to conclude that one-fifth of drugs for which the sponsor believes there is a favorable benefit–risk fail to reach approval. Those that do may have significantly curtailed usage as a result of restrictions placed by both regulatory and payer agencies. Detailed analyses by regulators such as that by Putzeist [18] discussed earlier and Sacks et al. [19] should be critically reviewed by all involved.

The approval rate in Japan appears to be slightly higher than that reported in the United States and EU, for 1050 applications submitted between 2004 and 2012: 79.1% (831) have been approved, 10.1% (106) either withdrawn or not approved, and with 10.8% (113) still under review [20]. Asada et al. [21] reviewed the causes of delay in the approval of NDAs in Japan in 2001 and 2011; since the reasons for withdrawal are not well described in the public domain, they analyzed the issues that delayed approval. This analysis considers 53 of 727 NDAs that had serious and identifiable issues prior to approval. Their conclusions are consistent with those described earlier for the United States and EU with the majority based on clinical data and study designs, although a particular feature of development strategies for Japan is highlighted in this review, that is, bridging strategies. Five of 12 classified as having an inadequate development strategy were the result of failed bridging studies. In a group of 16 applications where the dosing strategy was considered the primary cause of delay, the majority (12) had not adequately defined the dosing regimen.

As highlighted earlier, there has been a strong emphasis since the early 1990s on numbers of approvals and review times; consequently, there are regular analyses of the approval rates on an annual basis. Given the small numbers of approvals, it is likely that individual year data would provide limited insight to trends but a deeper analysis and review of individual case studies would be informative. Dowden et al. [22] reviewed the performance in 2012 for the United States, EU, Japan, and several emerging markets. The number of new molecular entity approvals in the United States (39) was the greatest since 1997 and was higher than the previous year in the markets in both EU and Japan (EU 33; Japan 45). The FDA achieved their lowest ever median approval times of 5.8 months (priority review) and 10 months (standard review) for applications received in 2011. Analysis of the 39 approvals by the FDA indicates that 21 received postmarketing requirements (PMR) or PMC a higher proportion in the case of standard review approvals. It will be important to consider the impact of these commitments on the business case for these products, and a similar trend is being observed in Europe; applicants will need to make an effort to coordinate these discussions with various regulators to maximize the return and utility of these studies in improving the understanding of benefit–risk. One important trend highlighted in this review [22] impacts primary care treatments in areas where there are existing therapies. In 2012, three such new drugs had regulatory setbacks: two antidiabetic agents (dapagliflozin and alogliptin) and an anticoagulant (apixaban) received complete response letters (comparable to a negative opinion from the CHMP) from the FDA. Each of these agents was subsequently approved, but the delays reflect both the importance and the challenge of demonstrating favorable benefit–risk for drugs intended to be used in diseases for which well-understood agents are already available. These examples also highlight the requirement for larger trials to support approval in many therapeutic areas.

6.2.2 Impact of Black Box Warnings

The "black box warning" is just one of the tools available to regulators (in this case the FDA) for advising and highlighting safety considerations of medicines primarily to prescribers but also to the wider community including patients. The "black box warning" is a relatively long-standing tool, and only limited research is available on its effectiveness. It raises a variety of issues, as do other aspects of warning statements in product labeling. When developing new agents, it is important that differentiation is demonstrated. It is often the case that significant class-labeling statements or the potential elimination of "black box warnings" will be a major focus of the development program. The likelihood of such changes and clarity of the scientific pathway to this goal will have an impact on the developer's decisions and consequently a major contribution to overall attrition.

In a study of the effectiveness of an FDA advisory, including "black box warning" in respect of the increased risk of mortality associated with the use of antipsychotic medicines in elderly dementia patients was reported by Dorsey et al. [23]. Examining the usage of atypical antipsychotic agents in a period before and after the issue of the FDA advisory in April 2005 that requested manufacturers include a "black box warning," Dorsey et al. demonstrate that there was a fall in use of atypical antipsychotic agents of

2% overall and 19% in patients with dementia. However, they note that atypical usage fell not only in the off-label use in dementia but also through 2008 in all populations. This analysis highlights a number of issues associated with the uncertain utility of labeling in assuring safe use of medicines; for example, the use of atypical antipsychotic in elderly dementia patients was still 9% in 2008. The authors note that the impact of this specific advisory would have been influenced by several case-specific issues: the usage in elderly dementia was off-label, lack of understanding of the benefits of medicines in behavior in elderly patients, and the safety risks associated with alternative treatments (e.g., typical antipsychotics). The FDA advisory was extended to include typical antipsychotic agents in 2008, although the authors had only noted a small increase in the use of these agents.

6.2.3   Importance and Impact of Pharmacovigilance

Pharmacovigilance is the monitoring of the safety of medicines following their introduction and throughout their marketed life [24]. The marketing authorization holder has the primary responsibility for monitoring the safety of the medicines through evaluation of spontaneous adverse event reporting, review of published literature, and continued studies. The development of proactive risk management strategies is required by regulators, as described in the ICH guidelines [25], and represents a major step in the evolution of pharmacovigilance from "passive" monitoring through reporting of adverse events and signal detection through review of adverse event databases held by regulators and companies to a "proactive" risk-based monitoring approach. The main focus of the ICH guideline is the development of a safety specification and pharmacovigilance plan and its incorporation in the CTD or MAA. The safety specification is an evaluation of the important identified risks from the development program; important potential risks, which may not have been fully evaluated in a preapproval development program; and important missing information. The pharmacovigilance plan is based on the safety specification and the benefit–risk assessment undertaken at the time of approval and would include a summary of the issues identified in the safety specification, the routine pharmacovigilance practices that will be used to evaluate benefit–risk, and an action plan to manage and evaluate the risks. There are substantive differences in the ways that these approaches are implemented by national regulators. In the United States, the FDA was granted additional powers by the US FDA Amendments Act of 2007 to enhance drug safety through the implementation of risk evaluation and mitigation strategies (REMS); but in contrast to the European approach, which requires risk management plans for all submissions, the FDA only requires REMS where there are specific questions relating to risk–benefit that might prevent approval [26]. The FDA provides a list of products that currently have REMS [27], as of February 2014 there were 86 individual product REMS in place with requirements ranging from publication of a medication guide to a full package (medication guide; elements to assure safe use (ETASU); communication plan; implementation plan) and various combinations.

Although there are procedural differences, as highlighted previously, between the FDA and EMA, the tools applied to pharmacovigilance and benefit–risk assessment are increasingly consistent in their development and application. Zomerdijk et al. [28] reviewed risk management activities associated with centralized EU approvals from January 1995 to January 2010. This period covers the earliest EU centralized approvals through the introduction of mandatory risk management plans; however, it provides a better comparison with the FDA approach to REMS. Additional requirements were identified for 58 of 391 active substances during the period under review, compared with the 86 individual product REMS reported by the FDA. The proportion of products requiring additional measures rises from 5% (11/227), prior to the introduction of risk management plans, to 29% (47/164). All (47) required additional physician or patient educational materials, 19 required additional patient screening or monitoring, and 10 included controlled distribution. This latter category was less frequent after the introduction of risk management plans, although given the relatively small numbers, this may be coincidental. Alternatively, it represents a better use of the available risk management strategies beyond restricted distribution. The opportunities presented by the increasingly proactive approach to pharmacovigilance should enable earlier approval of drugs and reduce the likelihood of market withdrawal. However, given the increased burden on the healthcare systems, companies, and the regulators, it is critical that these approaches are evaluated further.

6.2.4   Prospects of Market Withdrawals for New Drugs

Drug withdrawals following approval due to safety are relatively rare events that can have a disproportionate effect both on public perceptions and on regulatory approval pathways. Qureshi et al. [12] studied the withdrawal of NDAs approved between 1980 and 2009; in this analysis, withdrawals for all reasons are identified. Of 740 new drugs approved in this cohort, 118 (15.9%) were discontinued from the market. In 26 (3.5%) cases, these discontinuations were associated with significant safety findings, although a few of these examples were restricted following FDA action rather than withdrawn immediately. The median time to withdrawal was 5.9 years (range 0.3–18.2 years; 95% CI, 4.0–7.8 years). This analysis provides some evidence for a lower rate of market withdrawals due to post-marketing safety findings for more recent approvals, only 3 of the 26 were withdrawn in the 2000s (1980s, 9; 1990s, 14). This is also supported by Sacks et al. [19] who point out that only one (valdecoxib) of the 302 new drugs approved from 2000 to 2012 cohort was withdrawn due to safety concerns. It may take some further time to establish that this reduction in the withdrawal rate is sustained and associated with the proactive approach to risk management established over the past 10 years. The interpretation of this apparent trend should be considered in the context of clustering of the causes of withdrawal, either associated with a common toxicity (e.g., QT interval prolongation) or class of therapeutic agents (e.g., coxibs). Seven of the withdrawals during the 1990s and one in 2001 were associated with QT interval prolongation representing an extreme example of this source of bias [29]. Other leading contributing factors of drug withdrawals due to safety events include liver toxicity and hypersensitivity reactions. A review of withdrawals between 1960 and 1999 by Fung et al. [30] highlighted hepatic adverse events as the main cause of safety-related withdrawals (26.2%) over that period.

The development of a new class of antidiabetic agents provides an interesting case study; with the introduction of troglitazone and subsequent agents in the same class (PPAR-γ) including rosiglitazone and pioglitazone. The first of these agents (troglitazone) was withdrawn from the US market in March 2000, as a result of hepatotoxicity findings [31]. In 2007, an increased risk for myocardial infarction was identified for rosiglitazone (OR 1.43; 95% CI, 1.03–1.98), although no signal was identified for pioglitazone.

The manufacturers undertook further evaluation of the cardiovascular risk (RECORD for rosiglitazone; PROactive for pioglitazone), and following publication of the results of these randomized trials, the FDA responded to an advisory committee recommendation by implementing a black box warning for rosiglitazone to highlight the risk of myocardial infarction. In addition, the FDA requested a head-to-head trial to evaluate the relative risk associated with these agents. At that time, the PROactive trial showed no difference in risk versus placebo (HR 0.90; 95% CI, 0.80–1.02; P=0.095), whereas the RECORD trial did not conclusively show cardiovascular harm (HR 0.99; 95% CI, 0.85–1.16; P=0.93) but was considered underpowered. The regulatory action at the time by the FDA of the introduction of a black box warning and a requirement for further evaluation is based on this evidence. Of greater significance in this discussion than the specific case is the consequent increase in regulatory requirements for the development of antidiabetic agents resulting in FDA guidance issued in final version in December 2008 [32]. This guidance added a requirement for demonstration of cardiovascular safety in addition to previously required surrogate efficacy biomarkers demonstrating lower plasma glucose and hemoglobin A_{1c}. The EMA issued similar draft guidelines in 2010 [33]. This guidance was an update of a guideline issued in 2002 and incorporated the following statement: "For future developments, it is expected that the development programme provides sufficient data supporting the lack of a drug-induced excess cardiovascular risk both from a clinical and regulatory perspective." And "At the time of the MAA, the overall results of this safety program should be submitted and discussed in terms of internal and external validity and clinical justification of the safety outcome. Acceptability of the data presented will be decided based on its overall quality, the point and interval estimates obtained for the calculation of specific risks, including cardiovascular risk compared to controls, and the reliability of these estimations. A summary of what is known about CV risk should be proposed for the SPC." FDA guidance provides a greater level of specificity regarding the consideration of hazard ratios and the design of the studies, but it could be anticipated that the two agencies would make similar decisions based on similar results from the meta-analysis of phase II/III data. However, the EMA would appear to allow more flexibility in the interpretation of the benefit–risk at the time of initial approval. This guideline was finalized in 2012 [34] and included the following updated guidance: "In the past, the assessment of cardiovascular safety in the context of the clinical development of glucose lowering agents has not been possible; the generally benign baseline CV risk profile of patients recruited in confirmatory studies presented for licensure and the limited treatment or diabetes duration have played a major role. For future developments, it is expected that the development programme provides sufficient information supporting the lack of a drug-induced excess cardiovascular risk."

In their study of the impact of FDA guidelines on cardiovascular studies, Bethel and Sourji discussed potential implications for future diabetes treatments [31]. They observed that following publication of the draft FDA guidance, there was an increase in the number and size of cardiovascular outcome trials, illustrated by an increase in median number of participants by sixfold and also by changes in the nature of study outcomes including changes in the nature of the composite cardiovascular outcome to move toward "hard" cardiovascular outcomes and an increase in the proportion of studies that included as primary or major secondary endpoints. The regulatory guidelines also lead to the inclusion of higher-risk populations into phase III studies to avoid a greater escalation in the size of the studies required to enable approval. These trends are clearly desirable with respect to informing the benefit–risk of new agents although as seen in other areas (e.g., coxibs) the absence of reliable information on disease outcomes and the impact of older agents will present significant challenges to regulators. This is exemplified by the EMA review of the cardiovascular safety of ibuprofen [35] based on a meta-analysis of the vascular and gastrointestinal effects of nonsteroidal anti-inflammatory agents in randomized trials [36] published 8 years after the withdrawal of rofecoxib.

This short case study illustrates several concepts outlined in this chapter—not only the broad application of a new safety finding from a specific drug or class to the disease area in general but also the time taken to follow the "regulatory learning cycle." It will be of great interest to follow the effectiveness of these measures on the postmarketing safety experience of more recently approved antidiabetic agents that have been approved based on these guidelines. What cannot be in any doubt is the escalation in development requirements and consequent extended development time and cost. Fortunately, in the case of diabetes, the global epidemic does not limit the desire to develop new therapies in this arena.

Regulatory actions may not result in complete market withdrawal; nevertheless, even a partial withdrawal may have a substantial effect on both the individual product and the regulatory environment. In some cases, the resulting escalation in requirements may make it challenging to justify investment in certain therapeutic areas, even in the case of an obvious medical need. This can be illustrated by the escalation in FDA requirements following the limitation of indications for telithromycin, which had been approved in 2001 for community-acquired pneumonia (CAP), acute bacterial sinusitis (ABS), and acute bacterial exacerbations of chronic bronchitis (ABECB). In late 2006, an FDA Advisory Committee met to reconsider the benefit–risk of telithromycin, following postmarketing adverse events that included liver injury, and concluded that there was an unfavorable benefit–risk for both ABS and ABECB indications [37]. Antibacterial agents represent an unusual area for regulatory approvals since the efficacy assessment is typically based on noninferiority studies, which have always been controversial for both regulators and clinicians. Such studies represent many technical design challenges particularly the definition of the noninferiority margin. Although the trigger event for the review of the telithromycin indications appears to the change risk assessment, as a result of postmarketing safety data, the main driver is an evolving concern regarding the establishment of appropriate noninferiority margins. The original review had been based on the FDA policy in place at the time of the application. At the advisory committee, the FDA argued that "new science" suggested that the noninferiority study design for ABS was not valid, although Echols [37] argues it was really an alternative approach to the calculation of the margins. Following the outcome of the advisory committee advice, the FDA restricted the use of telithromycin to the CAP indication. Final guidance was issued by the FDA in October 2012 for clinical development of antibacterial agents in ABS confirming a requirement for superiority studies in this indication [38]. It is beyond the scope of this chapter to review the detailed evolution of antibacterial guidelines over the past decade. However, it is clear that the uncertainty presented by the evolving situation over a period of more than 10 years will have had adverse effect on business decisions in an important therapeutic area. Echols [37] reviews the outcome of regulatory actions in the 2006–2010 period following the telithromycin review and highlights a series of 10 NDAs or supplements that failed to reach approval (see Table 6.1). During this same period, only three new molecular entities (doripenem, telavancin, and ceftaroline) were approved by the FDA.

As a result of significant public debate both in Europe and the United States regarding the public health importance of developing new antibiotics, there is some cause for

TABLE 6.1 Recent Regulatory Actions for New NDAs and Supplemental NDAs Resulting in Nonapprovals[a]

Drug	Year	Indications	NDA outcome	Reason for outcome
Gemifloxacin[b]	2006	ABS	Negative AIDAC	Noninferiority margin justification
Faropenem	2006	ABS, ABECB, CABP	Not approved by FDA	FDA requests additional clinical studies
Garenoxacin	2006	ABS, ABECB, CABP	NDA withdrawn before AIDAC	Safety and study design
Dalbavancin[c]	2007	ABSSI	FDA issued approvable letter; sponsor withdraws NDA	Study design, additional studies required
Oritavancin[d]	2008	ABSSSI	Negative AIDAC; FDA issued complete response letter	Study design, data quality
Doripenem[b]	2008	NP	Negative AIDAC	Study design, data quality
Iclaprim	2009	ABSSSI	Negative AIDAC; FDA issued complete response letter	Inadequate efficacy, noninferiority margin justification
Cethromycin	2009	CABP	Negative AIDAC; FDA issued complete response letter	Study design, data quality, additional studies required
Ceftobiprole	2009	ABSSSI	No AIDAC; FDA issued complete response letter	Data quality; additional studies required
Telavancin[b,e]	2010	NP	No AIDAC; FDA requires additional studies	Study design

AIDAC, Anti-infective Drugs Advisory Committee; ABECB, acute bacterial exacerbation of chronic bronchitis; ABS, acute bacterial sinusitis; ABSSSI, acute bacterial skin and skin structure infections; CABP, community-acquired bacterial pneumonia; NP, nosocomial pneumonia.

[a] Adapted and updated from Echols [37].
[b] Previously approved for other indications.
[c] Approved May 23, 2014.
[d] Approved August 6, 2014, following priority review.
[e] Approved June 21, 2013, restricted indication.

optimism that this period of great uncertainty may be receding and regulatory pathways to approval are somewhat clearer. In 2014, four new antibacterial agents were approved by the FDA (dalbavancin, oritavancin, tedizolid for skin infections, and ceftolozane/tazobactam for complicated intra-abdominal and urinary tract infections) [39]. However, a consistent theme from recent guidelines and scientific advice requiring smaller noninferiority margins, more robust definitions of primary study populations, and the implementation of novel outcome measures has resulted in substantially larger study sizes and increased uncertainty of trial outcomes. Prior to 2006, studies were typically 300–600 patients, whereas based on these updated requirements, studies of 400–1800 patients will be required, typically two or three times the size of historical requirements [37]. The development of restricted use or indications may facilitate the approval of new antibacterial agents based on more limited databases [40].

The introduction of proactive risk management strategies and the greater transparency in the understanding of the benefit–risk assessment appears to have reduced the likelihood of market withdrawal of new drugs. Review of the regulatory history of natalizumab first approved in November 2004 for the treatment of patients with relapsing multiple sclerosis [41] provides an insight into the progression of a new drug after approval. Within a few months of marketing, cases of progressive multifocal leukoencephalopathy (PML) were reported, and the FDA issued a "public health advisory" highlighting the potential cases and their intention to review the possible risks to patients. The manufacturer voluntarily suspended marketing on February 28, 2005, and the FDA announced their intention to work with the Biogen Idec to assess the potential relationship between natalizumab and PML. The rapid action reflects both the seriousness of the events (one of the two cases was fatal) and the absence of any diagnostic or therapeutic interventions. In June 2006, the FDA approved the resumption of marketing of natalizumab through a "special distribution program"; this followed a consultation with an FDA Advisory Committee that recommended a risk minimization plan. In May 2010, the FDA issued a Drug Safety Communication to highlight new safety information on the risk of PML including additional information to patients and physicians. Based on a review of data received up to January 21, 2010, there were 31 confirmed cases of PML and that it was caused by the John Cunningham virus (JCV), a common childhood virus. Although most adults are infected with JCV but do not develop PML, and it appears that taking immunosuppressive drugs increases the risk. The number of patients treated with at least one infusion was 66,000, and the FDA estimated the rate of PML as 0.5 per 1000 patients, rising to 1.9 per 1000 patients that received at least 24 infusions. The FDA approved a test (Stratify JCV Antibody ELISA test) in January 2012 to support the risk assessment of patients intended to receive natalizumab. The risk factors identified include presence of anti-JCV antibodies, receiving natalizumab for longer than 2 years, and prior treatment with immunosuppressive agents. In the presence of all three risk factors, the risk of developing PML was increased to 11 per 1000 patients, and the label was updated to include the risk information. The use of the drug continues to be limited to patients in the restricted distribution program; however, based on the medical need for a treatment of multiple sclerosis, the rareness of the serious adverse events and the identification of risk factors have allowed the drug to remain on the market rather than being permanently withdrawn.

Qureshi [12] and Sacks [19] suggested that the prospect for withdrawal of new drugs appears to be lower in the past decade than prior to 2000, which may be associated with the more proactive approach to risk management. As highlighted earlier, one dominant and persistent source of attrition remains impacting both the preapproval and postapproval setting, that is, drug-induced liver injury (DILI) [42]. The challenges represented by this cause of attrition include both the diversity of mechanisms of toxicity and the detection and confirmation of the effects. In a review of DILI, Liese et al. [43] highlight several liver abnormality patterns and six histologic patterns of liver injury and more than 50 individual marketed drugs or classes associated with these adverse effects. The estimated annual incidence of DILI is around 19 in 100,000 inhabitants and approximately 10% of acute liver failure cases in the United States [43]. Drugs causing severe liver injury often have rates less than 1 in 10,000 patients and do not demonstrate notable findings in animal studies; as highlighted by FDA guidance, this presents a significant dilemma in that a highly risk averse analysis of findings in clinical studies is likely to prevent useful drugs reaching the market [42].

Ximelagatran was initially approved in nine European markets, including France and Germany, through mutual recognition for short-term use in hip and knee surgery for the prevention of venous thromboembolic events. An EU marketing authorization submission for longer-term use in prevention of cardiovascular events in patients with atrial fibrillation was withdrawn in February 2006, at the same time the product was withdrawn from existing markets [44]. According to Keisu and Andersson [45], short-term use did not indicate any hepatic injury potential and that a higher frequency of reports of elevated hepatic enzyme levels were first observed in long-term studies (>35 days). It appeared that severe hepatic injury could develop after completion of drug exposure and that liver function monitoring might not mitigate the risk of liver injury; as a consequence, the development program was terminated. Subsequent evaluation of the standard preclinical toxicology testing and further investigation of *in vitro* models were unable to define the mechanisms of toxicity observed in the long-term studies. The authors report that pharmacogenomic studies suggested a possible immunogenic pathogenesis, a retrospective analysis showing an association between patients carrying human leukocyte antigen (HLA) class II alleles DRB1*07 and DQA1*02 and elevated alanine aminotransferase (ALT) treated with ximelagatran. This finding has not been confirmed in prospective studies that would be required to validate a screening tool. Although ximelagatran was never approved in the United States, it does feature as one of the examples cited in the FDA guidance [42] highlighting a high rate of Hy's law cases with an apparent incidence of 1/500, suggesting a rate of severe hepatic injury of 1/5000. The challenges associated with understanding DILI have resulted in many efforts to develop additional models and potential biomarkers including the Critical Path Initiative, leading to 14 conferences from 1999 to 2014, and the IMI-SAFE-T Consortium developing DILI biomarkers [46, 47]. It seems likely that this will remain a significant cause of attrition both as a result of sponsor decisions and the increasing sophistication of preapproval safety evaluations.

6.2.5   What Are the Challenges for the Industry Given the Current Regulatory Environment?

Scannell et al. [48] discussed the decline in R&D productivity highlighting the decline in productivity measured in numbers of new drugs per billion US$ invested, indicating that the number of new drugs approved by the US FDA per billion US dollars (adjusted for inflation) spent on R&D has approximately halved every 9 years since 1950. Their analysis suggests four major causes for this long-term trend of declining productivity: "better than the Beatles" problem, the "cautious regulator" problem, the "throw money at it" tendency, and the "basic research–brute force" bias. Several aspects of the first two of these causes are relevant to the dilemmas discussed in this chapter.

The challenge of demonstrating improvement over existing therapies typically results in larger and longer studies, particularly when the disease has several existing treatments and is not associated with serious short-term outcomes. Several authors have highlighted the impact of this effect with a sixfold increase in the size of diabetes trials with the introduction of cardiovascular safety evaluations [31] and also, the increase in the size of studies of antibacterial agents resulting from the introduction of more stringent noninferiority margins [37]. Although the increase in size of studies required for regulatory approval is a general observation, the impact is greater when the objective is to provide greater assurance of safety, as a result of requirements to detect small differences. The requirement for larger preapproval studies is also associated with the availability of existing therapies as the regulator is less inclined to address these issues through PMC. In a small number of disease areas these trends have not yet had a marked impact, for example, oncology. Where it is possible to demonstrate substantial changes in outcomes in well-defined populations, for example, improved survival in a biomarker-selected population regulator, then accelerated approvals are much more likely and much of the burden associated with understanding the benefit–risk can be shifted to the postapproval arena. It must be appreciated that these apparent lower hurdles are transitory as they are dependent on the absence of adequate therapies for life-threatening situations. When these conditions no longer apply to a particular disease or class of treatments, the regulatory hurdle will return to the norm.

The impact of reimbursement or health technology assessment is not the focus of this discussion, but it should be noted that the interaction between payers and regulators is likely to lead to a further escalation in regulatory requirements and the cost of drug development. The EMA and EU network of health technology assessment bodies (EUnetHTA) announced a joint 3-year plan that will focus on early dialogue with sponsors, exchange of scientific and methodological guidance for clinical trials, development of methods for collection of postapproval data, and sharing information about orphan medical products [49]. This initiative is developed from a desire to improve the utilization of information developed during the regulatory review in the subsequent health technology assessment. This is clearly an important step forward in ensuring that potentially divergent interests of these reviewers are integrated for the benefit of patients and the wider healthcare systems. However, it represents a further example of the developer's dilemma as additional complexity is added to the process of drug development.

6.2.6   Future Challenges for Both Regulators and the Pharmaceutical Industry

The evolution of the regulatory environment over the past several years highlights some further challenges for both regulators and the drug developer. The trends that may be of particular significance include greater patient involvement, increased transparency, and a move from binary approval decisions. Each of these pressures will have an impact on areas of the drug development and approval process, some of which are discussed earlier including the efforts of regulators to provide greater transparency regarding their decisions and the data used to support approval.

The level of patient engagement in the development and approval of medicines follows the broader social trend for increased consumer involvement in many areas of public policy. Companies have sought to gain input from patients and patient groups to increase understanding of their condition or needs. Dewulf [50] describes the importance of gaining deeper understanding of patient needs and the challenges that the industry faces in developing a compliant process in shifting from a primary focus that sought insight from physicians. With the increased engagement of patients in both the design and interpretation of benefits and risks of new and existing treatments, regulators will need to adapt both guidelines and evaluation procedures. Eichler et al. [51] state that the demand from patients for access to medicines is not just focused on rapidly progressing or life-threatening conditions but applies equally to chronic conditions.

Regulators need to develop new approaches to balancing "unmet need" and "uncertainty" of evidence in order to address the demands of patients for earlier access to treatment when evaluating benefit–risk. Approaches to accelerated or conditional approval introduced by regulators have not addressed this demand, but have markedly changed the nature of recent approvals. This is illustrated by considering the 41 FDA approvals in 2014 [52]: 17 (41%) orphan drugs and 9 (22%) oncology drugs, 7 of these were subject to accelerated approval, and all oncology approvals had orphan designations. Some observers have suggested that this shift in favor of rare diseases and life-threatening conditions in the composition of regulatory approvals is driven not just by potential commercial value but also the reduced regulatory burden. The challenge for regulators and the industry is to bring forward treatments for long-term chronic conditions and early intervention in progressive diseases, for which the regulatory system is ill-prepared. The introduction of "adaptive pathways" discussed by Eichler et al. [51] recognizes this challenge, but it remains to be seen how both developers and regulators respond. This requires a move from binary approval decisions to "A life-span approach to bringing innovation to patients … to address the perceived access vs. evidence trade-off…." This approach is consistent with the modern approaches to risk management described earlier; however, it will require much greater levels of transparency and information management tools to provide a real-time evaluation of the benefit–risk of treatments following approval.

6.3 CONCLUSION

This consideration of the impact of the regulatory environment on attrition in drug development has focused primarily on the later stages of development and the postapproval experience. The insights gained should be applicable across the development continuum but are less easily separated from the multiplicity of factors that impact on decisions to progress a particular candidate. The history of drug regulation is the layering of learnings from mainly negative experiences associated with the safety of new and marketed drugs and a primary focus on avoiding the repetition of past mistakes. This is associated with an increased regulatory burden, and decreased productivity in the industry has led to a marked change in the portfolios of the pharmaceutical industry with a shift away from the chronic diseases that affect the largest proportion of patients. There are reasons to be hopeful that the development of new regulatory pathways and a greater emphasis on continual evaluation of benefit–risk and understanding of uncertainty will address some of these issues. However, these changes will require much greater transparency and collaboration between stakeholders in order to enable earlier but monitored access to medicines still in development and life cycle approach to regulation. It remains to be seen whether the increased emphasis on regulatory science and introduction of adaptive pathways can reduce the trend in recent decades of decreased productivity in pharmaceutical development, but the shared desire of all stakeholders to bring benefit to patients should provide the motivation.

REFERENCES

1   Eichler, H.-G., Pignatti, F., Flamion, B., Leufkens, H., Breckenridge, A. (2008). Balancing early market access to new drugs with the need for benefit/risk data: a mounting dilemma. *Nature Reviews Drug Discovery, 7*, 818–826.

2   Stockbridge, N., Morganroth, J., Shah, R.R., Garnett, C. (2013). Dealing with global safety issues: was the response to QT-liability of non-cardiac drugs well coordinated. *Drug Safety, 36*, 167–182.

3   Prescription Drug User Fee Act. www.fda.gov. http://www.fda.gov/ForIndustry/UserFees/ PrescriptionDrugUserFee/ucm272170.htm (accessed July 16, 2015).

4   ICH History. www.ich.org. http://www.ich.org/about/history.html (accessed July 16, 2015).

5   The value and benefits of ICH to drug regulatory authorities—advancing harmonization for better health. Geneva: ICH, 2010.

6   Critical Path Institute. Coalition against major diseases 2014. http://c-path.org/programs/ camd/ (accessed July 16, 2015).

7   Manolis, E., Vamvakas, S., Isaac, M. (2011). New pathway for qualification of novel method-,ologies in the European medicines agency. *Proteomics Clinical Applications, 5*, 248–255.

8   Guidance for Industry and FDA Staff: Qualification Process for Drug Development Tools. www. fda.gov. http://www.fda.gov/downloads/Drugs/GuidanceComplianceRegulatoryInformation/ Guidances/UCM230597.pdf (accessed July 16, 2015).

9   Prescrire Editorial Staff (2011). New drugs and indications in 2010: inadequate assessment; patients at risk. *Prescrire International, 20*, 105–107, 109–110.

10   Eichler, H.-G., Aronsson, B., Abadie, E., Salmonson, T. (2010). New drug approval success rate in Europe in 2009. *Nature Reviews Drug Discovery, 9*, 355–356.

11   Halsay, J.P., Cardoe, N. (1992). Benoxaprofen: side-effect profile in 300 patients. *British Medical Journal, 284*, 1365–1368.

12   Qureshi, Z.P., Seoane-Vazquez E., Rodriguez-Monguio R., Stevenson K.B., Szeinbach S.L., (2011). Market withdrawal of new molecular entities approved in the United States from 1980 to 2009. *Pharmacoepidemiology and Drug Safety, 20*, 772–777.

13   Eichler, H.-G., Abadie, E., Raine, J.M., Salmonson, T. (2009). Safe Drugs and the Cost of Good Intentions. *New Engl. J. Med., 360*, 1378–1380.

14   COMMISSION IMPLEMENTING REGULATION (EU) No 520/2012 of 19 June 2012 on the performance of pharmacovigilance activities provided for in Regulation (EC) No 726/2004 and Directive 2001/83/EC. *Official Journal of the European Union*, 2012, L *159*, 5–25.

15   Questions & Answers: Fees for Pharmacovigilance activities—European Commission— MEMO/14/314 16/04/2014. European Commission—Press Release Database. April 16, 2014. http://europa.eu/rapid/press-release_MEMO-14-314_en.htm (accessed July 16, 2015).

16 Rengstrom, J., Koenig, F., Aronsson, B., Reimer, T., Svendsen, K., Tsigkos, S., Flamion, B., Eichler, H.G., Vamvakas S. (2010). Factors associated with success of market authorisation applications for pharmaceutical drugs submitted to the European Medicines Agency. *Eur. J. Clin. Pharmacol., 66*, 39–48.

17 Czerepak, E.A., Ryser, S. (2008). Drug approvals and failures: implications for alliances. *Nature Reviews Drug Discovery, 7*, 197–198.

18 Putzeist, M., Mantel-Teeuwisse, A.K., Aronsson, B., Rowland, M., Gispen-de Wied, C.C., Vamvakas, S., Hoes, A.W., Leufkens, H.G., Eichler, H.G. (2012). Factors influencing non-approval of new drugs in Europe. *Nature Reviews Drug Discovery, 11*, 903–904.

19 Sacks, L.V., Shamsuddin, H.H., Yasinskaya, Y.I., Bouri, K., Lanthier, M.L., Sherman R.E. (2014). Scientific and regulatory reasons for delay and denial of FDA approval of initial applications for new drugs, 2000-2012. *JAMA, 311*, 378–374.

20 The Pharmaceuticals and Medical Devices Agency, Japan. Annual report by 2012.

21 Asada, R., Shimizu, S., Ono, S., Ito, T., Shimizu, A., Yamaguchi, T. (2013). Analysis of new drugs whose clinical development and regulatory approval were hampered during their introduction in Japan. *Journal of Clinical Pharmacy and Therapeutics, 38*, 309–313.

22 Dowden, H.M., Jahn, R., Catka, T., Jonsson, A., Michael, E., Miwa, Y., Zinkand, W. (2013). Industry and regulatory performance in 2012: a year in review. *Clinical Pharmacology & Therapeutics, 94*, 359–366.

23 Dorsey, E. R., Rabbani, A., Gallagher, S.A., Conti, R.M., Alexander, G.C. (2010). Impact of FDA black box advisory on antipsychotic medication use. *Arch. Intern. Med., 170*, 96–103.

24 Pharmacovigilance—how we monitor the safety of medicines. MHRA. http://www.mhra.gov.uk/Safetyinformation/Howwemonitorthesafetyofproducts/Medicines/Pharmacovigilance/index.htm (accessed July 16, 2015).

25 ICH Tripartite Guideline Pharmacovigilance Planning E2E. November 18, 2004. http://www.ich.org/fileadmin/Public_Web_Site/ICH_Products/Guidelines/Efficacy/E2E/Step4/E2E_Guideline.pdf (accessed July 16, 2015).

26 Nicholson, S.C., Peterson, J., Yektashenas, B. (2012). Risk evaluation and mitigation strategies (REMS) educating the prescriber. *Drug Safety, 35*, 91–104.

27 Approved Risk Evaluation and Mitigation Strategies (REMS). *FDA Drug Safety.* http://www.fda.gov/drugs/drugsafety/postmarketdrugsafetyinformationforpatientsandproviders/ucm111350.htm (accessed July 16, 2015).

28 Zomerdijk, I.M., Sayed-Tabatabaei, F.A., Trifirò, G., Blackburn, S.C., Sturkenboom, M.C., Straus, S.M. (2012). Risk minimisation activities of centrally authorised products in the EU. *Drug Safety, 35*, 299–314.

29 Fermini, B., Fossa, A.A. (2003). The impact of drug-induced QT interval prolongation on drug discovery and development. *Nature Reviews Drug Discovery, 2*, 438–447.

30 Fung, M., Thornton, A., Mybeck, K., Wu, J.H., Hornbuckle, K., Muniz, E. (2001). Evaluation of the characteristics of safety withdrawal of prescription drugs from worldwide pharmaceutical markets—1960 to 1999. *Drug Information Journal, 35*, 293–317.

31 Bethel, M.A., Sourji, H. (2012). Impact of FDA guidance for developing diabetes drugs on trial design: from policy to practice. *Curr. Cardiol. Rep., 14*, 59–69.

32 FDA Guidance for Industry: Diabetes Mellitus-Evaluating cardiovascular risk in new antidiabetic therapies to treat type 2 diabetes. U.S. Department of Health and Human Services Food and Drug Administration Center for Drug Evaluation and Research (CDER), December 2008.

33 EMA Guideline on clinical investigation of medicinal products in the treatment of diabetes mellitus. EMA, July 2010.

34   Guideline on clinical investigation of medicinal products in the treatment or prevention of diabetes mellitus. May 14, 2012. http://www.ema.europa.eu/docs/en_GB/document_library/Scientific_guideline/2012/06/WC500129256.pdf (accessed July 16, 2015).

35   Press Release European Medicines Agency starts review of ibuprofen medicines. Review to evaluate cardiovascular risk with high doses taken over long periods. European Medicines Agency. June 13, 2014. http://www.ema.europa.eu/ema/index.jsp?curl=pages/news_and_events/news/2014/06/news_detail_002125.jsp&mid=WC0b01ac058004d5c1 (accessed July 16, 2015).

36   Coxib and traditional NSAID Trialists' (CNT) Collaboration, Bhala, N., Emberson, J., Merhi, A., Abramson, S., Arber, N., Baron, J,A., Bombardier, C., Cannon, C., Farkouh, M.E., FitzGerald, G.A., Goss, P., Halls, H., Hawk, E., Hawkey, C., Hennekens, C., Hochberg, M., Holland, L.E., Kearney, P.M., Laine, L., Lanas, A., Lance, P., Laupacis, A., Oates, J., Patrono, C., Schnitzer, T.J., Solomon, S., Tugwell, P., Wilson, K., Wittes, J., Baigent, C. (2013). Vascular and upper gastrointestinal effects of non-steroidal anti-inflammatory drugs: meta-analyses of individual participant data from randomised trials. *Lancet, 382*, 769–89.

37   Echols, R.M. (2011). Understanding the regulatory hurdles for antibacterial drug development in the post-Ketek world. *Annals of the New York Academy of Sciences, 1241*, 153–161.

38   Guidance for Industry: Acute Bacterial Sinusitis: Developing Drugs for Treatment. October 2012. http://www.fda.gov/downloads/drugs/guidancecomplianceregulatoryinformation/guidances/ucm070939.pdf (accessed July 16, 2015).

39   FDA Approved Drugs by Therapeutic Area. Centerwatch. http://www.centerwatch.com/drug-information/fda-approved-drugs/therapeutic-areas (accessed July 16, 2015).

40   FDA Advisory Committee. December 4, 2014: Anti-infective drugs Advisory Committee Meeting Announcement. http://www.fda.gov/AdvisoryCommittees/Calendar/ucm420134.htm (accessed July 16, 2015).

41   Drugs@FDA   2014.   http://www.accessdata.fda.gov/scripts/cder/drugsatfda/index.cfm (accessed July 16, 2015).

42   Guidance for Industry Drug-Induced Liver Injury: Premarketing Clinical Evaluation. fda.gov. July 2009. http://www.fda.gov/downloads/Drugs/.../Guidances/UCM174090.pdf (accessed July 16, 2015).

43   Liese, M.D., Poterucha, J.J. Talwalkar, J.J. (2014). Drug-Induced Liver Injury. *Mayo Clinical Proceedings, 89*, 95–106.

44   Press Release AstraZeneca withdraws its application for Ximelagatran 36-mg film-coated tablets EMEA/57827/2006. http://www.ema.europa.eu/ema/ (February 16, 2006); http://www.ema.europa.eu/docs/en_GB/document_library/Press_release/2010/02/WC500074073.pdf (accessed July 16, 2015).

45   Keisu, M. Andersson, T.B. (2010). Drug-induced liver injury in humans: the case of ximelagatran. Uetrecht, J. (ed). *Adverse Drug Reactions, Handbook of Experimental pharmacology*, Springer-Verlag, Berlin, Vol. *196*, pp. 407–418.

46   FDA Science & Research—Drug Liver Safety. www.fda.gov. http://www.fda.gov/Drugs/scienceresearch/researchareas/ucm071471.htm (accessed July 16, 2015).

47   Safer and Faster Evidence-based Translation. *IMI-SAFE-T*. http://www.imi-safe-t.eu/ (accessed July 16, 2015).

48   Scannell, J.W., Blanckley, A., Boldon, H., Warrington, B. (2012). Diagnosing the decline in pharmaceutical R&D efficiency. *Nature Reviews Drug Discovery, 11*, 191–199.

49   European Medicines Agency and EUnetHTA agree joint work plan—Press Release European Medicines Agency (19 November 2013).

50   Dewulf, L. (2015). Patient engagement by pharma—why and how? A framework for compliant patient engagement. *Therapeutic Innovation & Regulatory Science, 49*, 9–16.

51   Eichler, H.-G., Baird, L.G., Barker, R., Bloechl-Daum, B., Børlum-Kristensen, F., Brown, J., Chua, R., Del Signore, S., Dugan, U., Ferguson, J., Garner, S., Goettsch, W., Haigh, J., Honig, P., Hoos, A., Huckle, P., Kondo, T., Le Cam, Y., Leufkens, H., Lim, R., Longson, C., Lumpkin, M., Maraganore, J., O'Rourke, B., Oye, K., Pezalla, E., Pignatti, F., Raine, J., Rasi, G., Salmonson, T., Samaha, D., Schneeweiss, S., Siviero, P.D., Skinner, M., Teagarden, J.R., Tominaga, T., Trusheim, M.R., Tunis, S., Unger T.F., Vamvakas, S., Hirsch, G. (2015). From adaptive licensing to adaptive pathways: Delivering a flexible life-span approach to bring new drugs to patients. *Clinical Pharmacology & Therapeutics, 97*, 234–246.

52   Mullard, A. (2015). 2014 FDA drug approvals. *Nature Reviews Drug Discovery, 14*, 77–81.

7

EXPERIMENTAL SCREENING STRATEGIES TO REDUCE ATTRITION RISK

MARIE-CLAIRE PEAKMAN, MATTHEW TROUTMAN, ROSALIA GONZALES AND ANNE SCHMIDT

Hit Discovery and Lead Profiling Group, Department of Pharmacokinetics, Dynamics and Metabolism, Pfizer Inc., Groton, CT, USA

7.1 INTRODUCTION

Productivity across the pharmaceutical industry has been the subject of intense scrutiny in recent years [1–6]. Estimates of productivity losses greater than 70% [1] and quotes ranging as high as $11 billion (USD) in spending per new drug approved [2] have fueled a sense of crisis in which various alternate strategies are expounded for their potential impact on recovery. In almost all cases, decreasing attrition at each stage of the portfolio and divesting programs at earlier stages of discovery and development factor as significant levers in the drive to improved returns on investment.

The design and application of screening strategies across the breadth of discovery disciplines provide key opportunities to impact attrition, particularly in the hit identification and hit-to-lead optimization stages, by guiding teams to therapeutics with improved pharmacodynamic, pharmacokinetic (PK), and safety profiles and a higher likelihood of translation into the patient. Hit identification and optimization strategies have evolved radically in the past 25 years. Advances in the molecular understanding of enzymes and receptors throughout the twentieth century [7, 8] together with the revelation of the human genomic sequence [9] and the concepts of drugability [10] heralded an era of molecular target-based drug discovery in the 1990s and the early 2000s. Concomitant technology advances in recombinant DNA and automation enabled the generation of new, miniaturized assays with increased throughput and improved efficiency. Compound files expanded with the fruit of combinatorial chemistry [11] and, to some extent, an

Attrition in the Pharmaceutical Industry: Reasons, Implications, and Pathways Forward,
First Edition. Edited by Alexander Alex, C. John Harris and Dennis A. Smith.
© 2016 John Wiley & Sons, Inc. Published 2016 by John Wiley & Sons, Inc.

exuberant philosophy held sway in which more compounds tested should yield more hits and subsequently enhance the identification of successful leads [12]. Although this "numbers approach" has been somewhat tempered, the advancement of screening capabilities has opened the doors to novel target space, expanded coverage of chemical space, precise measures of potency and selectivity, the ability to differentiate mechanism of action, greater understanding of desired physicochemical properties and ADME parameters, predictive *in silico* models, and early safety assessments. The key premise remains that the evolution of these approaches has increased the breadth of the discovery screening funnel, improved the filtering efficiency, and increased the quality of filtering and selection [13]; thus yielding a greater number of starting points from which to launch a chemistry design campaign, an improved ability to identify series with more favorable pharmacology, ADME and safety profiles, and consequently a reduced risk of attrition. Today, improved return on investment requires that hit identification and optimization screening strategies are designed with success across a portfolio of targets in mind. The specific challenges for each program demand that the opportunity for screening is over-laid with the biological relevance to disease and the precedence for chemistry success and is balanced against the cost investment for prosecution.

Similar evolutions of screening strategy are evident beyond lead identification. In addition to the technological and scientific advances described previously, safety and tox-icology strategies were impacted by the reduced amounts of compound being synthesized in combinatorial methodologies, the expanding diversity of potential drug candidates, and the higher number of series being pursued in parallel. This drove the requirement for higher-throughput, faster, predictive *in vitro* assays to prescreen compounds for progres-sion into *in vivo* studies [14]. Examples include hERG screening for prediction of QT prolongation [15], cytotoxicity measured in transformed human liver epithelial (THLE) cells as a general predictor of toxicology, glucose/galactose assay as a surrogate for mitochondrial toxicity, and *in vitro* micronucleus assessment for genetic toxicology [16].

The need for increased productivity and market share and the desire to avoid fiscal cliffs due to loss of exclusivity also drove the wave of mergers and acquisitions that has been seen across the pharmaceutical industry over the past 25 years [17, Table 7.1]. This has, in turn, impacted the implementation and prosecution of screening strategies. As organizations seek to increase efficiencies and reduce redundancy, high-throughput approaches, particularly those that are capital intensive, have routinely become centralized. Core centers of expertise frequently partner across multiple therapeutic areas, enabling streamlined global practices, strategic capital investments to capitalize on advancing technologies, platform-based approaches, knowledge-driven screening cascades, and focused computational support. From this culture of partnering across organizational lines, it has not been a significant jump, in the blossoming global climate, to build partnerships externally, particularly where this creates an opportunity to reduce fixed internal costs, operate flexibly in response to a dynamic portfolio, and benefit from the reduced share of capital costs. Consequently, many screening funnels that shape attrition across the portfolio are now conducted at global contract research organizations (CROs). Initially, design concepts for these strategies were maintained internally, and partnerships were built on a foundation of standard protocols and robust assays with significant periods of repetitive screening to justify the challenges of resource investment in assay transfer and infrastructure support. However, as the expe-rience at CROs evolves and relationships are solidified, there is an increasing trend to

TABLE 7.1 *Recent History of Large Pharmaceutical Mergers*
(Survivors are Ranked by 2010 Worldwide Sales)[a]

1. Pfizer
2009: Acquired Wyeth (which resulted from 1994 merger of American Cyanamid and American Home Products)
2003: Acquired Pharmacia (which acquired Upjohn in 1995)
2000: Acquired Warner-Lambert

2. Johnson & Johnson (no major mergers)

3. Novartis
2011: Acquired Alcon
1996: Resulted from merger of Ciba Geigy and Sandoz

4. Roche
2009: Acquisition of Genentech
1995: Acquired Syntex

5. Bayer (no major mergers)

6. Merck
2009: Acquired Schering-Plough

7. Sanofi-Aventis
2011: Acquired Genzyme
1999: Name changed after merger of Rhone-Poulenc and Hoechst
1995: Hoechst acquired Marion Merrell Dow
1995: Rhone-Poulenc acquired Fisons
1990: Rhone-Poulenc acquired Rorer

8. GlaxoSmithKline
2000: SmithKline Beecham merged with Glaxo
1995: Wellcome merged with Glaxo
1989: Beecham merged with SmithKline

9. Abbott (no major mergers)

10. Astra Zeneca
1999: Zeneca Group merged with Astra AB

11. Eli Lilly (no major mergers)

12. Bristol-Myers Squibb
2001: Acquired DuPont Pharmaceuticals
1989: Bristol-Myers and Squibb merged; name change

[a] Taken from Reference 17.

outsource assay development, more transient and complex protocols, and even to invest in risk-sharing deals in which the accountability for approaches that will ultimately shape the attrition profile of a portfolio is shared by multiple parties [18].

This chapter will provide a nonexhaustive historical perspective on the evolution of *in vitro* screening strategies in drug discovery and attempt to address some of the key drivers for current experimental strategies to reduce attrition. Specifically, we will focus on the stage of hit identification, considering concepts of biological and chemical space, and their superimposition through the use of high-throughput screening technologies. We will describe the progression of newly identified active compounds through screening efforts that provide hit validation and pharmacology optimization. Additionally, we will discuss screening strategies to optimize PK parameters and safety profiles, illustrating how early discovery disciplines can build a solid data package to mitigate risk of future clinical failure.

7.2 SCREENING STRATEGIES IN HIT IDENTIFICATION

High-throughput screening has proven a valuable path to the identification of lead material for small-molecule discovery programs and, at Pfizer, has historically yielded 30–40% of candidate compounds moving into development. However, across the industry, there tend to be disparate philosophies on hit identification strategies. While some pharma companies see the compound file itself as a key corporate advantage, others have collaborated in significant compound collection swaps (e.g., AstraZeneca and Bayer) [19, 20] presumably intending to drive competitive advantage through key decisions such as target selection, screening assay format, and lead series selection. One strategy to enable exposure of the greatest possible chemical space to a target of interest has led to screening in higher-density formats (1536, 2080, 3456 wells) and the use of a variety of compound mixture or compression algorithms. Conversely, knowledge-based subset screening strategies are often advocated for specific areas of biology space or well-precedented targets, particularly when strong computational chemistry support is available. Other approaches to lead material identification also exist, for example, knowledge-based design, virtual screening, and alternate screening approaches such as biophysical techniques (e.g., affinity selection mass spectrometry (MS)) and fragment screening, to name a few. In the current climate of low industry productivity and high budget pressure, hit identification strategies need to be considered in the light of investments across a portfolio of targets, and HTS should be balanced against other techniques taking into account cost, speed, and starting chemical equity. In our minds, there are three key drivers (Fig. 7.1) in the decision to use a classical HTS approach to screening: confidence in rationale (CIR) trajectory (or increasing likelihood of biology translation), chemistry need including the likelihood of equity, and screening risk (scalability, precedence). High CIR, such as a compelling case from human biology for causality and functionality in disease, a high chemical need for novel equity, such as an unprecedented target or lack of intellectual property space, and/or a well-precedented screening technology for valid hit identification (thus ameliorating screening risk) would all support investment in large-scale HTS. Typical reasons for low return on investment campaigns include running the screen too late in the project life cycle (i.e., when chemical lead matter is already available), running a screen too early in a project life cycle (i.e., when CIR has yet to be fully developed), and when the screening technology selected fails to deliver valid and viable hits for the mechanism in question. Once the decision has been made to engage in a hit identification campaign through HTS, the concepts of biological space, chemical space, and screen design become relevant. The next sections will address each of these aspects in turn.

7.2.1 Screening Strategies and Biology Space

Ultimately, attrition for target-based drug discovery approaches relies at the outset on the fundamental selection of appropriate target space. In the past decade, numerous attempts have been made to define the target space covered by currently marketed therapeutic agents [10, 21–23]. Determining the number of drug targets *per se* is challenging, since

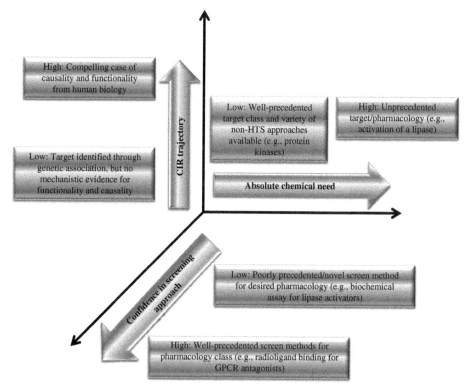

FIGURE 7.1 *Representation of 3 key drivers in the decision to use a classical HTS approach to hit identification*; CIR trajectory, absolute chemical need and confidence in screening approach. Figure illustrates examples of scenarios representing high and low ratings for each of these parameters. High ratings on two or three of these axes would suggest a high return on investment for a traditional high-throughput screening campaign.

some therapeutic agents, for example, display polypharmacology. In 1996, Drews and Ryser determined 483 targets engaged by the agents listed in the 9th edition of Goodman and Gilman's *The Pharmacological Basis of Therapeutics*. Furthermore, they postulated the existence of 5–10,000 targets based on the prevailing idea at the time that there were ~300,000 genes in the human genome, of which 5% represented drugable targets. In 2002, the size of the human genome was found to be ten times smaller than originally anticipated, ~30,000 genes, and new target number estimates arose taking into account both the size of the drugable genome and the probability of genetic linkage to disease. Hopkins and Groom [10] estimated 120 targets for the small-molecule drugs (rule of 5 compliant: [24]) listed in the Investigational Drug Database and Pharmaprojects (experimental and launched agents). Similarly, Overington et al. [27] estimated 266 human genome targets for drugs in the FDA Orange Book (small molecule) and Center for Biologics Evaluation and Research (CBER; biologics) databases. The anticipated number of desirable targets available to the pharmaceutical industry was reduced to just over 3000 [10].

An analysis of the targets by gene family showed that the largest class was G-protein-coupled receptors (GPCRs; 26% of drug targets), followed by nuclear hormone receptors (23%) and ligand- and voltage-gated ion channels (7.9 and 5.5%, respectively) with the remaining gene families representing less than few percent each [22]. Over 60% of the drugs acted through targets at the cell surface, while only ~22% of all proteins are cell surface located. A more recent analysis in 2011 by Rask-Andersen et al. illustrates the shift of target space across the industry. Using the DrugBank Database, these authors categorized 989 agents acting through 435 targets. Drugs acting at receptors still represented the largest drug class (193 or 44% of all targets, with GPCRs making up 36% of these); however, the next largest families were enzymes (124 or 29%) and transporters (67, 15%). The majority of enzyme targets (78%) were soluble, not membrane-bound, proteins. Hydrolases were the most common enzyme target (42%), followed by oxidoreductases and transferases. In spite of this shift, the majority of new approvals each year are for agents acting at precedented targets, and the rate of innovation (drugs acting at new target classes) over the last four decades has remained relatively constant at ~5 new targets per year, primarily outside the major disease indications. The top 4 therapeutic classes are antihypertensive, antineoplastic, anti-inflammatory, and hypnotic/sedative agents [22, 23].

Rask-Andersen et al. [23] created a computational drug–target interaction network map (interactome) for all the drugs in their dataset, the largest component representing drug–target interactions for 489 drugs (49% of the total) and 131 targets (30% of the total). New targets may be deduced from these types of analyses. Analyses of network pharmacology [25] may also lead to a better understanding of pathways underlying drug action and the contribution of multiple targets to drug efficacy. Hypotheses exist that highly selective drugs, targeting a single protein, may exhibit lower clinical efficacy. The integration of systems biology and polypharmacology may provide a path to greater clinical efficacy; however, polypharmacology in a single molecule poses significant challenges such as validating target combinations (ratio of contribution) and in optimizing activity at two or more pharmacophores. Combinations approaches may overcome some of these difficulties.

The next wave of new drugs is purportedly coming from close partnerships between academia and industry so an assessment of academic target space may provide insight into future portfolio directions and the challenges that will be faced in the design of appropriate screening strategies. Upon completion of human genome sequencing, the NIH created drug discovery screening centers as part of the NIH RoadMap initiative. These screening centers provide academic labs with access to the NIH's small-molecule libraries (Molecular Libraries Probe Production Centers Network—MLPCN library) for screening at proposed drug targets (precedented and novel). Numerous additional academic screening centers exist with over 70 in the United States alone. In a recent analysis [26], cancer and infectious diseases represent the highest areas of interest (86 and 71% of all centers, respectively) as well as areas not typically addressed by commercial drug research such as diseases prevalent in less developed countries and orphan diseases (30 and 36%, respectively). Expansion of this analysis to UK academic centers [27] showed similar trends but with higher levels of effort in gastrointestinal and respiratory diseases and less focus on orphan diseases. In contrast, pharma organizations tend to have higher efforts in psychiatric and neurodegenerative diseases. The target portfolios of academic centers are diverse and 20% of the centers perform phenotypic assays to identify compounds affecting a specific physiological outcome rather than compounds that bind to a defined protein target. The majority of targets under investigation (49%) have little or no validation in the literature; some preclinical validation is available in the literature for 27% of the targets, while only 18% of the targets being studied have clinical validation. In contrast to commercial drug research, academics are taking on portfolios with higher risk. In spite of this risk, the potential opportunity for competitive advantage, often combined with access to specific patient cohorts, makes academic or biomedical partnerships appealing for pharmaceutical organizations once biology validation of the target or pathway begins to emerge.

Target validation is a significant concern in compound attrition and, together with the predictive validity of animal models, has been called into question by recent clinical outcomes. In phase II clinical trials, success rates from 2008 to 2010 fell from 28 to 18% with efficacy being cited most often for failure [28]. While survival rates in phase III are up 7%, success rates in phase II remain under 20% for 2011 and 2012 [29]. In 2011, Bayer colleagues, Prinz et al. [30] reviewed 67 internal projects that had data, or were initiated on data, in the literature. Only 20–25% of project teams were able to fully replicate the published data, whereas in 67% of projects, encompassing oncology (70%), cardiovascular, and women's health, some aspects of the published data could not be reproduced. Consequently, this resulted in extended periods of target validation or ultimate termination of the projects. These findings are very troubling and not unique to Bayer. In a similar exercise at Amgen, only 11% of the published findings could be replicated [31]. Based on these and other recent reports, *Nature* Journals will require additional information on methodology, analysis, and reporting of data [32], and the NIH also plans to ensure reproducibility [33]. The bottom line is that solid evidence validating a target for a specific disease/indication is absolutely critical before launching into a drug discovery program that requires substantial investment.

Future directions aimed at improving target selection are expected to include advancing targets that have a high level of validation through linkage to disease and human genetic data. The use of human genetic data to increase confidence in target choice, instead of relying on animal models that do not always recapitulate aspects of human disease, may lead to a greater survival rates in phase II. Plenge et al. [34] provide a good review of learnings from naturally occurring mutations ("experiments of nature"), where there are clear function–phenotype findings. For example, several gene mutations have been identified that modulate levels of LDL and are associated with the risk of cardiovascular disease, including mutations in LDL receptor, PCSK9, and HMG-CoA. Gain of function mutations in the PCSK9 gene leads to increased levels of LDL and thus a higher risk for cardiovascular disease, whereas a loss of function mutation in this gene has the opposite effect suggesting a role for PCSK9 in modulating LDL levels. Accordingly, in the clinic, a monoclonal antibody against PCSK9 reduced LDL levels in patient populations [35]. Understanding how common genetic variants contribute to complex traits (diseases that do not segregate by families) such as type 2 diabetes and rheumatoid arthritis, where the underlying genetic cause is often polygenic and influenced by a multitude of factors, is much more difficult. Many of these alleles have been identified from genome-wide association studies (GWAS), and validation requires unraveling the biological mechanism contributing to disease phenotypes. Additional sequencing data from large patient cohorts are needed to help define the complete picture of genetic variation (from common to rare alleles). Preclinical screening strategies can support these efforts by generating data in human systems and comparing patient-derived tissue/cells with the normal state in order to validate the target and link to disease [34].

Trends also indicate a focus on specific gene families or target classes, and this is likely to improve the probability of progressing successful drug candidates since knowledge gained from one project can be applied to the next. For example, the number of drugs that target GPCRs has increased at a steady pace over the last decade or so. The kinase target class is also gaining momentum. Originally, kinases were thought less attractive and even "undrugable" by some, but as of September 2013, 24 small-molecule drugs targeting kinases have been approved by FDA (FDA Orange Book). Target classes that are likely to see similar focus include ion channels, transporters, non-GPCR receptors, and enzymes.

7.2.2   Screening Strategies and Chemical Space

The content, size, and quality of a compound collection used for hit identification approaches and HTS are also fundamental to the ultimate direction of a project and its future success or attrition. Even the most advanced screening technologies and the most physiologically relevant assays will be defeated by poor quality compound collections and poor decisions in the chemical matter that should be applied to the target. Historically, compound collections were assembled or grown in a number of ways:

1. Synthesis of proprietary compound designs
2. Acquisition of commercially available compounds
3. Assimilation with other collections as a consequence of mergers and acquisitions
4. Strategic collaborations and/or shared access

The synthesis of proprietary designs can be the result of ongoing or past drug discovery efforts and/or a deliberate effort to populate specific areas of chemical (e.g., macrocycles, steroids) or biological (e.g., GPCRs, kinases) space. Big pharma, then, by virtue of the relatively large numbers of targets they work on across different therapeutic areas, generally have larger proprietary HTS collections. The deliberate effort to synthesize proprietary compounds is an expensive proposition, although it does increase the likelihood of compound exclusivity. In the early to mid-2000s, several million dollars were spent across the industry on combinatorial chemistry technology to enrich collections in this way. At Pfizer, the HTS collection grew by over 2.0 million compounds through this effort [36]. A similar effort at Bayer resulted in their collection being comprised of two-thirds combichem molecules [37]. The acquisition of commercially available compounds is an ongoing approach for both larger pharma and smaller biotechs. For example, the "compound collection enhancement" at GSK and "file enrichment" efforts at Pfizer tap into the catalogues of a large number of suppliers and "aggregators" to enrich existing collections with specific chemistries. These purchases are often in line with emerging strategies within an organization, for example, a desire to move into new gene family targets or to explore specific physicochemical property space deemed to be advantageous. In the past, GSK, seeking to enhance compounds with good potential to be orally active, used Lipinski rule of 5 criteria (see Section 4 for more specifics on rule of 5, [24]) to select compounds for purchase [38]. The goal of deliberate compound acquisitions and chemical synthetic efforts is to have the highest possible coverage of chemical and biological space to maximize the probability of finding starting material [39]. This becomes more critical when working on unprecedented targets. Mergers and acquisitions can significantly increase the compound collection of the newly formed company. The Novartis compound file is, in part, a combination of the collections of Ciba, Geigy, and Sandoz. The same is true for GSK consisting of matter from legacy Glaxo, Burroughs Wellcome, SmithKline, and Beecham collections. Pfizer Inc.'s collection is the product of legacy Pfizer, Warner-Lambert, Pharmacia, and Wyeth collections. Among many other examples are the BMS collection that hails from Bristol Myers and Squibb and Merck's collection from Merck and Schering-Plough [17, 37, 40]. Usually, during a corporate integration, some element of streamlining these collections takes place to refine screening files, reduce redundancy, concord identical structures, and maintain collections thought most likely to be of value in hit identification. The redundancy of structures in merged collections is of concern particularly when a sizeable collection is being integrated. Bayer and Schering AG did an assessment of the combined collection and found a surprisingly low overlap in structure identity and similarity [37]. Pfizer did a similar exercise during the Wyeth merger and only added ~500,000 compounds to their existing collection [36]. Without that initiative, the combined Pfizer collection would have been in excess of four million compounds.

An increasing awareness of the costs required to expand and to maintain compound collections, challenged by a continuous drive to increase compound diversity, has led to the sharing or exchange of partial corporate collections while seeking ways to apply rational methods to random screening [36, 45]. This model is a good compromise between efforts that sacrifice intellectual exclusivity (i.e., through purchase of commercially available compounds) and cost (i.e., through the synthesis of proprietary designs). Lundbeck acquired over 300K compounds through a collaboration with Novo [41]. AstraZeneca and Bayer recently announced sharing of their collections in areas where they do not compete [20]. There are also initiatives that grant broader access to collections in the spirit of accelerating the discovery of drugs for diseases with the greatest unmet need. In the United States, The National Center for Advancing Translational Sciences (NCATS) at the NIH provides access to its pharmaceutical collection through the Therapeutics for Rare and Neglected Diseases (TRND) program and as part of the compound collection for the Tox21 Initiative, a collaborative effort for toxicity screening among several government agencies. All data generated through these efforts are made available in the PubChem database [42]. In Europe, the Innovative Medicines Initiative (IMI), partially funded by the EU, brought pharma members together to build the Joint European Compound Library (JECL), which is projected to be a 500K compound collection [43].

Determining the appropriate size of an HTS file is challenging and balances multiple factors, taking into account the desired outcomes and cost. The target space pursued has implications on the breadth of chemical space that needs to be covered. A portfolio exclusively focused on one target class with key structural requirements, for example, would in theory require a compound collection covering a narrower band of chemical space than a diverse portfolio of soluble kinases, membrane-bound GPCRs, and other enzyme classes. However, the density of chemical space coverage (i.e., the number of compounds populating a given dimension of space) also plays significantly into the overall collection size. It follows then that there is a balance between breadth of the space covered and the density of the compounds covering the space. Population of the entire area of theoretical chemical space has been estimated to require 166 billion small molecules [44]. Clearly, building and maintaining collections of this size is unfeasible, but even as file sizes grow beyond a few million compounds, the costs of screening become considerable and must

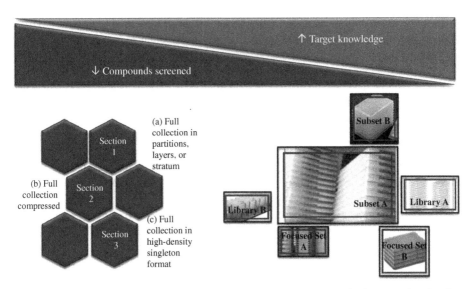

FIGURE 7.2 *The number of compounds screened against a target is highly correlated to how much is known about it.* Exposing unprecedented targets to the full compound collection (on the left) is more likely to yield a series of interest. Well-precedented targets exposed to focused sets (on the right) will quickly yield tool compounds that could be used to mine the full compound collection. The full screening collection can be physically accessed in a number of different ways: (a) as progressing partitions that are rationally grouped, (b) as compressed or multiplexed wells containing more than one compound per well, or (c) as one compound per well in high-density plates (e.g., 1536 well or higher) for speed and screening cost containment. Focused libraries or subsets are of varying sizes, depending on the richness of the full collection and historical knowledge about compounds therein. These sets may be enriched with the addition of commercially available compounds with well-characterized activities.

be weighed against the value of testing more chemical matter. Important principles to consider in this respect are that the primary goal of an HTS is to find starting templates to launch a medicinal chemistry program, not to define full structure–activity relationships (SAR) on the series of interest and that a key outcome is the knowledge gained of the target (and potentially other members of the target class) or pathway.

In essence, the measure of quality of a compound collection is how many series it yields given the investment made for an HTS campaign. One way to efficiently gain knowledge on a target is to screen smaller libraries focused on the target of interest. This is particularly helpful for well-precedented targets or when a tool compound is desired for further biological target validation. By their very nature, however, target-focused libraries are often designed using knowledge of previously active compounds, and when screening only known chemotypes, identification of completely novel active templates becomes a challenge (Fig. 7.2). Phenotypic screening approaches that are agnostic to target require diverse chemotypes but quite often utilize smaller well-annotated compound subsets that may help to define an effector pathway and thereby facilitate biological validation of the approach. Focused libraries or subsets can range in size from small collections of 50 compounds to 60K or more. Partitioning, stratifying, or layering large compound collections is another strategy that enables the search for a series of interest in a more cost-effective way.

In 2009, Pfizer redesigned its screening collection to select for compounds with drug-like or lead-like properties and introduced the concept of molecular redundancy to determine how many compounds in a given area of chemical space are required to find a hit [36]. Reducing the redundancy of structures is one way to reduce the number of wells screened while maintaining maximal diversity in the chosen set. A 150K subset of compounds was selected to represent the diversity of the four million collection. In addition, the full collection was split to form a Tier 1, ~3 million drug-like collection and a Tier 2 segment of less drug-like compounds that are screened more rarely. Tier 1 members were selected, in part, using an algorithm based on molecular redundancy, thereby limiting the density of compounds in a given section of chemical space. Tier 2 compounds often exceeded the limits of Lipinski's rule of 5 [24], contained undesirable functional groups, provided poor starting points for most programs seeking orally active medicines, and/or had extreme physicochemical properties. AstraZeneca did something along the same lines to increase the efficiency of screening their collection [45], while Sanofi maintains a 25K compound inner core collection [46]. Since the computational algorithms for diversity have evolved over time and, moreover, the concepts of and approach to diversity are often deeply embedded in an organization's culture and prior experience, diversity algorithms are likely unique to each company yielding different biases in physical collections per given area of chemical space.

Compound collection quality is measured at a more granular level by multiple parameters such as the confidence in the identity and purity of each compound in the collection. In the 1990s, follow-up on active hits from HTS campaigns was frequently derailed by the poor quality of compounds purchased from commercial sources. Today at Pfizer, we assess a sample both when it enters the collection and again when it is identified as a confirmed hit and is requested for further screening. Our QC fail rate has fallen dramatically to generally below 10%. Similar workflows are in place at GSK [47, 48] and Lundbeck, to name a few [41].

Stability of the compound in the solvent of choice (usually DMSO) is also key to minimizing precipitants that can hamper screening both by reducing the compound concentration accessible to the target and potentially by creating false positive readouts. Stability of any solution is influenced by the degree of saturation of the solute in the solvent at a given temperature. In the case of DMSO stocks, the hydrophilicity of DMSO adds another dimension to the issue, and there have been several studies to determine the optimal temperature, relative humidity, concentration, and time that ensure stability of solutions of compounds in DMSO [49, 50]. The concentration of compound stocks is also influenced by the effects of DMSO on biological systems, notably elevating cell stress pathways and interfering with compound binding to protein targets (see also Section 7.2.3). It is important to recognize that there is no set of conditions that will guarantee that 100% of the compounds in a collection will be in a homogeneous form; therefore, decisions on storage and handling conditions balance doability and scientific confidence based on data available. Some basic premises are universal:

1. Single-use storage containers are better than multiuse containers as one avoids any issues associated with freeze–thaw cycles.
2. Storing more dilute concentrations in DMSO is better than storing higher concentrations. Somewhat ironically, however, storing solids (i.e., no solvent at all) is better than storing liquids.

3. For DMSO stocks, minimizing contact with water is good.

4. Freshly made solutions are better than older solutions.

At Pfizer, the major liquid collection is stored as single-use tubes at −20°C, while lead optimization compounds are stored at room temperature for about 90 days under a nitrogen atmosphere. BMS stores long-term stocks at 4°C, while lead optimization compounds are kept at room temperature for 2 weeks [50], and Novartis maintains stocks in 10% water/DMSO stored at 4°C [39].

It should be appreciated that compound collection quality is equally maintained by the data systems that define the contents of a well. After all, small system errors such as an inversion of the compounds in a plate or the transfer of compounds in an S pattern instead of a Z pattern (e.g., when transferring from 96 to 384 well format) can render all data derived from those compounds meaningless. Compound management workflows, therefore, focus on both the sample and the data.

The stewards of a screening collection should be closely attuned to the trends used to identify lead matter. As target space of interest to the pharmaceutical industry expands, screening collections are enhanced to cover broader chemical space. Miniaturized screening technologies set the demand for nanoliter volume dispensing, and an increase in the number of assays using physiologically relevant systems with limited reagent availability may limit the number of wells that can be screened driving strategies for compound mixing or compression. Affinity selection MS assays have the ability to test hundreds of compounds in one experiment so highly compressed versions (upward of 100 compounds per well) of the screening collection are needed [51]. Nuclear magnetic resonance (NMR) screening often requires high concentration of fragments in deuterated DMSO. Acoustic dispensing technology has made custom compression or multiplexing more achievable than ever before. Hence, emerging compound management strategies continue to emphasize knowledge-based subset creation and custom-compressed compound sets tailored to the target of interest and the assay system in use.

7.2.3 High-Throughput Screening Technologies

As combinatorial fever swept the industry in the late 1990s, corporate compound files underwent rapid expansion [52, 53]. In 2001, the Pfizer HTS file size was somewhat shy of one million compounds and was predicted to increase to three million [54], a figure at which it currently hovers today. HTS facilities responded by searching for higher-density technologies, higher-throughput detection methods with increased signals, more automated instrumentation, particularly for liquid handling, and technologies for more robust assays. A worldwide study of 39 HTS providers in 2000 indicated that 384 well plate technologies were being used in approximately 50% of all screening campaigns and were on the increase [52, 53]. Accordingly, Pfizer transitioned almost 100% of screening campaigns from 96 well plates to 384 well plates between 1999 and 2003. Further, miniaturization to 1536 well formats was embraced in many facilities [55, 56] often with a drive to fully automated platforms. This was frequently the path of choice for groups with limited staffing but accessible financial support for capital investment. Conversely, workstation approaches tended to be adopted by institutions with higher staff counts, lower financial resources for capital investment, and/or those maintaining a significant diversity of detection technologies. More atypical higher-density formats (e.g., 3456 wells, 2080 well nanocarriers) were often associated with specific screening platforms and formed the basis for multiple technology partnerships [57–59].

The greater use of automation, higher costs of screening larger libraries, and the increased campaign length, requiring greater reagent stability, combined with increasing availability of cheaper, rapid fluorescent detection methodologies drove a significant shift away from radioactive readouts in the early 2000s [52, 53]. In parallel, compound management strategies aimed at decreasing costs and campaign times established new ways to screen in mixtures or compressed formats, while computational expertise was applied to creating smaller target-specific or diversity-based subsets that were representative of the chemical space desired [60]. The introduction of acoustic liquid handling technologies [61] has further enabled this space, improving data quality in miniaturized assays through accurate nanoliter dispensing and removal of intermediate compound handling. Additionally, acoustic dispensing has eliminated contamination potential, reduced consumables costs, and extended the lifetime of compound inventories.

Against this backdrop, recent retrospective analyses regarding the high attrition rate in phase II clinical trials have suggested that the "reductionist" approach to drug discovery taken over the past two to three decades may have contributed to failure to demonstrate efficacy of new medical entities (NMEs) [69, 70]. Prior to the molecular biological revolution, most screening assays utilized whole animals, animal or human tissue or immortalized cell lines to evaluate the effects of compounds on a specific phenotype of interest. Improved molecular biological techniques that allowed individual proteins of interest to be easily expressed in abundance subsequently led to the design of assays using isolated protein or proteins overexpressed in cell lines. Frequently, enzyme activity was measured using a truncated form of protein, such as the catalytic domain, in a noncellular, biochemical environment. While these systems allowed for the development of robust, high-throughput assays, the physiological relevance was knowingly questionable and had the potential to limit the type of chemical matter identified. Recent emergence of allosteric modulators, DFG-out kinase inhibitors, etc. has demonstrated the value of using full-length protein whenever possible in biochemical assays [62].

The drive for more physiologically relevant systems coupled with the patient-focused strategies of the precision medicine field has advanced more complex biological systems such as patient-relevant genotypes, primary cells [63], stem cell-derived models [64], or patient-derived cells [65]. Often, these reagents are difficult to procure or require considerable resource investments thereby limiting the scalability of screening campaigns and providing yet another driver for miniaturization and restricted compounds sets that are often custom compressed and tailored to the target of interest and the assay system in use. Recent examples from Pfizer include a 150K compound library screen of primary human T-cells [66, 67] and a two million compound file screen of mouse embryonic stem cells to identify potentiators of AMPA activity [68]. Although for targets with clinical precedent, recombinant systems may still be considered as "fit for purpose," for novel targets, a clear trend is to use the most physiologically relevant system possible for hit identification campaigns. Importantly, the use of such systems requires appropriate counterscreens and hit validation strategies to confirm that the activity is due to engagement of the target of interest and not to off-target effects.

The importance of screening strategies in attrition has also led to the resurgence of the phenotypic screening debate and the call for a more holistic system or pathway approach with clinically relevant endpoints [69, 70]. Swinney and coauthors [69] reviewed FDA approvals between 1999 and 2008 and determined 37% of "first-in-class" agents approved in this time period were discovered in phenotypic screens versus 23% identified in target-based screens. In this analysis, two therapeutic areas stood out, infectious diseases and central nervous system disorders, each with seven first-in-class approvals where agents were identified in phenotypic screens. Phenotypic screens measure changes in a specific, clinically relevant outcome (e.g., lipid lowering, abeta lowering, reduction of high blood pressure, etc.), where the screen is agnostic to the underlying targets and pathways. An important feature of the assay design is that it mimics the clinical disease state as closely as possible, using the most relevant tissues/cell source, to optimize the likelihood of translation of in vitro results to the clinic. A good example was described in 2011 by scientists at Vertex pharmaceuticals [65] who monitored chloride ion transport and epithelial cell function in patient-derived bronchial epithelial cells to successfully identify a new treatment for cystic fibrosis. In some cases, plate-based whole-animal screens are possible (e.g., C. elegans, D. melanogaster, zebrafish, X. laevis) and provide the advantage that, in addition to measuring efficacy, ADME and toxicity properties can be inferred, or measured directly, in parallel [71]. The challenge of these complex biological systems, however, is that lead matter may be missed due to poor physicochemical properties that lead to inadequate absorption, inadequate systemic stability, or inability to reach the target, such that no potential efficacious (or toxic) effect can be observed [70]. Phenotypic screens utilizing high-content imaging allow for visual monitoring of phenotypes such as autophagy, apoptosis, nuclear translocation, receptor internalization, and neurite outgrowth. Furthermore, high-content imaging enables the measurement of multiple phenotypes simultaneously, including potential toxicity measures through the recording of cell health parameters (e.g., cell membrane integrity, nucleus size and shape, DNA amounts, etc.) [70]. Feng et al. [72] highlight the benefits of phenotypic profiling early in discovery using a variety of methods including imaging. In particular, they note the potential power of integrating a phenotypic profile with knowledge of biochemical pathway interactions and functional validation data linking disease with drug action to improve lead series selection (e.g., increased efficacy, decreased toxicity).

Whether target based or phenotypic, several factors play a role in developing an assay for a successful high-throughput screening campaign, including "measuring the right signal," assay robustness and reproducibility, toleration to DMSO, ability to detect active moieties, and selection of the appropriate compound collection. Once an assay is designed to measure a desired endpoint in an appropriate source of protein, with an adequate signal window in the linear dynamic range, all of these parameters should be assessed to ensure confidence in data. HTS runs are frequently large and may take extended periods of time; therefore, reagent and signal stabilities need to be checked over this time period. Temperature sensitivity has become a more significant issue in our recent experience of working with CROs in global locations that experience significantly varying temperature conditions throughout the year. In one recent kinase assay, significant shifts in data were observed and were found to be caused by the practice of turning off the laboratory air conditioning during the nighttime periods. Even though the lab had sufficiently cooled by the time the assay was prosecuted, lab plastics and consumables were still at elevated temperatures and required prechilling to restore confidence in data. Most compound collections are solubilized in DMSO. In general, cell-based assays can tolerate up to 1% DMSO (although we have observed impacts of DMSO at concentrations as low as 0.05%, particularly in primary cell-based assays), whereas biochemical assays have been shown in our hands to tolerate up to 10% DMSO. Reproducibility from plate to plate in a single screening run is required, in addition to consistency across experiment days. Several useful measures of understanding and monitoring reproducibility include the Z factor ("statistical effect size"—reflective of both the assay signal dynamic range and the data variation associated with the signal measurements) [73], monitoring of control wells (zero effect and maximal effect), and the performance of standard compounds (if available). Custom visualizations of the data help to detect plate-level artifacts such as edge effects or pipetting errors, the latter of which can be minimized by routine quality control procedures on automated instrumentation, particularly liquid handlers.

Often, prior to the launch of a large-scale HTS, a pilot run using a small set of compounds is performed at a number of concentrations to predict hit rate, confirmation rate, and false positive/negative rates. During pilot runs, a greater understanding of the assay's ability to detect actives under different conditions (e.g., singleton versus compressed compound format) can be achieved by spiking wells with available, known actives (e.g., full and partial agonists) at varying concentrations. The results from these experiments enable final decisions on what compound format and concentration to use. In some cases, the choice is made to test compounds in a full concentration-response mode in the primary HTS that can ensure actives with varying effect profiles are identified (e.g., inverted U-shaped dose-response curves) [74] and may reduce, but not eliminate, the amount of follow-up.

7.2.4 Future Directions for High-Throughput Screening

The future evolution of high-throughput screening will be aimed at truly bringing the patient to the forefront of earliest discovery efforts, using clinically relevant cell systems (e.g., 3D cultures, cocultures, patient-derived stem cells) to create the virtual "patient in a dish." Miniaturization is likely to remain an opportunity for the maximal exploitation of these precious, often labor-intensive, patient-specific reagents. These systems will provide the means to explore novel targets, identified from genomic or phenotypic approaches, which lie outside current biology and chemistry space experience. Phenotypic readouts will demand a systematic approach to target or safety deconvolution in order to mitigate on- and off-target safety risks. As datasets are built across the industry from new screening paradigms, there should also arise a large opportunity to mine data within gene families to assess, and ideally predict, trends in pharmacological SAR.

7.3 SCREENING STRATEGIES IN HIT VALIDATION AND LEAD OPTIMIZATION

Once the primary HTS has identified active compounds, these hits are confirmed and validated, usually through a downstream *in vitro* screening funnel. Often, hits are retested in the same assay format at a single, high concentration. This can be performed in parallel with cytotoxicity assays, if the assay is cell based, and / or counterscreens to eliminate false positives. A common counterscreen for recombinant cell line-based assays is testing compounds for activity in the parental cell line. Confirmed hits progress to full concentration-response studies (typically 11 doses in our groups) for potency determination. The robustness of the assay will determine if this can be done in singlicate measurements as opposed to duplicate/triplicate wells per concentration. If each concentration is run in multiplicate, these replicates can be dosed within a plate or across multiple plates, again driven by statistical measures of reproducibility.

Before significant medicinal and synthetic chemistry is applied to pursue compounds identified through an HTS campaign, there is usually a substantial effort invested in hit validation and triage. Invariably, strategies involve exploring and expanding the developing SAR through file mining efforts to select more compounds for testing that show similarities to hits. This is particularly the case if similar compounds were not included in the original primary screen and subset screening often leads to iterative rounds of follow-up that can significantly increase the cost and time of a program. Structurally enabled programs or homology models may enable 3D pharmacophore searching that allows for leaps to structurally dissimilar series. Computational expertise can also be applied to weed out potential false positives from the wealth of data already in a corporate database. For example, some of the hit compounds may turn out to be promiscuous binders, the so-called "frequent hitters." In a recent study, colleagues of ours surveyed 96 HTS campaign runs within Pfizer (82 different targets, 66 gene families, ~300 million data points) and characterized the common features among frequent hitters [75]. The topology and physical attributes of these compounds were analyzed and compared with random sets of compounds. A fingerprint-based Bayesian model was built to predict frequent hitters and potentially reduce the time and cost of HTS triage and chemistry follow-up during hit-to-lead optimization. Currently, this model is used as part of Pfizer's tool box for HTS triage and hit follow-up. Multiple groups have published on the nonspecific inhibition of enzymes by compounds that form organic small-molecule aggregates, and more recently, Brian Shoichet's lab has attempted to describe the mechanism of action of aggregate-based promiscuous inhibitors as a partial protein unfolding (see Ref. 76). In the past, fluorescent compounds were often false positives in fluorescence-based assays and were triaged out of hit lists or passed through nonfluorescence-based assays for confirmation. Newer generations of fluorescent probes and fluorescent technologies, together with improved compound file characteristics, however, have reduced the impact of these compounds on screening campaigns.

The biology screening funnel generally incorporates an orthogonal assay technology to confirm biological activity, for example, hits from a biochemical assay may be tested in cell-based formats; GPCR signaling may be tested via second messenger accumulation and confirmed through GTPγS studies. Additionally, biophysical methods such as surface plasmon resonance (SPR), NMR, light scattering, MS, or calorimetry may be used to confirm binding to the protein target of interest and to understand conformational changes and the thermodynamic interaction.

Off-target effects, both within target class and across target classes, are assessed to understand the selectivity profile of hits. This can be accomplished by data mining for activity in other institutional or reported HTS and SAR screens, computational models to predict activity at targets, or by profiling compounds in selectivity panels. Many CROs now provide standard selectivity panels that provide some flexibility and reduce the necessity to build internal commitments to these ongoing investments [77]. Subsequent secondary and tertiary assays in the *in vitro* screening funnel often become more physiologically relevant, more reflective of the disease state, and are often more labor intensive.

Further, characterization of hit matter builds deeper knowledge of a compound's mode of action and can support routes to differentiation from existing clinical agents. Multiple assays assessing intracellular signaling can be used to understand any potential ligand bias of GPCR agonists. Binding to an enzyme target can be elucidated in terms of its competitive, noncompetitive, and uncompetitive kinetics. On and off rates and the reversible or irreversible nature of binding can all be ascertained and potency can be compared across relevant species (e.g., rats, dogs, monkeys, humans). These parameters are incorporated into computational models for translation from preclinical efficacy and safety studies to human dose and therapeutic index prediction. Chemical matter that is highly species specific raises downstream challenges, for example, in determining how to assess preclinical safety of small molecules that only interact with a human receptor.

Evaluating HTS hits from phenotypic-based screens is somewhat different from traditional target-based screening. Although knowledge of the biological target *per se* is not required for drug approval, provided that safety and efficacy are assured, the timelines for project progression can be significantly impacted by the path selected. In theory, in the absence of knowing the target, hits from high-throughput phenotypic screens could be triaged for potential safety and ADME liability through *in vitro* screens and *in silico* models and be characterized further in a multitude of cell health measures leading to advancement of the most attractive chemical matter for optimization. However, most medicinal chemists would prefer to work on target-based approaches because following SAR around a phenotype is complex. Additionally, a significant risk with this approach is that the potential inherent safety profile of the biological target/pathway, for example, indicated by the distribution of tissue expression of the target, is unknown. If safety findings occur at a later stage of development, then some level of deconvolution will have to take place in order to design mitigation strategies. Doing some level of target/pathway deconvolution as part of HTS hit triage and then pursuing the most promising target(s) or pathway(s) for SAR may enable better decision making; however, target/pathway deconvolution takes a significant time investment (months to years). The advantage is early safety assessment prior to launching a costly development program. Deconvolution strategies will be project specific and may include a variety of different approaches and technologies.

Chemical genomics and RNA interference studies are frequently part of deconvolution strategies for identifying potential targets and pathways underlying a phenotype. Classical chemogenomics aims to identify small molecules that can act as tools to identify the protein(s) involved in a particular phenotypic response. Chemogenomic subsets have been created using compounds with well-annotated pharmacological activities, often covering over 1000 mechanisms. Hits from these subsets with known common pharmacologies can be used to make and test hypothesis about the causative factors for phenotypic changes. Known targets of confirmed hits can be analyzed by computational pathway analyses such as a causal reasoning engine [78], which in turn leads to additional hypothesis testing of compounds that should also produce the desired phenotype. Many iterations of hypothesis testing are done to shore up targets and pathways involved in phenotype effects. While there are 1000–1500 mechanisms that can be interrogated using chemogenomics subsets, genome-wide RNAi libraries are available to both confirm mechanisms implicated by small-molecule approach and to test mechanisms for which there are no small molecules [79]. RNAi, which allows for selective knockdown of specific genes, is a posttranslational process triggered by the introduction of double-stranded RNA that leads to gene silencing in a sequence-specific manner. RNAi knockdown isn't always complete, however, and off-target effects can be produced. Recently, several newly developed genomic editing approaches such as zinc-finger nuclease (ZFN), transcription activator-like effector nucleases (TALEN), and clustered regulatory interspaced short palindromic repeat (CRISPR) have shown promise in improving knockdown efficiency, but there is limited experience with these novel technologies and they come with their own set of caveats [80]. Combining chemogenomics with specific gene knockdown technologies provides a powerful approach to interrogating a wide range of potential mechanisms underlying phenotypic endpoints. There are a number of available technologies to support target deconvolution including chemoproteomics, metabolomics, transcriptomics, and computational methods that are beyond the scope of this chapter, but see the following references for a more in-depth review [81, 82]. In addition to these approaches to biological validation and characterization, assessment of chemical matter for drug-like properties, *in vitro* ADME, and *in vitro* safety properties all contribute to the winnowing and selection of a chemical series that is optimal for progression to lead optimization.

7.4 SCREENING STRATEGIES FOR OPTIMIZING PK AND SAFETY

The importance of optimizing the PK profile and eliminating known safety issues for a compound as part of the process leading to the nomination of a clinical candidate has been clearly illustrated through analyses of factors contributing to clinical attrition. Indeed, in the 1990s, poor PK and biopharmaceutical properties were found to cause a high percentage of clinical candidate attrition [83–86]. Safety has historically been, and continues to be, a key attrition factor, though as mechanistic toxicology and adverse drug reactions (ADRs) become better understood, it has become possible to proactively avoid some recognized issues [87, 88]. Since we do not have a complete knowledge of the mechanisms contributing to either inadequate PK or safety profiles, it is unlikely that attrition due to these factors will ever be totally eliminated, particularly since experimental strategies to treat disease states constantly evolve and diversify. However, significant progress has been made to drive down attrition in these areas over the last two decades. One strategic approach that has produced positive results in reducing attrition has focused on clearly identifying and measuring key properties that convey the PK profile and eliminating known safety liabilities. Important to this has been establishing the means to profile these properties in early discovery while there is still opportunity to take action with respect to candidate design and selection. Practically, this has been realized as plate-based measurement of physical properties, *in vitro* absorption, distribution, metabolism and elimination (ADME) assays, and safety assays run at a throughput commensurate with discovery chemistry efforts. This enables the profiling of multiple series and the optimization of these factors in large numbers of compounds in parallel with efforts to improve potency. As time has evolved, databases have been created and populated, from which powerful *in silico* tools have been built. Many of these *in silico* tools have become integral to the drug design process [89–91] thus helping to avoid liabilities or to optimize desirable profiles even before compounds are made (a more thorough review of *in silico* models and their application to ADME and safety profiling and prediction is presented in Chapter 10). This section will review key advances, strategies, and technologies that are routinely employed in early discovery to optimize small-molecule PK profiles and eliminate known safety risks.

7.4.1   High-Throughput Optimization of PK/ADME Profiles

In modern drug discovery, optimization of ADME and physical properties now occurs at very early stages, in parallel with, or closely following, potency optimization. Along with directly reducing attrition due to poor PK profile, optimizing ADME properties early in lead molecules has positive indirect impact through enabling preclinical *in vivo* efficacy and safety testing. More specifically, lower, knowledge-based doses may be administered thus avoiding the need for large quantities of chemical matter and potential solubility challenges that may occur at higher doses [92, 93, 85]. The availability of high-volume ADME data also allows the creation of SAR and the possibility of probing broader dimensions of chemistry space for potential drugability. The rationale to conduct ADME profiling on a scale to support high-volume chemistry analoging and early potency optimization efforts is clear, though realizing the ability to do this has required several unique scientific and technical advancements along with strong sponsorship from senior leadership to resource and support focused efforts in this arena [85].

Perhaps, one of the key scientific advancements was the establishment of a relationship between ADME properties and the PK profile. Indeed, *in vivo* to *in vitro* correlation (IVIVC) is well understood for many properties and mechanisms though with varying degrees of sophistication from flagging or binning to prediction. Furthermore, the practice of predicting human PK profiles from modeling of *in vitro* ADME endpoints is now routinely performed preclinically, in preference to prediction from preclinical animal PK studies [94, 95]. Much work has been done in academia and industry to build relationship models from *in vivo* (whole organism) to *in situ* (organ), to *in vitro* (cell and tissue models), to physical properties, and in recent times to *in silico*. Certainly, this has driven greater clarity in experimental model selection due to a better understanding of the key factors to be measured. Our modern-day understanding of these mechanisms was often the consequence of studies aimed at understanding interesting drug behavior, and we now have many well-established compound sets that have great utility in establishing quantitative or semiquantitative translation from experimental models to potential clinical outcomes. It is beyond the scope of this chapter to exhaustively review the specific mechanisms and details underlying ADME properties, but further information can be found in Chapter 2 of this book.

Keeping pace with chemistry efforts aimed at optimizing potency, often times utilizing high-speed analogue and parallel medicinal chemistry approaches that could yield hundreds of compounds per cycle (albeit producing only small amounts of compound), required the creation of *in vitro* ADME models that could be performed quickly on plate-based scales utilizing minimal compound amounts. The widespread establishment of cell culture models (e.g., Caco-2, MDCK) and readily available reagents (e.g., human liver microsomes and hepatocytes) provided key tools for *in vitro* ADME assay development; indeed, these models and reagents are amenable to miniaturization, and their ease of use enables a high number of studies to be performed per given cycle. At present, a wide variety of high-quality and well-characterized *in vitro* ADME reagents and cell lines is now commercially available, and further expansion and establishment of *in vitro* ADME reagents remain an active area of research.

In parallel, and of equal importance to high-throughput *in vitro* model development, have been complementary technological advances in workflow and sample analysis. In most cases, the endpoint measured in ADME profiling is the compound of interest itself, rather than a reporter or downstream signaling event as is often the case in pharmacology and safety assays. This presented four major challenges to be overcome [92]. First, given hundreds to thousands of compounds requiring analysis, an ability to rapidly generate analytical methods was needed. The solution involved creating automated methods development, applicable on the plate scale with associated informatics capabilities to manage compound details and the resulting analytical conditions specific to each test compound [96–98]. The second challenge was that miniaturized assays using small amounts of compound dictated that highly sensitive instruments were used to perform sample analysis. In this case, the utilization of MS technology played a powerful role by providing the necessary sensitivity to quantify samples in the nanomolar to micromolar ranges. The third major challenge was a requirement for high analytical throughput and capacity since multiple assays, each yielding many sets of samples, are commonly performed per given compound. One approach to overcoming this challenge was the creation of high-throughput autosamplers, capable of handling many samples per minute and plates per hour (the autosampler takes samples from plates and introduces them to the analyzer, in this case the mass spectrometer) [97, 98]. The importance of tandem mass spectrometry (MS/MS) to overcoming the three aforementioned challenges should be noted. MS/MS allows for a rapid and highly specific, yet transferrable, analytical method to be created that can be utilized by multiple instruments. The high sensitivity and specificity greatly reduces the need to perform chromatographic separations for reliable analysis thus further increasing sample throughput. Finally, the forth challenge required the ability to rapidly handle a large volume of data, both in terms of reviewing each sample analysis and in constructing endpoints from multiple samples. Automated approaches to each problem have been created and typically involve laboratory information systems. Over time, many of these solutions have gone from being proprietary to being commercially available and include software products such as Discovery Quant Optimize and Analyze (AB Sciex) and Galileo (Thermo Scientific) and hardware platforms such as the RapidFire (Agilent) and Apricot Designs Dual-Arm Autosampler (Apricot). The creation of this high-throughput analytical platform represents a major step forward in enabling early ADME profiling both from the standpoint of meeting capacity and demand, along with being universally applicable as an approach to measure samples across a wide variety of ADME endpoints.

7.4.1.1 Commonly Employed HT ADME Endpoints in Early Discovery

7.4.1.1 Commonly Employed HT ADME Endpoints in Early Discovery The following parameters are assessed when optimizing a compound's PK profile: the compound's bioavailability and distribution, or ability to reach the target site following an acceptable dosing route (typically oral); the required residency in the body to yield efficacy (related to half-life and elimination, expressed as clearance); and the potential for drug–drug interactions (DDI). Assays and relationships to physical properties have been developed to evaluate each of these key aspects, and these are routinely employed and well documented in the literature [93]. Rather than provide an in-depth review of the existing suite of assays, the following section will highlight mechanistic elements of these critical processes of bioavailability, distribution, clearance, and DDI and provide some example systems that are used for profiling.

Most drugs are dosed orally and, as such, ensuring that the compound has the necessary biopharmaceutical properties to be absorbed from the gut lumen into the blood is critical. The basic underlying properties related to absorption are permeability and solubility [99]. These two properties can be described by more fundamental physical characteristics including lipophilicity, hydrogen bonding abilities (to donate and to accept), polar surface area, and molecular size. In fact, Lipinski's rule of 5 is rooted in these basic properties [24]. The central importance of optimizing these properties was illustrated in a review of commercially available oral drugs where a scant 12% did not obey this rule [100]. Most of these properties are calculated values, but often, lipophilicity is experimentally measured. Commonly utilized approaches include classic octanol–water shake-flask methods, immobilized artificial membrane or chromatographic approaches, and computational fragment-based approaches [101, 102]. It is also a common practice to experimentally determine solubility, and methods deployed in early discovery include kinetic approaches where predissolved compound in a universal solvent such as DMSO is left to equilibrate in the appropriate buffer. The dissolved concentration can then be determined by a variety of endpoint measurements including UV detection, nephelometry, turbidity, and MS [103, 104]. Additionally, permeability (notably, intestinal permeability) is commonly measured using artificial models and tissue culture models where the ability of a compound to partition into the membrane and subsequently diffuse across the membrane is determined by using bicameral systems with donor and acceptor chambers. The parallel artificial membrane permeability assay (PAMPA) utilizes common lipids (such as dodecane and phosphatidylcholine) or extracts (such as soy lecithin) to form the membrane [105, 106]. Key advantages of PAMPA include ease of generating and optimizing the reagent, whereas disadvantages include translation to true tissue permeability and series-specific behavior. Tissue culture models such as Caco-2 and MDCK, which spontaneously differentiate into polarized epithelium under normal culture conditions, have been widely utilized [107–109]. Both models allow for a multimechanistic study of transport processes across intestinal epithelium though both require cell culturing and can be labor intensive.

Once the compound is absorbed and present in the systemic circulation, the compound must have the ability to distribute to the site of action in order to elicit an effect. Again, permeability is a critical parameter that governs a compound's ability to access tissues. This is particularly relevant for targets in the brain where ability for the compound to cross the specialized blood–brain barrier (BBB) is a prerequisite to therapeutic effect. Many of the aforementioned permeability models have been used as surrogates to gauge tissue distribution since specialized permeability models for BBB transport have been slow to develop and difficult to use effectively in early discovery. The role that transporters play in the distribution of their substrates is, in some cases, well understood, and transporter impact has been clearly shown in relation to BBB penetration. Many of the ABC transporters, notably P-glycoprotein (P-gp), breast cancer resistance protein (BCRP), and multidrug-resistance protein (MRP), can efflux a wide spectrum of drugs and exist in barrier tissues throughout the body. It is thought that these transporters serve a protective role and limit exposure of these pharmacological sanctuaries to foreign compounds that are substrates. Drugs with greater lipophilicity are typically effluxed by P-gp, whereas drugs with intermediate lipophilicity and some conjugated drugs are substrates for BCRP and MRP. Assays to profile for efflux potential have tended to utilize Caco-2 cells (P-gp and BCRP), or transfected MDCK cells (P-gp, BCRP, MRP), and to determine transport polarity via comparing absorptive transport (apical to basolateral) versus secretory transport (basolateral to apical) [110, 111]. Another key factor involved in distribution is protein binding, notably plasma protein binding. While this parameter is useful in establishing the free drug concentration, which is necessary for PK predictions and is therefore routinely profiled, it is not a property that should be actively optimized against as it plays no role in drug action [85]. Commonly used methods to measure protein binding in higher-throughput formats include ultrafiltration and equilibrium dialysis approaches [112–114].

It is important to understand how long a drug can remain in the circulation, or conversely, the rate at which it is removed from the body, which is commonly expressed as the flow parameter, clearance. Major mechanisms that determine clearance are metabolic stability (metabolic clearance is typically observed for more lipophilic compounds) and biliary and renal excretion, which may involve passive and/or active transport from blood into bile and urine, respectively. The liver is the major organ active in drug metabolism, and commonly employed reagents used to gauge metabolic stability are derived from liver tissue including human liver microsomes and hepatocytes. Microsomes have been used with much success to profile phase I, or oxidative, metabolism where polarity is imparted to the parent compound [115]. Hepatocytes are a whole-cell system and can be used to assess both phase I and II metabolism. Phase II metabolism imparts hydrophilicity to the compound and includes synthesis and conjugation reactions such as glucuronidation and sulfation. Reports have shown that data generated using these assays can be used to inform PK predictions when metabolism is the rate-limiting event that determines clearance [94]. The amenability of these reagents to high-throughput approaches, translatability to clinical outcomes, and applicability to a large majority of drug-like chemical space have led to them being placed on the front lines of ADME profiling. The emerging importance of biliary and renal excretion as major pathways of clearance for metabolically stable compounds has led to the creation of models to characterize these events. When these pathways have been shown to be important contributors to overall clearance, active transport mediated by transporters has played a significant role. Whole-cell systems such as sandwich-cultured human hepatocytes have been used to determine biliary elimination [116, 117]. Additionally, cell systems transfected with transporters of interest, including OATP, OAT, and OCT isoforms, have also been used to gauge these processes [118, 119].

Finally, the potential for a drug to interact with another drug, resulting in significant changes to the PK profile, is commonly assessed early in discovery to identify any safety-related implications. These DDIs can result in changes to bioavailability, distribution, and/ or clearance parameters and thus impact the PK profile of the victim drug. In most cases, these interactions can be linked to specific biochemical interactions between a drug and enzymes, proteins, or transporters. Specifically, profiling for DDI potential with the major cytochrome P450 enzymes, CYP1A2, CYP2C, CYP2D6, and CYP3A4 using either recombinant CYP enzymes or microsomes together with marker substrates, is performed early in discovery in a high-throughput fashion due to numerous clinical reports that have revealed CYP-mediated DDI [120]. Some CYP enzymes are important to profile due to the large variety of substrates metabolized by the isoform (e.g., CYP3A4), while others, which may be more minor players in overall numbers of substrates metabolized, are important to profile due to the existence of polymorphisms within the population (e.g., CYP2D6). There are a variety of potential DDIs involving CYPs that need to be assessed including the potential for competitive inhibition and time-/mechanism-based inhibition [121]. Certain CYPs are also inducible, necessitating yet another DDI potential to be quantitated. In this case, experiments may study perpetrator propensity to bind and activate the nuclear receptors, PXR and CAR, or profile changes in CYP expression in whole-cell models such as primary hepatocytes [122]. These data are typically included in regulatory submissions; therefore, the degree of rigor of the assays typically increases as a candidate compound progresses from discovery into development. Recently, the potential of transporters to be involved in DDI has been appreciated. The impact of interactions between P-gp and digoxin has been seen clinically, and assays have been developed to profile these events [123, 124]. Assessments of the interactions with OATP and with renal transporters such as OATs are commonly performed, particularly for compounds that are primarily renally excreted [119]. It is likely that as more candidates, for which transporters play a major ADME role, reach the clinic, a greater need to profile transporter-mediated DDI will emerge.

7.4.2   Early Safety Profiling

Safety concerns remain a leading cause of attrition [87, 88]. While safety is always studied in preclinical models, profiling in whole-animal models has proven to be insufficient to achieving an optimal approach to safety profiling and optimization [125]. Indeed, these models, typically short-term, repeat dose studies, are intensive studies requiring large amounts of compound in addition to significant resources and planning and the use of animals. As such, they tend to be performed in the later stages of discovery when reoptimization of lead compounds becomes more challenging. Furthermore, there remains the issue of translation from animal species to humans. For these reasons, safety screening and profiling have been advanced to the early stages of drug discovery alongside potency and ADME optimization. The strategic approach for this early screening often biases toward allowing false negatives rather than false positives to avoid the risk of deselecting potentially useful compounds along with the reliance on downstream profiling to ultimately identify unsafe compounds [125]. While there is a reasonable breadth in the technologies employed to profile safety in early discovery, early safety profiling falls into two broad categories including general, and specific, or off-target pharmacology.

7.4.2.1   General Safety Profiling in Early Discovery   In general, safety profiling outcomes are typically profiled using whole-cell, tissue, and in some cases organ models. The endpoints of these assays tend to be complex and may be the result of multiple perturbations of the system. The Ames assay that addresses a compound's potential to cause genotoxicity is the most widely utilized assay and a prerequisite to dosing in humans [125]. Scaled down versions of the Ames test and surrogates, including the *in vitro* micronucleus assay, have been developed for high-throughput profiling [126, 127]. Recently, a high-throughput assay using THLE cells and monitoring ATP depletion was shown to successfully predict cytotoxicity [128].

An area of current active research involves optimizing emerging technologies such as microarray profiling and high-content screening to support early discovery efforts. Many microarrays now exist that include, but are not limited to, oxidative stress, mitochondrial dysfunction, differentiation, inflammation, and apoptosis pathways allowing a targeted profiling of these outcomes [125]. The effectiveness of these toxicogenomic tools is dependent on the genomic coverage of the microarray associated with a given outcome and the bioinformatics capabilities to perform inference analysis against curated databases and to provide direction on next steps to eliminate the toxic outcome [129–132]. In addition to the need to advance the science underlying these genomic-based technologies, key operational needs include cost reduction and throughput enhancement.

High-content screening has been employed successfully for early safety profiling. Much of the work has involved establishing models to profile liver toxicity, and in many cases, hepatocytes or surrogate systems have been utilized [128, 133]. The human hepatocellular liver carcinoma (HepG2) cell line has been used in a high-content format to profile cell number, nuclear area, calcium content, and mitochondrial membrane potential. Primary human hepatocytes have been used to determine potential liver injury with known toxic reference compounds [134]. A commercially available assay with these cells is offered by CellCiphr that includes markers for oxidative stress, organelle function, stress pathway activation, cytoskeletal integrity, cell cycle, and DNA damage. In addition to cell/organ high-content screens, assays exist to profile toxicity on the organelle level. Recently, Luxcel Biosciences has developed assays that can profile mitochondrial toxicity using soluble phosphorescence and fluorescence-based sensors for extracellular oxygen consumption. These assays can be adapted to the system of interest, from isolated mitochondria to organisms and tissues [135–137]. Key advances in high-content screening beyond continued ability to profile new endpoints include a push toward establishing physiologically relevant assay formats that more accurately recapitulate and translate to *in vivo* outcomes [125, 138].

7.4.2.2 Safety (Off-Target) Pharmacology in Early Discovery While there are many well-known Adverse Drug Reactions (ADRs) that have been established from published clinical results and from specific experiences, there is only one safety pharmacology assay required for regulatory submissions [88]. Specifically, profiling against the heterologously expressed, human voltage-gated potassium channel subfamily H member (hERG) is required due to the association of compounds that block this channel with potentially fatal arrhythmia (Torsades de Pointe). Typically, profiling includes the use of a binding assay, screening with astemizole or dofetilide, followed by patch-clamp assays [139, 140].

A recent publication from Bowes et al. [88] compares the collective safety experiences of four major pharma and suggests a strategic approach together with a baseline set of targets against which to profile, using *in vitro* models [88]. These authors suggest a five-tiered approach to safety that is enabled by *in vitro* pharmacology. The first phase is directed at early hazard identification and is performed mostly at the series level by screening against panels of molecular targets associated with known ADRs. The second phase is focused on hazard elimination in the lead optimization phase and utilizes many of the same screens as the first phase to enable compound optimization through the elimination of off-target pharmacology. Phase three is performed as part of the candidate nomination process and involves the synthesis of experimental data that can be used to guide compound selection and to design subsequent *in vivo* studies. As more data are gathered for a compound such as the ADME and PK profiles, the safety information can be placed in context with dose, expected free plasma concentrations, and efficacious concentration in phase four to enable risk assessment in early development. Finally, the fifth phase typically occurs in preclinical and clinical development when regulatory studies are conducted and the safety pharmacology data can be applied to an understanding of the mechanistic effects underlying clinical observations. In this way, *in vitro* safety pharmacology is driven simultaneously in both a feedforward (meeting the needs of more complex challenges as discovery and development progress) and a feedback fashion (where feedback is provided in the form of novel clinical ADRs).

The technologies utilized for *in vitro* safety pharmacology profiling are very much the same as those used in plate-based primary pharmacology and include ligand-binding approaches and functional approaches performed in single-concentration and dose-response formats to obtain an effect profile. *In vitro* safety pharmacology studies typically include panels of molecular targets including receptors, ion channels, enzymes, and transporters. The following is the baseline panel suggested by Bowes et al. (for more details, please reference Table 1 in the Bowes work, which lists key references that discuss at more length the association of the ADR with the molecular target [88]). GPCRs in both the neurotransmitter and peptidergic classes dominate the panel. Key examples profiled include isoforms of adenosine, adrenergic, muscarinic, opioid, and serotonergic GPCRs. The peptidergic receptors include those responsive to cholecystokinin, endothelin, and vasopressin. Members of the ion channel family that are often profiled include the acetylcholine receptor subunit $\alpha 1$ or $\alpha 4$, calcium voltage-gated channel Cav1.2, GABA receptors, the aforementioned hERG channel, the potassium voltage-gated channel KQT, NMDA receptors, 5-HT3 receptors, and the sodium voltage-gated channel Nav1.5. Enzymes include cyclooxygenases 1 and 2, monoamine oxidase A, phosphodiesterases 3 and 4, and lymphocyte-specific protein tyrosine kinase. In the transporter class, neurotransmitter transporters for dopamine, noradrenaline, and serotonin are included. Finally, nuclear hormone receptors profiled include androgen and glucocorticoid receptors.

Key challenges within this field include the continued expansion of molecular target space and the need to build computational abilities to inform compound design. It is likely that as newer target families including kinases and epigenetic targets gain more exposure in the clinic, novel ADRs will arise with a consequent need to establish the means to avoid them through *in vitro* screening. As the datasets of compounds profiled in safety pharmacology assays increase in size, corporate databases are being utilized in the creation of *in silico* models that can identify pharmacophores for undesired activity.

7.4.3 Future Directions for ADME and Safety in Lead Optimization

The clear impact that optimization of ADME and safety endpoints has had on attrition has solidified their place next to potency as integral to consider during early drug discovery. High-throughput ADME and safety screening has evolved considerably from a place of establishing a mechanistic understanding of key factors conferring desirable PK profile and eliminating avoidable safety risks in hand. The technical ability to profile these endpoints at scale has enabled the harmonization and standardization of conditions thus driving SAR and computational efforts. Currently, we continue to incorporate additional factors as ADME and safety sciences advance and related clinical outcomes become more clearly understood. Indeed, as the number of factors profiled increases, digesting and making decisions on this volume of multifactorial data in a high-paced early discovery environment is becoming more challenging. The focus on precision medicine approaches will also manifest increased complexity as many ADME and safety-relevant mechanisms show genetic variation across populations.

The future early discovery challenge for ADME will be how to quickly identify key factors that dictate the PK profile while creating the logistics and assay paradigms that allow rapid profiling against these factors to drive optimization and design efforts. The importance of physical properties in determining PK and ADME characteristics of compounds has been recognized and further utilized as a means to categorize and predict *in vivo* behavior [141–144]. Frameworks such as this could be employed for directed profiling. Future safety efforts in early discovery will build translatable safety endpoints and relationships (including *in silico* and modeling efforts) to predict safety margins, a step beyond the current capability to flag potential issues [88, 125]. The physiologically relevant and phenotypic assay trends hold true in this arena also, and induced pluripotent stem cells or patient-derived stem cells along with high-throughput whole-animal models (e.g., zebrafish and nematodes) will ultimately enable a more complex and biologically relevant safety profiling [144–146].

7.5 SUMMARY

Scientific and technological advances over the past 25 years together with the growing body of knowledge from clinical outcomes have led to a dramatic evolution of screening paradigms applied in early discovery. *In vitro*, high-throughput screening capabilities have been developed, implemented, and in many cases optimized across the breadth of early pipeline activities in an attempt to build a greater understanding of compound characteristics as early as possible in a program. The large volumes of data produced by these approaches have been ideal for the generation of *in silico* models that have further enabled the leap to predictive sciences. Jointly, these approaches drive iterative design hypothesis and enable parallel multiparameter optimization across specific endpoints in an attempt to achieve desired pharmacology, ADME/PK, and safety profiles that will ultimately translate successfully into the clinic.

Despite these tools at our fingertips, success for the pharma industry will still require careful selection among the multitude of options that arise at each stage of the program commencing with the identification of disease area in which to invest, the pathway or clinical phenotype to pursue, or the specific protein target to modulate. These decisions can be guided by known linkage to human genetics and any clinical- or literature-based prece- dence. Solid validation data are required early to build the confidence in biology necessary to ensure return on screening investments. Prosecuting a portfolio of programs in a cost- constrained environment requires the selection of the most appropriate hit identification strategy for each target and likely includes a balance of phenotypic and biochemical target- based approaches with an increasing emphasis on using more physiologically relevant systems such as stem cells, primary cells, or patient-derived cells. The ability to expose the most appropriate chemical equity to the disease model/target requires strategic library design in addition to flexible collection access in order to align the chemical matter screened with the desired outcome, whether it is a tool compound or a therapeutic lead. The format of compound presentation to the assay system (solvent, concentration, singleton, mixtures) is also a significant factor in ensuring the identification of active chemical matter for a program. In fact, these choices are generally made hand in hand with the decisions on screening methodologies that are possible and most suitable for the program. Assay development will aim to employ disease-relevant systems, stimuli, and endpoints while if possible, ensuring scalability to enable exposure to the greatest chemical space if needed.

Screening funnel design is critical to ensure assays generate data that will guide medicinal chemistry decision making in the subsequent design cycles. Funnels typically incorporate potency measures for selectivity and for species translation, which provide guidance in the suitability of preclinical models. In a similar timeframe, ADME screening panels assess parameters associated with bioavailability, distribution, clearance, and DDI. Early safety assessments have also become incorporated into early discovery activities to probe endpoints indicative of general toxicity (including genotoxicity, cytotoxicity, and some specific organ toxicity, e.g., liver and cardiac) and for specific off-target pharmacological safety issues through panel screening. The balance of profiles along with a clear understanding of medical need, development risks, and commercial viability will all play into the critical decisions around lead series selection and clinical candidate nomination at which point the attrition destiny is at least partly determined.

Even though we have achieved tremendous advances in screening strategies, the science and knowledge base continues to advance, especially as we gain experience with an ever-expanding target space. Future work will continue to progress high-content imaging capabilities, phenotypic deconvolution strategies, better predictive toxicity measures, an expansion into chemical space beyond rule of 5 compliant compounds, and will need to tease apart the role of transporters in tissue distribution for both an understanding of drug exposure to the target and also for predicting and optimizing clearance.

REFERENCES

1   Hewitt J., Campbell, J. D., Cacciotti, J. (2011). Beyond the shadow of a drought the need for a new mindset in pharma R&D. *Oliver Wyman Health and Life Sciences*, 1–20.

2   Herper, M. (2012). The truly staggering cost of inventing new drugs. *Forbes Magazine*, February 2012.

3   Paul, S.M., Mytelka, D.S., Dunwiddie, C.T., Persinger, C.C., Munos, B.H., Stacy R. Lindborg, S.R., Schacht, A.L. (2012). How to improve R&D productivity: the pharmaceutical industry's grand challenge. *Nature Reviews Drug Discovery 9*, 203–214.

4   Samanen, J. (2012). NME Output versus R&D Expense – Perhaps there is an explanation. Samanen Consulting. http://www.discoverymanagementsolutions.com/the-organization-of-biopharmaceutical-rd/common-goals-between-discovery-and-development/innovation/nme-output-versus-rd-expense-perhaps-there-is-an-explanation/

5   Tollman, P., Morieux, Y., Murphy, J.K., Schulze, U. (2011). Identifying R&D outliers. *Nature Reviews Drug Discovery, 10*, 653–654.

6   Munos, B. (2009). Lessons from 60 years of pharmaceutical innovation. *Nature Reviews Drug Discovery, 8*, 959–968

7   Johnson, K.A. (2013). A century of enzyme kinetic analysis, 1913 to 2013. *FEBS Letters 587*, 2753–2766.

8   Winquist, R.J., Mullane, K., Williams, M. (2014). The fall and rise of pharmacology—(Re-) defining the discipline? *Biochemical Pharmacology 87*, 4–24.

9   Venter, C. et al. (2001). The Sequence of the Human Genome. *Science, 291*, 1304–1351.

10 Hopkins, A.L., Groom, C.R. (2002). The Druggable Genome. *Nature Reviews Drug Discovery, 1*, 727–730.

11 Ratner, M. (2002). File enrichment: the way out of pharma's productivity crisis? StartUp Technology Strategies. Article # 2002900199.

12 Sundberg, S.A. (2000). High-throughput and ultra-high-throughput screening: solution- and cell-based approaches. *Current Opinion Biotechnology, 11*:47–53

13 Scannell, J.W., Blanckley, A., Boldon, H., Warrington, B. (2012). Diagnosing the decline in pharmaceutical R&D efficiency. *Nature Reviews Drug Discovery, 11*, 191–200.

14 Schoonen, W.G.E.J., Westerink, W.M.A., Horbach, G.J. (2009). High-throughput screening for analysis of in vitro toxicity. *Molecular, Clinical and Environmental Toxicology, Springer-Verlag Basel, 2009*, pp 401–452.

15 Recanatini, M., Poluzzi, E., Masetti, M., Cavali, A., De Pont, F. (2005). QT prolongation through hERG K$^+$ channel blockade: current knowledge and strategies for the early prediction during drug development. *Medicinal Research Reviews, 25*, 133–166.

16 Benbow, J., Aubrecht, J., Banker, M., Nettleton, D., Aleo, M.D. (2010). Predicting safety toleration of pharmaceutical chemical leads: cytotoxicity correlations to exploratory toxicity studies. *Toxicology Letters, 197*, 175–182.

17 Comanora, W.S., Schererb, F.M. (2013). Mergers and innovation in the pharmaceutical industry. *Journal of Health Economics, 32*, 106–113.

18 Elmer, D. J. (2008). Today's biopharma venture capital challenge: risk-sharing models versus risk avoidance. Drug Discovery News, p. 10, Online, http://www.ddn-news.com/index.php?newsarticle=2225 (accessed July 16, 2015).

19 News in Brief (2012). Nature Reviews Drug Discovery, *11*, 739.

20 Kogej, T., Blomberg, N., Greasley, P.J., Mundt, S., Vainio, M.J., Schamberger, J., Schmidt, G., Huser, J. (2012). Big Pharma screening collections: more of the same or unique libraries? The AstraZeneca-Bayer Pharma AG case. *Drug Discovery Today, 18*, 1-14-1024.

21 Drews, J., Ryser, S. (1997). Classic drug targets. *Nature Biotechnology, 15*,1318–1319.

22 Overington, J.P., Al-Lazikani, B., Hopkins, A.L. (2006). How many drug targets are there? *Nature Reviews Drug Discovery, 5*, 993–996.

23 Rask-Andersen, M., Almen, M.S., Schioth, H.B. (2011). Trends in the exploitation of novel drug targets. *Nature Reviews Drug Discovery, 10*, 579–590.

24 Lipinski, C.A.; Lombardo, F., Dominy, B.W., Feeney, P.J. (1997). Experimental and computational approaches to estimate solubility and permeability in drug discovery and development settings. *Advanced Drug Delivery Reviews, 23*, 3–25.

25 Hopkins, A.L. (2008). Network pharmacology: the next paradigm in drug discovery *Nature Chemical Biology, 4*, 682–690.

26 Frye, S., Crosby, M., Edwards, T., Juliano, R. (2011). US academic drug discovery. *Nature Reviews Drug Discovery, 10*, 409–410

27 Tralau-Stewart, C. Low, C.M., Martin, N. (2014). UK academic drug discovery. *Nature Reviews Drug Discovery, 13*, 15–16.

28 Arrowsmith, J. (2011). Trial watch: phase III and submission failures: 2007–2010. *Nature Reviews Drug Discovery, 10*, 87.

29 Arrowsmith, J., Miller, P. (2013). Trial watch: phase II and phase III attrition rates 2011–2012.*Nature Reviews Drug Discovery, 12*, 569.

30 Prinz, F. Schlange, T., Asadullah, K. (2011). Believe it or not: how much can we rely on published data on potential drug targets? *Nature Reviews Drug Discovery, 10*, 712.

31 Begley, C.G., Ellis, L.M. (2012). Drug development: raise standards for preclinical cancer research. *Nature, 483*, 531–533.

32 Facilitating reproducibility, (2013). editorial, *Nature Chemical Biology, 9*, 345.

33 Collins, F.S., Tabak, L.A. (2014). NIH plans to enhance reproducibility. *Nature, 505*, 612.

34 Plenge, R.M., Scolnick, E.M., Altschuler, D. (2013). Validating therapeutic targets through human genetics. *Nature Reviews Drug Discovery, 12*, 581–594.

35 Stein, E.A., Gipe, D., Bergeron, J. Gaudet, D., Weiss, R., Dufour, R., Wu, R. Pordy, R. (2012). Effect of a monoclonal antibody to PCSK9, REGN727/SAR236553, to reduce low-density lipoprotein cholesterol in patients with heterozygous familial hypercholesterolaemia

on stable statin dose with or without ezetimibe therapy: a phase 2 randomised controlled trial. *The Lancet, 380*, 29–36.

36　Bakken, G.A., Bell, A.S., Boehm, M., Everett, J.R., Gonzales, R., Hepworth, D., Klug-McLeod, J.L., Lanfear, J., Loesel, J., Mathias, J., Wood, T.P. (2012). Shaping a screening file for maximal lead discovery efficiency and effectiveness: elimination of molecular redundancy. *Journal of Chemical Information and Modeling, 52*, 2937–2949.

37　Schamberger, J., Grimm, M., Steinmeyer, A., Hillisch, A. (2011). Rendezvous in chemical space? Comparing the small molecule compound libraries of Bayer and Schering. *Drug Discovery Today, 16*, 636–641.

38　Pope, A. (2012). Screening Heuristics & Chemical Property Bias—New directions for Lead Identification and Optimization. Oral presentation at 2012 Meeting of the Society of Laboratory Automation and Screening (SLAS), San Diego, CA.

39　Jacoby, E., Schuffenhauer, A., Popov, M., Azzaoui, K., Havill, B., Schopfer, U., Engeloch, C., Stanek, J., Acklin, P., Rigollier, P., Stoll, F., Koch, G., Meier, P., Orain, D., Giger, R., Hinrichs, J., Malagu, K., Zimmermann, J., Roth, H.-J. (2005). Key aspects of the Novartis compound collection enhancement project for the compilation of a comprehensive chemogenomics drug discovery screening collection. *Current Topics in Medicinal Chemistry, 5*, 397–411.

40　LaMattina, J.L. (2011). The impact of mergers on pharmaceutical R&D. *Nature Reviews Drug Discovery, 10*, 559–560

41　Otterbein, L. (2013). Cost efficient and reliable management of compound quality by utilization of statistical methods. Oral presentation delivered at the 2013 IQPC Compound Management Meeting in London, UK, May 2013.

42　Roy, A., McDonald, P.R., Sittampalam, S., Chaguturu, R. (2010). Open access high throughput drug discovery in the public domain: a mount everest in the making. *Current Pharmaceutical Biotechnology, 11*, 764–778.

43　Boucharens, S. (2013). European Lead Factory: game changing for innovative medicine. Oral presentation delivered at the 2013 IQPC Compound Management Meeting in London, UK, May 2013.

44　Ruddigkeit L., van Deursen, R., Blum, L.C., Reymond, J.-L. (2012). Enumeration of 166 Billion Organic Small Molecules in the Chemical Universe Database GDB-17. *Journal of Chemical Information and Modeling, 52*, 2864–2875.

45　Nissink, J.W.M., Schmitt, S., Blackburn, S., Peters, S. (2014). Stratified High-Throughput Screening Sets Enable Flexible Screening Strategies from a Single Plated Collection. *Journal of Biomolecular Screening, 19*, 369–378.

46　Krass, N. (2013). Open innovation: new approaches supported by Compound Management. Oral presentation delivered at the 2013 IPQC Compound Management Meeting in London, UK, May 22, 2013.

47　Blaxill, Z., Holland-Crimmin, S., Lifely, R. (2009). Stability through the ages: the GSK experience. *Journal of Biomolecular Screening, 14*, 547–556.

48　Popa-Burke, I., Novick, S., Lane, C.A., Hogan, R., Torres-Saavedra, P., Hardy, B., Ray, B., Lindsay, M., Paulus, I., Miller, L. (2014). The effect of initial purity on the stability of solutions in storage. *Journal of Biomolecular Screening, 19*, 308–316.

49　Janzen, W.P., Popa-Burke, I.G. (2009). Advances in improving the quality and flexibility of compound management. *Journal of Biomolecular Screening, 14*, 444–451.

50　Matson, S.L., Chatterjee, M., Stock, D.A., Leet, J.E., Dumas, E.A., Ferrante, C.D., Monahan, W.E., Cook, L.S., Watson, J., Cloutier, N.J., Ferrante, M.A., Houston, J.G., Banks, M. N. (2009). Best practices in compound management for preserving compound integrity and accurately providing samples for assays. *Journal Biomolecular Screening, 14*, 476–484.

51 Bergsdorf, C., Ott, J. (2010). Affinity-based screening techniques: their impact and benefit to increase the number of high quality leads. *Expert Opinion on Drug Discovery*, *11*, 1095–1107.

52 Fox, S., Wang, H., Sopchak, L., Khoury, R. (2001). High throughput screening: early successes indicate a promising future. *Journal Biomolecular Screening*, *6*, 137–140.

53 Fox, S., Wang, H., Sopchak, L., Khoury, R. (2000). High throughput screening 2000: new trends and directions. HighTech Business Decisions. Available online at http://www.hitechbiz.com.

54 Jeffrey, P., Lock, R., Witherington, J. (2009). Highlights of the society for medicines research symposium: approaches to lead generation. *Drugs of the Future*, *34*, 501–508.

55 Gribbon, P., Lyons, R., Laflin, P., Bradley, J., Chambers, C., Williams, B.S., Keighley, W., Sewing, A. (2005). Evaluating real-life high-throughput screening data. *Journal of Biomolecular Screening*, *10*, 99–107.

56 Klumpp, M., Boettcher, A., Becker, D., Meder, G., Blank, J., Lederm, L., Forstner, M., Ott, J., Mayr, L.M. (2006). Readout technologies for highly miniaturized kinase assays applicable to high-throughput screening in a 1536-well format. *Journal of Biomolecular Screening*, *11*, 617–633.

57 Gribbon, P., Schaert, S., Wickenden, M., Williams, G., Grimley, R., Stuhmeier, F., Preckel, H., Eggeling, C., Kraemer, J., Everett, J. (2004). Experiences in implementing uHTS: cutting edge technology meets the real world. *Current Drug Discovery Technologies*, *1*, 27–35.

58 Boettcher, A., Mayr, L.M. (2006). Miniaturisation of assay development and screening. *Drug Discovery World Summer 2006*, 17–27. http://www.ddw-online.com/screening/p97061-miniaturisation-of-assay-development-and-screening-summer-2006.html

59 Brandish, P.E., Chiu, C-S., Schneeweis, J., Brandon, N.J., Leech, C.L., Kornienko, O., Scolnick, E.M., Strulovici, B., Zheng, W. (2006). A cell-based ultra-high-throughput screening assay for identifying inhibitors of D-amino acid oxidase. *Journal of Biomolecular Screening*, *11*, 481–487.

60 Vaschetto, M., Weisbrod, T., Bodle D., Güner, O. (2003). Enabling high-throughput discovery. *Current Opinion Drug Discovery & Development*, *6*, 377–383.

61 Merten, C.A. (2010). Screening Europe 2010: an update about the latest technologies and applications in high-throughput screening. *Expert Review of Molecular Diagnostics*, *10*, 559–563.

62 Gavrin, L.K., Saiah, E. (2013). Approaches to discover non-ATP site kinase inhibitors. *Medicinal Chemistry Communications*, *4*, 41–51.

63 Sharma, P., Ando, D.M., Daub, A. Kaye, J.A., Finkbeiner, S. (2012). High-throughput screening in primary neurons. *Methods in Enzymology*, *506*, 331–360.

64 Engle, S.J., Vincent, F. (2014). Small molecule screening in human induced pluripotent stem cell-derived terminal cell types. *Journal of Biological Chemistry*, *289*(8), 4562–4570.

65 Neuberger, T., Burton, B., Clark, H. van Goor, F. (2011). Use of primary cultures of human bronchial epithelial cells isolated from cystic fibrosis patients for the pre-clinical testing of CFTR modulators. *Cystic Fibrosis, Methods in Molecular Biology*, *741*, 39–54.

66 Dickinson, T., Banker, M.E., Bora, G., Salafia, M., Ren, Y., Rockwell, K., Vincent, F., Doyonnas, R., Hawrylik, S. (2014). Isolation/expansion of activated T cells to support physiological relevant HTS. Society for Laboratory Automation & Screening (SLAS). 2014 Annual Meeting Poster presentation, Washington, DC.

67 Banker, M.E., Bora, G., Salafia, M., Dickinson, T., Hawrylik, S. Vincent, F. (2014). Primary T-cell high throughput screen to identify novel lead matter. Society for Laboratory Automation & Screening (SLAS). 2014 Annual Meeting: Poster Presentation, Washington, DC.

68 McNeish, J., Roach, M., Hambor, J., Mather, R.J., Weibley, L., Lazzaro, J., Gazard, J., Schwarz, J., Volkmann, R.V., Machacek, D., Stice, S., Zawadzke, L., O'Donnell, C., Hurst, R. (2010).

High-throughput screening in embryonic stem cell-derived neurons identifies potentiators of α-amino-3-hydroxyl-5-methyl-4-isoxazolepropionate-type glutamate receptors. *Journal of Biological Chemistry*, *285*, 17209–17217.

69   Swinney, D.C., Anthony J. (2011). How were new medicines discovered? *Nature Reviews Drug Discovery*, *10*, 507–519.

70   Zheng, W., Thorne, N., McKew, J.C. (2013). Phenotypic screens as a renewed approach for drug discovery. *Drug Discovery Today*, *18*, 1067–1073.

71   Tan, J.L., Zon, L.I. (2011). Chemical screening in zebrafish for novel biological and therapeutic discovery. *Methods in Cell Biology*, *105*, 493–516.

72   Feng, Y., Mitchison, T.J., Bender, A., Young, D.W., Tallarico, J.A. (2009). Multi-parameter phenotypic profiling: using cellular effects to characterize small-molecule compounds. *Nature Reviews Drug Discovery*, *18*, 567–578.

73   Zhang, J.-H., Chung, T.D.Y., Oldenburg, K.R. (1999). A simple statistical parameter for use in evaluation and validation of high-throughput screening assays. *Journal of Biomolecular Screening*, *4*, 69–73.

74   Inglese, J., Auld, D.S., Jadhav, A., Johnson, R.L., Simeonov, A., Yasgar, A., Zheng, W., Austin, C.P. (2006). Quantitative high-throughput screening: A titration-based approach that efficiently identifies biological activities in large chemical libraries. *Proceedings of the National Academy of Sciences USA*, *103*, 11473–11478.

75   Song, K., Wang, J., Boehm, M., Mathiowetz, A., Klug-McLeod, J., Bhattacharya, S., Carpino, P. (2011). Analysis of Pfizer HTS promiscuous hits. Abstracts of Papers, 241st ACS National Meeting & Exposition, Anaheim, CA, Pages COMP-179.

76   Coan, K.E.D., Maltby, D.a., Burlingame, A.l., Shoichet, B. (2009). Promiscuous aggregate-based inhibitors promote enzyme unfolding. *Journal of Medicinal Chemistry*, *52*, 2067–2075.

77   Uitdehaag, J.C.M., Verkeer, F., Alwam, H., de Man, J.,Buijsman, R.C., Zaman, G.J.R. (2012). A guide to picking the most selective kinase inhibitor tool compounds for pharmacological validation of drug targets. *British Journal of Pharmacology*, *166*, 858–876.

78   Enayetallah, A.E., Ziemek, D., Leininger, M.T., Randhawa, R., Yang, J., Manion, T.B., Mather, D.E., Zavadoski, W.J., Kuhn, M., Treadway, J.L., des Etages, S.A.G., Gibbs, E.M., Greene, N., Steppan, C.M. (2011). Modeling the mechanism of action of a DGAT1 inhibiton using a causal reasoning platform. *PLoS ONE 6*, 1–10.

79   Yang, Y.M., Gupta, S.K., Kim, K.J., Powers, B.E., Cerqueira, A., Wainger, B.J., Ngo, H.D., Rosowski, K.A., Schein, P.A., Ackeifi, C.A., Arvanites, A.C., Davidow, L.S., Woolf, C.J., Rubin, L.L. (2013). A small molecule screen in stem-cell-derived motor neurons identifies a kinase inhibitor as a candidate therapeutic for ALS. *Cell Stem Cell*, *12*, 713–726.

80   Gaj, T., Gersbach, C.A., Barbas, C.F. (2013). ZFN, TALEN, and CRISPR/Cas-based methods for genome engineering. *Trends in Biotechnology*, *7*, 397–405.

81   Terstappen, G.C., Schlüpen, C., Raggiaschi, R., Gaviraghi, G. (2007). Target deconvolution strategies in drug discovery. *Nature Reviews Drug Discovery 6*, 891–903.

82   Schenone, M., Danick, V., Wagner, B.K., Clemons, P.A. (2013). Target identification and mechanism of action in chemical biology and drug discovery. *Nature Chemical Biology*, *9*, 232–240.

83   Baillie, T.A., Pearson, P.G. (2000). The impact of drug metabolism in contemporary drug discovery: new opportunities and challenges for mass spectrometry. In *Mass Spectrometry in Biology & Medicine* (Burlingame, A.L.; Carr, S.A.; Baldwin, M.A. Eds.), Humana Press, Totowa, NJ, pp. 481–496.

84   Kola, I., Landis, J. (2004). Can the pharmaceutical industry reduce attrition rates? *Nature Reviews Drug Discovery*, *3*, 711–715.

85 Smith, D.A., Early screening for ADMET properties. *Lead Generation Approaches in Drug Discovery*, John Wiley & Sons, Inc. New York, 2010, pp. 231–258.

86 Smith, D.A., Schmid, E.F. (2006). Drug withdrawals and the lessons within. *Current Opinion on Drug Discovery Development*, *9*, 38–46.

87 Kramer, J.A., Sagartz, J.E., Morris, D.E. (2007). The application of discovery toxicology and pathology towards the design of safer pharmaceutical lead candidates. *Nature Reviews Drug Discovery*, *6*, 636–649.

88 Bowes, J., Brown, A.J., Hamon, J., Jarolimek, W., Sridhar, A., Waldron, G., Whitebread, S. (2012). Reducing safety-related drug attrition: the use of in vitro pharmacological profiling. *Nature Reviews Drug Discovery*, *11*, 909–922.

89 Keefer, C.E., Kauffman, G.W., Gupta, R.R. (2013). Interpretable, probability-based confidence metric for continuous quantitative structure-activity relationship models. *Journal of Chemical Information and Modeling*, *53*, 368–383.

90 Keefer, C.E., Chang, G., Kaufman, G.W. (2011). Extraction of tacit knowledge from large ADME data sets via pairwise analysis. *Bioorganic and Medicinal Chemistry*, *19*, 3739–3749.

91 Honório, K.M., Moda, T.L., Andricopulo, A.D. (2013). Pharmacokinetic properties and in silico ADME modeling in drug discovery. *Medicinal Chemistry*, *9*, 163–76.

92 Hop, C.E., Cole, M.J., Davidson, R.E., Duignan, D.B., Federico, J., Janiszewski, J.S., Jenkins, K., Krueger, S., Lebowitz, R., Liston, T.E., Mitchell, W., Snyder, M., Steyn, S.J., Soglia, J.R., Taylor, C., Troutman, M.D., Umland, J., West, M., Whalen, K.M., Zelesky, V., Zhao, S.X. (2008). High throughput ADME screening: practical considerations, impact on the portfolio and enabler of in silico ADME models. *Current Drug Metabolism*, *9*, 847–853.

93 Di, L., Kerns, E.H. ADME properties of drugs. *Chemical Biology: Approaches to Drug Discovery and Development to Targeting Disease*, John Wiley & Sons, Inc. New York, 2012, pp. 101–114.

94 Hosea N.A., Collard W.T., Cole S., Maurer T.S., Fang R.X., Jones H., Kakar S.M., Nakai Y., Smith B.J., Webster R., Beaumont K. (2009). Prediction of human pharmacokinetics from preclinical information: comparative accuracy of quantitative prediction approaches. *Journal of Clinical Pharmacology*, *49*, 513–533.

95 Poulin P., Jones R.D., Jones H.M., Gibson C.R., Rowland M., Chien J.Y., Ring B.J., Adkison K.K., Ku M.S., He H., Vuppugalla R., Marathe P., Fischer V., Dutta S., Sinha V.K., Björnsson T., Lavé T., Yates J.W. (2011). PHRMA CPCDC initiative on predictive models of human pharmacokinetics, part 5: prediction of plasma concentration-time profiles in human by using the physiologically-based pharmacokinetic modeling approach. *Journal of Pharmaceutical Sciences*. *100*, 4127–4157.

96 Whalen, K., Gobey, J., Janiszewski, J. (2006). A centralized approach to tandem mass spectrometry method development for high-throughput ADME screening. *Rapid Communications in Mass Spectrometry*, *20*, 1497–1503.

97 Janiszewski, J.S., Rogers, K.J., Whalen, K.M., Cole, M.J., Liston, T.E., Duchoslay, E., Fouda, H.G. (2001). A high-capacity LC/MS system for the bioanalysis of samples generated from plate-based metabolic screening. *Analytical Chemistry*, *73*, 1495–1501.

98 Whalen, K.M., Rogers, K.J., Cole, M.J., Janiszewski, J.S. (2000). AutoScan: an automated workstation for rapid determination of mass and tandem mass spectrometry conditions for quantitative bioanalysis mass spectrometry conditions for quantitative bioanalysis mass spectrometry. *Rapid Communications in Mass Spectrometry*, *14*, 2074–2079.

99 Amidon, G.L., Lennernäs, H., Shah, V.P., Crison, J.R. (1995). A theoretical basis for a biopharmaceutic drug classification: the correlation of in vitro drug product dissolution and in vivo bioavailability. *Pharmaceutical Research*, *12*, 413–420.

100   Gleeson, M.P., Hersey, A., Montanari, D., Overington, J. (2011). Probing the links between in vitro potency, ADMET and physicochemical parameters. *Nature Reviews Drug Discovery*, *10*, 197–208.

101   van de Waterbeemd, H., Gifford, E. (2003). ADMET in silico modeling: towards prediction paradise? *Nature Reviews Drug Discovery*, *2*, 92–204.

102   Lombardo, F., Shalaeva, M.Y., Tupper, K.A., Gao, F. (2001). ElogD(oct): a tool for lipophilicity determination in drug discovery. 2. Basic and neutral compounds. *Journal of Medicinal Chemistry*, *44*, 2490–2497.

103   Avdeef, A. (2001). Physicochemical profiling (solubility, permeability and charge state). *Current Topics in Medicinal Chemistry*, *1*, 277–351.

104   Bevan, C.D., Lloyd, R.S. (2000). A high-throughput screening method for the determination of aqueous drug solubility using laser nephelometry in microtiter plates. *Analytical Chemistry*, *72*, 1781–1787.

105   Kansy, M., Senner, F., Gubernator, K. (1998). Physicochemical high throughput screening: parallel artificial membrane permeation assay in the description of passive absorption processes. *Journal Medicinal Chemistry*, *41*, 1007–1010.

106   Avdeef, A. (2005). The rise of PAMPA. *Expert Opinion on Drug Metabolism and Toxicology*, *1*, 325–342.

107   Artursson, P., Karlsson, J. (1991). Correlation between oral drug absorption in humans and apparent drug permeability coefficients in human intestinal epithelial (Caco-2) cells. *Biochemical and Biophysical Research Communications*, *175*, 880–885.

108   Cho, M.J., Thompson, D.P., Cramer, C.T., Vidmar, T.J., Scieszka, J.F. (1989). The Madin Darby canine kidney (MDCK) epithelial cell monolayer as a model cellular transport barrier. *Pharmaceutical Research*, *6*, 71–77.

109   Di, L., Whitney-Pickett, C., Umland, J.P., Zhang, H., Zhang, X., Gebhard, D.F., Lai, Y., Federico, J.J., Davidson, R.E., Smith, R., Reyner, E.L., Lee, C., Feng, B., Rotter, C., Varma, M.V., Kempshall, S., Fenner, K., El-Kattan, A.F., Liston, T.E., Troutman, M.D. (2011). Development of a new permeability assay using low-efflux MDCKII cells. *Journal of Pharmaceutical Sciences*, *100*, 4974–4985.

110   Augustijns, P.F., Bradshaw, T.P., Gan, L.S., Hendren, R.W., Thakker, D.R. (1993). Evidence for a polarized efflux system in CACO-2 cells capable of modulating cyclosporin A transport. *Biochemical and Biophysical Research Communications*, *197*, 360–365.

111   Troutman, M.D., Thakker, D.R. (2003). Novel experimental parameters to quantify the modulation of absorptive and secretory transport of compounds by P-glycoprotein in cell culture models of intestinal epithelium. *Pharmaceutical Research*, *20*, 1210–1224.

112   Banker, M.J., Clark, T.H., Williams, J.A. (2003). Development and validation of a 96-well equilibrium dialysis apparatus for measuring plasma protein binding. *Journal of Pharmaceutical Sciences*, *92*, 967–974.

113   van Liempd, S., Morrison, D., Sysmans, L., Nelis, P., Mortishire-Smith, R. (2011). Development and validation of a higher-throughput equilibrium dialysis assay for plasma protein binding. *Journal of Laboratory Automation*, *16*, 56–67.

114   Pacifici, G.M., Viani, A. (1992). Methods of determining plasma and tissue binding of drugs. Pharmacokinetic consequences. *Clinical Pharmacokinetics*, *23*, 449–468.

115   Obach, R.S., Baxter, J.G., Liston, T.E., Silber, B.M., Jones, B.C., MacIntyre, F., Rance, D.J., Wastall, P. (1997). The prediction of human pharmacokinetic parameters from preclinical and in vitro metabolism data. *Journal of Pharmacology and Experimental Therapeutics*, *283*, 46–58.

116 Bi, Y.A., Kazolias, D., Duignan, D.B. (2006). Use of cryopreserved human hepatocytes in sandwich culture to measure hepatobiliary transport. *Drug Metabolism and Disposition*, *34*, 1658–1665.

117 Hoffmaster, K.A., Turncliff, R.Z., LeCluyse, E.L., Kim, R.B., Meier, P.J., Brouwer, K.L. (2004). P-glycoprotein expression, localization, and function in sandwich-cultured primary rat and human hepatocytes: relevance to the hepatobiliary disposition of a model opioid peptide. *Pharmaceutical Research*, *21*, 1294–1302.

118 Kusuhara, H., Sugiyama, Y. (2009). In vitro-in vivo extrapolation of transporter-mediated clearance in the liver and kidney. *Drug Metabolism and Pharmacokinetics*, *24*, 37–52.

119 Feng, B., Hurst, S., Lu, Y., Varma, M.V., Rotter, C.J., El-Kattan, A., Lockwood, P., Corrigan, B. (2013). Quantitative prediction of renal transporter-mediated clinical drug-drug interactions. *Molecular Pharmacology*, *10*, 4207–4215.

120 Obach, R.S., Walsky, R.L., Venkatakrishnan, K., Gaman, E.A., Houston, J.B., Tremaine, L.M. (2006). The utility of in vitro cytochrome P450 inhibition data in the prediction of drug-drug interactions. *Journal of Pharmacology and Experimental Therapeutics*, *316*, 336–348.

121 Obach, R.S., Walsky, R.L., Venkatakrishnan, K. (2007). Mechanism-based inactivation of human cytochrome p450 enzymes and the prediction of drug-drug interactions. *Drug Metabolism and Disposition*, *35*, 246–255.

122 Fahmi, O.A., Ripp, S.L. (2010). Evaluation of models for predicting drug-drug interactions due to induction. *Expert Opinion on Drug Metabolism and Toxicology*, *6*, 1399–1416.

123 Fenner, K.S., Troutman, M.D., Kempshall, S., Cook, J.A., Ware, J.A., Smith, D.A., Lee, C.A. (2009). Drug-drug interactions mediated through P-glycoprotein: clinical relevance and in vitro-in vivo correlation using digoxin as a probe drug. *Clinical Pharmacology and Therapeutics*, *85*, 173–181.

124 Cook, J.A., Feng, B., Fenner, K.S., Kempshall, S., Liu, R., Rotter, C., Smith, D.A., Troutman, M.D., Ullah, M., Lee, C.A. (2010). Refining the in vitro and in vivo critical parameters for P-glycoprotein, [I]/IC50 and [I2]/IC50, that allow for the exclusion of drug candidates from clinical digoxin interaction studies. *Molecular Pharmacology*, *7*, 398–411.

125 Thomas, C.E., Will, Y. (2012). The impact of assay technology as applied to safety assessment in reducing compound attrition in drug discovery. *Expert Opinion on Drug Discovery*, *7*, 109–122.

126 Bryce, S.M., Avlasevich, S.L., Bemis, J.C., Phonethepswath, S., Dertinger, S.D. (2010). Miniaturized flow cytometric in vitro micronucleus assay represents an efficient tool for comprehensively characterizing genotoxicity dose-response relationships. *Mutation Research*, *703*, 191–199.

127 Mondal, M.S., Gabriels, J., McGinnis, C., Magnifico, M., Marsilje, T.H., Urban, L., Collis, A., Bojanic, D., Biller, S.A., Frieauff, W., Martus, H.J., Suter, W., Bentley, P. (2010). High-content micronucleus assay in genotoxicity profiling: initial-stage development and some applications in the investigative/lead-finding studies in drug discovery. *Toxicological Sciences*, *118*, 71–85.

128 Greene, N., Aleo, M.D., Louise-May, S., Price, D.A., Will, Y. (2010). Using an in vitro cytotoxicity assay to aid in compound selection for in vivo safety studies. *Bioorganic Medicinal Chemistry Letters*, *20*, 5308–5312.

129 Bureeva, S., Nikolsky, Y. (2011). Quantitative knowledge-based analysis in compound safety assessment. *Expert Opinion on Drug Metabolism and Toxicology*, *7*, 287–298.

130 Ganter, B., Zidek, N., Hewitt, P.R., Müller, D., Vladimirova, A. (2008). Pathway analysis tools and toxicogenomics reference databases for risk assessment. *Pharmacogenomics*, *9*, 35–54.

131 Fielden, M.R., Nie, A., McMillian, M., Elangbam, C.S., Trela, B.A., Yang, Y., Dunn, R.T. 2nd, Dragan, Y., Fransson-Stehen, R., Bogdanffy, M., Adams, S.P., Foster, W.R., Chen, S.J., Rossi, P., Kasper, P., Jacobson-Kram, D., Tatsuoka, K.S., Wier, P.J., Gollub, J., Halbert, D.N., Roter,

A., Young, J.K., Sina, J.F., Marlowe, J., Martus, H.J., Aubrecht, J., Olaharski, A.J., Roome, N., Nioi, P., Pardo, I., Snyder, R., Perry, R., Lord, P., Mattes, W., Car, B.D. (2008). Predictive Safety Testing Consortium; Carcinogenicity Working Group. Interlaboratory evaluation of genomic signatures for predicting carcinogenicity in the rat. *Toxicological Sciences*, *103*, 28–34.

132 Low, Y., Uehara, T., Minowa, Y., Yamada, H., Ohno, Y., Urushidani, T., Sedykh, A., Muratov, E., Kuz'min, V., Fourches, D., Zhu, H., Rusyn, I., Tropsha, A. (2011). Predicting drug-induced hepatotoxicity using QSAR and toxicogenomics approaches. *Chemical Research in Toxicology*, *24*, 1251–1262.

133 O'Brien, P.J., Irwin, W., Diaz, D., Howard-Cofield, E., Krejsa, C.M., Slaughter, M.R., Gao, B., Kaludercic, N., Angeline, A., Bernardi, P., Brain, P., Hougham, C. (2006). High concordance of drug-induced human hepatotoxicity with in vitro cytotoxicity measured in a novel cell-based model using high content screening. *Archives of Toxicology*, *80*, 580–604

134 Xu, J.J., Henstock, P.V., Dunn, M.C., Smith, A.R., Chabot, J.R., de Graaf, D. (2008). Cellular imaging predictions of clinical drug-induced liver injury. *Toxicological Sciences*, *105*, 97–105.

135 Will, Y., Hynes, J., Ogurtsov, V.I., Papkovsky, D.B. (2006). Analysis of mitochondrial function using phosphorescent oxygen-sensitive probes. *Nature Protocols*, *1*, 2563–2572.

136 Papkovsky, D.B., Hynes, J., Will, Y. (2006). Respirometric screening technology for ADME-tox studies. *Expert Opinion on Drug Metabolism Toxicology*, *2*, 313–323.

137 Hynes, J., Marroquin, L.D., Ogurtsov, V.I., Christiansen, K.N., Stevens, G.J., Papkovsky, D.B., Will, Y. (2006). Investigation of drug-induced mitochondrial toxicity using fluorescence-based oxygen-sensitive probes. *Toxicological Sciences*, *92*, 186–200.

138 Khetani, S.R., Bhatia, S.N. (2008). Microscale culture of human liver cells for drug development. *Nature Biotechnology*, *26*,120–126.

139 Chadwick, C.C., Ezrin, A.M., O'Connor, B., Volberg, W.A., Smith, D.I., Wedge, K.J., Hill, R.J., Briggs, G.M., Pagani, E.D., Silver, P.J., et al. Identification of a specific radioligand for the cardiac rapidly activating delayed rectifier K+ channel. *Circulation Research*, *72*, 707–714.

140 Chiu, P.J., Marcoe, K.F., Bounds, S.E., Lin, C.H., Feng, J.J., Lin, A., Cheng, F.C., Crumb, W.J., Mitchell, R. (2004). Validation of a [3H]astemizole binding assay in HEK293 cells expressing HERG K+ channels. *Journal of Pharmacological Sciences*, *95*, 311–319.

141 Wu, C.Y., Benet, L.Z. (2005). Predicting drug disposition via application of BCS: transport/ absorption/ elimination interplay and development of a biopharmaceutics drug disposition classification system. *Pharmaceutical Research*, *22*, 11–23.

142 Varma, M.V., Gardner, I., Steyn, S.J., Nkansah, P., Rotter, C.J., Whitney-Pickett, C., Zhang, H., Di, L., Cram, M., Fenner, K.S., El-Kattan, A.F. (2012). pH-Dependent solubility and permeability criteria for provisional biopharmaceutics classification (BCS and BDDCS) in early drug discovery. *Molecular Pharmacology*. *9*, 1199–1212.

143 Smith, D.A. (2013). Evolution of ADME science: where else can modeling and simulation contribute? *Molecular Pharmacology*, *10*, 1162–1170.

144 Wobus, A.M., Löser, P. (2011). Present state and future perspectives of using pluripotent stem cells in toxicology research. *Archives in Toxicology*, *85*, 79–117.

145 Bailey, J., Oliveri, A., Levin, E.D. (2013). Zebrafish model systems for developmental neurobehavioral toxicology. *Birth Defects Research Part C Embryo Today Reviews*, *99*, 14–23.

146 Jadiya, P., Nazir, A. (2012). Environmental toxicants as extrinsic epigenetic factors for parkinsonism: studies employing transgenic *C. elegans* model. *CNS and Neurological Disorders Drug Targets*, *11*, 976–983.

8

MEDICINAL CHEMISTRY STRATEGIES TO PREVENT COMPOUND ATTRITION

J. RICHARD MORPHY

Lilly Research Centre, Surrey, UK

8.1 INTRODUCTION

Historically, the main scientific causes of compound attrition in the clinic have been poor pharmacokinetics (PK), safety, and efficacy. Once a compound is selected for clinical development, the properties of the molecule are fixed and any liabilities that may reduce the probability of technical success (pTS) cannot be rectified by medicinal chemists. Therefore, it is crucial to select high-quality molecules that can probe a novel mechanism in the case of a first-in-class agent and compete against existing molecules for a best-in-class approach.

Over recent years, PK deficiencies have become a less common cause of attrition relative to efficacy and safety, as PK modeling using a variety of *in vitro* and *in vivo* data from preclinical rodent and nonrodent species has improved. However, the prediction of human PK remains an inexact science, so it is critical that medicinal chemists discover molecules that are "good enough" to allow for errors in these calculations.

Safety liabilities come in several manifestations, on-target, off-target, and idiosyncratic, and each will be considered in turn in the following sections. Lack of efficacy in a clinical proof-of-concept study is currently the predominant cause of compound attrition. In a large part, this is due to current shortcomings in target identification and validation, and the best efforts of medicinal chemists to obtain the right compound will be in vain unless working on the right target. By collaborating with biologists and geneticists, experienced medicinal chemists can, and indeed should, influence these decisions based on their detailed knowledge of particular disease areas.

Attrition in the Pharmaceutical Industry: Reasons, Implications, and Pathways Forward,
First Edition. Edited by Alexander Alex, C. John Harris and Dennis A. Smith.
© 2016 John Wiley & Sons, Inc. Published 2016 by John Wiley & Sons, Inc.

8.2 PICKING THE RIGHT TARGET

The earliest stages of drug discovery should be highly interdisciplinary and collaborative, and medicinal chemists can provide valuable input to the choice of target. Indeed, many companies now employ multidisciplinary teams, including chemists, *in vitro/in vivo* biologists, and geneticists, whose role is to select targets with an attractive combination of biological rationale and chemical tractability. Increasingly, human genetic data are being used to select targets, with some diseases being associated with polymorphisms/gene mutations, such as LRRK2 for Parkinson's disease, ApoE4 for Alzheimer's disease, and Nav1.7 for pain [1].

Medicinal chemists can help ensure that novel mechanisms are probed effectively by developing tool compounds with adequate potency, selectivity, and other properties appropriate for *in vitro* or *in vivo* experimentation. The desired properties of tool compounds have been reviewed extensively elsewhere [2]. Medicinal chemists are increasingly involved in the development of tracer molecules that are invaluable for probing target engagement (TE) in preclinical species and ultimately in humans. In numerous failed clinical studies, it is unknown if a novel target was engaged to the required extent since positron emission tomography (PET) ligands or alternative biomarkers were not available.

Whereas highly selective tool compounds are useful for target validation, efficacious drugs may require activity at more than one target [3]. Whether single target modulation will achieve the desired effect needs to be considered carefully, whenever novel mechanisms of action are being proposed. Due to the difficulty of identifying and validating individual targets, phenotypic screening is enjoying a resurgence in popularity [4]. Compounds that are active in a disease-relevant assay system can in some cases be successfully optimized to drugs without knowledge of their molecular mechanism of action, and this was the case for most drugs developed before the molecular biology revolution of the late twentieth century, such as many of the current drugs for treating depression and schizophrenia [5].

8.3 FINDING STARTING COMPOUNDS

The choice of starting compound is a critical decision point that profoundly influences the overall pTS. Historically, responsibility for the smaller teams at the lead generation stage has in many companies been assigned to less experienced medicinal chemists, with more experienced medicinal chemists being charged with delivering success in lead optimization. Medicinal chemists with deep experience in all phases of discovery need to be involved in hit selection to reduce attrition at a later stage.

Following a high-throughput screening (HTS) campaign, a rigorous process of active/ hit validation is now commonplace in many organizations to ensure first that any activity is real and second that hits with the best combination of properties are selected. Certain structural features in molecules often lead to false positives in screening assays, for example, by imparting fluorescence or causing aggregation. Databases of problematic groups that are enriched in these Pan Assay Interference Compounds (PAINS) have been published and can be used as filters if desired [6]. When selecting hits, it is important to bear in mind that the most successful hits with respect to multiparameter optimization are those with the best balance of properties, including potency, selectivity, physicochemical

TABLE 8.1 A Typical Hit Validation Checklist

Activity	Selectivity	DMPK/physchem	Toxicity
Affinity	Related targets	MW, LogP, PSA, etc.	CYP inhibition
Functional activity	Panel selectivity	Solubility	hERG inhibition
Native system		Permeability	Structural alerts
SAR (from analogues)		Chemical stability	
		Microsomal stability	

properties, PK, and toxicity profile (Table 8.1). This has required a cultural change among medicinal chemists from 10 years ago when an excessive focus on potency in the primary assay sometimes leads to suboptimal hits being progressed. Many medicinal chemists now view potency as one of the easier properties to optimize if other properties are reasonable. During optimization, enhancing *in vivo* effectiveness is a much more important goal than simple measures of *in vitro* potency [7]. Where possible, activity in a native cell or tissue should be confirmed at the hit stage to provide confidence that the activity is physiologically relevant. Where available, testing of analogues can provide useful SAR around the hit without the need for new compound synthesis.

Unsurprisingly, medicinal chemists are known to exhibit biases in hit selection based on their past experiences with particular structural types, so gathering opinions from multiple sources can ensure that potential issues with hit compounds are anticipated and suitable experiments are brought forward in the flowchart. The introduction of more objective measures of hit quality into decision making has been a major development in recent years. One of the earliest metrics of hit quality was ligand efficiency (LE = maximum potency from minimum heavy atoms) [8], and this was followed by lipophilic ligand efficiency (LLE = maximum potency from minimum lipophilicity) [9], and a multiplicity of other metrics with the most recent being a QED score as a measure of "chemical beauty" [10].

While HTS has become the default method of generating starting compounds, several other complementary methods can also be considered and can present advantages in terms of delivering starting compounds with more balanced properties. Fragment-based drug discovery (FBDD) has been claimed to more consistently deliver candidates with better physicochemical properties and PK profiles [11]. Since molecular weight (MW) and lipophilicity tend to increase during optimization, starting "small and polar" allows room for maneuver for the medicinal chemist.

Whereas HTS has been very successful for certain targets and target families, such as GPCRs, less success is apparent for newer target classes such as protein–protein interactions (PPIs), epigenetics, and transcription factors that are becoming increasingly important in drug discovery portfolios. This has spawned efforts to expand HTS libraries to cover this new chemical space using, for example, diversity-oriented synthesis, natural products, and peptidomimetics [12]. Again, fragment approaches can sometimes deliver starting compounds where HTS has failed since smaller compounds can more efficiently cover a wider region of chemical space. Inhibitors of PPIs derived from FBDD have now advanced into late-stage clinical testing such as the Bcl-xL inhibitor navitoclax, which is in phase II for leukemia (Fig. 8.1) [13]. In this case, the molecule has a high MW, but this is not uncommon for inhibitors of PPIs where binding sites can be large and shallow.

Over the past decade, FBDD has matured as a field and several compounds have reached clinical testing. Moreover, the first fragment-derived drug, vemurafenib, a B-Raf

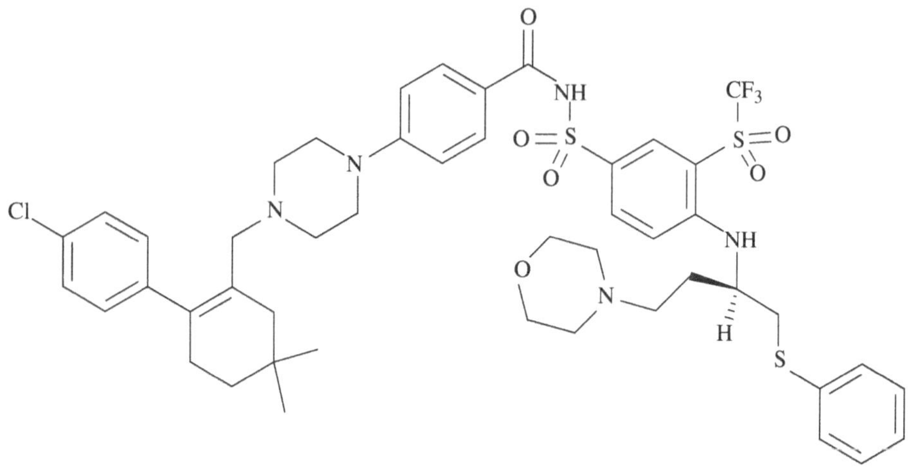

FIGURE 8.1 Structure of a protein–protein (Bcl-2/Bcl-xL) interaction inhibitor, navitoclax.

kinase inhibitor, was recently approved for metastatic melanoma [14]. Nonetheless, FBDD presents unique challenges to medicinal chemists less experienced in the field, and lessons learned from past projects can be especially valuable to avoid attrition [15].

Of course, screening of molecules of whatever size is not the only way to generate lead compounds. In the oft-quoted remark from Sir James Black, "The most fruitful basis of the discovery of a new drug is to start with an old drug." In part, old drugs can provide useful starting points for new efforts due to the fact that clinical PK and safety data are already available, reducing the risk of attrition from these sources. In addition, drugs often exhibit polypharmacology, and the selective optimization of side activities (SOSA) has been described as another approach to delivering new lead matter [16].

The rational design of compounds using the structures of natural ligands such as peptides and proteins (peptidomimetics) is another area that has recently regained interest among the medicinal chemistry community. To increase the chance of success in this challenging endeavor, crystallography is invaluable. Mimics of alpha-helices or beta-turns have been utilized in the design of PPI libraries [17].

8.4 COMPOUND OPTIMIZATION

8.4.1 Drug-Like Compounds

Medicinal chemists must focus on optimizing multiple parameters in parallel including potency, selectivity, physicochemical properties, PK, and toxicology. Attrition can be reduced by selecting starting compounds that have fewest independent parameters to optimize. The publication of the "rule of five" in 1997 first raised awareness of the need to focus on physicochemical properties in order to reduce attrition due to poor oral bioavailability [18]. Simple parameters such as LE and LLE have proved to be useful guidelines during the early stages of optimization. As projects progress to lead optimization, increasingly complex metrics have been employed such as Pfizer's CNS MPO approach, which combines 6 physicochemical properties proposed to be important when optimizing

CNS active agents [19]. The last fifteen years has seen a proliferation of guidelines intended to help medicinal chemists design "drug-like" compounds. A comparison of the relative value of these metrics has recently been published [20]. This showed that monitoring lipophilicity using the LLE parameter is perhaps the most robust guideline for chemists to follow. The importance of optimizing lipophilicity stems from its influence on many other molecular parameters such as potency, selectivity, PK, and toxicity. Increasing lipophilicity can be an extremely effective way of improving a compound's *in vitro* affinity, but often this strategy is counterproductive when *in vivo* potency is compromised due to lower solubility, resulting in poor oral absorption, and higher metabolism due to stronger interaction with cytochrome P450 (CYP) and other metabolizing enzymes. In addition, higher lipophilicity has been associated with greater promiscuity of binding to off-target proteins, increasing the risk of deleterious side effects [21].

8.4.2 Structure-Based Drug Design

Structure-based drug discovery (SBDD) has become an essential component of the medicinal chemist's drug discovery toolkit. An increasing number of drugs discovered at least in part using SBDD have reached the market. For example, the c-MET/ALK inhibitor, crizotinib, approved to treat non-small-cell lung cancer, was the product of the judicious use of SBDD to guide the reduction of MW and lipophilicity during lead optimization (Fig. 8.2) [22]. While in the past, crystallography came too late in many projects to have a major impact, now it is not unusual for screening hits to be crystallized before optimization starts, meaning that biostructural data can guide compound design from the very earliest stage. Crystallography data for a compound bound to off-targets can also be useful for designing out any activities that may increase the risk of undesired side effects.

In particular, crystallography of membrane-bound targets has expanded enormously over the past decade; most notably, a large number of high-resolution GPCR structures have been reported, including class A and more recently class B [23]. For the first time, this has allowed FBDD to be applied to GPCRs such as the beta-1 adrenergic receptor [24]. Where biostructural data is not available for a particular target of interest, homology models can be used for virtual screening and compound design, although the rate of success is highly variable.

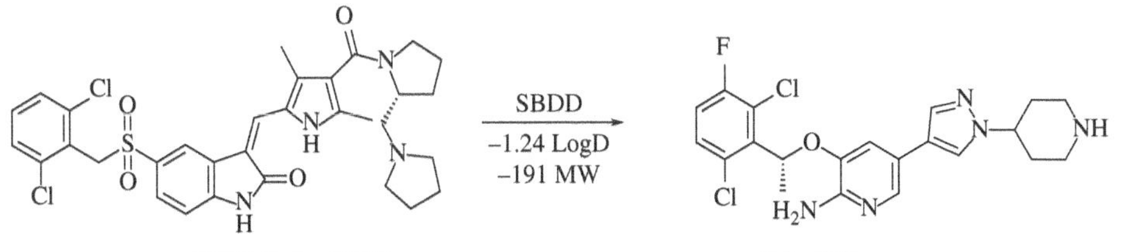

MW 641, LogD 3.2
c-Met cell IC$_{50}$ 9 nM

SBDD
−1.24 LogD
−191 MW

Crizotinib
MW 450, LogD 1.96
c-Met cell IC$_{50}$ 8 nM
ALK cell IC$_{50}$ 20 nM

FIGURE 8.2 Optimization of MW and lipophilicity during the discovery of crizotinib.

8.4.3   The Thermodynamics and Kinetics of Compound Optimization

Recently, the concept of enthalpic efficiency has also been introduced. A focus on optimizing the enthalpic contribution to binding, by forming specific interactions between ligand and protein, for example, optimized H bonds, may deliver potent ligands with improved properties compared to increasing potency entropically via increases in lipophilicity [25]. However, this is not an easy task and the generality of this approach to compound optimization, even when both isothermal calorimetry (ITC) and crystallography data are available, is unknown.

In addition to optimizing the thermodynamic components of binding, it has been recognized that the kinetics of binding can also be an important determinant of the ultimate success of a compound [26]. Depending upon the target, compounds with either slow or fast off-rates have been proposed to deliver safer compounds. Memantine, which is a weak partial agonist at the NMDA receptor, has a fast off-rate associated with reduced side effects [27]. In other cases, the efficacy and safety of a drug has been associated with a slow off-rate [28]. This can allow prolonged modulation of a target long after the compound has cleared from the systemic circulation, allowing efficacy from lower compound exposures and improved safety margins. Interest in measuring binding kinetics has increased as techniques such as surface plasmon resonance (SPR) have become more widely available within the drug discovery community.

8.4.4   PK

In recent years, medicinal chemists have become ever more proficient at optimizing oral PK profiles during hit and lead optimization. This has been achieved in part by an increased focus on physicochemical properties, minimizing MW and LogP, balancing solubility and permeability, and reducing metabolic turnover. The use of *in silico* and *in vitro* models of *in vivo* PK, as well as an earlier use of *in vivo* PK itself across multiple species, has facilitated this positive trend. Another key factor has been close collaboration between medicinal chemists and DMPK specialists, who are now firmly embedded within project teams, helping to plan the right PK experiments at the right time to ensure that the right compounds are progressed through the flow scheme. Compounds that appear most promising from early (*in vitro*) PK work may appear much less attractive once the next set of data become available. Gathering appropriate PK data in a timely manner reduces the amount of *in vivo* testing needed overall by "killing" poor compounds earlier. On the other hand, compounds with an apparently poor PK profile from initial testing can sometimes be "rescued" by testing in carefully chosen follow-up studies, for example, the liability may be specific to a single preclinical species.

Medicinal chemists increasingly generate *in silico* and *in vitro* PK data across a large number and structurally diverse range of project compounds (Table 8.2). Providing these data are shown to be relevant by correlating with *in vivo* data, this approach can be highly efficient in terms of time and cost. For example, if hepatic metabolism is the predominant mechanism of clearance, data from microsomes or hepatocytes can be very useful. If absorption is rate limiting, data from solubility and *in vitro* permeability assays can provide a more in-depth understanding of a particular compound's liabilities, although at an earlier stage of development, *in silico* models of metabolism using, for example, biostructural data for metabolizing enzymes and quantum mechanical calculations are being explored [29].

TABLE 8.2   Frequently Acquired Pharmacokinetic Data During Hit and Lead Optimization Projects

In silico	In vitro	In vivo
Microsomal turnover	Metabolic turnover	Rodent PK
Cytochrome P450 inhibition	Microsomes	i.v. clearance, $t_{1/2}$, volume
Solubility	Hepatocytes	p.o. F%, $t_{1/2}$, C_{max}, T_{max}
Brain penetration	Permeability	Higher species PK
MetID	MDCK-MDR1	Dog, monkey
	Caco-2	MetID
	CYP inhibition/induction	
	Metabolite identification	

Enhancing solubility remains a significant challenge for medicinal chemists, not least since solubility remains a difficult parameter to predict *in silico*. Poor solubility can cause compound attrition in multiple ways. If a compound is not in solution during an *in vitro* assay, potency at an off-target or the extent of metabolic turnover can be underestimated and suboptimal compounds can be progressed to later-stage testing only to fail one more time with loss of money that has been invested. Poor solubility can also lead to difficulties in formulating compounds for preclinical and clinical studies and ultimately reduce the chance of finding a suitable marketable formulation. Dose size is an important parameter for low solubility compounds. If there is insufficient margin between the maximum absorbable dose (MAD) and the efficacious dose, the chance of a compound successfully reaching the market will be much reduced. For low solubility compounds, achieving the high exposure levels typically required for toxicological evaluation can be especially challenging. Encouragingly, an increasing range of enabling formulations can sometimes help rescue low solubility compounds, but even with newer methods such as nanoparticles and surfactants, there is a limit to what can be achieved [30].

Poor intrinsic permeability can be an even more serious issue than solubility since it cannot be improved by any formulation tricks. *In vitro* models such as MDCK-MDR1 and Caco-2 monolayers should be used to identify low permeability compounds early in the optimization process.

According to the "free drug hypothesis," only compound that is unbound to tissue is available for interaction with its target [31]. Therefore, chemists need to optimize free (rather than total) drug levels in order to achieve sufficient TE. Excessive lipophilicity tends to decrease free drug concentrations by increasing tissue binding. This is particularly an issue for CNS targets where high lipophilicity causes nonspecific brain tissue binding. Extensive optimization can be required to ensure compounds have the best balance of properties to pass the blood–brain barrier (BBB) without efflux by BBB transporters such as p-glycoprotein (P-gp) while at the same time providing sufficient free drug for TE [32].

Some target classes continue to present medicinal chemists with significant challenges with respect to oral PK. For peptide-binding GPCRs and PPIs, ligands often have a high MW and are poorly absorbed, and for nuclear receptors, they often are highly lipophilic and poorly soluble [33]. Despite these challenges, medicinal chemists have achieved notable success in recent year's reducing PK-related attrition even for these difficult target classes [13]. For example, the PPI inhibitor, navitoclax (Fig. 8.1), achieved acceptable oral bioavailability across multiple species despite its high MW [13]. This discovery

TABLE 8.3   Typical Exploratory (non-GLP) Toxicology Assessment During Lead Optimization

General in vivo toxicology	Cardiovascular safety	Genetic toxicology
4–14-day rodent toxicology	hERG	Ames test
Life phase	Binding	
Histopathology	Electrophysiology	
4–14-day nonrodent toxicology	*In vivo* CV study (dog)	Clastogenicity assay
	Hemodynamic	
	Electrophysiology	

is remarkable in that it was rationally designed to bind to two PPI sites (Bcl-xL and Bcl-2) to enhance anticancer efficacy. Given provenance of the compound in fragment screening, the availability of biostructural information and an emphasis on maximizing ligand efficiency was undoubtedly helpful.

8.4.5   Toxicity

In contrast to the general level of success in reducing PK-related attrition, reducing attrition due to unfavorable compound toxicity remains a major challenge for medicinal chemists. An increasing number of safety assessment studies have been introduced into the lead optimization phase of projects in order to reduce attrition at the later (more costly) phases of drug development. Studies that are now run routinely in lead optimization include genetic toxicology, cardiovascular safety, and exploratory general toxicology across multiple species (Table 8.3). In some organizations, toxicology is done in nonrodent species prior to candidate selection. What constitutes an acceptable margin of safety between the exposures at the highest "clean" toxicology dose (NOAEL) and at the lowest dose required for efficacy will depend upon multiple factors such as the indication, acute versus chronic dosing, the monitorability of any side effects, and existing treatment options.

Safety-related attrition can be due to three main factors: primary ("on-target") pharmacology, secondary ("off-target") pharmacology, or idiosyncratic toxicity (where there is no obvious cause). On-target toxicity may or may not be predictable depending upon how much is known for the target of interest. In some cases, it may be possible to ameliorate on-target toxicity by adjusting potency/dose, route of administration, or by medicinal chemists optimizing the distribution of the compound so that it cannot access compartments of the body associated with toxicity rather than efficacy. To achieve efficacy, compound exposure should be optimized to a level needed for adequate TE, but not so high as to cause safety liabilities. For example, in the treatment of schizophrenia, dopamine D2 antagonists need to achieve receptor occupancy above 65% but less than 80% to avoid extrapyramidal side effects [34]. Another strategy that is rapidly gaining interest for reducing on-target toxicity is the discovery of biased ligands. Here, compounds that exhibit selectivity between signaling pathways associated with efficacy versus toxicity are sought. For example, compounds that are selective for the beta-arrestin opioid pathway may produce analgesic efficacy with reduced side effect liability [35].

For medicinal chemists, it is often clearer how to address off-target toxicity than on-target toxicity by seeking to design out activity at the culprit target. This of course assumes that the origin of the off-target toxicity is known. Cardiac toxicity is one of the most commonly observed toxicities in preclinical studies and has been associated with a diverse range of off-targets including the hERG potassium channel, $5HT_{2B}$ receptor, phosphodiesterase-3, and Abl kinase [36]. It is not uncommon for molecules to inhibit drug-metabolizing enzymes and transporters, and this can lead to liver toxicity and undesirable drug–drug interactions (DDIs). Compounds are profiled in diverse panels to identify binding to off-targets, typically at the lead compound stage, but increasingly at earlier stages of the optimization process. Some activities are regarded as so problematic, such as hERG, that cross-screening is often performed as early as the hit stage to reduce attrition later in discovery. *In silico* predictions of off-target binding may be useful for prioritizing experimental testing [37].

Selectivity panels need to be carefully selected so that relevant measures of selectivity are obtained [36]. In some cases, it could be sufficient to simply measure binding in a biochemical assay versus an isolated target, whereas in other cases, measuring functional activity in a whole-cell assay may be more relevant, especially in a more physiologically relevant native cell or tissue. Occasionally, panel screening indicates secondary pharmacology, which, rather than being detrimental, may actually be beneficial in terms of efficacy or safety [3].

Depending upon the similarity of the pharmacophores, it can be an intricate task for medicinal chemists to remove off-target activities from the candidate molecule while retaining the required activity. Achieving acceptable off-target selectivity can be particularly challenging for target families containing large numbers of homologous targets with similar binding sites, such as aminergic GPCRs and kinases. Pan-kinase inhibition has been associated with cardiac side effects, and it has been reported that two-thirds of kinase inhibitors bind to more than 10 kinases, illustrating the extent of the selectivity challenge in the kinase area [38]. Barely 20 years ago, the prospect of discovering even moderately selective kinase inhibitors was felt to be low and yet so many are now known. Efforts by medicinal chemists across the drug discovery community resulted in enormous progress in kinase research in recent years, and large numbers of kinase inhibitors are present in clinical development and increasingly as approved drugs [39].

As more data from panel screening becomes available, it is apparent that cross-reactivity across target families is surprisingly common. For example, molecules that have been carefully crafted as selective kinase inhibitors can still suffer from detrimental activities outside the kinome. A recently disclosed c-Met inhibitor was found to potently inhibit several phosphodiesterases, which lead to undesirable cardiac toxicity (Fig. 8.3) [40].

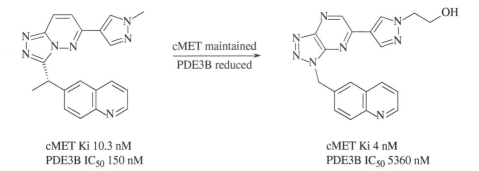

cMET maintained
PDE3B reduced

cMET Ki 10.3 nM
PDE3B IC_{50} 150 nM

cMET Ki 4 nM
PDE3B IC_{50} 5360 nM

FIGURE 8.3 Mitigation of off-target PDE3B activity during optimization of a c-MET inhibitor.

Fortunately, in this case, it was possible to mitigate the offending PDE3B activity during lead optimization, but this may not always be possible if the problem is scaffold wide. A historical tendency during the early stages of optimization projects has been to screen only at the most closely related targets for off-target activity. Cross-family binding site similarity also needs to be factored into the design of flow schemes. In the case of kinases and PDEs, both can bind adenine-containing endogenous ligands, so counterscreening against targets associated with known toxicology should be performed earlier. The pharmacophores for some proteins are so promiscuous that they often overlap with other target classes, for example, hERG and CYP activities overlapping with aminergic GPCRs. The 3-dimensional structures of these promiscuous proteins may in the future assist medicinal chemists to reduce the compound attrition they can cause [41].

Idiosyncratic toxicity describes toxicity whose cause is unknown and therefore is difficult, if not impossible to predict from preclinical studies. It has been associated with the generation of reactive metabolites, normally electrophilic species capable of reacting nonspecifically with nucleophilic groups on DNA, proteins, and other biomolecules. The presence of certain functionalities can predispose molecules toward bioactivation and the generation of reactive species [42]. For example, medicinal chemists will often seek to avoid anilines, which can be oxidized to nitroso groups and other reactive species; thiophenes, which can generate Michael acceptors; and alkenes, which can generate epoxides. The anti-inflammatory drug zomepirac, which was withdrawn from the market due to severe anaphylactic reactions, is bioactivated to an arene oxide species that forms glutathione conjugates (Fig. 8.4) [43].

Extensive lists of so-called structural alerts are consulted by medicinal chemists when prioritizing hits from screening [6]. Occasionally, hits containing structural alerts are progressed into optimization if the profile is otherwise attractive or no other hits are available. Emphasis is placed on replacing the offending group with a bioisostere that retains activity. Fortunately, medicinal chemists now have access to an extensive repertoire of bioisosteric replacements, and excellent reviews on this topic are available in the literature [44]. If the structural alert cannot be replaced, the compound should be prioritized

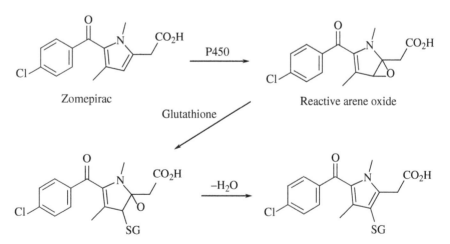

FIGURE 8.4   Bioactivation of zomepirac to a reactive arene oxide intermediate.

for testing in assays that measure the potential for bioactivation *in vitro* and ultimately *in vivo*. As a first-line screen, compounds may be incubated with microsomes or hepatocytes and conjugation to glutathione or some other nucleophilic trapping species is measured. Compounds that form conjugates can under certain circumstances still be developed if the risk–benefit analysis is perceived to be favorable [45]. The risk of bioactivation would be regarded as higher if the drug needed to be given chronically at a high dose and the disease is not regarded as life threatening. The importance of dose size is illustrated by the fact that no drugs that are given at doses less than 5 mg/day have been demonstrated to give rise to idiosyncratic toxicity [46].

Given the high rate of attrition due to compound toxicity, it is desirable to assess compounds at an earlier stage in the drug discovery process than was historically the case. The cost of *in vivo* testing, and the desire to reduce animal use, has led to interest in developing predictive *in vitro* toxicity models. For example, induced pluripotent stem cell-derived cardiomyocytes have been used to assess cardiac toxicity [47]. Approaches to predicting toxicity *in silico* are also being explored, and large investments are being made in this area such as the IMI eTOX consortium [48]. Simple rules of thumb have also been suggested, such as the "3/75" concept [49], proposing that compounds with LogP < 3 and PSA > 75 are significantly less likely to be associated with negative *in vivo* toxicity outcomes. Such guidelines can be useful to medicinal chemists during lead optimization projects, but their empirical nature and the frequency of exceptions mean that their use remains controversial and overly stringent use may lead to unwarranted compound attrition [20]. The search for more reliable predictive toxicology methods will no doubt continue, aided by an increased sharing of relevant experimental preclinical and clinical data across organizations.

8.5 SUMMARY

This chapter has highlighted some of the challenges faced by medicinal chemists with respect to reducing compound attrition as well as the opportunities provided by particular drug discovery strategies. It is a well-known fact that the pharmaceutical industry has faced many difficulties in recent years. Increased competition within the industry and price pressures from healthcare providers have led to a need to provide innovative medicines more quickly [50]. Medicinal chemists have not been immune to these pressures to increase the efficiency and reduce the cost of drug development and have faced a particularly difficult period [51]. Nonetheless, from target validation to the clinic, the role of the medicinal chemist in drug discovery is as crucial as it has ever been. A focus on developing the highest-quality molecules has already produced a positive impact in terms of reducing clinical attrition due to poor PK. An increased emphasis on assessing compound safety preclinically in the years ahead, for example, *in vitro* and *in silico* polypharmacology screening, should produce the same benefit in terms of reducing safety-related attrition in the clinic. Medicinal chemists need to focus on hypothesis-based design of the next generation of compounds to reduce the number of compounds and iterations required during a typical project [52]. Finding more efficient paths to marketed drugs, by reducing compound attrition, should be our clear goal. This will require judicious use of new technologies and computational methods and the full range of internal and external data in our decision making [53, 54].

REFERENCES

1 Plenge, R.M., Scolnick, E.M., Altshuler, D. (2013). Validating therapeutic targets through human genetics. *Nat Rev Drug Disc.,12*, 581–594.

2 Workman, P., Collins, I. Probing the probes: fitness factors for small molecule tools. (2010). *Chem Biol., 17*, 561–577.

3 Morphy, J. R., *Designing Multi-Target Drugs*. RSC Publishing, Cambridge, 2012, pp 141–154.

4 Swinney, D.C., Anthony, J. (2011). How were new medicines discovered. *Nat Rev Drug Disc., 10*, 507–519.

5 Ramachandraiah, C.T., Narayana Subramaniam, N., Tancer, M. The story of antipsychotics: past and present *Indian J Psychiatry., 51*, 324–326.

6 Baell, J.B., Holloway, G.A. (2010). New substructure filters for removal of pan assay interference compounds (PAINS) from screening libraries and for their exclusion in bioassays. *J Med Chem., 53*, 2719–2740.

7 Hann, M.M., Keserü, G.M. (2012). Finding the sweet spot: the role of nature and nurture in medicinal chemistry. *Nat Rev Drug Discov.,11*, 355–65.

8 Hopkins, A.L., Groom, C.R., Alex, A. (2004). Ligand efficiency: a useful metric for lead selection. *Drug Discov Today., 9*, 430–431.

9 Leeson, P.D., Springthorpe, B. (2007). The influence of drug-like concepts on decision-making in medicinal chemistry. *Nat Rev Drug Discov., 6*, 881–890.

10 Bickerton, G.R., Paolini, G.V., Besnard, J., Muresan, S., Hopkins, A.L. (2012). Quantifying the chemical beauty of drugs. *Nat Chem., 4*, 90–98.

11 Murray, C.W., Verdonk, M.L., Rees, D.C. (2013). Experiences in fragment-based drug discovery. *Trends Pharmacol Sci., 33*, 224–232.

12 Barker, A., Kettle, J.G., Nowak, T., Pease, J.E. (2013). Expanding medicinal chemistry space. *Drug Discov Today.,18*, 298–304.

13 Park, C.M., Bruncko, M., Adickes, J., Bauch, J., Ding, H., Kunzer, A., Marsh, K.C., Nimmer, P., Shoemaker, A.R., Song, X., Tahir, S.K., Tse, C., Wang, X., Wendt, M.D., Yang, X., Zhang, H., Fesik, S.W., Rosenberg, S.H., Elmore, S.W. (2008). Discovery of an orally bioavailable small molecule inhibitor of prosurvival B-cell lymphoma 2 proteins. *J. Med. Chem., 51*, 6902–6915.

14 Baker, M. (2013). Fragment-based lead discovery grows up. *Nat Rev Drug Discov., 12*, 5–7.

15 Davis, B.J., Erlanson, D.A. (2013). Learning from our mistakes: the 'unknown knowns' in fragment screening. *Bioorg Med Chem Lett.,15*, 2844–2852.

16 Wermuth, C.G., Selective optimization of side activities: the SOSA approach. *Drug Discov Today.* 2006 *11*(3–4), 160–164.

17 Fry, D., Huang, K.S., Di Lello, P., Mohr, P., Müller, K., So, S.S., Harada, T., Stahl, M., Vu, B., Mauser, H. (2013). Design of libraries targeting protein-protein interfaces. *ChemMedChem., 8*, 726–732.

18 Lipinski, C.A.. Lombardo, F., Dominy, B.W., Feeney, P.J. (1997). Experimental and computational approaches to estimate solubility and permeability in drug discovery and development settings. *Adv Drug Deliv Rev., 23*, 3–25.

19 Wager TT, Hou X, Verhoest PR, Villalobos A. (2010). Moving beyond rules: the development of a central nervous system multiparameter optimization (CNS MPO) approach to enable alignment of druglike properties. *ACS Chem Neurosci., 1*, 435–449.

20 Shultz, M.D. (2013). Setting expectations in molecular optimizations: strengths and limitations of commonly used composite parameters. *Bioorg Med Chem Lett., 23*(1), 5980–5991.

21 Tarcsay, A., Keseru, G.M. (2013). Contributions of molecular properties to drug promiscuity. *J Med Chem.*, *56*, 1789–1795.

22 Cui, J.J., Tran-Dubé, M., Shen, H., Nambu, M., Kung, P.P., Pairish, M., Jia, L., Meng, J., Funk, L., Botrous, I., McTigue, M., Grodsky, N., Ryan, K., Padrique, E., Alton, G., Timofeevski, S., Yamazaki, S., Li, Q., Zou, H., Christensen, J., Mroczkowski, B., Bender, S., Kania, R.S., Edwards, M.P. (2011). Structure based drug design of crizotinib (PF-02341066), a potent and selective dual inhibitor of mesenchymal–epithelial transition factor (c-MET) kinase and anaplastic lymphoma kinase (ALK). *J Med Chem.*, *54*, 6342–6363.

23 Tse, M.T. (2013). G protein-coupled receptors: two landmark class B GPCR structures unveiled. *Nat Rev Drug Discov.*,*12*, 579.

24 Christopher, J.A., Brown, J., Doré, A.S., Errey, J.C., Koglin, M., Marshall, F.H., Myszka, D.G., Rich, R.L., Tate, C.G., Tehan, B., Warne, T., Congreve, M. (2013). Biophysical fragment screening of the β1-adrenergic receptor: identification of high affinity arylpiperazine leads using structure-based drug design. *J Med Chem.*, *56*, 3446–3455.

25 Freire, E. (2008). Do enthalpy and entropy distinguish first in class from best in class? *Drug Discov Today.*,*13*, 869–874.

26 Copeland, R.A., Pompliano, D.L., Meek, T.D. (2006). Drug-target residence time and its implications for lead optimization. *Nat Rev Drug Discov.*, *5*, 730–739.

27 Copeland, R.A., Pompliano, D.L., Meek, T.D. (2006). Drug-target residence time and its implications for lead optimization. *Nat Rev Drug Discov.*, *5*, 730–739.

28 Glossop, P.A., Watson, C.A., Price, D.A., Bunnage, M.E., Middleton, D.S., Wood, A., James, K., Roberts, D., Strang, R.S., Yeadon, M., Perros-Huguet, C., Clarke, N.P., Trevethick, M.A., Machin, I., Stuart, E.G., Evans, S.M., Harrison, A.C., Fairman, D.A., Agoram, B., Burrows, J.L., Feeder, N., Fulton, C.K., Dillon, B.R., Entwistle, D.A., Spence, F.J. (2011). Inhalation by design: novel tertiary amine muscarinic M3 receptor antagonists with slow off-rate binding kinetics for inhaled once-daily treatment of chronic obstructive pulmonary disease. *J Med Chem.*, *54*, 6888–6904.

29 Sun, H., Scott, D.O. (2010). Structure-based drug metabolism predictions for drug design *Drug Des.*, *75*, 3–17.

30 Higgins, J., Cartwright, M.E., Templeton, A.C. (2012). Progressing preclinical drug candidates: strategies on preclinical safety studies and the quest for adequate exposure. *Drug Discov Today.*,*17*, 828–836.

31 Mariappan, T.T., Mandlekar, S., Marathe, P. (2013). Insight into tissue unbound concentration: utility in drug discovery and development. *Curr Drug Metab.*, *14*, 324–340.

32 Claffey, M.M., Helal, C.J., Verhoest, P.R., Kang, Z., Fors, K.S., Jung, S., Zhong, J., Bundesmann, M.W., Hou, X., Lui, S., Kleiman, R.J., Vanase-Frawley, M., Schmidt, A.W., Menniti, F., Schmidt, C.J., Hoffman, W.E., Hajos, M., McDowell, L., O'Connor, R.E., Macdougall-Murphy, M., Fonseca, K.R., Becker, S.L., Nelson, F.R., Liras, S. (2012). Application of structure-based drug design and parallel chemistry to identify selective, brain penetrant, in vivo active phosphodiesterase 9A inhibitors. *J Med Chem.*, *55*, 9055–9068.

33 Morphy, J.R. (2006). The influence of target family and functional activity on the physicochemical properties of pre-clinical compounds. *J Med Chem.*, *49*, 2969–2978.

34 Kapur, S., Zipursky, R., Jones, C., Remington, G., Houle, S. (2000). Relationship between dopamine D2 occupancy, clinical response, and side effects: a double-blind PET study of first-episode schizophrenia. *Am J Psychiatry.*, *157*, 514–520.

35 Violin, J.D., Lefkowitz, R.J. (2007). Beta-arrestin-biased ligands at seven-transmembrane receptors. *Trends Pharmacol Sci.*, *28*, 416–422.

36 Bowes, J., Brown, A.J., Hamon, J., Jarolimek, W., Sridhar, A., Waldron, G., Whitebread, S. (2012). Reducing safety-related drug attrition: the use of in vitro pharmacological profiling. *Nat Rev Drug Discov.*, *11*, 909–922.

37 Lounkine, E., Keiser, M.J., Whitebread, S., Mikhailov, D., Hamon, J., Jenkins, J.L., Lavan, P., Weber, E., Doak, A.K., Côté, S., Shoichet, B.K., Urban, L. (2012). Large-scale prediction and testing of drug activity on side-effect targets. *Nature.*, *486*, 361–367.

38 Bamborough, P., Drewry, D., Harper, G., Smith, GK., Schneider, K. (2008). Assessment of chemical coverage of kinome space and its implications for kinase drug discovery *J Med Chem.*, *51*, 7898–7914.

39 Cohen, P., Alessi, D.R. (2013). Kinase drug discovery—what's next in the field? *ACS Chem Biol.*, *8*, 96–104.

40 Aguirre, S.A., Heyen, J.R., Collette, W. (2010). Bobrowski W, Blasi ER. Cardiovascular effects in rats following exposure to a receptor tyrosine kinase inhibitor. *Toxicol Pathol.*, *38*, 416–428.

41 Stoll, F., Göller, A.H., Hillisch, A. (2011). Utility of protein structures in overcoming ADMET-related issues of drug-like compounds. *Drug Discov Today.*, *16*, 530–538.

42 Kalgutkar, A.S., Soglia, J.R. (2005). Minimising the potential for metabolic activation in drug discovery. *Expert Opin Drug Metab Toxicol.*, *1*, 91–142.

43 Chen, Q., Doss, G.A., Tung, E.C., Liu, W., Tang, Y.S., Braun, M.P., Didolkar, V., Strauss, J.R., Want, R.W., Stearns, R.A., Evans, D.C., Baillie, T.A., Tang, W. (2006). Evidence for the bioactivation of zomepirac and tolmetin by an oxidative pathway: identification of glutathione adducts in vitro in human liver microsomes and in vivo in rats. *Drug Metab Dispos.*, *34*, 145–151.

44 Meanwell, N.A. (2011). Synopsis of some recent tactical application of bioisosteres in drug design. *J Med Chem.*, *54*, 2529–2591.

45 Park, B.K., Boobis, A., Clarke, S., Goldring, C.E., Jones, D., Kenna, J.G., Lambert, C., Laverty, H.G., Naisbitt, D.J., Nelson, S., Nicoll-Griffith, D.A., Obach, R.S., Routledge, P., Smith, D.A., Tweedie, D.J., Vermeulen, N., Williams, D.P., Wilson, I.D., Baillie, T.A. (2011). Managing the challenge of chemically reactive metabolites in drug development. *Nat Rev Drug Discov.*, *10*, 292–306.

46 Stepan, A.F., Walker, D.P., Bauman, J., Price, D.A., Baillie, T.A., Kalgutkar, A.S., Aleo, M.D. (2011). Structural alert/reactive metabolite concept as applied in medicinal chemistry to mitigate the risk of idiosyncratic drug toxicity. *Chem Res Toxicol.* 24, 1345–1410.

47 Liang, P., Lan, F., Lee, A.S., Gong, T., Sanchez-Freire, V., Want, Y., Diecke, S., Sallam, K., Knowles, J.W., Want, P.J., Nguyen, P.K., Bers, D.M,, Robbins, R.C., Wu, J.C. (2013). Drug screening using a library of human induced pluripotent stem cell-derived cardiomyocytes reveals disease-specific patterns of cardiotoxicity. *Circulation.*, *127*, 1677–1691.

48 Cases, M., Pastor, M., Sanz, F. (2013). The eTOX library of public resources for in silico toxicity prediction. *Mol Informat.*, *32*, 24–35.

49 Hughes, J.D., Blagg, J., Price, D.A., Bailey, S., DeCrescenzo, G.A., Devraj, R.V., Ellsworth, E., Fobian, Y.M., Gibbs, M.E., Gilles, R.W., Greene, N., Huang, E., Krieger-Burke, T., Loesel, J., Wager, T., Whiteley, L., Zhang, Y. (2008). Physiochemical drug properties associated with in vivo toxicological outcomes. *Bioorg Med Chem Lett.* 18, 4872–4875.

50 Schulze, U., Ringel, M. (2013). What matters most in commercial success: first-in-class or best-in-class? *Nat Rev Neurol.*, *12*, 419–420.

51 Lowe, D. (2012). Nowhere to go but up: the return of medicinal chemistry. *ACS Med Chem Lett.*, *3*, 3–4.

52 Cumming, J.G., Winter, J., Poirrette, A. (2012). Better compounds faster: the development and exploitation of a desktop predictive chemistry toolkit. *Drug Discov Today.*, *17*, 923-927.

53 Beck, B. (2012). BioProfile—extract knowledge from corporate databases to assess cross-reactivities of compounds. *Bioorg Med Chem.*, *20*, 5428–5435.

54 Duffy, B.C., Zhu, L., Decornez, H., Kitchen, D.B. (2012). Early phase drug discovery: cheminformatics and computational techniques in identifying lead series. *Bioorg Med Chem.*, *20*, 5324–5342.

9

INFLUENCE OF PHENOTYPIC AND TARGET-BASED SCREENING STRATEGIES ON COMPOUND ATTRITION AND PROJECT CHOICE

ANDREW BELL[1], WOLFGANG FECKE[2] AND CHRISTINE WILLIAMS[3]

[1] *Institute of Chemical Biology, Department of Chemistry, Imperial College, London, UK*
[2] *VIB Discovery Sciences, Bio-Incubator, Leuven, Belgium*
[3] *Ipsen BioPharm Ltd, Global Project Management and Analytics, Slough, UK*

9.1 DRUG DISCOVERY APPROACHES: A HISTORICAL PERSPECTIVE

There are two types of drug discovery philosophy that have emerged over the last century. These have generally been referred to as (i) the "phenotypic" approach (or "forward pharmacology"/"classical pharmacology"/"function-first"), whereby molecules are initially assessed for their ability to evoke physiologically relevant effects in a system such as a tissue, organ, or organism, and (ii) the target-based approach (or "reverse pharmacology"/"target-first"), whereby molecules are initially assessed for their ability to induce an effect at a specific molecular target in an *in vitro* system (e.g., isolated protein, recombinant cell systems). The differences in these two approaches are discussed further later but are also summarized in Table 9.1.

9.1.1 Phenotypic Screening

The foundations for modern phenotypic drug discovery can arguably fall to the serendipitous detection of the antibiotic effect of penicillin by Fleming in the early twentieth century [1]. From his simple observation that there was a halo devoid of bacterial growth around a contaminating mold, he concluded that the mold was secreting a substance that repressed the growth of the bacteria. He subsequently isolated a culture of the contaminant and discovered it was a strain of *Penicillium*. In 1945, Crowfoot Hodgkin determined the chemical structure of penicillin [2] (see Fig. 9.1), and a method of mass production

Attrition in the Pharmaceutical Industry: Reasons, Implications, and Pathways Forward,
First Edition. Edited by Alexander Alex, C. John Harris and Dennis A. Smith.
© 2016 John Wiley & Sons, Inc. Published 2016 by John Wiley & Sons, Inc.

TABLE 9.1   Comparison of Target-based and Phenotypic Screening

	Target based	Phenotypic
Definition	Target or mechanism known	Target and mechanism unknown
Validation	Target or mechanism supported (e.g., genetics, animal models)	Disease relevance confirmed (e.g., physiological information)
Assay format	Binding or functional *in vitro* assays using expressed target proteins	Animals or primary cells

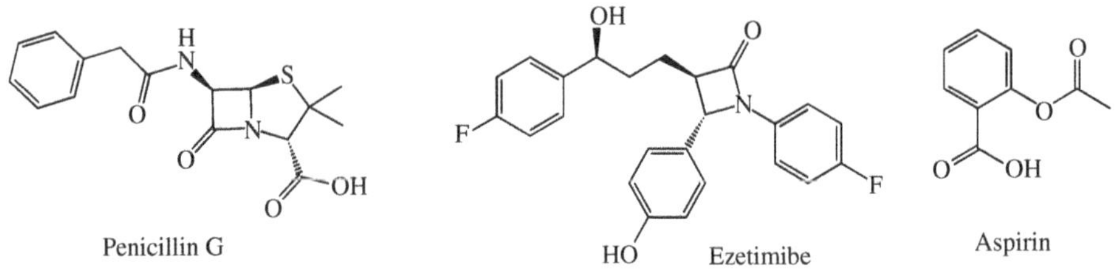

FIGURE 9.1   Structure of exemplar Drugs identified via phenotypic screens.

was developed by Florey and Chain [3]. As a consequence, the drug was used widely throughout the Second World War and had a significant impact on sepsis management. While a serendipitous discovery, this work highlighted that by looking directly at a physiologically meaningful, or phenotypic, endpoint (i.e., bacterial growth), a "drug" could be discovered that would have the desired effect, even when its exact mechanism was not understood. While such a lack of understanding clearly did not prevent its use, it is clear that efficacy had to be demonstrated before it was approved for use by the regulatory bodies of the time. Indeed, entities discovered using phenotypic means that have no identified mechanism of action at the time of approval still exist today. For example, ezetimibe (Zetia®) (Fig. 9.1) was discovered using an animal model with a high-cholesterol diet but, despite an unknown mechanism, was approved for use by the Food and Drug Administration (FDA) in 2002. It was only later that its molecular target was reported as the NPC1L1 transporter [4]. While in this example, there was a relatively short time between approval of the entity and the discovery of its target, historically, this has taken a considerable amount of time. A classic example of this is aspirin (Fig. 9.1), for which identification of the molecular target took close to 100 years [5]. Furthermore, this remains a key challenge for phenotypic approaches today and can prove a challenge in managing safety risks during clinical development [4].

9.1.2   Target-Based Screening

In contrast to phenotypic screening, the foundations for modern target-based drug discovery cannot be pinned to a single event. During the late nineteenth and early twentieth century, significant advances in chemistry were being made with the establishment of the periodic table and the proposal of the "Benzene Theory" [6]. Furthermore, our biological understanding of the molecular effects of drug interactions was evolving. Pivotal works by Ehrlich in 1913 [7] and Langley in 1906 [8] established modern-day receptor theory, and the expanding study of enzymes provided methodologies for calculating compound

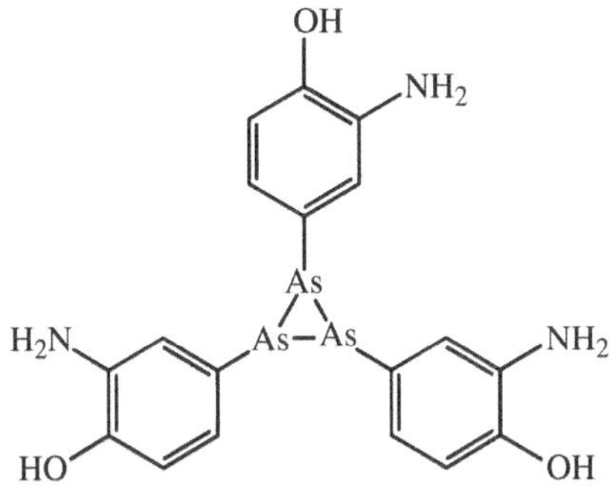

Arsphenamine (shown as trimeric structure)

FIGURE 9.2   The structure of arsphenamine.

potency and efficacy [9]. Cumulatively, these works facilitated the development of a rational synthetic drug design approach, exemplified by arsphenamine (Fig. 9.2), first marketed in 1910 [6]. However, it was only in the middle of the twentieth century when Watson and Crick discovered the molecular structure of DNA that such target-based drug design fully emerged. With the subsequent development of molecular biology techniques such as gene sequencing and recombinant expression, our understanding of the molecular interactions of cell surface receptors, intracellular signaling molecules, and enzymes increased and specific targets responsible for mediating biological effects were identified.

Throughout the late twentieth century, combinatorial chemistry [10] and a plethora of screening technologies for different types of target class [11, 12] and ADME/toxicology effects [13] promoted the synthesis and testing of a huge variety of molecules. This enabled rapid structure–activity relationships (SARs) to be identified and the tailored expansion of compound collections [14]. Furthermore, additional chemistry approaches such as natural product screening [15, 16] and fragment-based screening [17] provided the opportunity for developing molecules from a diversity of chemical space. This enabled the identification of both new entities but also tools to further dissect and understand the underlying mechanisms. Parallel developments in computational tools allowed the industry to tie together a wealth of chemical, biological, and structural information. Moreover, evolving systems biology analysis enabled us to apply that knowledge to disease mechanisms. With such tools to facilitate prosecution, it is no wonder that target-based drug discovery rapidly became the method of choice in the industry.

9.1.3   Recent Changes in Drug Discovery Approaches

There has clearly been a great deal of historical success with the target-based approach to drug discovery; however, costs have increased at an annual rate of 13.4%, while the number of new molecular entities (NMEs) have remained relatively constant [18]. Consequently, the industry and its predominantly target-based discovery paradigm have come under close scrutiny [19].

Analysis of the industry's productivity highlighted that clinical development represented 63% of the total costs and that only 8% of NMEs would survive the process [20], making attrition levels a key area of focus.

During the drug discovery process, attrition can be observed due to a lack of thorough relationship of target to disease mechanism, the failure to develop molecules with the required pharmacokinetic (PK) properties, an inability to translate target affinity to biological efficacy, poor therapeutic index over undesired side effects, and failure to differentiate over other available treatments. Inevitably, one of the first factors to be considered was safety or adverse events and as such thinking initially turned toward alternatives to small-molecule target-based therapies that may have less "off-target" effects. The hybridoma technologies emerging in the mid-1970s had greatly enhanced the use of monoclonal antibodies [21] and recombinant expression enabling the "artificial" production of natural proteins such as human insulin [22] enabled the development of recombinant biopharmaceuticals. With such new biological entities (NBEs), it appeared that their highly specific target-binding and "humanized" characteristics would reduce safety-based attrition [23]. However, recent analysis demonstrates that NBEs still only form a relatively small proportion of FDA approvals [18, 24]. Furthermore, in 2006, the adverse events associated with the clinical trial for TGN1412, intended for use in B-cell chronic lymphocyte leukemia and rheumatoid arthritis, highlighted that such approaches may not be without their own safety concerns [25, 26].

Interestingly, additional analysis of the industry's attrition rates indicated that safety was not the major cause of failure. Data indicated that in 1999 the major attrition factor appeared to be poor PK or bioavailability and that in 2000 the major contributor was clinical efficacy, with exact levels of attrition being dependent upon the specific therapeutic disease areas [27]. This emphasis on failure to achieve clinical efficacy was also highlighted by others [28, 29], and a detailed analysis of 44 phase II projects at Pfizer highlighted that although the majority of failure was caused by a lack of clinical efficacy, for 43% of projects, it was not even possible to confirm that the mechanism had been tested adequately [30].

The industry had already started to move away from the more affinity-based binding screens to more cell-based screens in the target-based model in order to limit early-stage attrition due to a lack of functional effects in *in vitro* systems and poor translation to animal models [31]. In addition, there was an increased focus on the development of preclinical biomarkers and better animal models [32, 33] to enhance the prediction of outcomes in human studies. However, it was an analysis of the first-in-class small-molecule drugs approved by the FDA between 1999 and 2008, conducted by Swinney and Anthony [24] that sparked a renewed interest in phenotypic screening. From this analysis, only 17 (38%) of the drugs approved during that period had been identified via a target-based approach compared to 28 (62%) via a phenotypic approach. This has naturally led to a renewed interest in the phenotypic approach, which was recently discussed at a Keystone meeting "Addressing the Challenges of Drug Discovery" in March 2012 [34]. It is clear from these discussions that the advantages and disadvantages of the approach are still in debate and that the industry as a whole is not completely aligned in its thinking on this subject or indeed the merits of such an approach on productivity and attrition. The benefits of phenotypic screening for complex diseases or areas in need of new mechanisms/targets were clearly stated. However, the challenges of target identification and clinical development remain contentious points [34]. Therefore, in this chapter, we intend to review recent successes from both phenotypic and target-based screening approaches. The review will focus on examples from the two therapeutic areas that appear to have adopted a more phenotypic approach in recent years, specifically anti-infectives and CNS drug discovery.

9.2 CURRENT PHENOTYPIC SCREENS

9.2.1 Definition of Phenotypic Screening

While the original definition of phenotypic screening seems clear, recent interpretation of what is phenotypic is becoming blurred with target-based screening. For example, signaling pathway assays that look at ligand-mediated transduction from cell surface through protein–protein interactions to transcriptional activation/repression have been described as partially phenotypic, even when the majority of the pathway is known in order to generate the high-throughput assay format [4]. Examples of recombinant systems that closely mimic physiological pathways have been developed. For example, a gp160 cell fusion assay that mimics HIV entry into cells [35], is this a target-based screening approach or phenotypic screening? There is clearly increased use of primary cell types in screen cascades [36] and examples where use of a functional assay has been pivotal to success, but is this phenotypic screening? When considering the discovery of cinacalcet (Sensipar®/Mimpara®), it is clear that a functional assay using a primary cell type was utilized, but the calcium receptor, albeit putative at the time, was the intended target [37]. Thus, while categorized as a phenotypic discovery by Swinney and Anthony [24], the mechanism was not completely undefined or unknown. Indeed, in his more recent discussion of the topic, Swinney has acknowledged that phenotypic success can be categorized as those discovered in animals, those in cells with a functional marker, those in cells with a known pathway or clinical serendipity [38]. This is something that will be questioned throughout the review, and for the purposes of this analysis, the following definition will be adopted. A phenotypic screen will be one in which the pathway from compound action to endpoint effect is, at least to a significant extent, undefined or unknown.

9.2.2 Recent Anti-infective Projects

The identification of novel classes of anti-infective medicines has been a major success story for phenotypic screening (see Table 9.2). Virtually all classes of antibacterial, antifungal, and antiparasitic drugs were first discovered by large-scale phenotypic screening, often using crude natural product extracts, with the biological target only emerging much later or, in some cases, remaining uncertain. Of the most recent drug approvals highlighted by Swinney and Anthony [24], the discovery of caspofungin (Cancidas®) is an illustrative example since its discovery and development spans the pre- and postgenomic period.

The approval of caspofungin—only antifungal agent as a first-in-class i.v.—took place in 2001 and was quickly followed by micafungin (Mycamine®, US approval 2005) and finally by anidulafungin (Eraxis®, 2006), the only member of the class suitable for oral delivery. The origin of each drug was the discovery of antifungal activity in extracts of another fungal species, *Aspergillus nidulans*, from which the active species echinocandin B was isolated [39]. A number of related natural products (pneumocandins, mulundocandins, and sporiofungins) were subsequently identified, all being characterized as cyclic hexapeptides with a long-chain carboxylic acid attached to the N-terminus. Replacement of the naturally occurring acid side chain led to the initial drug candidate, cilofungin [40], and after much research to the commercial drugs.

TABLE 9.2   Structures of Recent Anti-infective Drugs Identified Using Phenotypic Screens

Candin derivative	R_1	R_2	R_3	R_4	R_5
Echinocandin B		OH	Me	Me	H
Cilofungin		OH	Me	Me	H
Anidulafungin		OH	Me	Me	H
Pneumocandin B$_o$		OH	H	CH_2CONH_2	H
Caspofungin		$NHCH_2CH_2NH_2$	H	$CH_2CH_2NH_2$	H
Micafungin		OH	Me	CH_2CONH_2	SO_3H

The search for a mechanism of action for the fungin class is a classical example of scientific detective work. Although the drugs are predominantly known for their anti-*Candida* activity, much of the fundamental biology was carried out in the related yeast *Saccharomyces cerevisiae* due to the ease of genetic manipulation. The candins were known to act through cell wall biosynthesis, which was subsequently narrowed down to reduced levels of (1,3)-β-glucan. The responsible gene was first identified as KRE1 [41] and then as FKS1/ETG1 [42]. The delay in the identification of the target class was attributed to the weaker activity of the candin natural products against *Saccharomyces* than *Candida*, which was overcome by the identification of a strain resistant to the more potent analogue, caspofungin [43]. Final confirmation that the related *Candida* gene (CaFKS1) encoded the target of the candins required was not achieved until much later [44], due to the diploid nature of the target organism.

9.2.3 Recent CNS Projects

The first clinically efficacious modern drugs for diseases such as anxiety, bipolar, and psychotic disorders were serendipitous discoveries made during the 1950s to 1970s [45]. As these original discoveries showed significant adverse effects, the subsequent decades were characterized by the development of more selective drugs with fewer side effects. However, this was complicated to some extent by a lack of understanding of the biology and function of the molecular target and the precise mechanism of action of the drug.

Levetiracetam (Keppra®) (Fig. 9.3) is one of the most effective antiepileptic drugs in the clinic [46]. It was identified in an animal model of convulsions in the 1990s, appearing as a more effective analogue than a previously known compound [47]. The drug was

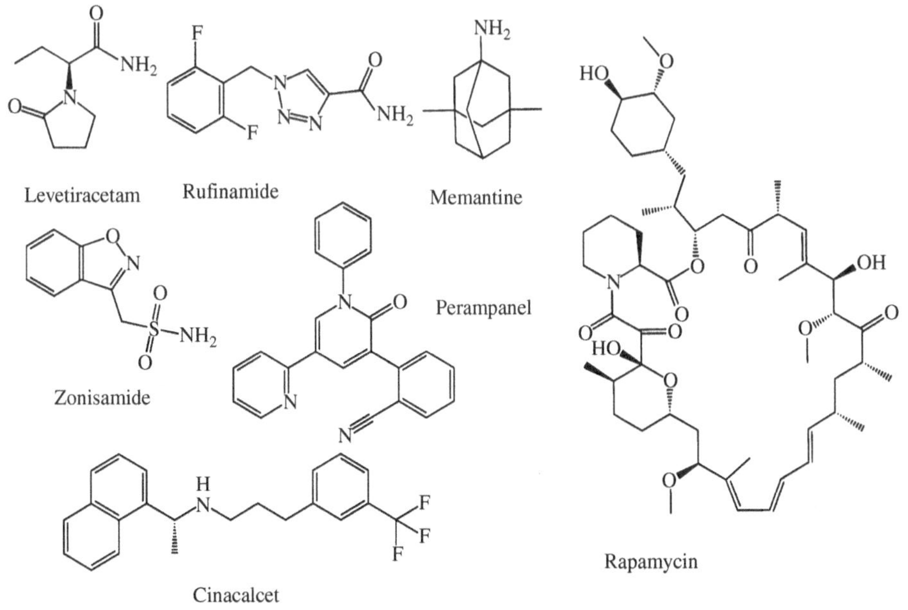

FIGURE 9.3 Structures of recent CNS drugs identified using phenotypic screens.

also shown to be active in other seizure models, although it was noted that its activity profile was quite different to other antiepileptic drugs of the time. For example, the compound was ineffective in acute seizure models such as maximal electroshock seizure (MES) and pentylenetetrazole (PTZ), tests which had been widely used in antiepileptic drug discovery. However, the value of the compound really became clear when it was tested in rodent models of spontaneous seizure occurrence, such as the 6 Hz mouse test [48], which better reflected human epilepsy compared to models that measure suppression of acute seizures. Although the synaptic vesicle protein 2A (SV2A) has now been firmly established as the molecular target of levetiracetam (Keppra®), the exact function of this protein in epilepsy still remains elusive [49].

The relative success rate of phenotypic screening campaigns for CNS diseases might also be related to the fact that the desired molecular properties of drug candidates simply cannot be identified in isolated, target-specific assays. It would appear that molecules need to interact in a receptor-dependent manner within the context of the brain or the whole animal for demonstration of superior therapeutic efficacy. The sodium channel blocker rufinamide (Banzel®/Inovelon®) (Fig. 9.3), which was identified in the MES and PTZ animal models as an antiepileptic drug, showed a very distinctive and highly protective anticonvulsant profile when compared to competitor compounds [50]. It is difficult to imagine how this information could have been obtained by simply using sodium channel-expressing cell lines. The sodium channel blocker zonisamide (Zonegran®) (Fig. 9.3) is another successful antiepileptic drug, which was discovered via screening in animal models [51]. However, despite being on the market for more than 13 years, its mechanism of action in epilepsy is still not clear.

Recently, cellular assays and pathway screens have demonstrated some success in replacing animal models as the primary assay for novel antiepileptic drugs. The drug perampanel (Fycompa®) (Fig. 9.3) was initially discovered in a cell death assay using cortical neurons, was then found to act as a glutamate receptor antagonist [52], and reached the market in 2012.

It is also interesting to note how unexpected phenotypes can be instrumental in identifying new treatments. For example, rapamycin (Sirolimus®) (Fig. 9.3) was discovered from a natural product screen as an antifungal antibiotic and was then found to have potent immunosuppressive qualities [53], mediated by binding to the regulatory kinase mammalian target of rapamycin (mTOR) (Sirolimus®) and inhibiting proliferation of T-cells. Following these initial discoveries, mTOR has been associated with a variety of genetic diseases [54], among them being tuberous sclerosis, which often results in intractable epilepsy. Consequently, mTOR is now seen as a promising new target for epilepsy [55].

Similar developments can also be seen in other CNS disease areas. For instance, memantine (Fig. 9.3) was first discovered more than 50 years ago in a phenotypic screen as an antidiabetic compound [56]. Its CNS activity was only established much later when it was found to act as an uncompetitive NMDA receptor antagonist with moderate affinity [57] but devoid of side effects that are common with other NMDA antagonists. Memantine has been shown to have a modest effect in moderate to severe Alzheimer's disease [58] as a symptomatic treatment and is currently being tested in many other CNS diseases, among them being neurodegenerative diseases (NDs) such as multiple sclerosis and amyotrophic lateral sclerosis but also epilepsy, bipolar disorders, and migraine.

9.3 CURRENT TARGETED SCREENING

9.3.1 Definition of Targeted Screening

As indicated earlier, the boundaries between phenotypic and target-based screening methodologies are blurring. There is evidence that the industry has shifted to cell-based technologies and indeed physiologically relevant cell types where possible early on in the screen cascade. However, if the project is aimed at modulating a specific target, or target-mediated pathway, the emphasis is on that target or pathway and not just the endpoint. Therefore, a target-based project will be defined as any screen (binding and/or functional) in which the target and/or pathway between compound action and endpoint is well understood.

9.3.2 Recent Anti-infective Projects

Since the Swinney and Anthony review [24] of first-in-class drugs breaks down new approvals by disease area, it is clear that target-based screening has been highly successful in certain disease areas. In particular, the approval of multiple anticancer drugs and anti-HIV therapies stand out as beacons of excellence. In this overview, we will focus on the latter area for a more in-depth analysis (see Fig. 9.4 for example structures).

The emergence of AIDS, and subsequent identification of the causative agent as the retrovirus HIV [59, 60], occurred in the same scientific era as the shift of emphasis from phenotypic to targeted screening. Since other retroviruses had already been characterized,

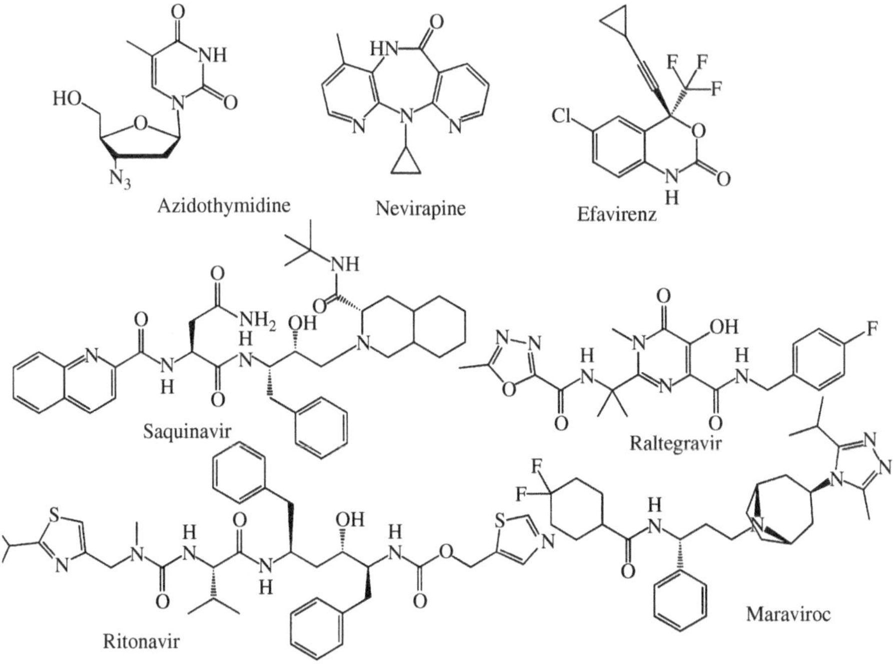

FIGURE 9.4 Structures of recent anti-infective drugs identified using target-based screens.

rapid advances resulted in the complete nucleotide sequence of HIV [61] being published within 2 years of its discovery. Simultaneously, cross-screening of inhibitors of other retroviral reverse transcriptases (RTs) resulted in the identification of analogues of thymidine as inhibitors of HIV-RT [62] and subsequent approval of the first drug, zidovudine (Retrovir®), in 1987.

With resistance to zidovudine already becoming apparent even during the incredibly short development period (24 months), the search for new antiviral agents with differing resistance profiles began in earnest. The second wave of anti-HIV research made use of two of the mainstay methods for first-in-class agents: wide high-throughput screening of random compound collections against known targets and knowledge-based design using inhibitors of members of the same class of enzymes. Both also benefitted from the availability of co-crystal structures enabling rapid optimization of enzyme affinity.

Since the principal objective was a reduction in spontaneous resistance, the finding that some non-nucleoside inhibitors of RT (NNRTIs) suffered an increased rate of resistance was an early disappointment. Fortunately, the binding region occupied by this class of inhibitors is much more promiscuous than the nucleoside binding site, and several different chemotypes were identified by screening. Subsequently, members of two structural classes (nevirapine/Viramune® [63] (1996) and efavirenz/Sustiva® [64] (1998)) with different resistance profiles were approved in the 1990s.

Following the publication of the complete nucleotide sequence of HIV, it became apparent that HIV infection resulted in expression of a protein sequence, known as the HIV propeptide, which was autocleaved by a protease enzyme into the required *gag*, *gag–pol*, and protease. The viral protease was found to be an aspartyl protease similar to the mammalian enzyme, renin, a long-standing target for the treatment of cardiovascular disease. Fortunately, medicinal chemistry programs targeting renin had identified a number of different dipeptide isosteres to mimic the transition state involved in peptide hydrolysis. Tailoring of these isosteres to match the substrate specificity of HIV protease rapidly lead to several series of potent inhibitors, including the first-in-class agent, saquinavir (Fortovase®) [65] (1995). However, despite exquisite enzyme affinity, translating this activity into antiviral efficacy both *in vitro* and *in vivo* emerged as a greater challenge. In a finding that was common with the renin inhibitor medicinal chemistry experience, the physicochemical properties required to inhibit the aspartyl protease were largely orthogonal to those for good cellular permeability or an acceptable PK profile. That said, for a life-threatening disease like AIDS, these issues could be addressed by using high-dose regimes or by compounds such as ritonavir (Norvir®) [66], which inhibits its own metabolism (resulting in drug–drug interactions that can be tolerated for this indication).

Although treatment of AIDS patients with combinations of the drugs discussed earlier (the HAART regimens) had been highly successful, further research led to two new classes of anti-HIV drugs based on targeted screening during the first decade of this century. In both cases, targeting HIV integrase inhibition and human CCR5 antagonism, the medicinal chemistry programs faced significant obstacles, which were aided by detailed knowledge of their biological targets.

The HIV integrase is the third of the enzymes to be coded as part of the *pol* gene by all retroviruses (along with RT and proteases). Unlike the other potential biological targets for treatment of HIV, there were no easy starting points for medicinal chemistry, despite the availability of several crystal structures of the three structural domains. However, these structures did confirm the common feature of all integrases, a triad of acidic residues (DDE), of which two are involved in coordination of a divalent metal (Mg^{2+} or Mn^{2+}). Since coordination of these metals can also be achieved with diacids, scientists at Shionogi were able to identify a number of diacid mimetics and to generate the first structure of an inhibitor bound to HIV integrase [67]. Despite selectivity problems with this approach, due to the propensity of inhibition of other metal-dependent enzymes, Merck was able to identify a number of alternate ligands through target-based high-throughput screening [68] and subsequently develop their leads to arrive at the first-in-class agent raltegravir (Isentress®).[69]

The focus on CCR5 antagonism was one of the first examples of a successful targeted discovery program based on genetic evidence. The finding that individuals with mutations in their CCR5 receptor structures could be carriers of HIV but remained symptom-free prompted a major investment in the search for a CCR5 antagonist drug from the late 1990s. Fortuitously, this G-protein-coupled receptor (GPCR) was amenable to the emerging technology of high-throughput screening, albeit using a surrogate assay based on binding of the peptide MIP-1β [70]. Although it was soon found to be a relatively simple task to achieve high levels of CCR5 antagonism and antiviral activity, the requirement of a basic center resulted in significant off-target effects, particularly cardiac effects mediated by the hERG channel [71]. As a result of the difficulties in combining both properties in the same molecule, only a single CCR5 antagonist has been approved (maraviroc/Selzentry® [72] (2007)).

9.3.3 Recent CNS Projects

A number of CNS therapeutic indications have seen recent success. As discussed before, a number of these were clearly identified via phenotypic means. However, two drugs, specifically aripiprazole (Abilify®) and varenicline (Champix®) (Fig. 9.5), were previously classified as phenotypic discoveries (Swinney & Anthony [24]), whereas here we suggest they were target-based approaches.

Inhibition of dopamine receptors is the most important mechanism supporting the efficacy of antipsychotic drugs but is also a key factor responsible for the sometimes significant side effects (such as Parkinson-like movement disorders). The antipsychotic

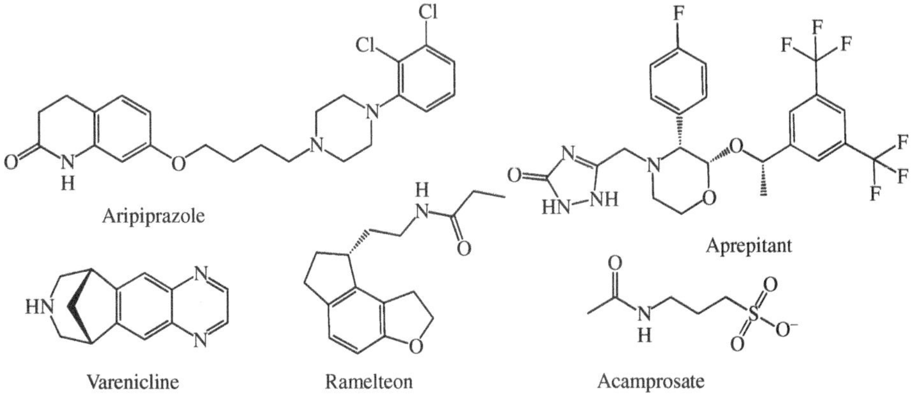

FIGURE 9.5 Structures of recent CNS drugs identified using target-based screens.

drug aripiprazole (Abilify®) was identified after testing a series of dopamine antagonists in animal models [73], which could be interpreted as a phenotypic screen; however, the approach was clearly driven through prior knowledge of the target (i.e., the dopamine D2 receptor) and the desire to differentiate the pharmacological profile. That said, the drug functions as a partial agonist of the dopamine D2 receptor, a mode of action that clearly differentiates it from several other antipsychotic drugs (which act purely as D2 antagonists) and one that was only identified after its initial identification [74]. Varenicline (Champix®), also a partial agonist but at the nicotinic acetylcholine receptors (nAChRs), was approved by the FDA as a smoking cessation aid in 2006. The drug was identified from a series of compounds based on (−)-cytisine, which was known to act as a partial agonist at nAChRs. The project was based on the premise that partial activation of nAChRs would produce a small but sustained increase in dopamine levels, which might overcome the craving and relapse observed in response to nicotine withdrawal (due to low dopamine levels observed in individuals attempting to quit) [75]. Clearly then, there was a target and mechanistic basis for the project.

Three other recent successes were identified via classical target-based approaches but coincidently also highlight three different approaches to identifying small molecules. The drug aprepitant (Emend®) (Fig. 9.5) was identified via a traditional high-throughput screening approach. As neurokinin 1 (NK1) receptors are highly expressed in the vomiting center of the brain and vomiting is a frequent and unpleasant side effect in patients undergoing chemotherapy, it was reasoned that an NK1 antagonist would be effective in preventing or delaying vomiting in cancer patients. The drug was approved in 2003 and so far remains the only NK1 antagonist on the market [76].

In contrast, the synthesis of a small set of compounds based on the endogenous receptor agonist melatonin, which were aimed at limiting the methoxy group conformational flexibility [77], was sufficient to discover the insomnia drug ramelteon (Rozerem®) (Fig. 9.5). The drug seems to have similar effects to melatonin for binding to its receptors and shows little/no significant side effects. In particular, the drug, which was brought to market in 2005, does not have any related dependence as is the case with GABA receptor modulators.

In the case of acamprosate (Campral®), a known starting point with appropriate pharmacological properties was used as the basis for the chemistry project. Specifically, ethanol binds to GABA$_A$ receptors in the CNS and acts as a positive allosteric modulator by increasing the inhibitory effects of the neurotransmitter GABA. In addition, ethanol inhibits NMDA receptors. Chronic alcohol consumption leads to tolerance and down-regulation of GABA receptors, whereas NMDA receptors are upregulated. Acamprosate (Campral®) is the acetylated form of homotaurine, which was known to antagonize NMDA receptors and activate GABA receptors, restoring the activity of neurotransmitters in the brain, which was disrupted by alcoholism [78]. Campral was officially approved by the FDA in 2004 but has been legal in Europe since 1989.

Finally, it is relatively rare that biologics are approved for CNS applications. However, natalizumab (Tysabri®) is a humanized monoclonal antibody to α4β1 integrin, which is a key molecule for the homing of leukocytes to inflamed brain regions. The antibody can block the interaction between the integrin proteins expressed on T-lymphocytes and receptors such as VCAM-1 on endothelial cells, leading to reduced inflammation in diseases such as multiple sclerosis [79]. The molecule was approved for multiple sclerosis in 2004' 12 years after its original discovery.

9.4 POTENTIAL ATTRITION FACTORS

While recent examples can be identified for both phenotypic and target-based screens, it is clear that overall productivity in either scenario can be influenced by attrition rates. As indicated earlier, the levels of attrition can be therapeutic area dependent, but many of the reasons remain common. Various attrition factors are thus discussed later, for both phenotypic and target-based approaches. A summary is also provided in Table 9.3.

9.4.1 Technical Doability and Hit Identification

9.4.1.1 Target-Based Projects   As indicated before, for target-based approaches, the choice of target is often determined by genetic evidence. Although this approach has generated much excitement, evidence for the relevance of a target is only one of the factors affecting the optimal choice of biological target. First and foremost is the ability to screen against a specific target in a biologically relevant manner. A plethora of assay technologies exist for the study of compound interactions at a wide variety of extracellular and intracellular targets, meaning this is rarely a limiting factor in the progression of target-based drug discovery. GPCRs and enzymes have classically dominated the portfolios of most major pharmaceutical companies, and as such, much of the new technology development over the last few years has been directed at these target classes. Historically, radiometric assays were employed for both GPCR and many enzyme inhibition assays, utilizing technologies such as scintillation proximity assays (SPA). However, recent years have seen a shift to more fluorescent technologies such as fluorescence polarization, fluorescence resonance energy transfer (FRET), homogeneous time-resolved fluorescence (HTRF), and fluorescence correlation spectroscopy (FCS). All of these technologies have been successfully applied to both GPCR and enzyme targets [11, 80]. However, recognition that the activities of compounds observed in these *in vitro* binding assay systems do not always translate through to target function has resulted in increasing complexity in the assays employed. Due to their weak potency, it is difficult to be certain that any cell-based activity is mechanism related. Chemical biology methods such as intracellular tagging [81] can be used to demonstrate that compounds are working on-target, but if independent target validation methods are unavailable, a substantial investment of medicinal chemistry resource may be required before cell-based activity is achieved (as was the case in the maraviroc example discussed earlier) [82]. Hence, in the GPCR field, the industry has seen a shift from binding to cellular assays [31], adoption of more high-content formats [83, 84], and a desire to use more physiologically relevant cell types [36, 85]. There are often so many technologies that enable the prosecution of the same type of assay that deciding which is the best to apply in any given setting can be daunting. For example, there are at least six different ways to measure cAMP accumulation and at least 4 commonly employed cAMP-responsive reporter gene systems for measuring GPCR signaling events that all have their own advantages and disadvantages [86].

In addition, it has been recognized that simple blockade or activation of responses may result in side effects or dramatic efficacy profiles, so with a desire to "fine-tune" drug profiles, screens for the identification of allosteric modulators are now also more prevalent [87, 88]. This is also true for kinase targets, which have seen a shift away from simple competition at the catalytic domain of the primary target and the development of coupled formats and the identification of noncompetitive compounds [89]. While these more

TABLE 9.3 Advantages and Disadvantages of Target-based and Phenotypic Screening Relative To Common Attrition Factors

Attrition factor	Target based		Phenotypic	
	Advantages	Disadvantages	Advantages	Disadvantages
Technical doability and hit identification	• Mostly high-throughput assays • False positives/negatives managed with identifiable counterscreen assays • Assay robustness (Z') typically high	• Careful choice of primary assay to maximize functionally active hits • Compound activity may not be relevant to disease	• Mostly medium-throughput assays	• False positives hard to identify/manage • Assay robustness can be challenging
Compound SAR and properties	• Easy to define tailored or focused collections • SAR readily defined due to direct target relationship • Additional crystallography information can drive design	• Polypharmacology can be overlooked	• Polypharmacology can be immediately identified if required for efficacy • Functionally active compounds identified immediately	• SAR can be complex to differentiate due to possible multiple targets • Compound permeability and metabolism can interfere with SAR
Safety	• Target- and mechanism-based toxicity can be readily assessed			• Target deconvolution preferred to manage target- and mechanism-based toxicity
Translation to models/clinic		• Target or mechanism may not demonstrate sufficient efficacy (e.g., redundancy, compensation)	• Efficacy determined at screen, if appropriately designed assay/model	

sophisticated approaches to classical targets are apparent, it is also clear that protein–protein interactions are also emerging more frequently as drug targets [90, 91]. These types of targets can be more challenging from a screening perspective as, although binding assays using isolated proteins may be simple to establish, the reality is that cellular context is often important and if the targets are intracellular, compound entry to the cell becomes a complicating factor.

Inevitably, high-throughput assays are a compromise between screen capacity and biological relevance. The choice of screen will also impact the quantity and quality of hit matter that will be obtained, with a desire to maximize chemical matter identified but to avoid false positives [92, 93]. One of the major advantages of targeted screening is the ability to define an appropriate selectivity target. The choice of selectivity target is also influenced by several factors including knowledge of the off-target pharmacology, target similarity, and the history of hits against the desired target. Ideally, the profile of hits against primary and selectivity target will be used to discard hits as false positives, but in most cases, many hits will be real but nonselective; hence, a structurally distinct target using the same assay format is used to eliminate hits as false positives. A further advantage of targeted screening is the ability to generate additional assays to confirm the validity of new hits and to further explore the properties of the hit matter. For well-defined targets amenable to structural biology, these can include NMR-based assays to show shifts in specific residues on binding or cocrystallization to determine which binding site is occupied by the new ligand. An interesting example illustrating this is the hepatitis C polymerase, which possesses at least 5 sites known to be amenable to inhibitor design [94]. Hits may be prioritized based on the properties of the site they occupy. With confirmed hit matter in hand, further assessment of off-target selectivity issues can be assessed, either through mining into the past screening history of the hits; by scoring of any hits through computational models; or by further wet screening against specific targets from the same gene family or through screening against a preselected diverse set of targets, often selected to include antitargets, which are known to result in negative effects seen clinically, for example, ion channels like hERG or sodium channels.

Hence, it is clear that to be successful it is important to consider the primary screen and the entire screen cascade carefully in the context of the desired biological outcome, whether it be for protein–protein interactions, enzymes, or even for classical targets such as GPCRs [11, 95].

9.4.1.2 Phenotypic Projects
Historically, phenotypic screening methods were slow, resource intensive, and relatively prohibitive for screening any more than a few compounds. However, there are now a wide variety of cells and organisms, which are being used successfully in medium- and higher-throughput phenotypic screens (see Table 9.4) [96–107]. The spectrum ranges from bacteria to yeast and mammalian cells up to small organisms (which are usually restricted to worms, flies, and zebrafish). More recently, primary human cells or even patient-derived and differentiated cell types obtained from induced pluripotent stem cells (iPSCs) have been more typically employed. In addition, advances in assay technologies, robotic platforms, and sensitive detection systems have improved the throughput of compound screening with cellular phenotypic assays. The most commonly adopted assays today measure cell viability (e.g., cancer cells), cell signaling events (e.g., reporter genes), or a disease-related endpoint (e.g., killing pathogens). In addition to these three dominating categories, there are also many other assay formats

TABLE 9.4 Representative Examples of Recent Phenotypic Screening Methods

Disease	Phenotype	Cell/organism	Readout	Reference
Huntington's disease	Huntingtin protein aggregation, cell viability	PC12 cells	2-step assay for protein secretion (luminescence) and aggregation (Htt-GFP, fluorescence)	Titus et al. [96]
Huntington's disease	Transcriptional dysregulation of CRE pathway	HEK293 cells	CRE luciferase reporter assay	Lazzeroni et al. [97]
Parkinson's disease	Inhibition of A-synuclein protein translation	H4 neuroglioma cells	A-Syn luciferase reporter assay	Ross et al. [98]
Parkinson's disease	Reduction of induced cytotoxicity	SH-SY5Y neuroblastoma cells	MTT absorbance assay	Zhao et al. [99]
Parkinson's disease	Protection of dopaminergic neurons	Caenorhabditis elegans	Automated recording of mobility and lethality, GFP intensity	Braungart et al. [100]
Alzheimer's disease	Reduction of secreted Aβ peptides	H4 neuroglioma cells	Aβ40/42 ELISA assay	Chakrabarti et al. [101]
Alzheimer's disease	Reduction of intracellular Tau protein	M17 neuroblastoma cells	In-cell Western assay with antibodies against Tau and GAPDH	Jones et al. [102]
Amyotrophic lateral sclerosis	Reduction of oxidative stress	NSC34 motor neurons	Fluorescence assay for reactive oxygen species	Barber et al. [103]
Malaria	Cell growth	Plasmodium falciparum	SYBR green assay for nucleic acid synthesis	Plouffe et al. [104] Gamo et al. [105]
Neglected diseases, inc leishmaniasis, schistosomiasis	Cell death	Leishmania donovani, T. brucei, T. Cruzi	Selectivity over human cell death	Nwaka et al. [106]
Tuberculosis	Cell death	Mycobacterium bovis	Mycobacterium tuberculosis activity confirmation	Ballell et al. [107]

that investigate a particular mechanism deemed to be important for a disease, such as cell migration, protein secretion or intracellular translocation, and neurite outgrowth. In addition to more conventional assay readouts such as fluorescent markers for cell death, assays based on fluorescence-activated cell sorting, and reporter gene technologies, novel methods increasingly comprise imaging technologies. Such high-content screening technologies, using automated microscopes and image analysis software, are particularly important for phenotypic assays as they enable the measurement of complex intracellular events, which were previously not accessible as readouts for higher-throughput assays [108]. Together, these developments mean that assays can be miniaturized in 384-well and even in some cases 1536-well formats and used to screen large-compound libraries, offering throughput more akin to target-based screening, albeit with the additional biological complexity of physiologically interacting proteins and cellular signaling networks.

However, despite all these options for setting up phenotypic screens, it is not surprising that a well-defined phenotype is the single most important factor in designing and executing a phenotypic screening assay. Only a good understanding of the phenotype, and the changes caused by small molecules, will allow correct interpretation of the results. This implies the use of appropriate controls and the statistical analysis of the screening data so that only the most promising compounds are selected for secondary assays and in-depth characterization while false positive and false negative rates are minimized. For instance, while many compounds affect cell viability or proliferation, only very few compounds from a screen will do so by interfering with a disease-relevant and targeted mechanism. Therefore, a phenotype should be described and evaluated by several parameters. It is often advisable to compare compound-related effects with perturbations caused by RNAi against proteins known to be involved in the particular signaling pathway or phenotype.

The success of phenotypic screening is also limited by the quality of the collection of compounds that are assayed, and given the virtually infinite potential of chemical space, all screening collections are a minute subset of the available space. For targeted screening, this limitation has been addressed through the screening of smaller collections of "fragment" molecules, but unfortunately, the technique is largely incompatible with phenotypic methods since the concentrations required to detect weakly active fragments often result in nonspecific toxicity in cell-based assays. Therefore, the actual screen might be performed with a set of approved and well-characterized drugs, with small focused libraries against particular target classes, which are thought to be involved in the phenotype, or with a large and diverse compound library. It is often advisable to exclude or filter out compounds with known unspecific cytotoxicity or by selecting against compounds with unwanted properties such as low solubility or known interference with the assay signal. However, despite using well-behaved and nontoxic compounds, phenotypic screens still often result in a fairly high number of hits. Some phenotypic screens may suffer with low assay robustness and data reproducibility due to the complexity of the cellular or organism-based assay, although there are reports that indicate that some assays are sensitive and robust [109]. Nevertheless, high hit rates can also be attributed to the poor specificity of early hits and the potential for modulating many cellular targets. It is therefore always advisable, similar to target-based screens, to have a rigorous hit selection strategy in place, which usually comprises the confirmation of the initial screening result by retesting the compounds in concentration response assays and by confirming the purity and chemical identity of the hits. Due to the sensitivity of phenotypic screens and the multitude of potential targets within cells, an independent chemical resynthesis of the most promising hits might be needed to exclude potential effects of impurities in the compound sample. Extracting further information from an image of cells incubated with a hit compound might also help, for instance, a change in cell numbers or morphology might indicate subtle toxic effects. The complexity of understanding SAR from potentially multiple mechanisms in a phenotypic screen is discussed later, but it is worth highlighting that these nonspecific effects represent a major challenge for both targeted and phenotypic screening.

9.4.2 Compound SAR and Properties

9.4.2.1 Target-Based Projects While target-based projects were initially favored due to their increased screening capacity relative to phenotypic screening, the need for large quantities of purified target protein has contributed to an exponential increase in the number of structure-guided drug design projects. Advances in structural biology have extended the range of target types from soluble enzymes toward multiple classes of GPCRs [110]. As more similar 3D structures become available, the quality of homology models has also improved, enabling application of drug design techniques to a wide range of related targets, either as the main target or as a selectivity target to avoid [111]. As a result, structures of multiple hits are often available at the start of a medicinal chemistry program, thus enabling the selection of chemotypes with appropriate physical properties to support delivery of improved drug candidates to ensure that the appropriate mechanism is tested [112].

Best practice examples, like the mGluR5 project highlighted before, do not lose sight of the need to improve potency without losing sight of the need to achieve a balance with appropriate physical properties. The Lipinski "rules of 5" were devised to address a trend toward poorer oral absorption of drug candidates, which has been separately attributed to the decline in phenotypic screening, the rise of combichem, and the shift to more challenging biological targets, or the need for greater selectivity over other biological targets.

In contrast, there has been less focus on the translation of target potency to cell-based activity. While cell type is undoubtedly a factor in certain organisms, levels can vary from complete translation of activity (1:1) to tens of 1000-fold translation. In rare cases, such as HIV protease inhibitors, poor translation is tolerated due to exquisite target potency [113], but more commonly compounds are inactive in the appropriate cell-based assay [114]. The latter example based on inhibition of a fungal N-myristoyltransferase is instructive since it involved depeptidization of an initial octapeptide inhibitor, with cell-based activity only being observed when the number of amide bonds had been reduced to one [115]. Inhibition of the orthologous enzyme in *Trypanosoma brucei* with a small-molecule inhibitor resulted in excellent translation [116]. Poor translation is often ascribed to low permeability, without any supporting evidence. In order for target-based screening to fulfill its potential, a greater focus on improving our understanding of this complex issue is required.

9.4.2.2 Phenotypic Projects For phenotypic screening hits, SARs can be developed in parallel with efforts to determine a biological target for the series of compounds or await assignment of a mechanism of action. If the latter approach is followed, the project

TABLE 9.5 Phenotypic Activity of Members of the Azole Class of Antifungal Agents

Compound	log *P*	Dose size (mg)	MIC *Candida albicans* (μg/ml)	MIC *Aspergillus fumigatus* (μg/ml)
Itraconazole	5.6	100	0.1	0.1
Fluconazole	0.4	200	1.0	>50
Voriconazole	1.0	200	0.03	0.09

can become operationally indistinguishable from a target-based approach, since issues relating to target selectivity and cellular translation can be tracked as part of the SAR studies. The former approach is more interesting since any SAR generated takes into account activity at the biological target and any permeability issues. The challenge for SAR development is to distinguish between structural modifications that increase permeability, for example, increased lipophilicity, fewer H bond donors, higher unionized drug levels, and increased intrinsic potency. Although increased cellular activity is desirable, it cannot be achieved without reference to free drug levels since increased clearance and lower free drug fractions are likely to be observed. In one of the more extremely differentiated examples, Table 9.5 shows the phenotypic activity of three members of the azole class of antifungal agents with the highest sales over the last 20 years.

Fluconazole (Diflucan®) and itraconazole were both developed to treat oral candidiasis (infection by *Candida* spp.), but despite widely differing *in vitro* activity, the active doses are very similar. Itraconazole is also approved for the treatment of aspergilloses (infection by *Aspergillus* spp.), which is consistent with its potent *in vitro* activity against both fungal species. Voriconazole has similar *in vitro* activity to itraconazole but is much more polar, resulting in a far higher free drug fraction (42% vs. 0.2%). Voriconazole has been shown to be superior to the previous standard agent (amphotericin B) in comparative trials [117] and has become the standard of care for the treatment of invasive aspergillosis.

While the antifungal armamentarium includes several classes of "drug-like" small molecules, the treatment of bacterial infections is heavily reliant on natural product-derived starting points. Of the major classes of antibacterial, only the quinolones, sulfonamides, and oxazolidinones can be considered as typical small-molecule drugs. In a further contrast to the antifungal arena, the range of *in vitro* activity achieved by antibacterials tends to be much narrower. The discovery of new classes of antibacterial continues to be an elusive objective for both targeted and phenotypic activity. In a recent example, scientists from GSK discovered a series of bactericidal inhibitors of *Mycobacterium tuberculosis Inh A*. In an effort to capitalize on the mechanism of action of the frontline antitubercular medicine, isoniazid, they used encoded library technology to discover a novel series of inhibitors of Inh A. Despite potent enzyme activity, limited whole-cell activity was achieved and their lead compound failed to reduce the bacterial load in an acute TB infection model [118]. This paper also raises the interplay between potency and lipophilicity and suggests that a measure of lipophilic efficiency could be important in optimizing the activity of novel anti-infectives, consistent with the azole example discussed earlier.

A further advantage of phenotypic screening is its ability to detect hits with polypharmacology, where effects at a combination of targets are required in order to achieve cellular activity. This situation is more likely to be successful when the biological targets are closely related; otherwise, combining positive effects at two or more targets but avoiding negative effects at other similar targets can be a major challenge.

There are a number of relevant examples in the field of malaria research due to the publication of sets of antimalarials arising from large-scale screening by pharmaceutical companies. The biological targets of several of the hit series have been determined, and in the case of PI(4) kinase, hits were found to be selective over the closest human orthologues [119], while a series of plasmepsin hits were found to have similar activity against the closest human orthologue, cathepsin D [120].

9.4.3 Safety

Safety-related attrition remains a major factor in the drug development process. Emphasis is therefore being put on identifying safety issues of drug candidates as early as reasonably possible in order to avoid attrition in expensive late-stage clinical trials. The safety profile of candidate compounds is defined by predicted or measured adverse drug reactions (ADRs), which can either be caused by the action of the compound on the intended drug target (mechanism-based toxicity) or by its interaction with other molecules or mechanisms (off-target toxicity) [121]. Both target and phenotypic drug programs face the challenge of off-target ADRs, and the approaches for limiting safety attrition due to these drug interactions are therefore also very similar for the two screening strategies.

9.4.3.1 Target-Based Projects   In target-based drug discovery programs, the available information on the (patho)physiological role of the target, the molecular mode of action, effects of known mutations from human and animal studies, protein overexpression and knockout data in mice or other organisms, and gene expression profiles in different human tissues are usually sufficient to predict potential target-related ADRs. This information is then used to design appropriate safety studies in rodents or other species, which help to define an acceptable balance between drug efficacy and adverse effects in the clinic. Companies seem to employ different strategies related to the best time point for these studies during drug development, but mechanism-based toxicity will be investigated for each target discovery program at some point. For instance, it might be beneficial to start with an initial safety assessment as early as in the target validation phase of a project, provided that suitable tool compounds are available, if knowledge on a new target is limited, or if there is a clear rationale for unwanted side effects. On the other hand, available clinical data with approved drugs or failed candidates for a particular target might not require an extensive mechanism-based safety assessment if the new molecule belongs to just another chemical series but works through an established mechanism.

Usually, the safety assessment starts with an *in vitro* profiling of lead molecules from different series against a range of well-characterized enzymes, receptors, ion channels, and transporters, while tightly regulated *in vivo* safety tests are reserved for advanced compounds, which are ready to be evaluated in the clinic. The only required *in vitro* assay is a test for hERG inhibition, due to the potential serious consequences of modulating that target [122]. Another example of an off-target modulation to be avoided is the serotonin 2B receptor (5-HT$_{2B}$), which is implicated in the function of the heart valves [123]. The human targets in a profiling panel belong to classes that are known to play an important role in the regulation of the cardiovascular, respiratory, and nervous systems, based on evidence from clinical safety data. Most companies use combinations of ligand binding and functional assays to define their individual panels, which are made up of between

40 and 120 individual assays, depending also on the time and exact purpose for what they are used.

Overall, there are several positive effects of *in vitro* profiling on compound attrition rates. When applied early in drug discovery, an assessment of compound promiscuity (% target binding compared to total number of targets tested) can help in the selection of the most promising lead series for further optimization. Profiling data are also used as starting points for SARs on the most important off-targets so that these activities can be designed out of the series while efficacy is retained. Later on, they can pave the way for integrated *in vivo* safety assessments by predicting the required free plasma concentration for the off-target ADR so that a safety window can be calculated for a given indication. Finally, the data can be used for understanding the biological effects seen *in vivo*, which should help in a risk assessment when patients are selected for clinical trials. However, it has to be kept in mind that the correlation between rodent and human affinity data is not always good and often depends on the target class; for instance, interspecies correlation for chemokine receptors is much weaker than for muscarinic acetylcholine receptors.

It is possible that advancements in high-throughput technologies will not only lead to a broadening of the common *in vitro* profiling panels, as more drug targets are being discovered [124] and currently underrepresented target classes such as kinases can be included, but also lead to their use earlier in the drug discovery process.

9.4.3.2 Phenotypic Projects One of the historical challenges with phenotypic screening approaches is the lack of an identified molecular target. While knowing the target is not a prerequisite for FDA approval, there is a drive in the industry to identify the target(s) and understand the mechanism of action, in order to mitigate any toxicity-related risks. As such, the project shifts to a more target-based approach and is handled similarly to that discussed earlier. However, to achieve this level of clarity, a potentially significant investment is required and as such is worth discussing further here. The process of target deconvolution, as it is commonly referred, is more subtle than simple target identification as the vast majority of compounds will have multiple-target interactions (on average, drug molecules have six target interactions) [125]. Therefore, once potential targets have been identified, it is also important to confirm the target as a key mediator of the biological effect.

A number of technologies have been applied successfully to the target deconvolution process over the last decade, with chemical proteomics being the most frequently adopted. This term refers to a subset of proteomics in which a small molecule is directly utilized to filter an entire proteome for proteins, which interact with that molecule. This direct approach requires that the small molecule can be tagged (with various probes, such as photoaffinity labels) and immobilized [126]. Affinity purification/pulldown is then used to isolate binding partners, which can be eluted, separated, and identified via mass spectrometry, immunoblotting, or other molecular identification techniques. Clearly, the approach requires that any tags and labels do not prevent any of the molecular interactions and that any biological extracts must also contain the target in a form that retains its biologically relevant conformation. However, it is also interesting to note that the specific immobilization/capture method can influence the efficiency of purification and the degree of false positives. In particular, magnetic nanobeads are emerging as a tool of choice, which eliminate the need for multiple wash and separation steps [127, 128]. Recently, label-free techniques have also emerged, which prevent the need for labeling of the active compound. These approaches depend on the changes in thermodynamic stability of proteins in an unbound versus bound state. Specifically, drug affinity responsive target stability (DARTS) exploits the fact that a protein demonstrates less conformational flexibility and thus susceptibility to proteolysis when bound to a small molecule. Stability of proteins from rates of oxidation (SPROX) considers the same point but under oxidative conditions (in the presence of H_2O_2 and a denaturant). The resulting peptide analysis from complex mixtures is achieved by using a tandem LC–MS/MS technique and hence has the potential for more direct global analysis of drug–protein interactions [128].

In addition to affinity purification, phage display, yeast three-hybrid, and protein microarrays can be applied to direct target identification [126–128]. All of these techniques require the use of a tagged small molecule; however, such expression cloning techniques can be advantageous if the targets are likely to be of low abundance or unstable [128]. In addition, phage display enables iterative affinity enrichment, via multiple rounds of selection, competition with free ligand, and monovalent rather than polyvalent display, which may all enhance identification of specific interaction [126].

In contrast to these direct methodologies, indirect methods using *in silico* or systems biology approaches have been applied to target identification. Databases of public or private information on the mRNA, protein, or metabolite expression profile of a compound can aid in the identification of specific targets. There are three types of approach that have been employed. Firstly, the changes observed with a compound can be combined with pathway analysis knowledge to select possible targets; secondly, the "fingerprint" or "signature" of a compound can be compared with the profile of drug with known molecular interactions; and finally, the profile of an unknown can be compared systematically with the profiles generated when specific proteins are removed (via gene deletion or RNAi). Such approaches have proven successful [126]; indeed, gene expression signatures could even be utilized to explore compound effects in the context of a specific genotype; however, they are all means of identifying possible targets rather than practical evidence of the molecular interaction.

In reality, a combination of both the direct chemical proteomics and *in silico* pathway analysis can be informative in the target deconvolution process. Regardless of methodology, possible target interactions need to be confirmed in biological studies such as cellular activity assays or molecular binding assays, making the process complex, potentially time consuming, and with no guarantee of success. At a recent meeting discussing this topic, Mark Fishman highlighted that the use of genetics and chemical proteomics had resulted in a target identification success rate of approximately 40% at the Novartis Institutes for BioMedical Research [34]. In contrast, Timothy Willson highlighted that with recent investment in chemical proteomics at GlaxoSmithKline, they were able to identify targets in 4–6 weeks for 70% of projects [34]. Regardless of the success rate, however, this can be one of the greatest hurdles with prosecution of phenotypic projects in organizations that have a predominantly target-based approach and where target-based and mechanism-based toxicology is a key "go/no go" decision factor, indicating that the organization philosophy to discovery and risk management can play a key role.

9.4.4 Translation to the Clinic

A crucial step in any drug discovery program is the selection of suitable animal models for a particular human disease so that the function of a new drug target can be assessed during the target validation phase of a project and promising lead candidates be tested in terms of target engagement and biological efficacy. It is therefore no surprise that deficits in validity and predictive power of a given animal model for a disease represent a major attrition factor for drug development and translation of compounds into the clinic. In most cases, it might not even be possible to initiate a discovery program for a new drug target if a model that links the target to a disease is not available. The efficacy of drug candidates emerging from both target and phenotypic screening approaches would need to be tested in animal disease models unless the model itself was used as the first assay in a phenotypic discovery program, as was discussed earlier, for instance, for epilepsy drug discovery projects. Therefore, we assume that issues of project selection and high attrition related to these models are very similar for both approaches and can be discussed together. It is beyond the scope of this book to explain in detail the measures used for evaluating the validity of animal models of human diseases. Instead, typical examples from CNS and infectious disease areas will be used to exemplify some issues researchers are facing when a good model needs to be selected.

NDs are often complex disorders, with multiple (sometimes still unknown) genetic and environmental factors contributing to the pathology, which makes the design and selection of animal models very challenging. Indeed, even for models of Huntington's disease (HD), which is caused by a CAG expansion of more than 36 repeats in the N-terminal exon of the Huntingtin (HTT) gene [129], the development of predictive models is not straightforward.

Since the discovery of the HTT mutation more than 20 years ago, HD was modeled in various animal species, from worms, fruit flies, mice, and rats up to sheep, pigs, and monkeys. Early animal models were created by injecting neurotoxins such as quinolinic acid into the striatum of rodents [130], leading to lesions and loss of GABAergic neurons, which is also observed in HD patients. However, these models are limited by the acute nature of the lesions, which appear within days, while HD typically takes decades for clinical manifestation, and by the obvious lack of the mutated HTT gene, which is ubiquitously expressed in patients and affects also the function of other brain and peripheral areas apart from the striatum. Working with the large (67 exons) HTT gene proved to be technically difficult, with error-free cloning, insertion into genomes, and obtaining high enough expression levels of the mutated gene compared to the endogenous one proving challenging. In addition, the extended CAG repeat is unstable *in vivo*, often leading to a drift and phenotypic differences in generations of the same animal model [131]. Apart from the use of either the mutated full-length or the N-terminal fragment of the gene, the models can be further distinguished by the exact length of the CAG repeat, overexpressed transgene or knockin expression into the endogenous gene locus, use of the human or animal form of the HTT gene, cDNA versus genomic DNA, and different promoter constructs. All these factors can affect the time of onset, severity, and type of physiological readouts and therefore also the validity of the models in respect to HD in patients.

In addition, the species choice can impact not only the throughput but also the translation. Whereas worm or fly models can also be used for higher-throughput screening at an earlier stage in HD discovery programs, there are now also more than 20 different genetic rodent models and several larger animal models of the disease available for preclinical studies. It could be argued that a nonhuman primate model should have the highest value in terms of predicting compound efficacy in HD patients, and a rhesus monkey model was indeed published several years ago [132]. However, although three animals were born alive, expressing the mutated HTT gene and demonstrating severe, clinically relevant phenotypes (dystonia and chorea), unfortunately none of them survived for longer than

6 months. Thus, although several rodent models effectively mimic certain aspects of the disease, for example, striatal atrophy in N-terminal fragment [133] and full-length mutated HTT models [134], none of the models can truly reflect the full spectrum of the human condition in terms of motor symptoms and cognitive or behavioral abnormalities.

Such difficulties are common and in fact expecting one animal model to fully predict the patient population seems unrealistic. Indeed, even apparently identical model systems can result in differences. This point is illustrated by a recent investigation into the behavior of different mouse strains toward cocaine and methamphetamine. It was found that two C57BL/6 mouse strains with nearly identical genetic background showed quite different responses to the same drug. In fact, the authors were able to track multiple phenotypic differences to a single point mutation in the CYFIP2 protein, leading to a destabilization of that protein (which is not even a direct target of the drug) and a sensitized response to cocaine [135]. The C57BL/6 mice are in widespread use in preclinical drug discovery as a strain of choice for protein overexpression and gene knockout models. If these inbred mouse strains can demonstrate such a remarkable amount of variability, it must be asked how predictive such animal models can ever be for a genetically diverse patient population. Thus, it is for these reasons that many organizations choose more than one model to assess the value of new compounds in complex CNS disease areas.

In contrast to the difficulty in devising appropriate models for CNS drugs, the picture for anti-infective research continues to be more promising. Animal models for bacterial, fungal, and parasitic diseases are well established, and as a result, there is an excellent correlation with the clinical outcome. As a result, the attrition rates for anti-infective drug candidates tend to be much lower than for other therapeutic areas. This success rate even extends to antiviral drug research, where development of animal models is more challenging. To take the examples of CCR5 antagonists, there have only been three drug candidates given USAN names (a measure of compounds reaching phase II trials), resulting in a 33% success rate to market following the launch of maraviroc (Selzentry®). The failed candidates (aplaviroc and vicriviroc) were halted due to unpredictable safety concerns (liver toxicity and increased cancer rate, respectively).

9.5   SUMMARY AND FUTURE DIRECTIONS

9.5.1   Summary of Impact of Current Approaches

From Swinney's original analysis [24], it could be interpreted that a project could be more successful if prosecuted via a phenotypic approach. More recent analysis by Swinney [38] scrutinizes this proposal further and highlights that for many of the phenotypic approaches, some mechanistic understanding was required to define the assays used or the compounds tested. Indeed, we would propose that the use of even complex functional systems is not truly phenotypic, as they are often defined or selected on the basis of known target interactions and mechanisms. We have acknowledged this difference in perspective throughout this chapter and would highlight that when the recent successes are viewed in this way, the relative impact of phenotypic versus target-based approaches becomes less clear.

During our analysis, we have focused on the CNS and anti-infective therapeutic areas, and it is clear from this that both phenotypic and target-based approaches have been successfully applied. However, when looking at discovery processes as a whole, it is also clear that both approaches can be subject to attrition. Furthermore, while the advantages and disadvantages of each approach may suggest that there will be differences in attrition levels and factors (see Table 9.3), even in this respect the picture is not always this clear cut.

A historical contributor to attrition in the target-based approach is a lack of clinical efficacy, often based on a lack of translation between initial *in vitro* screens, *in vivo* biology, and clinical efficacy. While it is perceived to be less likely for phenotypic screens to have lack of efficacy and to have higher levels of translatability between functional assay systems, the situation is not always that simple. For example, while the fungin class of anti-infectives is a clear success story for phenotypic screening, the identification of the eventual drug candidates utilized screening against both the target (*Candida* glucan synthase) and phenotypic screening in a range of whole-cell antifungal assays. These *in vitro* assays were followed up by *in vivo* efficacy studies in animal infection models. However, scientists at Lilly and Merck observed different correlations of activity. In the series of echinocandin analogues pioneered by Lilly scientists, activity in the phenotypic assay spanned several orders of magnitude and predicted well for *in vivo* activity, despite marginal differences in effects on glucan synthesis [136]. In contrast, caspofungin (Cancidas®) was identified as a potent inhibitor of glucan synthesis *in vitro*, with fungicidal effects in animal models, although a narrow potency range was observed in whole-cell assays [137]. Since neither group offered a rationale for the differences in correlation between systems that they had observed, it is difficult to make any firm conclusions on whether the phenotypic screens resulted in a higher level of translatability between systems. Moreover, given how these projects were prosecuted, it could be interpreted that both targeted and phenotypic screens are required to really understand the translation and maximize success. This complexity is also true when other classes of antifungal medicines are considered. The discovery of fluconazole (Diflucan®) provides an extreme example, since this series of polar molecules were inactive under 30 different conditions yet were among the most potent antifungal agents tested in animal models at the time they were tested [138]. Only after extensive investigation were conditions identified to obtain *in vitro* data that correlated with the observed efficacy. For the azole class, this required minimal media [139] or the presence of protein synthesis inhibitors [140]. Furthermore, this is not limited to the anti-infective area, as illustrated for levetiracetam (Keppra®). Although identified via phenotypic methods, levetiracetam (Keppra®) showed an apparent lack of efficacy in classical animal models of epilepsy.

Another well-cited advantage of the phenotypic screening approach compared to targeted screening is the ability to identify hits that act via different mechanisms. However, the use of the phenotypic endpoint does not necessarily mean all/the best start points will be found. For example, there have been no examples of validated antiviral CCR5 antagonists or antiparasitic N-myristoyltransferase inhibitors being discovered by phenotypic screening. Possible explanations for this include the dissimilarity of the structure of these mechanistic targets from those addressed by previous medicinal chemistry programs or the lack of overlap of their target space with either combichem chemical space or natural product chemical space. An additional risk is for any phenotypic hits to be excluded due to their lack of selectivity over other cell lines/assays. In addition, since selectivity at the target level is often amplified as potency improves, there is a real danger that useful hits against new targets might be discarded. As an example, it is doubtful whether phenotypic screening would have discovered inhibitors of *Plasmodium* kinases [141] since the hit series were derived from inhibitors of human kinases.

9.5.2   Future Directions

When considering target-based approaches, there is a distinct advantage for disease areas in which there is well-founded genetic information that confirms the critical role for a target and mechanism. While the discovery of the CCR5 antagonist maraviroc (Selzentry®) was one of the first examples of this approach, it is becoming much more common and is particularly prevalent in the cancer field. A recent success supporting this approach is the ALK inhibitor, crizotinib (Xalkori®) [142], which is a treatment for cancer patients that demonstrate a mutation in the ALK gene. In addition, the discovery of the role of the sodium channel (NaV1.7) in pain has sparked a worldwide search for selective channel blockers. The important role of the NaV1.7 channel in mediating pain was unknown until the finding that mutations in the SCN9a resulted in the complete inability to feel pain [143]. Less clinically useful, but equally interesting, was the finding that the mutations also caused loss of smell [144]. Following this initial discovery, over a 100 patents referring to NaV1.7 have been filed and several companies (Pfizer, Convergence, Xenon) are reported to have had clinical candidates based on this target. A recent paper provides evidence that the Xenon compound (XEN402) reduced pain in a clinical situation [145], which is again encouraging for this genetically founded approach to target-based screening.

Regardless, it still remains true that early use of functional testing and assessment of the phenotype is paramount for target-based approaches. Current technologies enabling high-content screening of physiologically relevant cell types that can be adopted at the front line of the discovery process as well as the development of biomarkers that test target engagement and mechanistic impact of the target in disease will remain key areas of focus in order to mitigate attrition risks. Therefore, the future of this approach can arguably remain dependent upon developments in genetic understanding, patient segregation, and the technology advancements required to measure biological activities in patients, patient samples, patient-derived cells, and physiologically relevant cell types. However, it is also clear that for more complex disease areas such as oncology, fibrosis, and CNS, in which disorders can be caused by a combination of genetic and environmental factors and might require the simultaneous modulation of several molecular targets or pathways, therapeutic approaches that focus on biological networks and a more phenotypic approach are more likely to succeed.

In the case of phenotypic approaches, it is ostensibly those screens that have adopted a more *in vivo* approach that have proven most successful. This is exemplified by phenotypic screening campaigns from the epilepsy field where several thousand compounds have been systematically screened in animal models or were simply found by serendipity [146]. As already indicated, though, this is predicated on the use of fully validated animal models that represent the complexity of the disease being targeted, and therefore, this must remain a key focus for these projects. Recent developments in the technologies supporting target deconvolution, undoubtedly a historical concern and bottleneck, are arguably less of an issue from a practical perspective today. Indeed, many would argue that any pain involved in this process is worth the gain [34]. However, what does play a significant part is the development of robust SAR and the ability to truly realize the perceived benefits of screening compounds versus multiple pathways and targets in the primary readout. The difficulty in designing molecules when combining multiple factors such as permeability, potency at the target, or indeed potency at multiple targets is a challenge and will depend upon the mindset of the chemists involved as well as the organizational philosophy as to when and if to deconvolute the target(s). Moreover, the choice of samples to screen needs to be carefully considered in order to maximize the likelihood of success. Thus, methods for differentiating pharmacological profiles and the creation of compound collections specifically aimed at phenotypic screens are likely to be of continued interest in this area.

9.5.3 Conclusion

While we have summarized the perceived advantages and disadvantages of each approach in Table 9.3, it is clear that each project has to be judged on its own merits in the context of the disease area. In fact, our analysis leads us to the conclusion that to be successful a combination of the phenotypic and target-based approach is required to minimize the risks of attrition, particularly due to translation and safety. Specifically, if a target-based approach is adopted, the philosophy should be to assess phenotype as early as possible, and if a phenotypic approach is adopted, the target or mechanism should be identified as early as possible. Therefore, it is likely that both target-based and phenotypic approaches will continue to be adopted in the industry, albeit the distinction between the two approaches may become less apparent. In addition, it is likely that what will dictate the approach selected will be dependent upon the level of disease biology knowledge, the tools available, and the fundamental philosophies of the organizations making the decisions.

REFERENCES

1 Drews, J. (2000). Drug discovery: a historical perspective. *Science, 287*, 1960–1964.

2 Glusker, J.P. (1994). Dorothy Crowfoot-Hodgkin (1910-1994). *Protein Sci., 3*, 2465–2469.

3 Florey, H.W. (1944). Penicillin: a survey. *Br Med J., 2*, 169–171.

4 Zheng, W., Thorne, N., and McKew, J.C. (2013). Phenotypic screens as a renewed approach for drug discovery. *Drug Discov Today, 18*, 1067–1073.

5 Vane, J.R., Botting, R.M. (2003). The mechanism of action of aspirin. *Thromb Res., 110*, 255–258.

6 Pina, A.S, Hussain, A., and Roque, A.C.A. (2009). An historical overview of drug discovery. *Methods Mol. Biol., 572*, 3–12.

7 Ehrlich, P. (1913). Chemotherapeutics: scientific principles, methods and results. *Lancet 2*, 445–451.

8 Langley, J.N. (1906). On nerve endings and on special excitable substances in cells. *Proc R. Soc., B78*, 170–194.

9 Segel, I.H. *Enzyme kinetics; behaviour and analysis of rapid equilibrium and steady-state enzyme systems*. Wiley 1975.

10 Moos, W.H., Hurt, C.R., Morales, G.A. (2009). Combinatorial chemistry: oh what a decade or two can do. *Mol Divers., 13*, 241–245.

11 McLoughlin, D.J., Bertelli, F., Williams C. (2007). The A,B,C's of G-protein coupled receptor pharmacology in assay development for HTS. *Expert Opin Drug Disc., 2*, 603–619.

12 Glickman, J.F., Schmid, A., Ferrand, S. (2008). Scintillation proximity assays in high-throughput screening. *Assay Drug Dev Technol., 6*, 433–455.

13 Pereira, D.A, Williams, J.A. (2007). Origin and evolution of high throughput screening. *Br J Pharmacol., 152*, 53–61.

14 Diller, D.J. (2008). The synergy between combinatorial chemistry and high-throughput screening. *Curr Opin Drug Discov Develop., 11*, 346–355.

15 Ganesan, A. (2008). The impact of natural products upon modern drug discovery. *Curr Opin Chem Biol. 12*, 306–317.

16 Newman, D.J., Cragg, G.M. (2012). Natural products as sources of new drugs over the 30 years from 1981 to 2010. *J Nat Prod., 75*, 311–335.

17 Albert, J.S., Blomberg, N., Breeze, A.L., Brown, A.J., Burrows, J.N., Edwards, P.D., Folmer, R.H., Geschwindner, S., Griffen, E.J., Kenny, P.W., Nowak, T., Olsson, L.L., Sanganee, H., Shapiro, A.B. (2007). An integrated approach to fragment-based lead generation: philosophy, strategy and case studies from AstraZeneca's drug discovery programmes. *Curr Top Med Chem., 7*, 1600–1629.

18 Munos, B. (2009). Lessons from 60 years of pharmaceutical innovation. *Nat Rev Drug Discov., 8*, 959–968.

19 Scannell, J.W., Blanckley, A., Boldon, H., Warrington, B. (2012). Diagnosing the decline in pharmaceutical R&D efficiency. *Nat Rev Drug Discov., 11*, 191–200.

20 Paul, S.M., Mytelka, D.S., Dunwiddie, C.T., Persinger, C.C., Munos, B.H., Lindborg, S.R., Schacht, A.L. (2010). How to improve R&D productivity: the pharmaceutical industry's grand challenge. *Nat Rev Drug Discov., 9*, 203–214.

21 Gershell, L.J., Atkins, J.H. (2003). A brief history of novel drug discovery technologies. *Nat Rev Drug Discov., 2*, 321–327.

22 Kayser, O., Muller, R.H. *Pharmaceutical Biotechnology*, Drug Discovery and Clinical applications. Wiley Press 2004.

23 DiMasi, J. (2007). An audience with... Joseph DiMasi. *Nat Rev Drug Discov., 6*, 512.

24 Swinney, D.C., Anthony, J. (2011). How were new medicines discovered? *Nat Rev Drug Discov., 10*, 507–19.

25 Goodyear, M. (2006). Learning from the TGN1412 trial. *Brit Med J., 332*, 677–678.

26 Danilenko, D.M., Wang, H. (2012). The yin and yang of immunomodulatory biologics: assessing the delicate balance between benefit and risk. *Toxicol Pathol., 40*, 272–287.

27 Kola, I., Landis, J. (2004). Can the pharmaceutical industry reduce attrition rates? *Nat Rev Drug Discov. 3*, 711–715.

28 Arrowsmith, J. (2011). Phase II failures: 2008-2010. *Nat Rev Drug Discov., 10*, 328.

29 DiMasi, J.A., Feldman, L., Seckler, A., Wilson, A. (2010). Trends in risks associated with new drug development: success rates for investigational drugs. *Clin Pharmacol Ther., 87*, 272–277.

30 Morgan, P., Van Der Graaf, P.H., Arrowsmith, J., Feltner, D.E., Drummond, K.S., Wegner, C.D., Street, S.D. (2012). Can the flow of medicines be improved? Fundamental pharmacokinetic and pharmacological principles toward improving Phase II survival. *Drug Discov Today., 17*, 419–424.

31 Williams, C., Sewing, A. (2005). G-protein coupled receptor assays: to measure affinity or efficacy that is the question. *Comb Chem High Throughput Screen., 8*, 285–292.

32 Wendler, A., Wehling, M. (2010). The translatability of animal models for clinical development: biomarkers and disease models. *Curr Opin Pharmacol., 10*, 601–606.

33 Gobburu, J.V., Lesko, L.J. (2009). Quantitative disease, drug, and trial models. *Annu Rev Pharmacol Toxicol., 49*, 291–301.

34 Kotz, J. (2012). Phenotypic screening, take two. *SciBX, 5*, 1–3.

35 Bradley, J., Gill, J., Bertelli, F., Letafat, S., Corbau, R., Hayter, P., Harrison, P., Tee, A., Keighley, W., Perros, M., Ciaramella, G., Sewing, A., Williams, C. (2004). Development and automation of a 384-well cell fusion assay to identify inhibitors of CCR5/CD4-mediated HIV virus entry. *J Biomol Screen., 9*, 516–524.

36 Eglen, R., Reisine, T. (2011). Primary cells and stem cells in drug discovery: emerging tools for high-throughput screening. *Assay Drug Dev Technol., 9*, 108–124.

37 Nemeth, E.F. (2006). Misconceptions about calcimimetics. *Ann NY Acad Sci.*, *1068*, 471–476.

38 Swinney, D.C. (2013). The Contribution of Mechanistic Understanding to Phenotypic Screening for First-in-Class Medicines. *J Biomol Screen.*, *18*, 1186–1192.

39 Benz, F., Knüsel, F., Nüesch, J., Treichler, H., Voser, W., Nyfeler, R. Keller-Schierlein, W. (1974). Stoffwechselprodukte von Mikroorganismen 143. Mitteilung. Echinocandin B, ein neuartiges Polypeptid-Antibioticum aus Aspergillus nidulans var. echinulatus: Isolierung und Bausteine. *Helvetica Chim Acta*, *57*, 2459–2477.

40 Debono, M., Abbott, B.J., Turner, J.R., Howard, L.C., Gordee, R.S., Hunt, A.S., Barnhart, M., Molloy, R.M., Willard, K.E., Fukuda, D. and et al. (1988). Synthesis and evaluation of LY121019, a member of a series of semisynthetic analogues of the antifungal lipopeptide echinocandin B. *Ann NY Acad Sci.*, *544*, 152–167.

41 Boone, C., Sommer, S.S., Hensel, A. and Bussey, H. (1990). Yeast KRE genes provide evidence for a pathway of cell wall beta-glucan assembly. *J Cell Biol.*, *110*, 1833–1843.

42 Douglas, C.M., Foor, F., Marrinan, J.A., Morin, N., Nielsen, J.B., Dahl, A.M., Mazur, P., Baginsky, W., Li, W., el-Sherbeini, M. (1994). The Saccharomyces cerevisiae FKS1 (ETG1) gene encodes an integral membrane protein which is a subunit of 1,3-beta-D-glucan synthase. *Proc Natl Acad Sci USA.*, *91*, 12907–12911.

43 Douglas, C.M., Marrinan, J.A., Li, W. and Kurtz, M. B. (1994). A Saccharomyces cerevisiae mutant with echinocandin-resistant 1,3-beta-D-glucan synthase. *J Bacteriol.*, *176*, 5686–5696.

44 Douglas, C.M., D'Ippolito, J.A., Shei, G.J., Meinz, M., Onishi, J., Marrinan, J.A., Li, W., Abruzzo, G.K., Flattery, A., Bartizal, K., Mitchell, A. and Kurtz, M.B. (1997). Identification of the FKS1 gene of Candida albicans as the essential target of 1,3-beta-D-glucan synthase inhibitors. *Antimicrob Agents Chemother.*, *41*, 2471–2479.

45 Preskorn, S.H. (2010). CNS Drug Development: Part I: The Early Period of CNS Drugs. Psychopharmacology. *J Psychiatr Pract.*, *16*, 334–339.

46 Klitgaard, H., Verdru, P. (2008). Levetiracetam—the first SV2A ligand for the treatment of epilepsy. *Expert Opin Drug Discov.*, *2*, 1537–1545.

47 Gower, A.J., Noyer, M., Verloes, R., Gobert, J., Wülfert, E. (1992). UCB L059, a novel anticonvulsant drug: pharmacological profile in animals. *Eur J Pharmacol.*, *222*, 193–203.

48 Barton, M.E., Klein, B.D., Wolf, H.H., White, H.S. (2001). Pharmacological characterization of the 6 Hz psychomotor seizure model of partial epilepsy. *Epilepsy Res.*, *47*, 217–228.

49 Kaminski, R.M., Gillard, M., Klitgaard, H. Targeting SV2A for discovery of antiepileptic drugs. In: Noebels, J.L., Avoli, M., Rogawski, M.A., Olsen, R.W., Delgado-Escueta, A.V., editors. *Jasper's Basic Mechanisms of the Epilepsies [Internet].* 4th edition. Bethesda (MD): National Center for Biotechnology Information (US); 2012.

50 Jain, K.K. (2000). An assessment of rufinamide as an anti-epileptic in comparison with other drugs in clinical development. *Expert Opin Investig Drugs*, *9*, 829–840.

51 Masuda, Y. (1980). 3-Sulfamoylmethyl-1,2-benzisoxazole, a new type of anti-convulsant drug: pharmacological profile. *Arzneimittelforschung*, *30*, 477–483.

52 Rogawski, M.A. (2011). Revisiting AMPA receptors as an antiepileptic drug target. *Epilepsy Curr.*, *11*, 56–63.

53 Martel, R.R., Klicius, J., Galet, S. (1977). Inhibition of immune response by rapamycin, a new antifungal antibiotic. *Can J Physiol Pharmacol.*, *55*, 48–51.

54 Inoki, K., Corradetti, M.N., Guan, K.L. (2005). Dysregulation of the TSC-mTOR pathway in human disease. *Nat Genet.*, *37*, 19–24.

55 Ryther, R.C., Wong, M. (2012). Mammalian target of rapamycin (mTOR) inhibition: potential for antiseizure, antiepileptogenic, and epileptostatic therapy. *Curr Neurol Neurosci Rep.*, *12*, 410–418.

56 Gerzon, K., Krumkalns, E.V., Brindle, R.L., Marshall, F.J., Root, M.A. (1963). The adamantyl group in medicinal agents I. Hypoglycemic N-arylsulfonyl-N'-adamantylureas. *J Med Chem.*, 6, 760–763.

57 Bormann, J. (1989) Memantine is a potent blocker of N-methyl-D-aspartate (NMDA) receptor channels. *Eur J Pharmacol.*, 166, 591–592.

58 Reisberg, B., Doody, R., Stöffler, A., Schmitt, F., Ferris, S., Möbius, H.-J. (2003). Memantine in moderate-to-severe Alzheimer's disease. *New Engl J Med.*, 348, 1333–1341.

59 Barre-Sinoussi, F., Chermann, J., Rey, F., Nugeyre, M., Chamaret, S., Gruest, J., Dauguet, C., Axler-Blin, C., Vezinet-Brun, F., Rouzioux, C., Rozenbaum, W. and Montagnier, L. (1983). Isolation of a T-lymphotropic retrovirus from a patient at risk for acquired immune deficiency syndrome (AIDS). *Science*, 220, 868–871.

60 Gallo, R., Sarin, P., Gelmann, E., Robert-Guroff, M., Richardson, E., Kalyanaraman, V., Mann, D., Sidhu, G., Stahl, R., Zolla-Pazner, S., Leibowitch, J. and Popovic, M. (1983). Isolation of human T-cell leukemia virus in acquired immune deficiency syndrome (AIDS). *Science*, 220, 865–867.

61 Ratner, L., Haseltine, W., Patarca, R., Livak, K. J., Starcich, B., Josephs, S.F., Doran, E.R., Rafalski, J.A., Whitehorn, E.A., Baumeister, K. and et al. (1985). Complete nucleotide sequence of the AIDS virus, HTLV-III. *Nature*, 313, 277–284.

62 Mitsuya, H., Weinhold, K.J., Furman, P.A., St Clair, M.H., Lehrman, S.N., Gallo, R.C., Bolognesi, D., Barry, D.W. and Broder, S. (1985). 3'-Azido-3'-deoxythymidine (BW A509U): an antiviral agent that inhibits the infectivity and cytopathic effect of human T-lymphotropic virus type III/lymphadenopathy-associated virus in vitro. *Proc Natl Acad Sci USA.*, 82, 7096–7100.

63 Smerdon, S.J., Jäger, J., Wang, J., Kohlstaedt, L.A., Chirino, A. J., Friedman, J. M., Rice, P.A., Steitz, T.A. (1994). Structure of the binding site for nonnucleoside inhibitors of the reverse transcriptase of human immunodeficiency virus type 1. *Proc Natl Acad Sci USA.*, 91, 3911–3915.

64 Lindberg, J., Sigurðsson, S., Löwgren, S. O., Andersson, H., Sahlberg, C., Noréen, R., Fridborg, K., Zhang, H. and Unge, T. (2002). Structural basis for the inhibitory efficacy of efavirenz (DMP-266), MSC194 and PNU142721 towards the HIV-1 RT K103N mutant. *Eur J Biochem.*, 269, 1670–1677.

65 King, F. D. *The identification of the HIV protease inhibitor saquinavir.* pp. 397–406. Royal Society of Chemistry 2002.

66 Kempf, D.J., Sham, H.L., Marsh, K.C., Flentge, C.A., Betebenner, D., Green, B. E., McDonald, E., Vasavanonda, S., Saldivar, A., Wideburg, N.E., Kati, W.M., Ruiz, L., Zhao, C., Fino, L., Patterson, J., Molla, A., Plattner, J.J., Norbeck, D.W. (1998). Discovery of ritonavir, a potent inhibitor of HIV protease with high oral bioavailability and clinical efficacy. *J Med Chem.*, 41, 602–617.

67 Goldgur, Y., Craigie, R., Cohen, G.H., Fujiwara, T., Yoshinaga, T., Fujishita, T., Sugimoto, H., Endo, T., Murai, H., Davies, D.R. (1999). Structure of the HIV-1 integrase catalytic domain complexed with an inhibitor: a platform for antiviral drug design. *Proc Natl Acad Sci USA.*, 96, 13040–13043.

68 Pommier, Y., Johnson, A.A. and Marchand, C. (2005). Integrase inhibitors to treat HIV/AIDS. *Nat Rev Drug Discov.*, 4, 236–248.

69 Deeks, S.G., Kar, S., Gubernick, S.I. and Kirkpatrick, P. (2008). Raltegravir. *Nat Rev Drug Discov.*, 7, 117–118.

70 Dorr, P., Westby, M., Dobbs, S., Griffin, P., Irvine, B., Macartney, M., Mori, J., Rickett, G., Smith-Burchnell, C., Napier, C., Webster, R., Armour, D., Price, D., Stammen, B., Wood, A. and Perros, M. (2005). Maraviroc (UK-427,857), a potent, orally bioavailable, and selective small-molecule inhibitor of chemokine receptor CCR5 with broad-spectrum anti-human immunodeficiency virus type 1 activity. *Antimicrob Agents Chemother.*, 49, 4721–4732.

71 Price, D.A., Armour, D., de Groot, M., Leishman, D., Napier, C., Perros, M., Stammen, B. L. and Wood, A. (2006). Overcoming HERG affinity in the discovery of the CCR5 antagonist maraviroc. *Bioorg Med Chem Lett.*, *16*, 4633–4637.

72 Kuritzkes, D., Kar, S. and Kirkpatrick, P. (2008). Maraviroc. *Nat Rev Drug Discov.*, *7*, 15–16.

73 Kikuchi, T., Tottori, K., Uwahodo, Y., Hirose, T., Miwa, T., Oshiro, Y., Morita, S. (1995). 7-(4-[4-(2,3-Dichlorophenyl)-1-piperazinyl]butyloxy)-3,4-dihydro-2(1H)-quinolinone (OPC-14597), a new putative antipsychotic drug with both presynaptic dopamine autoreceptor agonistic activity and postsynaptic D2 receptor antagonistic activity. *J Pharmacol Exp Ther.*, *274*, 329–336.

74 Shapiro, D.A., Renock, S., Arrington, E., Chiodo, L.A., Liu, L.X., Sibley, D.R., Roth, B.L., Mailman, R. (2003). Aripiprazole, A Novel Atypical Antipsychotic Drug with a Unique and Robust Pharmacology. *Neuropsychopharmacology*, *28*, 1400–1411.

75 Coe, J.E., Brooks, P.R., Vetelino, M.G., Wirtz, M.C., Arnold, E.P., Huang, J. (2005). Varenicline: an alpha4beta2 nicotinic receptor partial agonist for smoking cessation. *J Med Chem.*, *48*, 3474–3477.

76 Alvaro, G., Di Fabio, R. (2007). Neurokinin 1 receptor antagonists—current prospects. *Curr Opin Drug Discov Develop.*, *10*, 613–621.

77 Uchikawa, O., Fukatsu, K., Tokunoh, R., Kawada, M., Matsumoto, K., Imai, Y., Hinuma, S., Kato, K., Nishikawa, H., Hirai, K., Miyamoto, M., Ohkawa, S. (2002). Synthesis of a novel series of tricyclic indan derivatives as melatonin receptor agonists. *J Med Chem.*, *45*, 4222–4239.

78 Boismare, F., Daoust, M., Moor, N., Saligaut, C., Lhuintre, J.P., Chretien, P., Durlach, J. (1984). A homotaurine derivative reduces the voluntary intake of ethanol by rats: are cerebral GABA receptors involved? *Pharmacol Biochem Behav.*, *21*, 787–789.

79 Yednock, T.A. Cannon, C., Fritz, L.C., Sanchez-Madrid, F., Steinman, L., Karin, N. (1992). Prevention of experimental autoimmune encephalomyelitis by antibodies against alpha 4 beta 1 integrin. *Nature*, *356*, 63–66.

80 Sundberg S.A. (2000). High-throughput and ultra-high-throughput screening: solution- and cell-based approaches. *Curr Opin Biotechnol.*, *11*. 47–53.

81 Wright, M.H., Clough, B., Rackham, M.D., Rangachari, K., Brannigan, J.A., Grainger, M., Moss, D.K., Bottrill, A.R., Heal, W.P., Broncel, M., Serwa, R.A., Brady, D., Mann, D.J., Leatherbarrow, R.J., Tewari, R., Wilkinson, A.J., Holder, A.A. and Tate, E.W. (2013). Validation of N-myristoyltransferase as an antimalarial drug target using an integrated chemical biology approach. *Nat Chem.*, *6*, 112–121.

82 Armour, D., de 14;Groot, M.J., Edwards, M., Perros, M., Price, D.A., Stammen, B. L., Wood, A. (2006). The discovery of CCR5 Receptor antagonists for the treatment of HIV infection: hit-to-lead studies. *ChemMedChem*, *1*, 706–709.

83 Lang, P., Yeow, K., Nichols, A., Scheer, A. (2006). Cellular imaging in drug discovery. *Nat Rev Drug Discov.*, *5*, 343–356.

84 Bickle, M. (2010). The beautiful cell: high-content screening in drug discovery. *Anal Bioanal Chem.*, *398*, 219–226.

85 Haggarty, S.J., Perlis, R.H. (2014). Translation: Screening for novel therapeutics with disease-relevant cell types derived from human stem cell models. *Biol Psychiatry*, *75*, 952–960.

86 Williams, C. (2004). cAMP detection methods in HTS: selecting the best from the rest. *Nat Rev Drug Discov.*, *3*, 125–135.

87 Langmead, C.J., Christopolous, A. (2006). Allosteric agonists of 7TM receptors: expanding the pharmacological toolbox. *Trends Pharmacol Sci.*, *27*, 475–481.

88 Jensen, A.A., Spalding, T.A. (2004). Allosteric modulation of G-protein coupled receptors. *Eur J Pharm Sci.*, *21*, 407–420.

89 Eglen, R., Reisine, T. (2011). Drug Discovery and the human kinome: recent trends. *Pharmcol Ther.*, *130*, 144–156.

90   London, N., Raveh, B., Schueler-Furman, O. (2013). Druggable protein-protein interactions—from hot spots to hot segments. *Curr Opin Chem Biol.*, *17*, 952–959.

91   Fuller, J.C., Burgoyne, N.J., Jackson, R.M. (2009). Predicting druggable binding sites at the protein-protein interface. *Drug Discov Today.*, *14*, 155–61.

92   McGovern, S.L., Helfand, B.T., Feng, B., Shoichet, B.K. (2003). A specific mechanism of nonspecific inhibition. *J Med Chem.*, *46*, 4265–4272.

93   Jadhav, A., Ferreira, R. S., Klumpp, C., Mott, B.T., Austin, C.P., Inglese, J., Thomas, C.J., Maloney, D.J., Shoichet, B.K., Simeonov, A. (2009). Quantitative analyses of aggregation, autofluorescence, and reactivity artifacts in a screen for inhibitors of a thiol protease. *J Med Chem.*, *53*, 37–51.

94   Talele, T.T. (2008). Multiple allosteric pockets of HCV NS5B polymerase and its inhibitors: a structure based insight. *Curr Bioact Compd.*, *4*, 86–109.

95   Williams, C, Hill, S.J. (2009). GPCR signaling: understanding the pathway to successful drug discovery. *Methods Mol Biol.*, *552*, 39–50.

96   Titus, S.A., Southall, N., Marugan, J., Austin, C.P., Zheng, W. (2012). High-throughput multiplexed quantitation of protein aggregation and cytotoxicity in a Huntington's disease model. *Curr ChemGenom.*, *6.* 79–86.

97   Lazzeroni, G., Benicchi, T., Heitz, F., Magnoni, L., Diamanti, D., Rossini, L., Massai, L., Federico, C., Fecke, W., Caricasole, A., La Rosa, S., Porcari, V. (2013). A phenotypic screening assay for modulators of huntingtin-induced transcriptional dysregulation. *J BiomolScreen.*, *18*, 984–996.

98   Ross, N.T., Metkar, S.R., Le, H., Burbank, J., Cahill, C., Germain, A., MacPherson, L., Bittker, J., Palmer, M., Rogers, J., Schreiber, S.L. Identification of a small molecule that selectively inhibits alpha-synuclein translational expression. In *Probe Reports from the NIH Molecular Libraries Program (Internet), Bethesda (MD): National Center for Biotechnology Information (US),* 2011.

99   Zhao, D.L., Zou, L.B., Zhou, L.F., Zhu, P., Zhu, H.B. (2007). A cell-based model of alpha-synucleinopathy for screening compounds with therapeutic potential of Parkinson's disease. *Acta Pharmacol Sin.*, *28*, 616–626.

100 Braungart, E., Gerlach, M., Riederer, P., Baumeister, R., Hoener, M.C. (2004). *Caenorhabditis elegans* MPP+ model of Parkinson's disease for high-throughput drug screenings. *Neurodegenerative Dis.*, *1*, 175–183.

101 Chakrabarti, E., Smith, J.D. (2005). Drug library screen to identify compounds that decrease secreted Abeta from a human cell line. *Curr Alzheimer Res.*, *2*, 255–259.

102 Jones, J.R., Lebar, M.D., Jinwal, U.K., Abisambra, J.F., Koren, J. 3rd, Blair, L., O'Leary, J.C., Davey, Z., Trotter, J., Johnson, A.G., Weeber, E., Echman, C.B., Baker, B.J., Dickey, C.A. (2011). The diarylheptanoid (+)-aR,11S-myricanol and two flavones from bayberry (Myrica cerifera) destabilize the microtubule-associated protein tau. *J Nat Prod.*, *74*, 38–44.

103 Barber, S.C., Higginbottom, A., Mead, R.J., Barber, S., Shaw, P.J. (2009). An in vitro screening cascade to identify neuroprotective antioxidants in ALS. *Free Radic BiolMed.*, *46*, 1127–1138.

104 Plouffe, D., Brinker, A., McNamara, C., Henson, K., Kato, N., Kuhen, K., Nagle, A., Adrián, F., Matzen, J.T., Anderson, P., Nam, T.-G., Gray, N.S., Chatterjee, A., Janes, J., Yan, S.F., Trager, R., Caldwell, J.S., Schultz, P.G., Zhou, Y., Winzeler, E.A. (2008). In silico activity profiling reveals the mechanism of action of antimalarials discovered in a high-throughput screen. *Proc Natl Acad Sci USA.*, *105*, 9059–9064.

105 Gamo, F.J., Sanz, L.M., Vidal, J., de Cozar, C., Alvarez, E., Lavandera, J.L., Vanderwall, D.E., Green, D.V., Kumar, V., Hasan, S., Brown, J.R., Peishoff, C.E., Cardon, L.R., Garcia-Bustos, J.F. (2010). Thousands of chemical starting points for antimalarial lead identification. *Nature*, *465*, 305–310.

106 Nwaka, S., Besson, D., Ramirez, B., Maes, L., Matheeussen, A., Bickle, Q., Mansour, N.R., Yousif, F., Townson, S., Gokool, S., Cho-Ngwa, F., Samje, M., Misra-Bhattacharya, S., Murthy, P.K., Fakorede, F., Paris, J.-M., Yeates, C., Ridley, R., Van Voorhis, W.C., Geary, T. (2011). Integrated dataset of screening hits against multiple neglected disease pathogens. *PLoS Negl Trop Dis.*, *5*, e1412.

107 Ballell, L., Bates, R.H., Young, R.J., Alvarez-Gomez, D., Alvarez-Ruiz, E., Barroso, V., Blanco, D., Crespo, B., Escribano, J., González, R., Lozano, S., Huss, S., Santos-Villarejo, A., Martín-Plaza, J.J., Mendoza, A., Rebollo-Lopez, M. J., Remuiñan-Blanco, M., Lavandera, J.L., Pérez-Herran, E., Gamo-Benito, F.J., García-Bustos, J.F., Barros, D., Castro, J.P. and Cammack, N. (2013). Fueling Open-Source Drug Discovery: 177 Small-Molecule Leads against Tuberculosis. *ChemMedChem*, *8*, 313–321.

108 Zanella, F., Lorens, J.B., Link, W. (2010). High content screening: seeing is believing. *Trends Biotechnol.*, *28*, 237–245.

109 Low, J., Stancato. L., Lee, J., Sutherland J.J. (2008). Prioritizing hits from phenotypic high-content screens. *Curr Opin Dug Disc Dev.*, *11*, 338.

110 Andrews, S.P., Brown, G.A., Christopher, J.A. (2014). Structure-Based and Fragment-Based GPCR Drug Discovery. *ChemMedChem*, *9*, 256–275,

111 Meng, F., Hou, J., Shao, Y.X., Wu, P.Y., Huang, M., Zhu, X., Cai, Y., Li, Z., Xu, J., Liu, P., Luo, H. B., Wan, Y. and Ke, H. (2012). Structure-based discovery of highly selective phosphodies-terase-9A inhibitors and implications for inhibitor design. *J Med Chem.*, *55*, 8549–8558.

112 Zhang, L., Balan, G., Barreiro, G., Boscoe, B. P., Chenard, L.K., Cianfrogna, J., Claffey, M.M., Chen, L., Coffman, K.J., Drozda, S.E., Dunetz, J. R., Fonseca, K.R., Galatsis, P., Grimwood, S., Lazzaro, J.T., Mancuso, J.Y., Miller, E.L., Reese, M.R., Rogers, B.N., Sakurada, I., Skaddan, M., Smith, D.L., Stepan, A.F., Trapa, P., Tuttle, J.B., Verhoest, P.R., Walker, D.P., Wright, A.S., Zaleska, M.M., Zasadny, K., Shaffer, C.L. (2014). The discovery and preclinical characteriza-tion of 1-methyl-3-(4-methylpyridin-3-yl)-6-(pyridin-2-ylmethoxy)-1H-pyrazolo[3,4-b]pyr-azine (PF470), a highly potent, selective and efficacious metabotropic glutamate receptor 5 (mGluR5) negative allosteric modulator. *J Med Chem.*, *57*, 861–877.

113 Kaltenbach, R.F., Klabe, R.M., Cordova, B.C., Seitz, S.P. (1999). Increased antiviral activity of cyclic urea HIV protease inhibitors by modifying the P1/P1' substituents. *Bioorg Med Chem Lett.*, *9*, 2259–2262.

114 Devadas, B., Freeman, S.K., Zupec, M.E., Lu, H.F., Nagarajan, S.R., Kishore, N.S., Lodge, J.K., Kuneman, D.W., McWherter, C.A., Vinjamoori, D.V., Getman, D.P., Gordon, J.I., Sikorski, J.A. (1997). Design and synthesis of novel imidazole-substituted dipeptide amides as potent and selective inhibitors of Candida albicans myristoylCoA: protein N-myristoyl-transferase and identification of related tripeptide inhibitors with mechanism-based anti-fungal activity. *J Med Chem.*, *40*, 2609–2625.

115 Devadas, B., Freeman, S.K., McWherter, C.A., Kishore, N.S., Lodge, J.K., Jackson-Machelski, E., Gordon, J.I., Sikorski, J.A. (1998). Novel biologically active nonpeptidic inhibitors of myristoylCoA: protein N-myristoyltransferase. *J Med Chem.*, *41*, 996–1000.

116 Frearson, J.A., Brand, S., McElroy, S.P., Cleghorn, L.A.T., Smid, O., Stojanovski, L., Price, H.P., Guther, M.L.S., Torrie, L.S., Robinson, D.A., Hallyburton, I., Mpamhanga, C.P., Brannigan, J.A., Wilkinson, A.J., Hodgkinson, M., Hui, R., Qiu, W., Raimi, O.G., van Aalten, D.M.F., Brenk, R., Gilbert, I.H., Read, K.D., Fairlamb, A.H., Ferguson, M.A.J., Smith, D.F. and Wyatt, P.G. (2010). N-myristoyltransferase inhibitors as new leads to treat sleeping sickness. *Nature*, *464*, 728–732.

117 Herbrecht, R., Denning, D.W., Patterson, T.F., Bennett, J.E., Greene, R.E., Oestmann, J.-W., Kern, W.V., Marr, K.A., Ribaud, P., Lortholary, O., Sylvester, R., Rubin, R.H., Wingard, J.R., Stark, P., Durand, C., Caillot, D., Thiel, E., Chandrasekar, P.H., Hodges, M.R., Schlamm, H.T., Troke, P.F., de Pauw, B. (2002). Voriconazole versus Amphotericin B for Primary Therapy of Invasive Aspergillosis. *New Engl J Med.*, *347*, 408–415.

118   Encinas, L., O'Keefe, H., Neu, M., Remuiñán, M.J., Patel, A.M., Guardia, A., Davie, C.P., Pérez-Macías, N., Yang, H., Convery, M.A., Messer, J.A., Pérez-Herrán, E., Centrella, P.A., Álvarez-Gómez, D., Clark, M.A., Huss, S., O'Donovan, G.K., Ortega-Muro, F., McDowell, W., Castañeda, P., Arico-Muendel, C.C., Pajk, S., Rullás, J., Angulo-Barturen, I., Álvarez-Ruíz, E., Mendoza-Losana, A., Pages, L.B., Castro-Pichel, J., Evindar, G. (2014). Encoded library technology as a source of hits for the discovery and lead optimization of a potent and selective class of bactericidal direct inhibitors of Mycobacterium tuberculosis InhA. *J Med Chem.*, *57*, 1276–1288.

119   McNamara, C.W., Lee, M.C.S., Lim, C.S., Lim, S.H., Roland, J., Nagle, A., Simon, O., Yeung, B.K.S., Chatterjee, A.K., McCormack, S.L., Manary, M.J., Zeeman, A.-M., Dechering, K.J., Kumar, T.R.S., Henrich, P.P., Gagaring, K., Ibanez, M., Kato, N., Kuhen, K.L., Fischli, C., Rottmann, M., Plouffe, D.M., Bursulaya, B., Meister, S., Rameh, L., Trappe, J., Haasen, D., Timmerman, M., Sauerwein, R.W., Suwanarusk, R., Russell, B., Renia, L., Nosten, F., Tully, D.C., Kocken, C.H.M., Glynne, R. J., Bodenreider, C., Fidock, D.A., Diagana, T.T., Winzeler, E.A. (2013). Targeting Plasmodium PI(4)K to eliminate malaria. *Nature*, *504*, 248–253.

120   Jaudzems, K., Tars, K., Maurops, G., Ivdra, N., Otikovs, M., Leitans, J., Kanepe-Lapsa, I., Domraceva, I., Mutule, I., Trapencieris, P., Blackman, M.J., Jirgensons, A. (2014). Plasmepsin inhibitory activity and structure-guided optimization of a potent hydroxyethylamine-based antimalarial hit. *ACS Med Chem Lett.*, *5*, 373–377.

121   Stevens, J.L. (2006). Future of toxicology – mechanisms of toxicity and drug safety: where do we go from here? *Chem Res Toxicol.*, *19*, 1393–1401.

122   Sanguinetti, M.C., Jiang, C., Curran, M.E., Keating, M.T.A. (1995). A mechanistic link between an inherited and an acquired cardiac arrhythmia: HERG encodes the I_{Kr} potassium channel. *Cell*, *81*, 299–307.

123   Huang, X.-P., Setola, V., Yadav,P.N., Allen, J.A., Rogan, S.C., Hanson, B.J., Revankar, C., Robers, M., Doucette, C., Roth, B.L. (2009). Parallel functional activity profiling reveals valvulopathogens are potent 5-hydroxytryptamine$_{2B}$ receptor agonists: implications for drug safety assessment. *Mol Pharmacol.*, *76*, 710–722.

124   Overington, J.P., Al-Lazikani, B., Hopkins, A.L. (2006). How many drug targets are there? *Nat Rev Drug Discov.*, *5*, 993–996.

125   Mestres, J., Gregori-Puigjané, E., Valverde, S., Solé, R.V. (2009). The topology of drug-target interaction networks: implicit dependence on drug properties and target families. *Mol Biosyst.*, *5*, 1051–1057.

126   Hart, C.P. (2005). Finding the target after screening the phenotype. *Drug Discov Today.*, *10*, 513–519.

127   Cho, Y.S., Kwon, H.J. (2012). Identification and validation of bioactive small molecule target through phenotypic screening. *Bioorg Med Chem.*, *20*, 1922–1928.

128   Lee, J., Bogyo, M. (2013). Target deconvolution techniques in modern phenotypic profiling. *Curr Opin Chem Biol.*, *17*, 118–126.

129   Walker, F.O. (2007). Huntington's disease. *Lancet*, *369*, 218–228.

130   Beal, M.F., Kowall, N.W., Ellison, D.W., Mazurek, M.F., Swartz, K.J., Martin, J.B. (1986). Replication of the neurochemical characteristics of Huntington's disease by quinolinic acid. *Nature*, *321*, 168–171.

131   Menalled, L., El-Khodor, B.F., Patry, M., Suárez-Fariñas, M., Orenstein, S.J., Zahasky, B., Leahy, C., Wheeler, V., Yang, X.W., MacDonald, M., Morton, A.J., Bates, G., Leeds, J., Park, L., Howland, D., Signer, E., Tobin, A., Brunner, D. (2009). Systematic behavioral evaluation of Huntington's disease transgenic and knock-in mouse models. *Neurobiol Dis.*, *35*, 319–336.

132   Yang, S.H., Cheng, P.H., Banta, H., Piotrowska-Nitsche, K., Yang, J.J., Cheng, E.C., Snyder, B., Larkin, K., Liu, J., Orkin, J., Fang, Z.H., Smith, Y., Bachevalier, J., Zola, S.M., Li, S.H.,

Li, X.J., Chan, A.W. (2008). Towards a transgenic model of Huntington's disease in a non-human primate. *Nature, 453,* 921.

133   Aggarwal, M., Duan, W., Hou, Z., Rakesh, N., Ross, C.A., Miller, M.I., Mori, S., Zhang, J. (2012). Spatiotemporal mapping of brain atrophy in mouse models of Huntington's disease using longitudinal in vivo magnetic resonance imaging. *Neuroimage, 60,* 2086–2095.

134   Carroll, J.B., Lerch, J.P., Franciosi, S., Spreeuw, A., Bissada, N., Henkelman, R.M., Hayden, M.R. (2011). Natural history of disease in the YAC128 mouse reveals a discrete signature of pathology in Huntington disease. *Neurobiol Dis., 43,* 257–265.

135   Kumar V. (2013). C57BL/6N mutation in cytoplasmic FMRP interacting protein 2 regulates cocaine response. *Science, 342,* 1508–1512.

136   Debono, M., Turner, W.W., LaGrandeur, L., Burkhardt, F.J., Nissen, J. S., Nichols, K.K., Rodriguez, M.J., Zweifel, M.J., Zeckner, D.J. (1995). Semisynthetic chemical modification of the antifungal lipopeptide echinocandin B (ECB): structure-activity studies of the lipophilic and geometric parameters of polyarylated acyl analogs of ECB. *J Med Chem., 38,* 3271–3281.

137   Bouffard, F.A., Zambias, R.A., Dropinski, J.F., Balkovec, J.M., Hammond, M.L., Abruzzo, G.K., Bartizal, K.F., Marrinan, J.A. and Kurtz, M.B. (1994). Synthesis and antifungal activity of novel cationic pneumocandin B_0 derivatives. *J Med Chem., 37,* 222–225.

138   Troke, P.F., Marriott, M.S., Richardson, K., Tarbit, M.H. (1988). In vitro potency and in vivo activity of azoles. *Ann NY Acad Sci., 544,* 284–293.

139   Odds, F.C., Cheesman, S.L. and Abbott, A.B. (1986). Antifungal effects of fluconazole (UK 49858), a new triazole antifungal, in vitro. *J Antimicrob Chemother., 18,* 473–478.

140   Odds, F.C., Abbott, A.B., Pye,G., Troke, P.F. (1986). Improved method for estimation of azole antifungal inhibitory concentrations against Candida species, based on azole/antibiotic interactions. *J Med Vet Mycol., 24,* 305–311.

141   Chapman, T.M., Osborne, S.A., Bouloc, N., Large, J.M., Wallace, C., Birchall, K., Ansell, K.H., Jones, H.M., Taylor, D., Clough, B., Green, J.L., Holder, A.A. (2013). Substituted imidazopyridazines are potent and selective inhibitors of Plasmodium falciparum calcium-dependent protein kinase 1 (PfCDPK1). *Bioorg MedChem Lett., 23,* 3064–3069.

142   Cui, J.J., Tran-Dubé, M., Shen, H., Nambu, M., Kung, P.-P., Pairish, M., Jia, L., Meng, J., Funk, L., Botrous, I., McTigue, M., Grodsky, N., Ryan, K., Padrique, E., Alton, G., Timofeevski, S., Yamazaki, S., Li, Q., Zou, H., Christensen, J., Mroczkowski, B., Bender, S., Kania, R.S. and Edwards, M.P. (2011). Structure Based Drug Design of Crizotinib (PF-02341066), a Potent and Selective Dual Inhibitor of Mesenchymal–Epithelial Transition Factor (c-MET) Kinase and Anaplastic Lymphoma Kinase (ALK). *J Med Chem., 54,* 6342–6363.

143   Cox, J.J., Reimann, F., Nicholas, A.K., Thornton, G., Roberts, E., Springell, K., Karbani, G., Jafri, H., Mannan, J., Raashid, Y., Al-Gazali, L., Hamamy, H., Valente, E.M., Gorman, S., Williams, R., McHale, D.P., Wood, J. N., Gribble, F.M., Woods, C.G. (2006). An SCN9A channelopathy causes congenital inability to experience pain. *Nature, 444,* 894–898.

144   Weiss, J., Pyrski, M., Jacobi, E., Bufe, B., Willnecker, V., Schick, B., Zizzari, P., Gossage, S.J., Greer, C.A., Leinders-Zufall, T., Woods, C.G., Wood, J. N., Zufall, F. (2011). Loss-of-function mutations in sodium channel Nav1.7 cause anosmia. *Nature, 472,* 186–190.

145   Goldberg, Y.P., Price, N., Namdari, R., Cohen, C.J., Lamers, M.H., Winters, C., Price, J., Young, C.E., Verschoof, H., Sherrington, R., Pimstone, S. N., Hayden, M.R. (2012). Treatment of Nav1.7-mediated pain in inherited erythromelalgia using a novel sodium channel blocker. *Pain, 153,* 80–85.

146   Brodie, M.J. (2010). Antiepileptic drug therapy: The story so far. *Seizure, 19,* 650–655.

10

IN SILICO APPROACHES TO ADDRESS COMPOUND ATTRITION

PETER GEDECK[1], CHRISTIAN KRAMER[2] AND RICHARD LEWIS[3]

[1] *Novartis Institute for Tropical Diseases Pte Ltd, Singapore*
[2] *Roche Pharmaceutical Research and Early Development, Molecular Design and Chemical Biology, Roche Innovation Center, Basel, Switzerland*
[3] *Novartis Pharma AG, Basel, Switzerland*

10.1   *IN SILICO* MODELS HELP TO ALLEVIATE THE PROCESS OF FINDING BOTH SAFE AND EFFICACIOUS DRUGS

There are a multitude of biological reasons for compound attrition that can roughly be divided into toxicity (TOX) and failure of efficacy due to poor pharmacokinetics (PK) and medically insignificant targets. The medicinal significance of targets is subject to a lot of basic research and differs in every single case. Most mechanisms of TOX and poor PK however are target independent, and knowledge gained about those can be accumulated and transferred between projects. Every large pharmaceutical company now possesses a knowledge base of compounds that failed due to PK and TOX reasons, and it is *in silico* models that capture this knowledge and extract chemical patterns that can be used to reduce attrition in future projects.

In 2004, Kola and Landis [1] showed that poor PK properties were one of the major reasons for drug attrition at the end of the last century. The pharmaceutical industry reacted and introduced early assessment of drug metabolism and PK into the discovery process. In addition, the industry has seen a strong focus on understanding the physicochemical compound properties that underpin phenomena like solubility and passive transport.

By their very nature, *in silico* approaches extract patterns from chemical activity or property datasets and based on these patterns predict the activity or property of novel compounds. Since these are statistical approaches, there are always elements of uncertainty, and none of these predictions is irrevocably correct. However, the approaches can

Attrition in the Pharmaceutical Industry: Reasons, Implications, and Pathways Forward,
First Edition. Edited by Alexander Alex, C. John Harris and Dennis A. Smith.
© 2016 John Wiley & Sons, Inc. Published 2016 by John Wiley & Sons, Inc.

be very useful for selecting the most promising sets of compounds and indicating the most critical PK/TOX parameters that should be optimized first.

In the following text, we will review different classes and target fields of *in silico* approaches that help reduce attrition. Since the field of *in silico* models for reducing attrition is quite large, we cannot cover it exhaustively. We rather focus on (a) approaches that we have found to be robust and reliable, (b) approaches that are particularly controversial, and (c) novel developments that we find promising and that might become part of the standard *in silico* toolbox for reducing compound attrition.

10.2 USE OF *IN SILICO* APPROACHES TO REDUCE ATTRITION RISK AT THE DISCOVERY STAGE

The impact of *in silico* approaches on attrition depends on the degree of acceptance of the models. The predicted properties for a query compound should be consistent and reliable. This means that the accuracy of the model and the associated error and confidence should be quantifiable, if the primary role of the model is to predict. We cover this topic under domain application and error prediction. Models can also be more descriptive and interpretable, sacrificing some prediction quality, in order to clearly indicate a strategy that can alleviate an issue (an example would be the rule of 5, which is covered later). There are scenarios throughout the discovery process where predictive or descriptive models can be used. If there is high confidence in a predictive model, it can be used to help triage hits generated from screening. However, as hits are only crude first structures, it could be argued that the offending parts of the molecule can be removed as part of the first cycles of SAR, using a descriptive model for guidance. Once a lead series has been established, the SAR direction is clear, in which case a predictive model might be more useful to detect small changes in an off-target activity as part of a multiobjective fine-tuning. These principles are the basis for the REACH criteria for a QSAR model to be an acceptable surrogate [2].

10.3 LIGAND-BASED AND STRUCTURE-BASED MODELS

Usually, attrition is mainly linked to failure of drug candidates in development and in the clinic. However, this concept can be extended to research as well, and we can look at the attrition of hits to leads to candidates and further into the clinic. Baell and Holloway [3] analyzed six HTS campaigns for frequent hitters. They identified common structural features and highlighted a number of Pan Assay Interference Compounds (PAINS) such as quinones, rhodanines, and 2-aminothiophenes (Fig. 10.1).

These structural features also show up frequently in the literature as potential starting points for lead programs. It is worrying how much time and money is wasted following up these inconspicuous hits. In particular, the rhodanines seem to have sparked a lot of follow-up as demonstrated by more than 150 publications citing this compound class as screening hits and 60 patents in 2010. A more recent analysis [4] from 2014 confirmed that these numbers continue to increase. Identifying these types of *frequent hitters* early on will certainly reduce attrition at a very early stage. Some caution should be exercised, particularly if the hit rate is low and only compounds containing PAINS substructures are

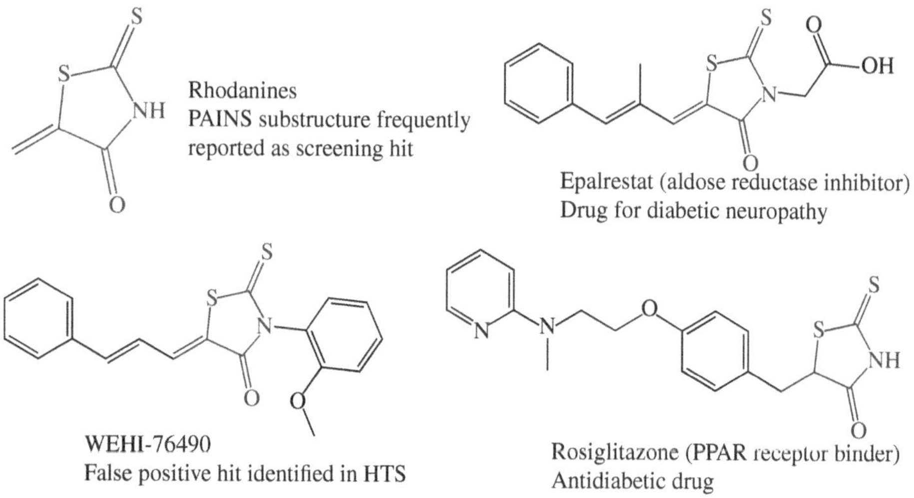

Rhodanines
PAINS substructure frequently
reported as screening hit

Epalrestat (aldose reductase inhibitor)
Drug for diabetic neuropathy

WEHI-76490
False positive hit identified in HTS

Rosiglitazone (PPAR receptor binder)
Antidiabetic drug

FIGURE 10.1   PAINS [3, 4] substructure for rhodanine-like substructures. This structural feature is susceptible to nucleophilic attack, and can act as a metal chelator, and compounds containing this motive are often colored. These effects can lead to high hit rates in biological assays. However, some marketed drugs contain this structural feature.

found. A good counterassay is required to distinguish true from false positives. Certain substructures also have graded responses, depending on their local environment, for example, Michael acceptors, whereas PAINS filters are binary. It has been shown that even groups that contain covalent warheads interact very specifically with protein targets [5]. Hu and Bajorath [6] demonstrated that some descriptors associated with promiscuity can be shown to be artifacts, if more stringent data selection criteria are used.

An integral part of medicinal chemistry knowledge is the substructures that can cause various toxic effects and should therefore be avoided. In 2007, Kazius *et al.* [7] published a list of manually curated chemical substructures that have a high propensity to cause AMES mutagenicity (Fig. 10.2).

There was a strong connection to chemical reactivity, so the focus was on isolated substructures. Since there are many different toxicological endpoints and the datasets have become rather large, manual curation is not an option for today's datasets. Therefore, a number of programs have been developed that automatically mine large datasets for substructures that have a higher propensity to cause various toxicological effects. The first commercial programs that were developed for this task at the beginning of the 1990s include DEREK [8] and MULTICASE [9]. Recently, the description of an implementation for a substructure mining program based on open-source software has been published by Ahlberg *et al.* [10]. The approaches differ in the way that substructures are identified, thus leading to strong performance differences and slightly different substructures for the same datasets. The modern approach recognizes that it is a combination of substructure and local environment, leading to better predictions but perhaps a slightly lower degree of interpretability. By contrast, these methods seem to be as good as and much cheaper than QM approaches that do indeed consider the whole molecule environment [11].

These topics also illustrate the role of risk in attrition. If a very risk-averse philosophy is adopted, then it is important to identify any possible cause of a toxicological issue, even

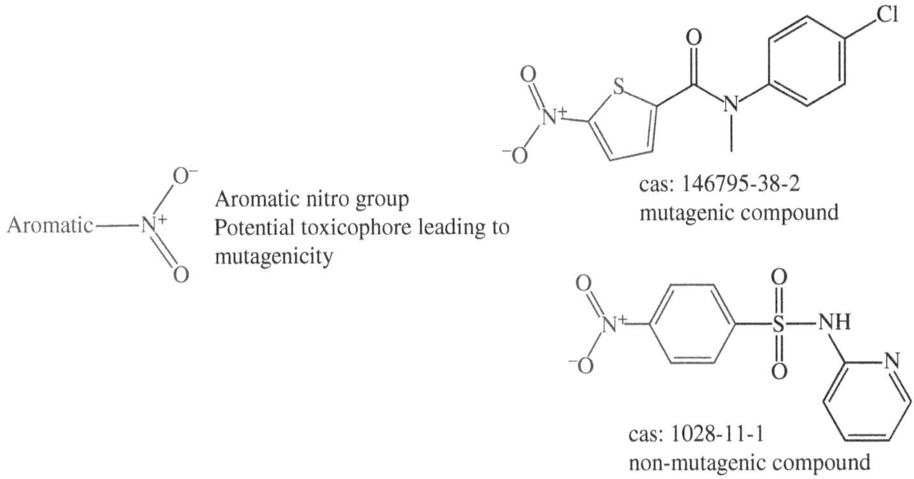

FIGURE 10.2 Substructure found in mutagenic compounds [7]. Aromatic nitro groups are often found in mutagenic compounds. The sulfonamide substituent removes the mutagenic potential. Other substituents that were found to detoxify the aromatic nitro groups are trifluoromethyl, sulfonic acid, and arylsulfonyl groups.

at the cost of discarding many innocent compounds. Some groups and software programs do take this approach, so it is necessary to be aware of the model building strategy when employing these as filters. For example, a model built for a very low false-negative rate on a permissive activity threshold for hERG will probably eliminate many known and safe drugs on the market.

In addition to ligand-based methods, it is also possible to utilize the many structures of key proteins responsible for attrition. X-ray structures of cytochrome P450s (CYP450s), PGP, nuclear hormone receptors, and many others are available, as well as homology models of hERG. The concept is to use docking to identify strong binders to these targets and thereby eliminate them from further consideration (or, as a binding model is extremely rich in description, to identify modifications that might reduce binding). There are flaws in this strategy. Some proteins have evolved to be promiscuous rather than specific, for example, CYP450 3A4. It is possible to dock almost anything into this enzyme, without being able to say with confidence if the compound is really a good ligand/substrate/inhibitor (and this is leaving aside any issues about binding site plasticity). If the site of metabolism is known, and the isoform responsible (and there may be more than one), it is possible to use this as a constraint in docking. Sadly, experience has shown that blocking one site of metabolism does not reduce clearance, as the body finds another pathway of destruction. The early expectations that were driven by the first CYP450 structures [12] (Fig. 10.3) have not been fulfilled.

Even if the site is more specific, inaccuracies in scoring functions do not allow one to make attrition decisions with confidence in this case. Many of the key targets have somewhat hydrophobic binding pockets, so that there are many equally feasible binding poses possible. hERG is another interesting case. Even though we talk about hERG pharmacophores, this is illusory, as the binding site in hERG is there by accident and has not been evolved to select for any particular pharmacophore. Homology models of hERG have

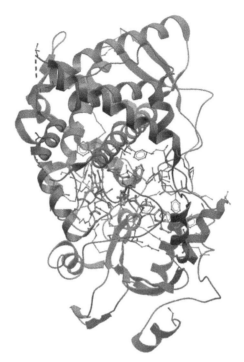

FIGURE 10.3   Crystal structure 3NXU [13] of cytochrome P450 3A4 with inhibitor ritonavir.

been built [14] and refined [15–17], but these have revealed that the linkage between binding and the phenotype of prolonged QT interval is not at all straightforward.

Structure-based methods are mainly useful when linked to experimental data such as binding affinity or sites of metabolism, which may allow the development of an accurate docking pose. Then established design methods can be used to reduce binding through addition of bulk, reduction in bond lability, or changes in polarity. The resulting designs can be made and tested, and the cycle repeated. If a large amount of data is available, ligand-based and structure-based methods can be combined [http://www.biograf.ch/index.php?id=projects&subid=virtualtoxlab]. The structural model is used to generate an overlay of the set of assayed molecules through docking. This can be used as input to generate a 3D QSAR model, whose quality can be assessed in the usual ways. New molecules can be docked by the same protocol and then scored by the 3D QSAR model to give a more accurate assessment of binding potential and hence liability. This approach has found more favor in academic labs; pharmaceutical industry teams prefer to work more with screening data, given the biases they have about the quality of the models.

10.4   DATA QUALITY

Being statistical models, the quality of *in silico* models' predictions depends on the quality of the underlying data. If the biological data is very noisy or full of errors, the models based on this data cannot be expected to be very reliable. Thus, it is very important to curate data carefully before training models.

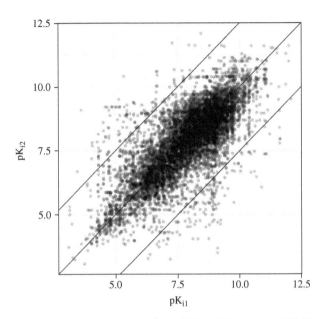

FIGURE 10.4   Reproducibility of pK_i values from different laboratories [18]. The average reproducibility of pK_i values measured in different laboratories and reported in the ChEMBL database is 0.54 log units. Reprinted with permission from Kramer et al. [18]. Copyright 2012 American Chemical Society.

There are different sources of errors and noise in the data. Assay readouts depend on the number of repeats, temperature, pressure, quality of the biological material, and solubility, purity, and correct concentration of the ligands tested. These have an influence on the values reported, and variation in those is usually referred to as experimental uncertainty. There is a marked difference in experimental uncertainty when comparing values measured in the same laboratory with values measured in different laboratories. Kramer *et al.* [18] showed that the average experimental uncertainty for pK_i values measured in different laboratories and reported in the ChEMBL database is 0.54 log units (Fig. 10.4).

In contrast, the experimental uncertainty of activity values measured in the same laboratory has been estimated to be around 0.2 log units [19]. pIC_{50} values measured in different laboratories on average differ with a standard deviation of 0.69 log units. Although in principle pIC_{50} values are assay specific and therefore not comparable between different assays, in practice they are compared very often, for example, when examining specificity between different kinase subtypes.

Biological activity data stored in in-house and public databases also contains other kinds of errors: in the ChEMBL database, we found that the most common errors are unit conversion errors, ignorance of qualifiers (activities measured as ">X μM"), and wrong or absent stereochemistry [20, 21]. These errors are not necessarily introduced when entering data into the database: we also found that the same types of errors are introduced when data is copied from one publication to another. Another worrying trend is if authors convert single % inhibition values to pIC_{50} values using highly idealized assumptions about the Hill equation. This clearly should not be done, since the resulting pIC_{50} values can be very far from the true pIC_{50} values.

While errors introduced when entering data into the database can often be found by recourse to the original publication of these data points, experimental uncertainty cannot be removed. This has to be taken into consideration when analyzing the performance of models. *In silico* models can only predict the signal in the data. As a consequence, the maximum R^2 that can be achieved when comparing predicted values to measured values is never 1.0, but must be smaller due to the experimental error [18]. If there is a lot of experimental uncertainty and little true biochemical variation in the assay readout, such datasets might not be amenable to *in silico* analysis.

In the context of data quality, it is instructive to look closer at the history of phospholipidosis datasets and models; Goracci *et al.* [22] give a good overview on this. Over the years, a considerable amount of data was collated by several groups and published in various papers. The largest dataset was collected by Orogo *et al.* [23]. While it contains in principle 743 compounds, about 300 of these were not disclosed, leaving just 447 compounds for comparative studies. More recently, Goracci *et al.* [22] extended this dataset with 19 public data points giving a dataset of 466 compounds. However, following a thorough analysis of the data quality, this left a curated dataset of 331 PLD data points. Only after this careful curation of data, it was possible to get good models for phospholipidosis. The curated dataset was recently used by Przybylak *et al.* [24] who derived SMARTS patterns for a number of substructures typically found in phospholipidosis compounds.

10.5 PREDICTING MODEL ERRORS

The reliability of individual predictions from a model can differ substantially. In fact, models very often perform well for some classes of compounds, whereas they fail for other classes. Therefore, judging the usefulness of a model by its average performance is not a good strategy if it is a few individual predictions that matter. Several strategies have been developed to estimate the reliability of each single prediction. Usually, *in silico* models are good at predicting new compounds that are similar and bad at predicting compounds that are very different from the training set compounds. This realization has led to the concept of applicability domain [25]: for every new compound, the similarity to the training set is calculated. Based on this similarity, the reliability of every single prediction is judged as reliable/unreliable (Fig. 10.5). The applicability domain concept has also been adapted in the European REACH legislation.

While the applicability domain is a good starting point for analyzing the reliability of individual predictions, it is still quite crude since it only gives a yes/no decision. Therefore, other approaches that yield reliabilities on a continuous scale have been developed: for classification tasks, probabilities of belonging to a specific class can be calculated from the distribution of prediction ensembles; Gaussian processes have been used to generate 2-class classification models to predict metabolic stability [26] and various other absorption/distribution/metabolism/excretion (ADME) and activity datasets [27] with reliable class-membership probabilities. A metaclassifier strategy that unites the predictions from several different machine learning approaches has been developed to filter out compounds that are likely to give wrong assay results due to insufficient solubility [28]. Norinder *et al.* [29] have recently introduced the concept of conformal prediction and showed that the fraction of decision trees in a random forest that vote for a specific class can be interpreted as class probabilities.

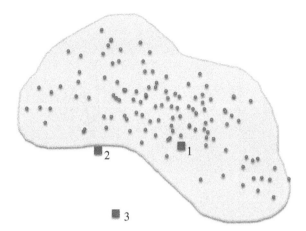

FIGURE 10.5 Schematic representation of the applicability domain of QSAR models. The points represent the compounds in the training set in a property space. (1) Predictions for compounds that are within the applicability domain should be reliable within the quality of the model (interpolation). (2) Predictions for compounds near the applicability domain should still be reliable but will have a higher predictive error (extrapolation). (3) Compounds that are far from the applicability domain will not be well predicted by the QSAR model.

In regression, the confidence in specific predictions can be expressed as standard deviations of the predicted values (Fig. 10.6).

Clark and coworkers [30] have developed neural network ensembles to predict boiling points, vapor pressure, and log$P$ with confidence intervals for every single prediction and showed that these correlate well with observed prediction accuracy. Schwaighofer *et al.* [31] showed that Gaussian process regression models could give very reliable individual estimates of individual prediction accuracies for solubility. Clark [32] developed a method to estimate prediction uncertainty based on the accuracy of the prediction of the most similar compound and the distance to this compound. Sheridan [33] showed that the similarity to other training compounds, the variation of predicted values among the trees, and the overall predicted value itself are useful criteria to estimate the reliability of individual predictions. In a follow-up work, Sheridan [34] showed that based on these three descriptors, a second random forest model can be built that gives highly reliable estimates of the accuracy of the first random forests' predictions.

While it is harder to interpret prediction ranges instead of single predicted values, the ranges have a number of striking advantages: First, ranges are more correct than single values since there is an uncertainty to every prediction. Keeping this uncertainty in mind can reduce frustration about "wrong" predictions. Second, ranges can be used to filter out compounds that have a certain probability of lying below a threshold. While users then have to shift their perspective to thinking in terms of probabilities, this again is scientifically more correct. Since the use of ranges allows calculating the expected number of misclassifications in advance, the frustration about wrong classifications can be reduced if the ranges are reliable. When analyzing prediction ranges instead of individual predictions, novel metrics for assessing model quality are needed. Wood *et al.* [35] recently showed that the Kullback–Leibler divergence is a metric that captures both the accuracy and the predicted uncertainty of individual predictions and should be used for assessing regression model performance instead of the standard

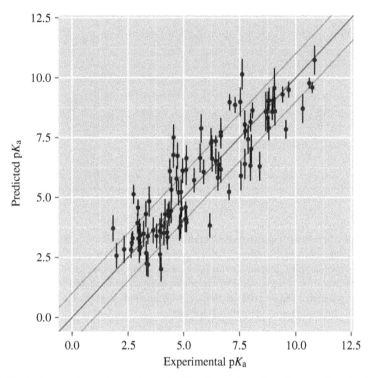

FIGURE 10.6   Comparison of experimental pK_a values with pK_a values predicted using Moka. The error bars show the error estimated by the QSAR model. The diagonal lines are the line of identity and plus and minus one pK_a unit.

RMSD. This topic is particularly important when semiautomated model building tools are available to the naive user [36]. As part of the European Innovative Medicines Initiative, a group was set up to develop models for toxicological endpoints. One of the first products was an Applicability Domain Analysis [37] that provided six criteria for comparing a query compound to the training set to provide an understandable measure of prediction reliability.

10.6   MOLECULAR PROPERTIES AND THEIR IMPACT ON ATTRITION

It all started with Lipinski's rule of 5 [38] published in 1997. This rule highlighted that the structural properties of most permeable and soluble drugs fall within certain ranges. As it was simple and easy to apply, it quickly spread through the industry achieving more than 1000 citations just 7 years later in 2004 [39]. The whole field of correlating ADMET properties with structural features flourished, and new rules popped up everywhere pushing into areas beyond physicochemical parameters. For example, Egan et al. [40] suggested a combination of polar surface area (PSA) and logP as a predictor for drug absorption. Gleeson [41] looked at the influence of molecular weight, logP, and ionization state on a wide variety of ADME properties: bioavailability, volume of distribution, plasma protein binding, and CNS penetration, to name just a few. Recently, Tarcsay and Keserű [42] gave an overview of correlations of molecular properties with drug promiscuity.

We all fell for the temptation to reduce success in drug discovery to a few simple rules and started to filter hit lists early on based on molecular weight and other simple descriptors. Because, if we believe these analyses, drug discovery should be really simple and analyzing hit lists reduces to filtering compounds out that don't fall into the right property space. However, drug discovery is not simple and rules always have exceptions. For example, 10% of all oral drugs fall outside the molecular weight cutoff of the rule of 5. The same holds for $\log P$ and the other rule-of-5 properties. All in all, we can estimate that at least 25% of all drugs violate at least one of the properties. While there is enough rationale [43, 44] to avoid large and lipophilic compounds, it is clear from this number that the rules must not be seen as hard cutoffs. It might be better to replace the hard cutoffs with smoother boundaries as done by Bickerton *et al.* [45]. In addition, Egan's analysis [40] showed that bioavailability can also be affected by active transport, the use of prodrugs, or internal hydrogen bonding (in the case of natural products).

While various rules were taken up widely, they started to receive criticism over the years as well. Kenny and Montanari [46] highlighted the fact that many of the drug property correlation stated in the literature is based on averaging properties within groups, for example, looking at the number of targets hit by compounds within certain molecular weight ranges. This averaging may reveal a trend based on the correlation of the average; however, the variation within each of the groups is so large that considering individual cases, the differences between the groups are statistically irrelevant. As an example, Kenny and Montanari looked at the correlation of solubility with the fraction of sp^3 atoms in molecule. For averaged values, the correlation is very strong; however, considering all data points individually, the data show basically no correlation, indicating that the fraction of sp^3 explains only 6% of the variation in solubility; this is an insufficient number to base a drug discovery project on.

Many of the rules also break down when applied to new data. In a recent study, Muthas *et al.* [47] analyzed 150 AstraZeneca development compounds and applied a number of common guidelines. Their results clearly showed that these common guidelines might not be commonly applicable. For example, the 3/75 guideline from Pfizer [48], which states that compounds with a tPSA < 75 and a $c\log P$ > 3 are more likely to fail due to toxicology reasons, failed when applied to AstraZeneca's compounds (Fig. 10.7). For them, the compounds had a higher likelihood to fail if their $c\log P$ was less than 3, so leading to the completely opposite conclusion to Pfizer's experience.

Roche looked at their safety panel data to see if there were any strong correlations with promiscuous compounds and adverse drug reactions [49]. They saw that compounds with a basic center were more likely to be promiscuous, but with the caveat that many of the endogenous ligands for the assays come from the aminergic family and thus the panel will be biased for bases. They also acknowledge that complete selectivity should not be a stringent criterion for the selection of a development candidate as long as the off-target effects do not lead to adverse effects *in vivo*. This is often a complex balance between desired pharmacodynamic (PD) actions, adverse actions, and PK.

Similarly, the fraction of sp^3 carbons, as a surrogate for molecular complexity, did not correlate with TOX in contrast to what was observed by Lovering *et al.* [50]. Muthas *et al.* conclude their analysis with the warning that all these guidelines must not be treated as hard cutoffs. If this is done, many important new drugs would have been eliminated before they even enter development, a warning we can only wholeheartedly support here.

In another study, Lloyd *et al.* [51] compared the properties of oncology drugs with the rule-of-5 and rule-of-3 properties and highlighted that a lot of interesting compounds would

Pfizer

	Clogp < 3	Clogp > 3
TPSA > 75	0.4	0.8
TPSA < 75	0.5	2.5

AstraZeneca

	Clogp < 3	Clogp > 3
TPSA > 75	2.1	0.5
TPSA < 75	1.8	0.4

FIGURE 10.7   Comparison of the 3/75 rule applied to development compounds from Pfizer and AstraZeneca. The numbers are the toxicity odds taken from Figure 10.1 in [47]. While the Pfizer results clearly show that $c$logP greater than 3 and tPSA less than 75 lead to a higher probability of toxicity, results obtained at AstraZeneca would lead to the opposite conclusion.

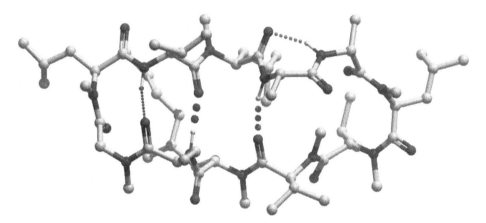

FIGURE 10.8   X-ray structure of cyclosporine A crystallized in carbon tetrachloride (CSD identifier: DEKSAN). The structure shows the formation of four intramolecular hydrogen bonds effectively reducing the polarity of the peptide and this way improving membrane permeability [57].

be missed if the rules were applied as hard cutoffs. A similar result was obtained recently by Doak *et al.* [52]. They focused on drugs that fall outside the property space of the rule of 5, in particular high molecular weight compounds. The analysis shows that it is possible to develop drug candidates successfully, which fall in the drug space beyond the rule of 5. For other studies, see Abad-Zapatero [53], Zhang and Wilkinson [54], and Walters [55].

   One chemical explanation for why rules fail can be found from considering that rules are usually based on simple 2D descriptors, whereas protein–ligand binding takes place in 3D. Due to the 2D–3D extrapolation problem, simple counts of hydrogen bond donors and acceptors can be misleading if molecules have the capability to form intramolecular hydrogen bonds. Kuhn *et al.* [56] analyzed preferred motifs for intramolecular hydrogen bonds and showed that those can have a substantial nonadditive effect on physicochemical properties. If a hydrogen bond donor and an acceptor form an intramolecular hydrogen bond, they mask each other relative to the surrounding, thus rendering molecules a lot more lipophilic than would be expected from just counting donors and acceptors (Fig. 10.8).

Switchable intramolecular hydrogen bonds are also considered to be the main reason that facilitates membrane permeability of small cyclic peptides [57]. Small cyclic peptides are clearly outside standard ADME space, since they contain too many hydrogen bond donors/acceptors and a too high PSA. Nevertheless, they are able to cross cellular membranes. It is highly likely that cell-permeable cyclic peptides exist in at least two conformations: one conformation inside the membrane when the compounds hide their donors/acceptors with intramolecular hydrogen bonds and a different conformation in solution presenting their donors/acceptors to the aqueous medium.

10.7 MODELING OF ADME PROPERTIES AND THEIR IMPACT OF REDUCING ATTRITION IN THE LAST TWO DECADES

ADME has become increasingly important to medicinal chemistry thinking in the last 20 years. Not only are many more assays available, but they also have increasing throughput and require smaller amounts of material. ADME groups have been reorganized to be better aligned with chemistry, and ADME experts are often key members of discovery teams, rather than outside experts. This change in culture is shown through a better awareness of risk factors and a desire to tackle them at an earlier stage. At the stage of characterization of a lead series, it can soon be shown if the front runner has a hERG liability using a high-throughput assay. Where a model can help is in assessing whether the liability is inherent to the series or there is a possibility that it can be designed out. A judicious balance between experimental and modeling results is required, in case the series is outside of the model domain, and hence poorly predicted. We would not advocate that the models be used on their own as decision-making tools for the reasons discussed in the previous section. Experimental data is needed to build trust first.

The impact of ADME and the change in culture on attrition can be seen in industry surveys of attrition rates across the various phases. Some companies have been able to reduce the attrition rate going from clinical candidate to phase 1 by 15% by bringing ADME into the preclinical lead optimization phase. It is clear more can be done: an analysis of project failures by AstraZeneca [58] indicated that 82% of discontinuations were due to safety, and the authors concluded that "our analysis demonstrated an intuitive but crucial need for teams to pay attention to preclinical safety signals, and also highlighted that safety signals become more problematic as a project progresses, resulting in project delays. It also indicated the progression of molecules that could have been stopped earlier through the application of more robust criteria."

It is harder to quantify the impact of models in terms of return on investment. As more assays are outsourced, there is pressure to reduce the number of compounds assayed by prescreening with ADME models. Informal reports suggest that the availability of a trusted and reliable model can reduce screening demand by at least 10%, in some cases much more. It is reported that the Roche model for phospholipidosis [59] has completely replaced an assay, but this is an extreme case. The drive to reduce the number of animal experiments by using *ex vivo* or *in silico* surrogates will increase this trend. Predictive models are being used as part of multiparameter optimization tools such as the Predictive Chemistry Network (PCN) [60] or StarDrop [61] that are routinely used by chemists during the compound design process. A survey of the usage of the PCN showed that "we learned that 89% of the users who responded believed that the PCN tools and approaches had had a large positive (26%), moderate positive (41%) or small positive (22%) effect on decision making in their projects." Discovery targets that are currently being pursued often have difficult characteristics, for example, protein–protein interactions. Perhaps the next step is to use models not just to identify poor series but also poor targets whose pharmacophores are inextricably linked to ADME issues.

10.8   APPROACHES TO MODELING OF TOX

Sutter *et al.* [62] analyzed a variety of *in silico* approaches to predict mutagenicity using various proprietary datasets from a large number of pharmaceutical companies. They found that DEREK predicts about 80% of mutagenic compounds correctly. Similar results were found when training QSAR models on AMES mutagenicity datasets [63, 64]. The availability of a curated public dataset [65] of considerable size helped in the development of different models.

The problem of drug TOX affecting the skin has been well described in the past; see Stein and Scheinfeld [66] or Elkeeb *et al.* [67] for recent reviews. Today, development candidates are regularly screened for their phototoxic potential using a hierarchy of experimental models [68]. While this helps identify the risk early, these experiments will become a bottleneck if applied early on in the research process. In order to assess the risk for a large number of structures, computational models were proposed based on the calculated HOMO–LUMO gap thresholds [69]. There is anecdotal evidence [70] that looking at this property can help estimate the risk and remove phototoxicity issues from a lead structure. We recently revisited the HOMO–LUMO model using experimental results from the last years. Based on our results [71], the proposed thresholds either have a small false-negative rate but lead to almost all compounds in a drug discovery context predicted as being potentially phototoxic or miss a large number of phototoxic compounds, yet the threshold will still lead to a considerable number of compound structures predicted as potentially phototoxic (60%). What these models can do however is to provide a probability or risk assessment of phototoxic potential. The quality of the models certainly will not justify their use as yes/no filters, however, they may be used as research stage-dependent filters for moving to a costly experiment early on.

Some workers have advocated the use of docking models to identify toxic compounds, particularly for nuclear receptor targets [72]. While this may be valid for removing compounds that cannot bind (and even that might be debatable for a binding site with high plasticity), the scoring functions currently available are not reliable enough to discriminate clearly between potent binders and weak binders. The validity of the docking models is not well established, and the prediction quality is not strong enough to make classifications, even though the applicability domain is very broad. We would not currently recommend that docking models be used as a filter by themselves but as an adjunct to other methods.

10.9   MODELING PK AND PD AND DOSE PREDICTION

Many of the approaches described in this chapter have been directed toward a PD endpoint, for example, predicting binding in a dofetilide displacement assay and assuming that the assay is a suitable surrogate for *in vivo* actions at the hERG channel. In almost all situations, we know that this assumption is flawed, but the hope is that the model errs on the side of caution. This also illustrates the need for secondary assays when the prediction is borderline, for example, using dog telemetry in the case of hERG. There is continual need to be aware of this balance between models as filters and the possibility of discarding a good compound. Some of this risk tolerance will be a function of the organization and its reward structure and its recent experiences with development compounds.

Another critical factor is the PK profile of the compounds and possibly its metabolites. The aim is to have sufficient concentration of drug at the site of action for it to be effective (PD) and for that concentration to be maintained for the desired amount of time (PK). If a compound has a low safety margin, then it is likely not to be given as a once-a-day dose, as the peak concentration may well stray into the range where TOX is observed. If a compound has a very long half-life, concentrations may build up, or slow toxicities such as time-dependent inhibition of CYP450s or 5HT-2b agonism-related valvulopathies (as seen with pergolide) may occur. Lipophilic compounds can also form depots in membranes, which, when combined with promiscuous activity, may lead to off-target effects. Some compounds may cause liabilities by inducing metabolic enzymes, leading to drug–drug interactions. So far, we have assumed fast equilibration. While we can predict off-target binding with reasonable confidence [73], we need to combine this with a consideration of the PK/PD profile of a compound to understand the safety margin of compounds with lower toxicities.

If the compound has slow rates of dissociation, one may see the PD effects long after the plasma concentration has gone to near zero. Almorexant (an orexin antagonist) is a good example of this. If there are no toxicities associated with the on-target effect, this can be a desirable characteristic [74]. If the off-targets are in fast equilibrium, then the clearance of the compound will reduce the effect here much faster than at the slowly dissociating target. We have also assumed that a compound is distributed to all body compartments. Partitioning can have both positive and negative effects: if there is an off-target effect associated with a target in the brain, it is wise to design a compound with low brain penetration. There have also been cases where compounds with cardiac toxicities have concentrated right in the heart, reducing the apparent safety margin.

Just as with off-target activities, models are being developed to predict PK profiles based on solubility, pK_a, rate of metabolism, and other parameters (physiology-based pharmacokinetic (PBPK) models [75, 76]). Appropriate use of the models allows one to assess the behaviors of a number of clinical candidates and to select the best ones to take into *in vivo*. One of the key features is a sensitivity analysis: one series might be controlled by solubility and another by metabolism. The techniques are being used more commonly, but with mixed results, indicating that there is still more research to done to build reliable models.

Dose prediction in humans can also have effects on attrition. Based on animal data, one must try to predict the key PK properties in man and again determine the safety margin. Most preclinical data comes from microsomes, which may not be predictive of intrinsic clearance. However, with the right estimates (e.g., from allometric scaling) of the PK parameters, PBPK models have been used successfully to reduce attrition by quickly identifying problem compounds that may need to be reformulated or the dose schedule changed.

10.10   NOVEL *IN SILICO* APPROACHES TO REDUCE ATTRITION RISK

The basic concept of computational models to reduce attrition risk is always the same: all approaches are based on a statistical analysis of databases and the identification of patterns that are correlated with attrition risk. There are developments in all aspects of this endeavor, including initiatives to significantly increase the data pool, better assessment of the data quality, novel statistical techniques, and the inclusion of novel data types in which patterns are sought.

All *in silico* attrition models are statistical models, and generally, statistical models become more powerful with an increasing dataset. Therefore, a number of data collection and sharing initiatives have been started: the well-known Protein Data Bank (PDB) [77] (www.rcsb.org) contains now more than 100 000 structures, and this number continues to grow exponentially (Fig. 10.9).

Since a couple of years, large public databases such as ChEMBL [78] and Binding DB [79] are being built up that collect bioactivity data from the public literature. Those databases and many other sources of bioactivity data are available via the Open PHACTS platform—a common interface to public drug discovery data [80]. There have been several attempts to directly share ADME data between companies. However, here the problem is the coding of molecules such that meaningful QSAR models can be built, but the original chemical structures cannot be traced back [81]. Finally, there are also approaches to collect company data via trustworthy third-party organizations, a system proposed by Dearden

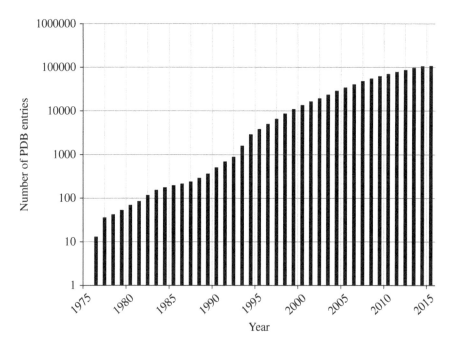

FIGURE 10.9   Increase of X-ray structure in the public domain database PDB. Note the logarithmic scale of the vertical axis. This shows that the number of structures increases exponentially. Data loaded from http://www.rcsb.org/pdb/static.do?p=general_information/pdb_statistics/index.html

[82]. Examples in case are the VITIC NEXUS database from Lhasa Limited [83] and the SALT MINER initiative based on matched molecular pair analysis [84] from MedChemica. Since the database is now becoming so large and heterogeneous in the case of the earlier examples, data quality and comparability are pressing questions. While the assembly of best practice rules for the generation of QSAR models has been a long-standing topic and covered in several reviews [85–88], it is becoming more and more evident that the quality of the underlying data itself can also be a major limiting component [18, 20, 21, 89]. For example, any model can only reproduce the signal in the data, not the noise. If there is a significant component of noise in the data, this obviously cannot be predicted. The authors, for example, have seen published QSAR models where the total range of measured bioactivity is smaller than 2 log units. For such datasets, a high R^2 cannot be expected, since a significant fraction of the total readout variability will likely be due to noise. Further improvement of existing models for the prediction of attrition will probably come from understanding experimental factors that contribute to systematic errors in the data. For example, it has been shown that errors in dilution series can have a dramatic impact on the measured bioactivity and the structural interpretation based on such data [90]. Assessment of the data quality and potential limitations requires expert knowledge in the experimental setup used to generate the data and a tight collaboration between experimentalists and people creating *in silico* models.

Based on the large datasets, which are nowadays available, novel statistical techniques like matched molecular pairs [91] or molecular transformations [92] have received a lot of attention [93]. They were applied early on as guides in the optimization of pharmaceutical properties like biological activities, aqueous solubility, plasma protein binding, and oral exposure (Fig. 10.10).

While this type of analysis was relatively successful when used for analyzing ADME/TOX datasets, results were less promising when applied to biological activities. For

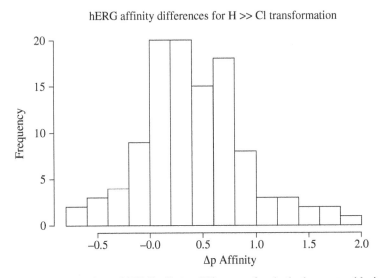

FIGURE 10.10   Distribution of hERG affinity differences for the hydrogen to chlorine transformation. Reprinted with permission from Kramer *et al.* [94]. Copyright 2014 American Chemical Society.

diverse SAR datasets, changes will be more inconsistent and will depend largely on the specific protein–ligand interactions that will be different for each target. To address this issue, O'Boyle *et al.* [95] recently extended the matched molecular pair approach to matched molecular series. So far, this has only been applied to SAR datasets; however, it will be interesting to see how this new idea will translate to ADME/TOX properties like CYP450 inhibition and other target-specific properties. As with standard QSAR models, experimental uncertainty needs to be taken into account to focus on significant matched molecular pairs [94].

Traditionally, calculated small-molecule chemical descriptors are used to model the similarity between compounds and predict biochemical properties. In recent years, a number of studies have shown that affinities to other targets and similarities between their binding sites can be very useful descriptors. Cortez-Ciriano *et al.* [96] have shown that in a proteochemometric framework, the affinity of small molecules toward different receptor subtypes can be predicted from chemical structures and the known affinity to other receptor subtypes. Petrone *et al.* [97] showed that similarities in activity patterns obtained from high-throughput screens (HTS fingerprints) could be used to predict novel bioactivities. Sedykh *et al.* [98] and Rusyn *et al.* [99] showed that the combination of classical chemical descriptors and *in vitro* assay results significantly improves the prediction of *in vivo* toxicological endpoints, when being compared to predictions based on any of the individual classes only. Biological descriptors are a lot more expensive than computed chemical descriptors, since they have to be measured. However, a good *in vitro–in vivo* correlation is essential for useful early warning signs, and therefore, the combination of chemical descriptors with fast *in vitro* measurements is a strong perspective.

10.11   CONCLUSIONS

Considerable progress has been made in the understanding of the causes of TOX and the development of assays that model the toxicities and hence in the amount of reliable data that is available for model building. QSAR models have also advanced, with progress in the fields of domain applicability, error estimates, and confidence predictions. If these are combined with an industry-wide desire to "fail fast, fail cheap," there is now a willingness to accept and use *in silico* approaches as tools to reduce attrition. TOX cannot be described by simple rules, so there is a need for users to understand what is being modeled, how it is being modeled, and how much to trust the result. As initiatives to share data more freely and to curate better what we do have (e.g., the European REACH initiative) take off, the number and quality of *in silico* approaches can only increase, providing a valuable tool set to assist the medicinal chemist.

REFERENCES

1   Kola, I., Landis, J. (2004) Can the Pharmaceutical Industry Reduce Attrition Rates? *Nat. Rev. Drug Discov.*, *3*, 711–716. DOI: 10.1038/nrd1470.
2   Nicolotti, O., Benfenati, E., Carotti, A., Gadaleta, D., Gissi, A., Mangiatordi, G. F., Novellino, E. (2014) REACH and In Silico Methods: An Attractive Opportunity for Medicinal Chemists. *Drug Discov. Today*, *19*, 1757–1768. DOI: 10.1016/j.drudis.2014.06.027.

3 Baell, J. B., Holloway, G. A. (2010) New Substructure Filters for Removal of Pan Assay Interference Compounds (PAINS) from Screening Libraries and for Their Exclusion in Bioassays. *J. Med. Chem.*, *53*, 2719–2740. DOI: 10.1021/jm901137j.

4 Baell, J., Walters, M. A. (2014) Chemistry: Chemical Con Artists Foil Drug Discovery. *Nature*, *513*, 481–483. DOI: 10.1038/513481a.

5 Flanagan, M. E., Abramite, J. A., Anderson, D. P., Aulabaugh, A., Dahal, U. P., Gilbert, A. M., Li, C., Montgomery, J., Oppenheimer, S. R., Ryder, T., Schuff, B. P., Uccello, D. P., Walker, G. S., Wu, Y., Brown, M. F., Chen, J. M., Hayward, M. M., Noe, M. C., Obach, R. S., Philippe, L., Shanmugasundaram, V., Shapiro, M. J., Starr, J., Stroh, J., Che, Y. (2014) Chemical and Computational Methods for the Characterization of Covalent Reactive Groups for the Prospective Design of Irreversible Inhibitors. *J. Med. Chem.*, *57*, 10072–10079. DOI: 10.1021/jm501412a.

6 Hu, Y., Bajorath, J. (2014) Influence of Search Parameters and Criteria on Compound Selection, Promiscuity, and Pan Assay Interference Characteristics. *J. Chem. Inf. Model.*, *54*, 3056–3066. DOI: 10.1021/ci5005509.

7 Kazius, J., McGuire, R., Bursi, R. (2004) Derivation and Validation of Toxicophores for Mutagenicity Prediction. *J. Med. Chem.*, *48*, 312–320. DOI: 10.1021/jm040835a.

8 Sanderson, D. M., Earnshaw, C. G. (1991) Computer Prediction of Possible Toxic Action from Chemical Structure; The DEREK System. *Hum. Exp. Toxicol.*, *10*, 261–273. DOI: 10.1177/096032719101000405.

9 Klopman, G. (1992) MULTICASE 1 A Hierarchical Computer Automated Structure Evaluation Program. *Quant. Struct.-Act. Relationships*, *11*, 176–184. DOI: 10.1002/qsar.19920110208.

10 Ahlberg, E., Carlsson, L., Boyer, S. (2014) Computational Derivation of Structural Alerts from Large Toxicology Data Sets. *J. Chem. Inf. Model.*, *54*, 2945–2952. DOI: 10.1021/ci500314a.

11 Leach, A. G., Cann, R., Tomasi, S. (2009) Reaction Energies Computed with Density Functional Theory Correspond with a Whole Organism Effect; Modelling the Ames Test for Mutagenicity. *Chem. Commun.*, *9*, 1094. DOI: 10.1039/b818744d.

12 Williams, P. A., Cosme, J., Ward, A., Angove, H. C., Matak Vinković, D., Jhoti, H. (2003) Crystal Structure of Human Cytochrome P450 2C9 with Bound Warfarin. *Nature*, *424*, 464–468. DOI: 10.1038/nature01862.

13 Sevrioukova, I. F., Poulos, T. L. (2010) Structure and Mechanism of the Complex Between Cytochrome P4503A4 and Ritonavir. *Proc. Natl. Acad. Sci.*, *107*, 18422–18427. DOI: 10.1073/pnas.1010693107.

14 Sánchez-Chapula, J. A., Navarro-Polanco, R. A., Culberson, C., Chen, J., Sanguinetti, M. C. (2002) Molecular Determinants of Voltage-dependent Human Ether-a-Go-Go Related Gene (HERG) K+ Channel Block. *J. Biol. Chem.*, *277*, 23587–23595.

15 Farid, R., Day, T., Friesner, R. A., Pearlstein, R. A. (2006) New Insights About HERG Blockade Obtained from Protein Modeling, Potential Energy Mapping, and Docking Studies. *Bioorg. Med. Chem.*, *14*, 3160–3173. DOI: 10.1016/j.bmc.2005.12.032.

16 Pearlstein, R. A., Vaz, R. J., Kang, J., Chen, X.-L., Preobrazhenskaya, M., Shchekotikhin, A. E., Korolev, A. M., Lysenkova, L. N., Miroshnikova, O. V., Hendrix, J., Rampe, D. (2003) Characterization of HERG Potassium Channel Inhibition Using CoMSiA 3D QSAR and Homology Modeling Approaches. *Bioorg. Med. Chem. Lett.*, *13*, 1829–1835. DOI: 10.1016/S0960-894X(03)00196-3.

17 Stansfeld, P. J., Gedeck, P., Gosling, M., Cox, B., Mitcheson, J. S., Sutcliffe, M. J. (2007) Drug Block of the hERG Potassium Channel: Insight from Modeling. *Proteins Struct. Funct. Bioinforma.*, *68*, 568–580. DOI: 10.1002/prot.21400.

18 Kramer, C., Kalliokoski, T., Gedeck, P., Vulpetti, A. (2012) The Experimental Uncertainty of Heterogeneous Public Ki Data. *J. Med. Chem.*, *55*, 5165–5173. DOI: 10.1021/jm300131x.

19 Kalliokoski, T., Kramer, C., Vulpetti, A., Gedeck, P. (2013) Comparability of Mixed IC50 Data—A Statistical Analysis. *PLoS ONE*, *8*, e61007. DOI: 10.1371/journal.pone.0061007.

20 Kramer, C., Lewis, R. (2012) QSARs, Data and Error in the Modern Age of Drug Discovery. *Curr. Top. Med. Chem.*, *12*, 1896–1902. DOI: 10.2174/156802612804547380.

21 Kalliokoski, T., Kramer, C., Vulpetti, A. (2013) Quality Issues with Public Domain Chemogenomics Data. *Mol. Informat.*, *32*, 898–905. DOI: 10.1002/minf.201300051.

22 Goracci, L., Ceccarelli, M., Bonelli, D., Cruciani, G. (2013) Modeling Phospholipidosis Induction: Reliability and Warnings. *J. Chem. Inf. Model.*, *53*, 1436–1446. DOI: 10.1021/ci400113t.

23 Orogo, A. M., Choi, S. S., Minnier, B. L., Kruhlak, N. L. (2012) Construction and Consensus Performance of (Q)SAR Models for Predicting Phospholipidosis Using a Dataset of 743 Compounds. *Mol. Informat.*, *31*, 725–739. DOI: 10.1002/minf.201200048.

24 Przybylak, K. R., Alzahrani, A. R., Cronin, M. T. D. (2014) How Does the Quality of Phospholipidosis Data Influence the Predictivity of Structural Alerts? *J. Chem. Inf. Model.*, *54*, 2224–2232. DOI: 10.1021/ci500233k.

25 Netzeva, T. I., Worth, A., Aldenberg, T., Benigni, R., Cronin, M. T. D., Gramatica, P., Jaworska, J. S., Kahn, S., Klopman, G., Marchant, C. A., Myatt, G., Nikolova-Jeliazkova, N., Patlewicz, G. Y., Perkins, R., Roberts, D., Schultz, T., Stanton, D. W., van de Sandt, J. J. M., Tong, W., Veith, G., Yang, C. (2005) Current Status of Methods for Defining the Applicability Domain of (quantitative) Structure-activity Relationships. The Report and Recommendations of ECVAM Workshop 52. *Altern. Lab. Anim. (ATLA)*, *33*, 155–173.

26 Schwaighofer, A., Schroeter, T., Mika, S., Hansen, K., ter Laak, A., Lienau, P., Reichel, A., Heinrich, N., Müller, K.-R. (2008) A Probabilistic Approach to Classifying Metabolic Stability. *J. Chem. Inf. Model.*, *48*, 785–796. DOI: 10.1021/ci700142c.

27 Obrezanova, O., Segall, M. D. (2010) Gaussian Processes for Classification: QSAR Modeling of ADMET and Target Activity. *J. Chem. Inf. Model.*, *50*, 1053–1061. DOI: 10.1021/ci900406x.

28 Kramer, C., Beck, B., Clark, T. (2010) Insolubility Classification with Accurate Prediction Probabilities Using a MetaClassifier. *J. Chem. Inf. Model.*, *50*, 404–414. DOI: 10.1021/ci900377e.

29 Norinder, U., Carlsson, L., Boyer, S., Eklund, M. (2014) Introducing Conformal Prediction in Predictive Modeling. A Transparent and Flexible Alternative to Applicability Domain Determination. *J. Chem. Inf. Model.*, *54*, 1596–1603. DOI: 10.1021/ci5001168.

30 Chalk, A. J., Beck, B., Clark, T. (2001) A Quantum Mechanical/neural Net Model for Boiling Points with Error Estimation. *J. Chem. Inf. Comput. Sci.*, *41*, 457–462.

31 Schwaighofer, A., Schroeter, T., Mika, S., Laub, J., ter Laak, A., Sülzle, D., Ganzer, U., Heinrich, N., Müller, K.-R. (2007) Accurate Solubility Prediction with Error Bars for Electrolytes: A Machine Learning Approach. *J. Chem. Inf. Model.*, *47*, 407–424. DOI: 10.1021/ci600205g.

32 Clark, R. D. (2009) DPRESS: Localizing Estimates of Predictive Uncertainty. *J. Cheminformat.*, *1*, 11. DOI: 10.1186/1758-2946-1-11.

33 Sheridan, R. P. (2012) Three Useful Dimensions for Domain Applicability in QSAR Models Using Random Forest. *J. Chem. Inf. Model.*, *52*, 814–823. DOI: 10.1021/ci300004n.

34 Sheridan, R. P. (2013) Using Random Forest to Model the Domain Applicability of Another Random Forest Model. *J. Chem. Inf. Model.*, *53*, 2837–2850. DOI: 10.1021/ci400482e.

35 Wood, D. J., Carlsson, L., Eklund, M., Norinder, U., Stålring, J. (2013) QSAR with Experimental and Predictive Distributions: An Information Theoretic Approach for Assessing Model Quality. *J. Comput. Aided Mol. Des.*, *27*, 203–219. DOI: 10.1007/s10822-013-9639-5.

36 Anger, L. T., Wolf, A., Schleifer, K.-J., Schrenk, D., Rohrer, S. G. (2014) Generalized Workflow for Generating Highly Predictive in Silico Off-Target Activity Models. *J. Chem. Inf. Model.*, *54*, 2411–2422. DOI: 10.1021/ci500342q.

37 Carrió, P., Pinto, M., Ecker, G., Sanz, F., Pastor, M. (2014) Applicability Domain Analysis (ADAN): A Robust Method for Assessing the Reliability of Drug Property Predictions. *J. Chem. Inf. Model.*, *54*, 1500–1511. DOI: 10.1021/ci500172z.

38 Lipinski, C. A., Lombardo, F., Dominy, B. W., Feeney, P. J. (1997) Experimental and Computational Approaches to Estimate Solubility and Permeability in Drug Discovery and Development Settings. *Adv. Drug Deliv. Rev.*, *23*, 3–25. DOI: 10.1016/S0169-409X(96)00423-1.

39 Lipinski, C. A. (2004) Lead- and Drug-like Compounds: The Rule-of-five Revolution. *Drug Discov. Today Technol.*, *1*, 337–341. DOI: 10.1016/j.ddtec.2004.11.007.

40 Egan, W. J., Merz, Baldwin, J. J. (2000) Prediction of Drug Absorption Using Multivariate Statistics. *J. Med. Chem.*, *43*, 3867–3877. DOI: 10.1021/jm000292e.

41 Gleeson, M. P. (2008) Generation of a Set of Simple, Interpretable ADMET Rules of Thumb. *J. Med. Chem.*, *51*, 817–834. DOI: 10.1021/jm701122q.

42 Tarcsay, Á., Keserű, G. M. (2013) Contributions of Molecular Properties to Drug Promiscuity. *J. Med. Chem.*, *56*, 1789–1795. DOI: 10.1021/jm301514n.

43 Hann, M. M., Leach, A. R., Harper, G. (2001) Molecular Complexity and Its Impact on the Probability of Finding Leads for Drug Discovery. *J. Chem. Inf. Comput. Sci.*, *41*, 856–864.

44 Michael M. Hann. (2011) Molecular Obesity, Potency and Other Addictions in Drug Discovery. *MedChemComm*, *2*, 349. DOI: 10.1039/c1md00017a.

45 Bickerton, G. R., Paolini, G. V., Besnard, J., Muresan, S., Hopkins, A. L. (2012) Quantifying the Chemical Beauty of Drugs. *Nat. Chem.*, *4*, 90–98. DOI: 10.1038/nchem.1243.

46 Kenny, P. W., Montanari, C. A. (2013) Inflation of Correlation in the Pursuit of Drug-likeness. *J. Comput. Aided Mol. Des.*, *27*, 1–13. DOI: 10.1007/s10822-012-9631-5.

47 Muthas, D., Boyer, S., Hasselgren, C. (2013) A Critical Assessment of Modeling Safety-related Drug Attrition. *Med. Chem. Commun.*, *4*, 1058–1065. DOI: 10.1039/C3MD00072A.

48 Hughes, J. D., Blagg, J., Price, D. A., Bailey, S., DeCrescenzo, G. A., Devraj, R. V., Ellsworth, E., Fobian, Y. M., Gibbs, M. E., Gilles, R. W., Greene, N., Huang, E., Krieger-Burke, T., Loesel, J., Wager, T., Whiteley, L., Zhang, Y. (2008) Physiochemical Drug Properties Associated with in Vivo Toxicological Outcomes. *Bioorg. Med. Chem. Lett.*, *18*, 4872–4875. DOI: 10.1016/j.bmcl.2008.07.071.

49 Peters, J.-U., Hert, J., Bissantz, C., Hillebrecht, A., Gerebtzoff, G., Bendels, S., Tillier, F., Migeon, J., Fischer, H., Guba, W., Kansy, M. (2012) Can We Discover Pharmacological Promiscuity Early in the Drug Discovery Process? *Drug Discov. Today*, *17*, 325–335. DOI: 10.1016/j.drudis.2012.01.001.

50 Lovering, F., Bikker, J., Humblet, C. (2009) Escape from Flatland: Increasing Saturation as an Approach to Improving Clinical Success. *J. Med. Chem.*, *52*, 6752–6756. DOI: 10.1021/jm901241e.

51 Lloyd, D. G., Golfis, G., Knox, A. J. S., Fayne, D., Meegan, M. J., Oprea, T. I. (2006) Oncology Exploration: Charting Cancer Medicinal Chemistry Space. *Drug Discov. Today*, *11*, 149–159. DOI: 10.1016/S1359-6446(05)03688-3.

52 Doak, B. C., Over, B., Giordanetto, F., Kihlberg, J. (2014) Oral Druggable Space Beyond the Rule of 5: Insights from Drugs and Clinical Candidates. *Chem. Biol.*, *21*, 1115–1142. DOI: 10.1016/j.chembiol.2014.08.013.

53 Abad-Zapatero, C. (2007) A Sorcerer's Apprentice and The Rule of Five: From Rule-of-thumb to Commandment and Beyond. *Drug Discov. Today*, *12*, 995–997. DOI: 10.1016/j.drudis.2007.10.022.

54 Zhang, M.-Q., Wilkinson, B. (2007) Drug Discovery Beyond the "rule-of-five." *Curr. Opin. Biotechnol.*, *18*, 478–488. DOI: 10.1016/j.copbio.2007.10.005.

55 Walters, W. P. (2012) Going Further Than Lipinski's Rule in Drug Design. *Expert Opin. Drug Discov.*, *7*, 99–107. DOI: 10.1517/17460441.2012.648612.

56 Kuhn, B., Mohr, P., Stahl, M. (2010) Intramolecular Hydrogen Bonding in Medicinal Chemistry. *J. Med. Chem.*, *53*, 2601–2611. DOI: 10.1021/jm100087s.

57 Alex, A., Millan, D. S., Perez, M., Wakenhut, F., Whitlock, G. A. (2011) Intramolecular Hydrogen Bonding to Improve Membrane Permeability and Absorption in Beyond Rule of Five Chemical Space. *Med. Chem. Commun.*, *2*, 669–674. DOI: 10.1039/C1MD00093D.

58 Cook, D., Brown, D., Alexander, R., March, R., Morgan, P., Satterthwaite, G., Pangalos, M. N. (2014) Lessons Learned from the Fate of AstraZeneca's Drug Pipeline: a Five-dimensional Framework. *Nat. Rev. Drug Discov.*, *13*, 419–431. DOI: 10.1038/nrd4309.

59 Fischer, H., Atzpodien, E.-A., Csato, M., Doessegger, L., Lenz, B., Schmitt, G., Singer, T. (2012) In Silico Assay for Assessing Phospholipidosis Potential of Small Druglike Molecules: Training, Validation, and Refinement Using Several Data Sets. *J. Med. Chem.*, *55*, 126–139. DOI: 10.1021/jm201082a.

60 Cumming, J. G., Winter, J., Poirrette, A. (2012) Better Compounds Faster: The Development and Exploitation of a Desktop Predictive Chemistry Toolkit. *Drug Discov. Today*, *17*, 923–927. DOI: 10.1016/j.drudis.2012.03.003.

61 Optibrium (2014) StarDrop; Optibrium.

62 Sutter, A., Amberg, A., Boyer, S., Brigo, A., Contrera, J. F., Custer, L. L., Dobo, K. L., Gervais, V., Glowienke, S., Gompel, J. van, Greene, N., Muster, W., Nicolette, J., Reddy, M. V., Thybaud, V., Vock, E., White, A. T., Müller, L. (2013) Use of in Silico Systems and Expert Knowledge for Structure-based Assessment of Potentially Mutagenic Impurities. *Regul. Toxicol. Pharmacol.*, *67*, 39–52. DOI: 10.1016/j.yrtph.2013.05.001.

63 McCarren, P., Bebernitz, G. R., Gedeck, P., Glowienke, S., Grondine, M. S., Kirman, L. C., Klickstein, J., Schuster, H. F., Whitehead, L. (2011) Avoidance of the Ames Test Liability for Aryl–Amines via Computation. *Bioorg. Med. Chem.*, *19*, 3173–3182. DOI: 10.1016/j.bmc.2011.03.066.

64 McCarren, P., Springer, C., Whitehead, L. (2011) An Investigation into Pharmaceutically Relevant Mutagenicity Data and the Influence on Ames Predictive Potential. *J. Cheminformat.*, *3*, 51. DOI: 10.1186/1758-2946-3-51.

65 Hansen, K., Mika, S., Schroeter, T., Sutter, A., ter Laak, A., Steger-Hartmann, T., Heinrich, N., Müller, K.-R. (2009) Benchmark Data Set for in Silico Prediction of Ames Mutagenicity. *J. Chem. Inf. Model.*, *49*, 2077–2081. DOI: 10.1021/ci900161g.

66 Stein, K. R., Scheinfeld, N. S. (2007) Drug-induced Photoallergic and Phototoxic Reactions. *Expert Opin. Drug Saf.*, *6*, 431–443. DOI: 10.1517/14740338.6.4.431.

67 Elkeeb, D., Elkeeb, L., Maibach, H. (2012) Photosensitivity: A Current Biological Overview. *Cutan. Ocul. Toxicol.*, *31*, 263–272. DOI: 10.3109/15569527.2012.656293.

68 Bauer, D., Averett, L. A., De Smedt, A., Kleinman, M. H., Muster, W., Pettersen, B. A., Robles, C. (2014) Standardized UV–Vis Spectra as the Foundation for a Threshold-based, Integrated Photosafety Evaluation. *Regul. Toxicol. Pharmacol.*, *68*, 70–75. DOI: 10.1016/j.yrtph.2013.11.007.

69 De Lima Ribeiro, F. A., Ferreira, M. M. C. (2005) QSAR Model of the Phototoxicity of Polycyclic Aromatic Hydrocarbons. *J. Mol. Struct. Theochem*, *719*, 191–200. DOI: 10.1016/j.theochem.2005.01.026.

70 Peukert, S., Nunez, J., He, F., Dai, M., Yusuff, N., DiPesa, A., Miller-Moslin, K., Karki, R., Lagu, B., Harwell, C., Zhang, Y., Bauer, D., Kelleher, J. F., Egan, W. (2011) A Method for Estimating the Risk of Drug-induced Phototoxicity and Its Application to Smoothened Inhibitors. *Med. Chem. Commun.*, *2*, 973–976. DOI: 10.1039/C1MD00144B.

71 Gedeck, P., Bauer, D. (2015) unpublished observation.

72 Nashev, L. G., Vuorinen, A., Praxmarer, L., Chantong, B., Cereghetti, D., Winiger, R., Schuster, D., Odermatt, A. (2012) Virtual Screening as a Strategy for the Identification of Xenobiotics Disrupting Corticosteroid Action. *PLoS ONE*, 7, e46958. DOI: 10.1371/journal.pone.0046958.

73 Nigsch, F., Lounkine, E., McCarren, P., Cornett, B., Glick, M., Azzaoui, K., Urban, L., Marc, P., Müller, A., Hahne, F., Heard, D. J., Jenkins, J. L. (2011) Computational Methods for Early Predictive Safety Assessment from Biological and Chemical Data. *Expert Opin. Drug Metab. Toxicol.*, 7, 1497–1511. DOI: 10.1517/17425255.2011.632632.

74 Copeland, R. A., Pompliano, D. L., Meek, T. D. (2006) Drug–Target Residence Time and Its Implications for Lead Optimization. *Nat. Rev. Drug Discov.*, 5, 730–739. DOI: 10.1038/nrd2082.

75 Gombar, V. K., Hall, S. D. (2013) Quantitative Structure–Activity Relationship Models of Clinical Pharmacokinetics: Clearance and Volume of Distribution. *J. Chem. Inf. Model.*, 53, 948–957. DOI: 10.1021/ci400001u.

76 Agoram, B., Woltosz, W. S., Bolger, M. B. (2001) Predicting the Impact of Physiological and Biochemical Processes on Oral Drug Bioavailability. *Adv. Drug Deliv. Rev.*, 50, Supplement 1, S41–S67. DOI: 10.1016/S0169-409X(01)00179-X.

77 Berman, H. M., Westbrook, J., Feng, Z., Gilliland, G., Bhat, T. N., Weissig, H., Shindyalov, I. N., Bourne, P. E. (2000) The Protein Data Bank. *Nucleic Acids Res.*, 28, 235–242. DOI: 10.1093/nar/28.1.235.

78 Gaulton, A., Bellis, L. J., Bento, A. P., Chambers, J., Davies, M., Hersey, A., Light, Y., McGlinchey, S., Michalovich, D., Al-Lazikani, B., Overington, J. P. (2012) ChEMBL: A Large-scale Bioactivity Database for Drug Discovery. *Nucleic Acids Res.*, 40 (Database issue), D1100–D1107. DOI: 10.1093/nar/gkr777.

79 Liu, T., Lin, Y., Wen, X., Jorissen, R. N., Gilson, M. K. (2007) BindingDB: A Web-accessible Database of Experimentally Determined Protein-ligand Binding Affinities. *Nucleic Acids Res.*, 35 (suppl_1), D198–201. DOI: 10.1093/nar/gkl999.

80 Williams, A. J., Harland, L., Groth, P., Pettifer, S., Chichester, C., Willighagen, E. L., Evelo, C. T., Blomberg, N., Ecker, G., Goble, C., Mons, B. (2012) Open PHACTS: Semantic Interoperability for Drug Discovery. *Drug Discov. Today*, 17, 1188–1198. DOI: 10.1016/j.drudis.2012.05.016.

81 Matlock, M., Swamidass, S. J. (2013) Sharing Chemical Relationships Does Not Reveal Structures. *J. Chem. Inf. Model.*, 54, 37–48. DOI: 10.1021/ci400399a.

82 Dearden, J. C. (2007) In Silico Prediction of ADMET Properties: How Far Have We Come? *Expert Opin. Drug Metab. Toxicol.*, 3, 635–639. DOI: 10.1517/17425255.3.5.635.

83 Briggs, K., Cases, M., Heard, D. J., Pastor, M., Pognan, F., Sanz, F., Schwab, C. H., Steger-Hartmann, T., Sutter, A., Watson, D. K., Wichard, J. D. (2012) Inroads to Predict In Vivo Toxicology—An Introduction to the eTOX Project. *Int. J. Mol. Sci.*, 13, 3820–3846. DOI: 10.3390/ijms13033820.

84 Dossetter, A. G., Griffen, E. J., Leach, A. G. (2013) Matched Molecular Pair Analysis in Drug Discovery. *Drug Discov. Today*, 18, 724–731. DOI: 10.1016/j.drudis.2013.03.003.

85 Fourches, D., Muratov, E., Tropsha, A. (2010) Trust, But Verify: On the Importance of Chemical Structure Curation in Cheminformatics and QSAR Modeling Research. *J. Chem. Inf. Model*, 50, 1189–1204. DOI: 10.1021/ci100176x.

86 Dearden, J. C., Cronin, M. T. D., Kaiser, K. L. E. (2009) How Not to Develop a Quantitative Structure–Activity or Structure–Property Relationship (QSAR/QSPR). *SAR–QSAR Environ. Res.*, 20, 241. DOI: 10.1080/10629360902949567.

87 Tropsha, A. (2010) Best Practices for QSAR Model Development, Validation, and Exploitation. *Mol. Informat.*, 29, 476–488. DOI: 10.1002/minf.201000061.

88   Krstajic, D., Buturovic, L. J., Leahy, D. E., Thomas, S. (2014) Cross-validation Pitfalls When Selecting and Assessing Regression and Classification Models. *J. Cheminformat.*, *6*, 10. DOI: 10.1186/1758-2946-6-10.

89   Stouch, T. R., Kenyon, J. R., Johnson, S. R., Chen, X.-Q., Doweyko, A., Li, Y. (2003) In Silico ADME/Tox: Why Models Fail. *J. Comput. Aided Mol. Des.*, *17*, 83–92.

90   Ekins, S., Olechno, J., Williams, A. J. (2013) Dispensing Processes Impact Apparent Biological Activity as Determined by Computational and Statistical Analyses. *PLoS ONE*, *8*, e62325. DOI: 10.1371/journal.pone.0062325.

91   Leach, A. G., Jones, H. D., Cosgrove, D. A., Kenny, P. W., Ruston, L., MacFaul, P., Wood, J. M., Colclough, N., Law, B. (2006) Matched Molecular Pairs as a Guide in the Optimization of Pharmaceutical Properties; A Study of Aqueous Solubility, Plasma Protein Binding and Oral Exposure. *J. Med. Chem.*, *49*, 6672–6682. DOI: 10.1021/jm0605233.

92   Sheridan, R. P., Hunt, P., Culberson, J. C. (2006) Molecular Transformations as a Way of Finding and Exploiting Consistent Local QSAR. *J. Chem. Inf. Model.*, *46*, 180–192. DOI: 10.1021/ci0503208.

93   Griffen, E., Leach, A. G., Robb, G. R., Warner, D. J. (2011) Matched Molecular Pairs as a Medicinal Chemistry Tool. *J. Med. Chem.*, *54*, 7739–7750. DOI: 10.1021/jm200452d.

94   Kramer, C., Fuchs, J. E., Whitebread, S., Gedeck, P., Liedl, K. R. (2014) Matched Molecular Pair Analysis: Significance and the Impact of Experimental Uncertainty. *J. Med. Chem.*, *57*, 3786–3802. DOI: 10.1021/jm500317a.

95   O'Boyle, N. M., Boström, J., Sayle, R. A., Gill, A. (2014) Using Matched Molecular Series as a Predictive Tool To Optimize Biological Activity. *J. Med. Chem.*, *57*, 2704–2713. DOI: 10.1021/jm500022q.

96   Cortes-Ciriano, I., van Westen, G. J., Lenselink, E. B., Murrell, D. S., Bender, A., Malliavin, T. (2014) Proteochemometric Modeling in a Bayesian Framework. *J. Cheminformat.*, *6*, 35. DOI: 10.1186/1758-2946-6-35.

97   Petrone, P. M., Simms, B., Nigsch, F., Lounkine, E., Kutchukian, P., Cornett, A., Deng, Z., Davies, J. W., Jenkins, J. L., Glick, M. (2012) Rethinking Molecular Similarity: Comparing Compounds on the Basis of Biological Activity. *ACS Chem. Biol.*, *7*, 1399–1409. DOI: 10.1021/cb3001028.

98   Sedykh, A., Zhu, H., Tang, H., Zhang, L., Richard, A., Rusyn, I., Tropsha, A. (2010) Use of in Vitro HTS-Derived Concentration–Response Data as Biological Descriptors Improves the Accuracy of QSAR Models of in Vivo Toxicity. *Environ. Health Perspect.*, *119*, 364–370. DOI: 10.1289/ehp.1002476.

99   Rusyn, I., Sedykh, A., Low, Y., Guyton, K. Z., Tropsha, A. (2012) Predictive Modeling of Chemical Hazard by Integrating Numerical Descriptors of Chemical Structures and Short-term Toxicity Assay Data. *Toxicol. Sci.*, *127*, 1–9. DOI: 10.1093/toxsci/kfs095.

11

CURRENT AND FUTURE STRATEGIES FOR IMPROVING DRUG DISCOVERY EFFICIENCY

PETER MBUGUA NJOGU AND KELLY CHIBALE

Department of Chemistry, University of Cape Town, Rondebosch, South Africa

11.1 GENERAL INTRODUCTION

The twentieth century will undoubtedly go down in the annals of history as the most spectacular century yet of the modern world. It heralded remarkable progress in several spheres of human life. One of the most enviable legacies of the century was the improvement in the overall health conditions of most of humanity [1]. Improved sanitation, increase in scientific knowledge of the pathogenesis of various diseases, discovery of more efficacious drugs, and the advent of immunization enabled global eradication and/or suppression of some of the worst human afflictions such as smallpox, plague, poliomyelitis, measles, and leprosy [2, 3]. This is attested to by a sharp decline in global infectious disease-related mortality, betterment in the quality of life, steady increase in life expectancy, and accelerated economic growth in the middle- and low-income countries. No scientific discipline has played a greater role in this commendable transformation than the field of drug discovery spearheaded largely by the big pharmaceutical companies and to a lesser extent by the biotechnology firms and academic research institutes [4].

The pharmaceutical industry recorded such impressive strides during the latter half of the twentieth century that it is currently regarded as one of the most successful human enterprises [5]. Supported by powerful and elegant new automated technologies such as high-throughput screening (HTS), computer-aided drug design (CADD), and combinatorial chemistry, as well as increased knowledge of molecular and cellular biology, the process of drug discovery has undergone rapid revolution [6, 7]. In parallel, the industry has embraced and applied newer concepts such as genomics, proteomics,

Attrition in the Pharmaceutical Industry: Reasons, Implications, and Pathways Forward,
First Edition. Edited by Alexander Alex, C. John Harris and Dennis A. Smith.
© 2016 John Wiley & Sons, Inc. Published 2016 by John Wiley & Sons, Inc.

and bioinformatics in the value chain of drug development [8, 9]. The results have increased rationality in drug design, synthesis, and clinical evaluation [10].

Paradoxically, in spite of substantial investments on innovative technologies, the number of new molecular entities (NMEs) entering the market has not increased proportionately [11, 12]. This apparent disconnection between investment and output has been attributed to several key factors. First is the observation that the entry bar for new drugs into the clinical market is considerably higher [13]. This is because more often than not, new drugs are competing with enhanced standards of care. A new molecule has to show marked benefits over existing drugs in terms of efficacy, side effects profile, and compliance. This is compounded by the perception that the national agencies tasked with evaluation of NMEs such as the US Food and Drug Administration (FDA) and the European Medicines Agency (EMA) have tightened the regulatory requirements for approval [14, 15]. Second is the fact that most easily druggable and disease-relevant targets have been picked, characterized, and addressed [16]. Unavoidably, the industry is currently tackling diseases of greater complexity, yet innovation has not kept pace with ever increasing need for newer therapeutically viable targets [13].

A third possible reason for the decline in pharmaceutical research and development (R&D) productivity is the industry's heavy reliance on new high-technology platforms and alchemy, which have yet to deliver on their promise. The added cost and challenges associated with integration of these technologies in the pharmaceutical R&D have disrupted coherence in the drug discovery and development processes [17]. Fourth, combinatorial chemistry and HTS platforms have tended to favor chemical libraries composed of compounds with both high molecular weight and lipophilicity [18]. Research has demonstrated that these larger, greasier molecules have poor absorption, distribution, metabolism, excretion, and toxicity (ADMET) properties [19, 20]. They therefore pose greater development challenges and are more liable to attrition during lead optimization due to their low "drug-likeness" [18, 21].

11.2 SCOPE

It is evident that the field of drug discovery experienced exponential growth and phenomenal developments in the latter half of the twentieth century. Over the same time frame, the pharmaceutical industry has impacted positively on the health and life span of humans across the globe. Unfortunately, the R&D productivity and the fiscal returns of the pharma industry in the first decade of the twenty-first century have been disappointing [22]. A number of unprecedented challenges currently confronting the pharmaceutical industry have been indicted as major drivers of this dismal performance. Foremost among these are the industry's unexpectedly low number of NME approvals, dry late-stage R&D pipelines prevalent throughout the industry, and dwindling revenue owing to patent expirations of the 1980s and 1990s' blockbuster drugs, leading to low returns on investment and poor shareholder stock performance [13, 17]. Further, agitation for cost containment of drugs and medical supplies advocated by governments, commercial players, and civil society activism has weighed heavily on industry's profit margins.

The inefficiency in the drug discovery process becomes even more apparent when one considers the large number of compounds that undergo attrition during preclinical research and the fact that only three out of ten drugs that make it to the clinic recover their capital investment [13]. It is imperative to note that efficiency in drug discovery is necessary for sustained profitability of the pharmaceutical industry, which will in turn support further reinvestment in R&D. This way, the industry can play its rightful role of providing much-needed innovative therapies for the existing and emerging unmet medical needs. Thus, the issue that begs answers is how the pharmaceutical industry can turn around its fortunes and regain its luster in innovative drug discovery.

The scope of this chapter is to highlight notable current and future strategies that, when properly harnessed, could bolster drug discovery efficiency. Here, using tropical diseases as an example, we argue that there are some niche disease areas and approaches that have hitherto been underexploited and/or only marginally applied as growth levers yet are likely to be valuable for the industry's present and future viability. As has been argued before, we support the assertion that future sustainability of the pharma industry lies not in a business model focused on discovering big blockbuster drugs but on one that will provide adequate profits from "not-so-profitable" disease areas. Thus, if another golden era in drug discovery is to happen, first, the pharmaceutical industry has to be restructured in a manner as to make profits from smaller patient population markets, and, second, newer approaches need to be inculcated in the pharma industry to increase efficiency in the drug discovery process. The discussion is presented under the following subtopics: neglected diseases, precompetitive drug discovery, exploitation of genomics, outsourcing strategies, multitarget drug design and discovery, and drug repositioning and repurposing.

11.3 NEGLECTED DISEASES

11.3.1 Introduction

Neglected diseases are primarily tropical infectious diseases that disproportionately afflict the poor and marginalized populations in sub-Saharan Africa, Southeast Asia, and Latin America [23]. As classified by the World Health Organization (WHO), neglected tropical diseases (NTDs) comprise 17 infectious ailments—namely, dengue, rabies, trachoma, Buruli ulcer (*Mycobacterium ulcerans* infection), endemic treponematoses, leprosy (Hansen's disease), Chagas disease (American trypanosomiasis), human African trypanosomiasis (HAT, sleeping sickness), leishmaniasis, cysticercosis, dracunculiasis (guinea worm disease), echinococcosis, foodborne trematode infections, lymphatic filariasis, onchocerciasis (river blindness), schistosomiasis (bilharziasis), and soil-transmitted helminthiases [24]. Malaria, although no longer classified as an NTD, is the preeminent tropical parasitic disease and one of the top three killers among communicable diseases [25, 26].

The WHO contends that although the NTDs have low visibility globally, they blight the lives of approximately one billion people and threaten the health of millions more. These ancient companions of poverty weaken impoverished populations, frustrate the attainment of the first six Millennium Development Goals and impede global development initiatives [27]. This chapter focuses largely on malaria, Chagas disease, HAT, leishmaniasis, onchocerciasis, lymphatic filariasis, and schistosomiasis. The extent of coverage will differ according to developments in each disease field.

11.3.2   Control of NTDs

Despite differences in their etiology and epidemiology, similar strategies have been adopted in the control and management of NTDs. Abundant clinical and experimental evidence indicates that vaccination still remains the best theoretical option in the fight against parasitism in both humans and animals [28]. Accordingly, vaccine R&D for NTDs has been attempted for decades with varied degrees of success [29–33]. Unfortunately up to date, there is no licensed vaccine against any NTD because of the extreme degree of antigenic variation exhibited by these parasites and lack of resources for translational work and large vaccine trials [28, 31]. However, a promising new malaria vaccine candidate, RTS,S (Mosquirix™), was announced in a press release on July 24, 2015. By GlaxoSmithKline (GSK) after positive appraisal by the EMA and is due for policy review by the WHO [34]. The vaccine candidate is based on plasmodial circumsporozoite protein and has been developed by GSK in partnership with the Malaria Vaccine Initiative (MVI) of the Program for Appropriate Technology in Health (PATH) with grant funding from the Bill & Melinda Gates Foundation. Results of a large-scale phase III trial have shown that, after 18 months of follow-up, the vaccine candidate reduces clinical malaria cases in infants by 31% and by 56% in young children [35].

Vector control strategies, through environmental management to eliminate vector breeding sites, indoor residual spraying and use of insecticide-impregnated bed nets to reduce human–vector contacts, and use of larvicides to interrupt parasites' life cycles, have saved millions of lives worldwide [36–38]. However, logistical complexity, cost, and moderate efficacy of vector control measures make this strategy largely ineffective in regions of high endemicity [39]. So far, case management through chemotherapy remains the cornerstone of successful control of tropical neglected infections [40]. Thus, NTDs represent one disease area in which the pharma industry could boost its R&D productivity.

11.3.3   Drug Discovery Potential of Neglected Diseases

Due to their occurrence among poor populations in the less developed regions of the world, NTDs do not have high visibility in the pharma industry since the prospects for financial returns on investment are too low to support market-driven drug development [41, 42]. Consequently, drug discovery efforts in this field are largely lacking, clearly evidenced by a dearth of new drugs entering clinical development for tropical diseases [43]. For example, of the 1,556 drugs granted market authorization between 1975 and 2004, only 18 (1.3%) were for tropical diseases, including 8 for malaria [44]. Not surprising, the pharmacopoeia for majority of the NTDs has remained essentially unchanged for decades; the available drugs are limited in efficacy, plagued by severe toxicities, associated with poor patient compliance, and hamstrung by drug resistance [45, 46]. Notable liabilities of the currently available drugs for management of NTDs are briefly presented in Table 11.1.

It is clear that the currently available chemotherapeutic armamentarium for NTDs is severely limited, convincingly revealing a public policy failure in drug discovery for tropical diseases. The challenges inherent in the field of NTDs present potential opportunities to the pharmaceutical industry to reengineer its dwindling innovation and productivity pipeline in several ways. First, reengagement of the pharmaceutical industry with the tropical disease research will improve its public relations. Such engagement will most likely occur through precompetitive research consortia in which there are shared resources and costs, thus mitigating the risks involved in drug discovery.

TABLE 11.1 Current Drugs for Some Neglected Tropical Diseases and their Liabilities[a]

Infection	Some widely used or recently introduced drugs (year first used)	Liabilities
Malaria	Chloroquine (1945); sulfadoxine/pyrimethamine (1961); mefloquine (1984); artemisinins (1994); artemether/ lumefantrine (1999); atovaquone/proguanil (1999); chlorproguanil/dapsone (2003)	Drug resistance is widespread to chloroquine and sulfadoxine/pyrimethamine, growing to mefloquine, and is a threat to other compounds; adverse effects to chloroquine, mefloquine, and proguanil; cost is an issue to most drugs and drug combinations; availability of artemisinins is problematic, and development of resistance is now a realistic possibility
HAT	Suramin (1920); pentamidine (1939); melarsoprol (1949); eflornithine (1991)	Risk of severe adverse effects with all drugs; suramin and pentamidine not effective in late-stage disease; eflornithine is expensive and only effective against *T.b. gambiense.*
Chagas disease	Nifurtimox (1970); benznidazole (1974)	Long treatment courses and adverse effects limit compliance; not effective in late-stage disease
Leishmaniasis	Pentamidine (1939); pentavalent antimonials (1950); liposomal amphotericin B (1990); miltefosine (2002)	Efficacy loss/drug resistance for pentamidine and antimonials; high cost for liposomal amphotericin B; adverse effects described for other drugs; miltefosine is contraindicated in women of childbearing age
Lymphatic filariases	Diethylcarbamazine (DEC, 1949); ivermectin (1989); albendazole/DEC; albendazole/ivermectin	DEC cannot be used in *Onchocerca volvulus*-endemic areas (risk of adverse effects); albendazole only used in combination; ivermectin does not eliminate adult worms
Schistosomiasis	Oxamniquine (1967); praziquantel (1975)	Oxamniquine is only effective against *Schistosoma mansoni*; praziquantel does not eliminate immature worms; possible resistance reported
Onchocerciasis	Ivermectin (1989)	Does not eliminate adult worms

[a]Adopted from Ref. 41.
HAT, human African trypanosomiasis.

Second, being a relatively underresearched field, it may qualify as a "niche market" in that the prospects for discovering a druggable target are far much higher than in other therapeutic areas. This prospect is likely through careful exploitation of genomic knowledge. Further, research in tropical diseases may provide a launch pad for lead discovery in profitable therapeutic areas. For example, a "hit-to-lead" program targeting a parasite enzyme might help a company build a chemical library around the lead for testing against a human isoenzyme amenable to commercialization [47]. Third, precompetitive research collaborations with other stakeholders in the industry such as biotechnology firms and academic institutions may open up further platforms for partnering. Inevitably, these partnerships could act as avenues for outsourcing strategies and sharing ideas on emerging drug discovery paradigms such as multitarget drug design and discovery, drug repositioning and repurposing, and other innovative strategies for improving drug discovery efficiency.

11.4 PRECOMPETITIVE DRUG DISCOVERY

11.4.1 Introduction

The pharmaceutical industry has had a long-standing history of research collaboration with academic institutions, biotechnology firms, and public healthcare organizations on various projects addressing issues of mutual concern. Traditionally, most collaborations have taken the form of a closed collaborative model. In this model, a partnership between a single company and a single academic researcher and/or host institution is established focusing on a specific research question in which the academic partner has significant expertise. The academic researcher gains by way of funding, and occasionally royalties, while the company retains all intellectual property rights [48]. The traditional model may be broadened into a closed consortium comprising one or several companies and one or more academic institutions. In this model, data is shared openly among the partners within the project but not to outsiders to protect any potential intellectual property rights [48]. Since the industry engages the academia on "as-needed" basis, the closed model is likely to create a huge translational gap and thus impede innovation [49]. Further, although a closed model in its various forms has had its successes in the past that was characterized by a scarcity of information and fierce competition among the major pharmaceutical companies, it is no longer tenable in the current information age, hence necessitating advocacy for establishment of open consortium networks [50, 51].

The open consortium model involves joint collaboration between the industry, academia, and public-funded organizations in a broad research area of mutual interest. The defining features of this model are the following: (1) the consortium members share the funding burden, resources, and expertise collectively, thus mitigating against the risks involved in drug discovery; (2) research activities are coordinated and usually carried out at one of the host institutions, or an established independent legal entity; (3) data are shared openly among consortium members; and (4) data are made publicly available with no restriction on use [48]. One of the consequences of the open consortium collaborative model is the emergence of precompetitive drug discovery.

Precompetitive research refers to multidisciplinary collaborative research practiced by organizations, which ordinarily are commercial competitors. This form of collaboration is deeply rooted in other industries such as electronics, telecommunications, and automotive engineering, but is still in its infancy in the pharmaceutical sector. It is modeled on an open-access platform. This emerging trend follows from an observation that the practice of keeping secret basic research findings so as to retain potential, and usually imaginary, commercial benefits does not stimulate scientific discourse and is ultimately detrimental to the public good [48]. In complete departure from the guarded nature of commercial scientific findings, it is recommended that the results of precompetitive research be made publicly available and subjected to scientific scrutiny providing avenues for cross-fertilization and refinement of new knowledge. Application of the shared insights will then greatly improve the prospects for invention-based competition. For this reason, it is sometimes dubbed "procompetitive" research [49]. The precompetitive drug discovery space may include determination of disease pathways and essential biomarkers for prediction of efficacy and safety, development of assay and diagnostic tools, development of predictive animal models of human disease, and in some cases target identification and validation.

An increasing collaborative landscape is emerging. Currently, there are several pre-competitive consortia that have been formed in specialized healthcare-related sectors. These include the Structural Genomics Consortium focusing on medically relevant proteins, Cancer Therapy Evaluation Program focusing on cancer therapies, and Innovative Medicines Initiative mandated to strengthen competitiveness of pharma sector in Europe [52]. In the field of NTDs, there exists no centralized open-access consortium. Much of the current discovery efforts are driven by the public–private partnerships (PPPs), each focusing on either one or two tropical infections (see Section 11.4.2). And there lies the disconnect. Due to the disjointed nature of these ventures, they are unable to benefit from economies of scale in terms of resources, expertise, and reduction in duplication of efforts. Although major tropical infections such as malaria, tuberculosis, and trypanoso-miasis are relatively well addressed, many others such as lymphatic filariasis, onchocerciasis, and schistosomiasis are underserved [53]. Therefore, centralized coordination of the various PPPs through a joint precompetitive consortium would have greater impact on drug discovery efficiency than the currently fragmented small-scale ventures [43]. Nevertheless, these initiatives have inspired formation of virtual discovery networks, improved collaborations between big pharma and academia, establishment of centers of excellence (CoE) and incubators, and enhanced screening data and compound file sharing.

11.4.2 Virtual Discovery Organizations

The drug discovery process is a long, torturous, expensive, and high-risk endeavor. The likelihood of success for any particular project is low, with industry metrics approximating that only one in 15 projects delivers a product to the clinic [54]. Due to the perceived absence of financial returns from the NTDs coupled with the demands on pharmaceutical companies to meet their business objectives, the big pharma started scaling down their tropical disease R&D portfolio in the early 1970s such that by the mid-1990s, this disengagement was virtually complete [55]. Since 1975, any mean-ingful research in the field of NTD has been provided by the WHO-sponsored Special Programme for Research and Training in Tropical Diseases (WHO/TDR) created through World Health Assembly resolution [56]. It was mandated to develop new tools and methodologies for control of ten target diseases and to strengthen research capacity in disease-endemic countries [57].

The WHO/TDR is credited with enabling the pharmaceutical industry to retain some level of foothold in tropical disease R&D. Some of its clinically marketed drugs successes are shown in Figure 11.1: eflornithine with Marion Merrell Dow for HAT, chlorproguanil/dapsone combination (Lapdap®) with GSK for malaria, ivermectin with Merck for oncho-cerciasis, praziquantel with Bayer for schistosomiasis, injectable artemether with Rhône-Poulenc Rorer and injectable β-arteether with Artecef for malaria, and miltefosine with Zentaris and liposomal amphotericin B with Nexstar for visceral leishmaniasis [57]. The WHO/TDR program is cosponsored by the World Bank, the United Nations Development Bank (UNDP), and the United Nations Children Fund (UNICEF) and funded through contributions from international, governmental, and philanthropic organizations.

Inspired by the successes of the WHO/TDR and using its organizational and funding model, numerous other PPPs focusing on NTDs have been established over the last slightly more than one decade. They include the MVI, the Medicines for Malaria Venture (MMV), the Institute for OneWorld Health (IOWH), the Drugs for Neglected Diseases

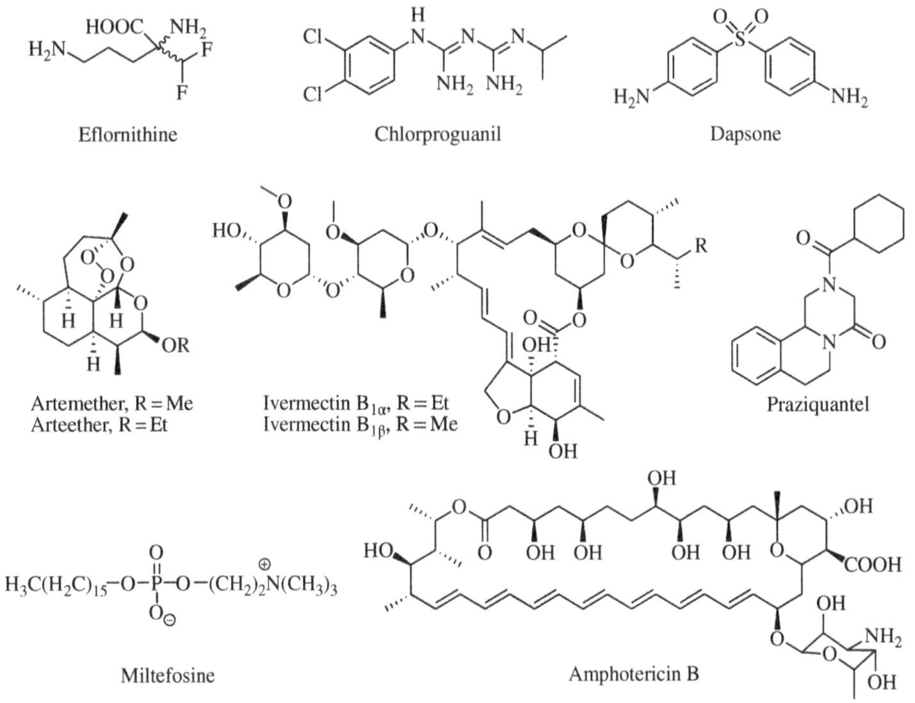

FIGURE 11.1 Selected drugs developed for tropical diseases through the aegis of the WHO/TDR program.

Initiative (DNDi), and the Global Alliance for Tuberculosis Drug Development (GATB). The PPPs are credited with reinvigorating drug discovery efforts in tropical infections evidenced by recent increase in the number of drug candidates for NTDs reaching clinical development [58, 59]. The success registered by the PPPs indicates that a consolidated public–private and philanthropic approach can stimulate R&D in neglected diseases, thus compensating for market failure by reducing the costs and risks involved for both public and private sector partners [60].

The MMV was the first organization to put the theory of PPP-driven virtual drug discovery into practice in a concerted manner and has several ongoing virtual discovery projects at various stages of development. However, our observations indicate there is a need for centrally coordinated virtual drug discovery networks among the various PPPs. Such coordination could markedly scale up drug discovery and translation research for NTDs. First, such networks would provide a framework of engagement through which partnering pharmaceutical companies, academic institutes, and philanthropic organizations can contribute their knowledge, expertise, compound libraries, and other infrastructural support, thus reducing overhead costs of running separate programs. Second, the networks can be mandated to develop joint portfolio of projects indicating priority areas. In this way, no disease area will be underserved as is the situation now. For several years, the WHO/TDR consortium had been the mainstay of systematic screening programs for most of neglected diseases [61]. Pharmaceutical companies provided access to a subset of their extensive libraries of proprietary chemical compounds that were screened across a broad range of tropical diseases. The TDR had partnerships with Pfizer, Merck Serono, and Chemtura that exemplify this approach [43].

11.4.3 Collaborations with Academic Laboratories

Pure basic research is a critical prelude to innovative drug discovery. Several studies in the field of drug discovery over the past 10 years reveal that the big pharma has not generally been a fertile ground for innovation. Rather, pharmaceutical companies have been largely dependent upon academia and biotechnology companies to fuel their drug pipelines [16, 62]. This is no coincidence. Academia has traditionally been the powerhouse of basic research and has been identifying targets, molecules, and disease models that feed into drug discovery and development for decades [63]. Subsequent technical development is left to the pharmaceutical industry [59]. However, the extent to which pharma has looked to academia for discovering ideas has varied and usually is ad hoc or opportunistic whenever circumstances required. Even when this happens, the extent to which the idea is pursued is a sole prerogative of the pharmaceutical company whose priority may be dictated by business considerations at the expense of good science. That is no accident but rather perpetuation of a survival culture: the purpose of academic, biological, and medical research is to pursue innovation and further knowledge about disease, in which the ultimate reward is journal publication, while the overriding motivation for the industry is to convert innovative ideas into novel therapeutics that address unmet medical needs whose ultimate reward is increased shareholder value [54]. This creates a huge translational gap between a laboratory discovery and a clinical therapy.

Further, current estimates indicate that the industry's failure rate of drugs against novel targets is at least 50% greater than against precedented and validated targets [64]. Thus, the big pharma is strategically retreating from fundamental exploration of novel therapeutic targets and early-stage drug discovery and instead focusing on clinically validated targets and later stages of the drug discovery process. Therefore, there is a compelling need for the industry to enhance its collaboration with the academia to invigorate new discoveries [63, 65]. In recognition of the need for closer bilateral partnerships between the industry and the academia, there has been an increase in formal agreements for collaboration between the pharma and academic institutions. As shown in Table 11.2, most of these academic–pharma bilateral collaborations are in the developed world.

Collaborative arrangements between the pharma and the academia have potential to boost drug discovery in three major ways. First, by tapping into the vast intellectual resources of academia, a great initial drug discovery engine can be unleashed, at great savings to the cost of drug discovery [65]. In addition to basic science academic

TABLE 11.2   Selected Examples of Big Pharma–Academia Bilateral Collaborations as of 2008[a]

Pharmaceutical partner	Academic institution	Focus area
GlaxoSmithKline	Immune Disease Initiative	Immunoinflammatory diseases
AstraZeneca	Columbia University Medical Center	Diabetes and obesity
Pfizer	University of California in San Francisco	Diverse disease areas

[a] Adapted from Ref. 63.

TABLE 11.3 Selected Examples of Academic Drug Discovery Units in Tropical Infections as of 2008[a]

Academic unit	Host academic institution	Focus area
Wellcome Trust	University of Dundee	Early drug discovery for HAT
Sandler Foundation	University of California, San Francisco	Early drug discovery for malaria, Chagas disease, and schistosomiasis
Broad Institute	Harvard University	Malaria drug discovery

[a] From Ref. 53.

investigators who, by serendipity and careful observation, develop new ideas for therapy, there are preclinical and clinical academic investigators whose research focus is to find and develop new therapies. Collaborative arrangements with these academicians could provide relatively inexpensive drug candidates for the pharmaceutical industry.

With increasing awareness of the commercial value of successfully translated drug discovery, several universities, especially in the developed world, have actively initiated and/or supported establishment of academic drug discovery units as a strategic source of income. These units are operated on a scaled-down industry management style and comprise most of the functions required for small-molecule drug discovery such as synthetic chemistry, HTS, and ADMET platforms [54]. Since the academic discovery units are established in such a manner as to enhance linkages with the pharma through collaborations, licensing, and spin-out activities, they provide an invaluable bridge for translation of university-based research innovation to the market. A few of the active representative academic drug discovery units in the field of neglected diseases are shown in Table 11.3. Thirdly, mixed university–industry teams present new opportunities for drug discovery and development. For example, they are able to address disease groups that are smaller and less commercially attractive such as NTDs, orphan diseases, and antibiotic research, as well as provide new innovative environments for the many highly trained pharma scientists made redundant by management restructuring in the big pharma [54].

11.4.4 CoE and Incubators

It has been rightly noted that a country's capacity to respond to the threat of disease is closely linked to its research capacity [66]. This creates a compelling need for the active involvement of the countries most affected by the NTDs, the disease-endemic countries, in the search for solutions, including research into new and better treatment options [41]. In any case, for sustainability and greater responsiveness to public health needs, the leadership to implement local agenda for R&D must come from within the disease-endemic countries [67]. Transfer of technology, tools, knowledge, and trained personnel are clearly important building blocks for advancing research capacity and establishment of vibrant local scientific communities [68]. The participation of institutions from developing countries in a precompetitive consortium ensures that the expertise is imported and built into the region where knowledge and drug discovery tools are needed most, adding a novel capacity-building dimension [47].

CoE and incubators domiciled within the disease-endemic countries are hallmarks of successful technological transfer. Apart from its long-term effect of fostering research culture, this strategy has an attendant effect of securing political will from governments concerned on the need for indigenization of tailored drug discovery efforts. One excellent example of a CoE is the University of Cape Town Drug Discovery and Development Centre, H3-D (http://www.h3-d.uct.ac.za). The H3-D is Africa's first integrated modern drug discovery center and is already providing a portal of collaborations to major global companies and not-for-profit organizations [69].

11.4.5 Screening Data and Compound File Sharing

Various research groups working in the field of NTD are releasing large datasets ranging from *in vitro* and *in vivo* efficacy data to ADMET and pharmacokinetic profiles of positive hits against tropical infectious microbes through publication, website posting, and other means [70, 71]. There is a great risk that most of this information will not be adequately prioritized and utilized [72]. However, since the various initiatives have a shared goal to discover new leads or drug candidates for tropical diseases, a coordination strategy that enhances networking and exchange of information between the research groups is not difficult to initiate. Central coordination would help to track, monitor, and prioritize research needs in any given NTD and also avoid unnecessary duplication of efforts.

One possible way of enhancing information flow is to set up a website on which investigators are encouraged to record what HTS campaigns they have conducted or which are ongoing [47]. The screening centers should also be encouraged and facilitated to communicate with each other and share lessons, standard operating procedures, and reagents as well as data. Where need be, multiple centers within the network can be engaged to perform *in vitro* or *in vivo* screens for similar pathogens in such a manner as to obtain internal cross validation of data within the compound screening network [53].

11.5 EXPLOITATION OF GENOMICS

11.5.1 Introduction

One of the often-quoted reasons for stagnation in R&D productivity is a diminution in the number of new druggable targets. It has been observed that although the big pharma still commits huge budgets to R&D, much of the funding goes toward later-stage development, with very minimal allocation to the riskier early stages of target identification and validation [54]. Hence, most discovery efforts have focused on a small subset of druggable targets that may have reached saturation levels especially for profitable therapeutic areas due to availability of several follow-on "me-too" drugs and generic versions.

Luckily for the industry, the Human Genome Project and the mapping of related genomes have made it possible to access a wide array of potential drug targets that could herald the beginning of a new golden era in drug discovery. It has been estimated that there may be as many as 3,000–10,000 new targets within the genome [65, 73]. This is an enormous treasure trove considering that, historically, all commercial drugs have addressed a total of approximately 500 molecular targets [74]. The next big challenge for the industry in general, and for tropical diseases research in particular, is determining the best way to translate the insights obtained from genomics into new, robust chemical leads that can form the basis of innovative drug discovery [41].

11.5.2 Target Identification and Validation

Pharmaceutical innovation occurs when biology has matured to a point where a viable druggable target emerges, setting the stage for product invention [75]. However, it is crucially important that validity of any putative target is fully demonstrated through data-driven proof-of-principle studies before committing substantial investments on a discovery project. Analysis of phase II and phase III attrition rates identified lack of efficacy as the most common reason for inefficiency in drug discovery during the 2011–2012 two-year period accounting for up to 56% of drug candidates' failure in clinical trials [76]. This failure rate was much higher than previous years [77]. In part, this implies that the validity of the targets is usually not sufficiently interrogated or the validation methods are deficient. One author has attributed the increased failure rate on the high-risk, poorly validated genomic targets that have been dominating the discovery pipelines over the last one decade, which are far less well characterized than previous targets of the 1990s [17].

Considering that many pharmaceutical companies tend to pursue same targets simultaneously, the industry suffers collective astronomical losses by way of wasted resources and lost opportunities whenever a putative target fails [78]. This shows that there is merit in the argument that precompetitive space be expanded to include target identification and validation [79]. Nonetheless, we reckon that efficacy-related failures could be markedly decreased by validation of putative targets through well-coordinated interfacing of genomics, proteomics, molecular genetics, biochemistry, physiology, and computer-based analysis [65]. Scrupulous identification and validation of drug targets could potentially facilitate more efficient drug discovery through (1) target-based drug discovery, (2) reverse determination of mode of action from whole-cell screening, and (3) individualized therapies for special patient populations.

Surprisingly, a very small number of truly innovative drug targets, typically two to three, are commercialized each year by the entire pharmaceutical industry [74]. Although this small number is sobering, the bright side is that these innovator targets provide tremendous potential for the industry and their effects are long term. Breakthrough targets deliver completely new mechanisms for treating disease and can therefore rapidly create large new medical markets. Prototypical examples of breakthrough targets include cyclo-oxygenase 2 (COX-2) that enabled development of selective COX-2 inhibitors such as celecoxib, phosphodiesterase type 5 (PDE-5) that is spurring development of PDE-5 inhibitors such as sildenafil, and Bcr-Abl fusion kinase that has paved way for development of tyrosine kinase receptor inhibitor antineoplastics such as imatinib [78].

11.5.3 Target-Based Drug Discovery

The two complementary drug discovery strategies, namely, target-based (protein screening) and phenotypic-based (whole-cell screening) approaches, have been reviewed extensively [80–82]. More recently and within the context of antiparasitic agents, these approaches have been compared and contrasted [83]. Although most authors contend that empirical phenotypic screening approaches have been far more successful than hypothesis-driven target-based approaches especially in the field of NTDs, drug discovery efforts are presently largely target driven [82, 84]. The target-based drug discovery strategy, also referred to as "one molecule–one target–one disease" paradigm, has its basis in the "lock and key" receptor concept first espoused by Ehrlich approximately one century ago [85]. The key principle of this approach is that there exists an identifiable molecular target that underlies pathogenesis of a particular disease. The molecular target is usually either a gene or a gene product such as a protein or an enzyme in an essential biochemical pathway unique to the pathophysiology of the offending disease. Compounds are then designed or screened for specific interaction with the identified therapeutic target. As such, a target-directed approach facilitates development of distinguishable rational therapies with well-founded molecular basis.

Based on the fact that most anti-infective agents exert therapeutic effects by binding to and regulating biological activity of a particular mechanistic target in the pathogenic microbe (see Table 11.4), target-driven drug discovery has been emphasized in recent years as a way of harvesting the significant investment made in parasite genomic programs [47]. Typically, target-based drug discovery may be achieved using one or a combination of three approaches: (1) de novo drug design, (2) *in silico* structure-based virtual screening, and (3) *in vitro* HTS of compound libraries. The de novo drug design is highly resource demanding and thus only feasible in therapeutic areas with viable commercial markets. Due to their relative ease of set up and low cost of goods, the *in silico* structure-based virtual screening and *in vitro* HTS render themselves suitable for target-directed discovery of drugs for NTDs. Compared with phenotypic whole-cell screening, mechanistic targets such as purified enzymes and receptors are amenable to rapid screening using the libraries of many hundreds of thousands of compounds that are available either commercially or in-kind donations from academic institutes, start-up ventures, and pharmaceutical companies. Valuable lead compounds can be missed in whole-organism assays when they are rapidly metabolized or are unable to cross cell membranes to reach their mechanistic target [43].

Besides delivering a wide array of new targets, genomic information has built a case for pursuing known but underexploited targets through "orthologue searching." Using this strategy, an existing focused library consisting of both active and inactive compounds against a host target is screened against a parasite target [86]. This approach could potentially deliver lead compounds that are selective for the parasite target if structural differences between the host and parasite targets can be exploited [61]. One appealing target is the kinome. Human protein kinases have proved a druggable target class especially in cases where the role of a particular protein kinase has a direct causal relevance to the diseased state through its inappropriate activation [87]. Interestingly, for the NTDs, protein kinases have been recognized as potential targets for antiparasitics [88]. By use of genome data, new kinase parasitic targets can easily be identified using already existing kinase inhibitor compound libraries. The cell cycle kinases offer a good starting point because of the similarities in the cell cycle of a rapidly replicating parasite and that of a rapidly replicating tumor cell [46]. Arguably, protein kinases have provided the pharmaceutical industry with an opportunity to play a pivotal role in NTD drug discovery by provision of kinase inhibitor libraries for screening against tropical parasites.

Another molecular target amenable to orthologue-based screening in NTD lead discovery is dihydroorotate dehydrogenase (DHODH), a flavin-dependent enzyme that catalyzes oxidation of dihydroorotate to orotate, a crucial fourth step in the de novo pyrimidine biosynthesis. Pyrimidines are essential building blocks for DNA and RNA biosynthesis in both eukaryotes and prokaryotes. Previous studies have shown that species-selective inhibition of DHODH is feasible [89]. Already, the *Plasmodium falciparum* DHODH is a proven drug target in the treatment of malaria, and the search is on

TABLE 11.4 Mechanistic Targets of Some Drugs for Tropical Infections as of 2005[a]

Target	Biochemical pathway	Parasite	Drug
Dihydrofolate reductase	Folate biosynthesis	*Plasmodium falciparum*	Pyrimethamine; cycloguanil
Dihydropteroate synthase	Folate biosynthesis	*P. falciparum*	Sulfonamides/sulfones, for example, sulfamethoxazole and dapsone
Cytochrome B	Electron transport	*P. falciparum*	Atovaquone
Ornithine decarboxylase	Polyamine biosynthesis	*Trypanosoma brucei gambiense*	Eflornithine
Sterol 14α-demethylase	Sterol biosynthesis	*T. cruzi*	Antifungal triazoles, for example, posaconazole
Farnesyl pyrophosphate synthase	Polyisoprene biosynthesis	*P. falciparum*; kinetoplastids	Bisphosphonates, for example, risedronate
Nicotinic acetylcholine receptors	Neurotransmission	Nematodes	Levamisole
Glutamate-gated chloride channels	Neurotransmission	Nematodes	Ivermectin
Tubulin	Cytoskeleton	Nematodes	Albendazole

[a]Adapted from Ref. 41.

for clinical candidates [90]. Independent studies have suggested that *Trypanosoma cruzi* and *Trypanosoma brucei* are susceptible to DHODH inhibition [91]. Yet another study has revealed high sequence and structural similarity between trypanosomal DHODH and leishmanial DHODH, suggesting that a common strategy of structure-based inhibitor design can be used to validate DHODH as a druggable target against trypanosomatid NTDs [92]. Such lead discovery efforts would greatly benefit from the already existing antiproliferative and parasitic DHODH inhibitor libraries.

Prenylated proteins play important signal transduction roles in all eukaryotes. Accordingly, protein prenyltransferases have been proved attractive targets for development of a number of therapeutics [93]. One of the most well-characterized prenyltransferase is the protein farnesyltransferase (FTase)—an emerging target for cancer chemotherapy [94]. Studies have shown that pathogenic microorganisms such as trypanosomes, leishmania, and plasmodia rely on protein farnesylation more heavily than their mammalian hosts. Indeed, several research groups have demonstrated susceptibility of trypanosomatids and plasmodia to available FTase inhibitor libraries, indicating that FTase is a potential target for NTD drug discovery [95]. Selectivity for parasitic FTase could be obtained by exploiting differences in amino acid sequence between parasite FTase and mammalian FTase, thus enabling synthesis of species-specific inhibitors.

We are yet to witness new drugs for NTDs that have resulted from genomics-driven target-based HTS campaigns. However, considering that NTDs have only recently started to benefit from HTS technologies implemented in the pharma industry for commercially attractive indications two decades ago, the strategy is still valid but needs to be augmented by use of validated molecular targets and high-quality compound libraries in the screening campaigns [53]. In any case, there are a number of anti-infectives that entered clinical market through target-based drug discovery approaches. Notable examples include human immunodeficiency virus (HIV) protease inhibitors such as zidovudine and stavudine, alpha-difluoromethylornithine that inhibits ornithine decarboxylase and used for HAT, and the neuraminidase inhibitors zanamivir and oseltamivir for influenza viruses. Target-based lead discovery for the NTD could greatly be augmented by a shared open-access repository of potential drug targets for tropical disease pathogens. In recognition of this, the TDR developed a drug target portfolio network—the TDR Targets Database [96]. The database is aimed at providing a central facility for information on drug targets. Various parasitic targets are ranked according to their druggability, essentiality to the organism, and selectivity index over related human macromolecules [96]. The synergistic, overlapping, and coordinated activities of partnering research teams would provide an opportunity for building a chemoinformatics and *in silico* drug discovery platform for tropical diseases [97].

11.5.4 Phenotypic Whole-Cell Screening

The advantages and disadvantages of target-based and phenotypic-based drug discovery strategies are briefly presented in Table 11.5. In relative terms and as already alluded to in Section 11.5.3, empirical phenotypic cell-based screening has been far more successful than target-based approach in delivering selective antiparasitic drug leads [98]. This observation may also hold true for other therapeutic areas. Analysis of NME approvals by the FDA between 1999 and 2008 revealed that the contribution of phenotypic screening to the discovery of first-in-class small-molecule medicines exceeded that of target-centric approaches, although target-based molecular strategies were more successful for

TABLE 11.5   Key Advantages and Disadvantages of Phenotypic and Target-based Screening[a]

Advantages	Disadvantages
Target-based screening	
• Hypothesis driven; capitalizes on genome advances	• Prebiased approach; focused on a relatively small "biological space"
• Target specific; mechanism of action understood	• "Artificial" *in vitro* environment
• Amenable for HTS	• Translation to *in vivo* disease state could be challenging
• Allow for structural biology and computational strategies in lead optimization and SAR development	• *In vivo* redundancy may null activity at a specific target
Phenotypic whole-cell screening	
• Unbiased approach; focused on a relatively large "biological space"	• Lead optimization could be hampered by poorly understood mechanism of action
• Allows for direct interrogation of complex biological systems	• Assays are mostly low throughput
• Activity may be better translatable to human disease state	• Assay variability may affect statistical robustness
• Hits likely to possess "drug-like" features	• Multiple assays may be required to fully understand spectrum of activity
• Potential toxicity issues associated with hit scaffolds could emerge earlier	

[a] From Ref. 82.

followers [99]. The apparent success of phenotypic screening has been attributed to its large "biological space" that provides a huge and unbiased target opportunity allowing for preclinical serendipity in drug discovery [82, 100]. In recent years, several research groups have used phenotypic whole-cell screening approaches to deliver preclinical antiparasitic compounds [101–103].

It is worth noting that majority of approved anti-infectives to date were discovered via nontarget-based approaches [83]. So far, the target-based HTS approach has yielded very few success stories. In part, this reflects the high rate of attrition in the process of progression from early-stage biochemical hits to robust lead compounds. Many compounds active in protein-based assays are inactive in whole cells. This can be due to failure to enter intact cells but can also occur because the chosen molecular target is not essential to the microbes. This indicates that for greater productivity in NTD lead discovery, the target-centric approach should be used complementarily with and not as a substitute for phenotypic cell-based screening. For example, the more "holistic" phenotypic screening may be used to identify initial hits that are potentially more useful starting points for medicinal chemistry lead generation and the target-based screening used to retrospectively identify molecular targets of phenotypic hits during lead optimization.

11.5.5   Individualized Therapy and Therapies for Special Patient Populations

In addition to uncovering a range of novel targets for drug discovery, the knowledge gained from sequencing of the genome can be exploited in two other therapeutically viable ways. First, correlation of gene haplotypes with individual drug responses could widen the scope of individualized drug treatment, which is very desirable from both medical and commercial perspectives [16]. Second is the identification of alleles that predispose people to complex genetic diseases such as diabetes, schizophrenia, arteriosclerosis, osteoporosis, hypertension, hypercholesterolemia, and cancer. This in turn will offer greater understanding of the pathophysiological mechanisms and hence possible therapeutic interventions for special patient populations affected by these diseases.

The practice of individualized therapy, also variably referred to as precision medicine, personalized medicine, targeted medicine, stratified medicine, and pharmacogenomics, enables the targeting of treatments specifically to patient subpopulations that are more likely to respond to a particular treatment [104]. It requires stratification of individuals into subpopulations that differ in their susceptibility to a particular disease or their response to a specific treatment. Critically, it also requires the development of companion diagnostics so as to focus drug therapy on appropriate patients and spare unresponsive individuals from fruitless treatment and debilitating drug toxicities. Equally crucial is the need to carefully analyze and select the patient population for clinical trials in order to demonstrate clear benefits for the target patient subpopulation. Analysis of data for the entire patient population will be unable to statistically demonstrate clinical benefits. Personalized medicine offers new approaches to drug discovery, replacing the one-drug-fits-all blockbuster model of the twentieth century. Chemical structures of representative small-molecule drugs in current clinical market are shown in Figure 11.2.

The development and marketing of Genentech's trastuzumab (Herceptin®) is a classic model of a true pharmacogenomic–diagnostic-led personalized medicine drug discovery. Trastuzumab, the first genetically guided therapy, is a HER2 dimerization inhibitor antineoplastic humanized monoclonal antibody. The human epidermal growth factor receptor 2 (HER2), also known as neu, ErbB2, CD340, or p185, is a proto-oncogenic transmembrane

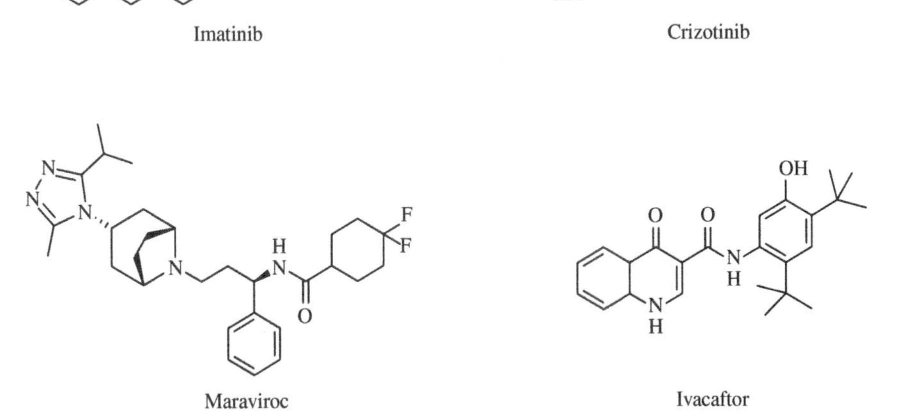

FIGURE 11.2 Chemical structures of some clinically used small-molecule personalized drugs.

receptor tyrosine kinase whose overexpression has been shown to correlate with pathogenesis and aggressive progression of metastatic breast cancer [105]. Dimerization of this protein is essential for initiation of its signaling cascade, and only approximately 20–25% of breast cancer cases exhibit aberrant HER2 signaling [104]. Thus, marketing authorization of the HER2 dimerization inhibitors trastuzumab and pertuzumab (Perjeta®) required concurrent development of companion diagnostic tests to identify a subset of HER2-positive breast cancer patients most likely to benefit from the drugs. Similarly, imatinib (Gleevec®) that is most effective for Philadelphia chromosome-positive neoplasms such as chronic myelogenous leukemia and crizotinib (Xalkori®) that works primarily in anaplastic lymphoma kinase gene-related cancers are effective in patient subpopulations with these biomarkers.

The discovery of maraviroc (Celsentri®), a selective C–C chemokine receptor 5 (CCR5) antagonist with potent activity against human immunodeficiency virus type 1 (HIV-1), is yet another proof of the value of expert exploitation of genomics in enhancing drug discovery for individualized medicines. Genomic analysis revealed that a rare mutation in the CCR5 gene led to a defective CCR5 coreceptor, thus conferring resistance against HIV infection in a patient subpopulation with this rare genetic polymorphism. It was hypothesized that pharmacological modulation of the CCR5 could reproduce this genetic polymorphism [75]. Maraviroc binds to the CCR5 coreceptor and blocks entry of HIV into the cell. However, since HIV can also use other coreceptors such as CXCR4, an HIV tropism test such as a Trofile assay must be performed to determine potential efficacy of the drug. Likewise, skilful application of genomic science led to the discovery and eventual clinical use of ivacaftor (Kalydeco®), a small-molecule potentiator of the cystic fibrosis transmembrane conductance regulator (CFTR) anion channel. Although there are hundreds of known CFTR gene mutations that can lead to cystic fibrosis (CF), ivacaftor is only effective in a subset of approximately 4% of CF cases where the disease is due to G551D mutation [106].

Because of its high genetic component, most of the currently available molecular personalized medicines are cancer chemotherapeutics. However, the discovery of HIV fusion inhibitor maraviroc shows that personalized medicines could become a reality in the design of anti-infective agents. Indeed, application of similar theranostics approaches in neglected disease research has the potential to improve NTD drug discovery efficiency, reduce healthcare costs, and provide improved treatment benefits to the patients [107]. While the near-term costs of introducing new personalized medicine technologies may be high, the long-term outlook is more effective, less costly, and less toxic medical care.

11.6 OUTSOURCING STRATEGIES

11.6.1 Introduction

In business terms, outsourcing may be defined as "a strategic decision that entails external contracting of determined non-strategic activities or business processes necessary for the manufacture of goods, or the provision of services, by means of agreements or contracts with higher capability firms to undertake those activities or business processes, with the aim of improving competitive advantage" [108]. The two main reasons for outsourcing are cost reduction and performance enhancement. The drive for greater efficiencies and cost reductions has compelled many organizations to specialize in a limited number of key areas referred to as core competencies while outsourcing other activities and services not deemed core to their operations.

The pharma industry has always worked with third parties to access specific technology, expertise, and other requirements. For example, pharmaceutical companies provide sponsorships for early-stage drug discovery and high-throughput combinatorial chemistry projects as well as engaging in in-licensing deals with biotechnology firms for development candidates and acquisition of companies that have resources they need. Due to continued restructuring in the pharmaceutical sector, most companies have scaled down on core activities, thus expanding the range of outsourcing requirements. Currently, these may include identification and validation of novel targets, determination of signaling pathways, sourcing for human cells and tissues for target validation studies, screening of suitable volunteers for safety and efficacy studies, sourcing for relevant platform technologies, novel chemistries, disease expertise, and pharmacology biomarker development and biotechnology processing [54].

Big pharma has increasingly turned to in-licensing to fill its pipeline and market portfolio [17]. However, most of this licensing activity tends to be late stage and expensive; early-stage in-licensing has been traditionally underappreciated and underleveraged by big pharma. Faced with high growth aspirations, pharmaceutical companies are competing to capture a disproportionate share of the industry's new products to meet their market expectations. To do this, big pharma will need to move aggressively for early-stage candidates. In addition to being an important source of growth and innovation, in-licensing is also productivity enhancing. Studies have shown that in-licensed candidates have not only had more success in the clinic but that early-stage candidates have also been less expensive than internally derived new development candidates [17]. It also costs much less to in-license an advanced therapeutic candidate for a neglected disease than would be required for other indications [109].

11.6.2 Research Contracting in Drug Discovery

According to the independent Tufts Center for the Study of Drug Development, clinical trials conducted by contract research organizations (CROs) are completed an average of 30% faster than the in-house conducted trials. It is estimated that this time-saving translates to approximately $120–150 million in increased revenue [110]. Further, small- and medium-sized biotechnology and pharmaceutical enterprises with closer links between research and development have been more successful than their big sisters at moving candidates through the development pipeline. In particular, they have been better at producing biological products, such as monoclonal antibodies, vaccines, and peptides, to tackle unmet needs in the market [54].

11.7 MULTITARGET DRUG DESIGN AND DISCOVERY

11.7.1 Introduction

As already alluded to, drug discovery is currently skewed in favor of the target-based approach due to its ability to allow for high-throughput technologies and development of rational medicines. However, despite the best efforts in rationally designed drugs, targeted therapies are satisfactory in very few diseases, while many others remain inadequately treated. This is due to a number of reasons. First, the inherent redundancy and robustness that exist in biological systems imply that functions of a disrupted molecular target can easily be compensated by overlapping pathways [111]. Second, selective pressure exerted by monotherapies is permissive to emergence of drug resistance. Third, high doses used to compensate for low efficacy may predispose to toxicity. Therefore, monotargeted therapies might not sufficiently achieve desirable efficacy especially for complex disorders [112].

Systems biology has shown that although many biological networks are typically characterized by robustness and redundancy, they are vulnerable to multiple hits, implying that modulation of a multiplicity of targets can be an asset in treating complex diseases [113, 114]. Further, modeling of network behavior has shown that the partial inhibition of several targets can be more efficient than complete inhibition of a single target [115]. This is supported by the fact that monotarget medicines are rarely used alone in clinical settings. Rather, they are used in combination regimens to improve therapeutic outcomes [116]. This creates a compelling justification for systems-oriented multitarget drug design and development.

11.7.2   Rationale for Multitargeted Drugs

Clinical experience has demonstrated that no matter how effective when first introduced, the therapeutic life span of most drugs is inevitably curtailed by the emergence of drug resistance [117]. This is especially so when used as monotherapeutic agents. Thus, combination of two or more complementary drugs with different modes of action in the treatment regimens is presently advocated for many disease conditions [118]. Combination therapy offers improved therapeutic outcomes in a number of ways. First, the drugs in combination may produce additive, or more likely, synergistic action by targeting two or more mechanisms involved in the pathogenesis of a particular disease [61]. This leads to improved treatment efficacy. Second, the synergy obtained allows lower doses to be used and hence reduction in propensity for dose-dependent drug toxicity with a resultant wider therapeutic window [119]. Third, combination therapy reduces the probability of selecting for resistant clones. This is because the drugs in a combination regimen may protect each other against spread of resistance by killing selected clone populations in the patient that are unresponsive to either one of the drugs [119].

The practice of combination therapy entails two major approaches: drug cocktails and fixed-dose combinations (FDCs) [120]. Most frequently, a cocktail of drugs is administered in the form of two or more individual pills. For example, a combination of pyrimethamine/sulfadoxine and amodiaquine in malaria treatment and nifurtimox–eflornithine combination therapy (NECT) in the second stage of *T.b. gambiense* sleeping sickness have been shown to improve therapeutic outcomes [121, 122]. Similarly, various combination regimens have provided better remissions than monotherapy in the treatment of leishmaniasis [123]. However, the benefits of cocktail polypharmacy are often compromised by poor patient compliance due to heavy pill burden [120]. Recently, coformulation of two or more individual drugs in a single pill as FDCs has gained currency as a means of simplifying dose regimes and hence improvement in patient compliance. For example, artemisinin-based FDCs such as Coartem® (artemether/lumefantrine), Artekin® (dihydroartemisinin/piperaquine), and Artequin® (artesunate/mefloquine) are currently the mainstay of antimalarial therapy [86].

In view of the obvious emphasis on FDCs, drug discoverers are increasingly considering the concept of multitarget molecules [124, 125]. In this approach, two or more pharmacophoric units of different known bioactive molecules are covalently linked into a single chemical entity so as to exert dual or multiple drug action [126]. Briefly, multitargeted compounds have the potential to (1) improve treatment efficacy, (2) reduce emergence and spread of drug resistance, (3) allow for improvement in pharmacokinetic profile of bioactive motifs, (4) decrease dose-dependent drug toxicity and drug–drug interactions, (5) allow for optimization of receptor affinity, and (6) hasten drug discovery efforts at a lower cost [114, 124]. If the number of publications and reviews in this field is anything to go by, current drug discovery efforts seem to endorse multitargeted drug molecules as the next-generation therapeutic agents. This section focuses mainly on the application of multitarget drug design and discovery in the field of NTDs. For additional details on this subject, readers are referred to excellent reviews available in a special issue of the *Current Medicinal Chemistry* journal titled "Complexity against Complexity" [127, 128].

11.7.3 Designed Multitarget Compounds for Neglected Diseases

11.7.3.1 Antimalarial Hybrids Among the tropical infectious diseases, malaria has registered the highest number of designed multitarget compounds. In their design, two major scaffolds have been utilized, namely, the 1,2,4-trioxane and the quinoline motifs, which are the biologically active pharmacophores in artemisinins and quinoline antimalarials, respectively. So far, trioxaquines and trioxolaquines are the most extensively pursued and most successful artemisinin-based antimalarial hybrids [129]. These hybrids are designed in such a manner as to contain a synthetic peroxide, either a trioxane or a trioxolane motif, linked to an aminoquinoline scaffold, via a noncleavable diaminoalkyl chain or a cyclohexyl ring. The clinical candidate PA1103 (Fig. 11.3) best exemplifies the design strategy and potential clinical success of trioxaquines as next-generation artemisinin-based antimalarials. This molecule has high *in vitro* activity on both chloroquine-sensitive and chloroquine-resistant strains and multidrug-resistant fresh isolates of *P. falciparum*. Based on its good *in vivo* efficacy in humanized malaria mouse models and favorable pharmacokinetic profile, it was selected for clinical development and is currently in late-stage preclinical evaluation [130].

Another class of antimalarial hybrids pursued to a great extent is the quinoline-chemosensitizer conjugates, also referred to as reversed chloroquines (RCQs). Therapeutic efficacy of quinoline antimalarials is attributed to their inhibition of the heme detoxification process in the plasmodial digestive vacuole—a target site where the drugs accumulate by pH trapping [131]. Studies have shown that plasmodial resistance to quinolines does not involve any change in the sensitivity of plasmodia to heme accumulation, but seems to be associated with enhanced efflux of the drugs from the digestive vacuole, hence lowering drug concentration at the site of action [132]. Thus, the target remains vulnerable, and the parasite is susceptible, so long as the intravacuolar drug accumulation can be restored through inhibition of the efflux pathways [128]. Several structurally diverse molecules termed quinoline-resistance reversal agents (or quinoline chemosensitizers) capable of inhibiting quinoline efflux pathways have been identified and include the antihypertensive Ca^{2+} channel blocker verapamil and the antidepressant imipramine [133]. Consequently, several hybrid molecules incorporating a quinoline motif and a reversal agent have been

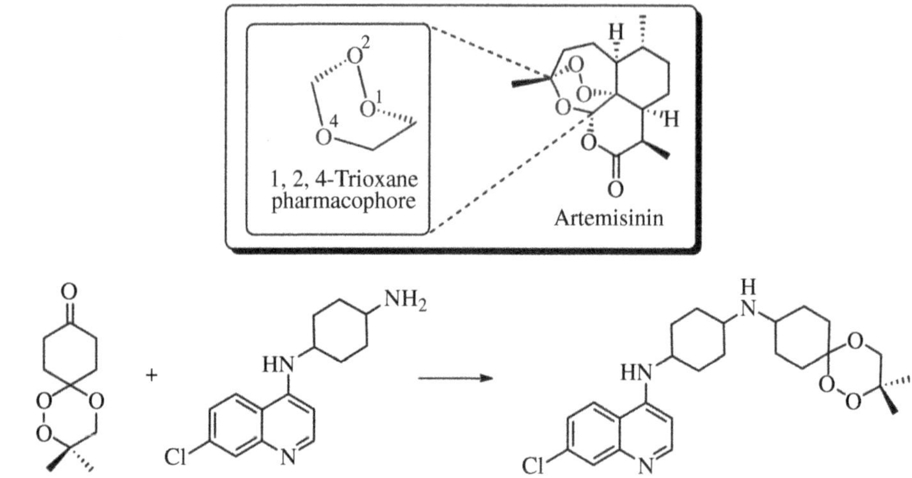

1, 2, 4-Trioxane 4-Aminoquinoline moiety PA1103

FIGURE 11.3 The design of trioxaquine preclinical candidate PA1103. *Inset*: The structure of artemisinin showing the 1,2,4-trioxane pharmacophore.

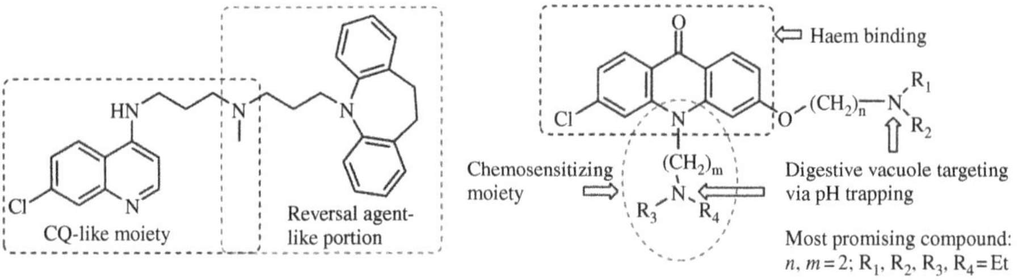

Chloroquine–imipramine hybrid Dual-function acridones

FIGURE 11.4 Structures and design of chloroquine–imipramine hybrid and dual-function acridones.

designed in such a manner as to retain both the antimalarial efficacy and quinoline-chemosensitization activity. These include the chloroquine–imipramine hybrid and the dual-function acridones (Fig. 11.4). The hybrids exhibit superior *in vitro* and *in vivo* efficacy against plasmodial parasites in comparison to chloroquine [134, 135].

The 8-aminoquinoline class, typified by primaquine and the investigational drug tafenoquine, is an important arsenal in the antimalarial artillery due to its unique properties. Primaquine is the only antimalarial drug in current clinical use with hypnozoitocidal activity and therefore capable of effecting radical cure of *Plasmodium vivax* and *Plasmodium ovale* infections [136]. It also has high tissue schizontocidal activity against preerythrocytic stages of *P. falciparum*, *P. vivax*, and *P. ovale*, thus being able to offer excellent causal prophylaxis [137]. Unfortunately, therapeutic utility of primaquine is severely limited by its prolonged 14-day treatment course that compromises patient adherence, rapid oxidative deamination to the inactive carboxy primaquine metabolite, and propensity for hemotoxicity especially hemolytic anemia in glucose-6-phosphate dehydrogenase-deficient individuals.

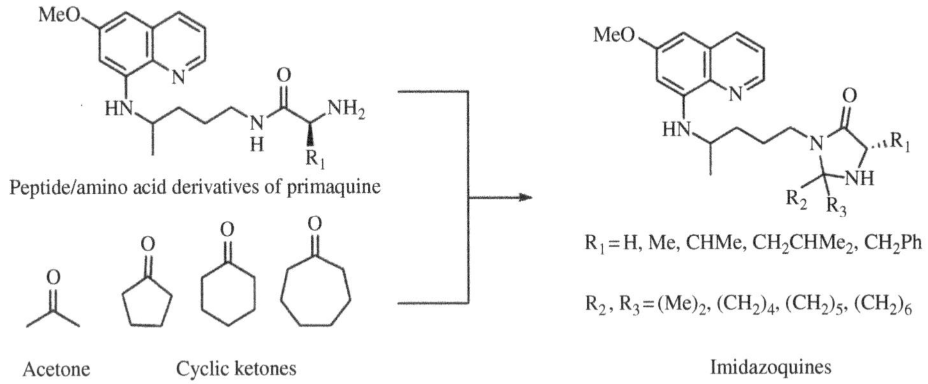

Peptide/amino acid derivatives of primaquine

Acetone   Cyclic ketones

$R_1 = H, Me, CHMe, CH_2CHMe_2, CH_2Ph$

$R_2, R_3 = (Me)_2, (CH_2)_4, (CH_2)_5, (CH_2)_6$

Imidazoquines

FIGURE 11.5   Design of imidazoquines from primaquine peptidomimetics [139].

In a bid to enhance therapeutic profile of primaquine while conserving its valuable antimalarial features, two classes of primaquine-based antimalarial hybrids have been designed: the imidazoquines and the primacenes. Precedent studies had revealed that peptidomimetic derivatives of primaquine have superior activity/toxicity profile compared to primaquine, although limited by possible hydrolytic and enzymatic instability of peptide bonds [138, 139]. Moreira and coworkers theorized that the masking of the *N*-terminal end amino group by incorporation in an imidazolidin-4-one moiety could improve their hydrolytic and enzymatic stability. Subsequent hybridization of the primaquine peptidomimetics and appropriate ketones as shown in Figure 11.5 yielded imidazoquines. Although they were less active against hypnozoites than primaquine *in vitro*, they exhibited improved *in vivo* gametocidal properties [139]. Their greater stability against both oxidative deamination and proteolytic degradation *in vivo* suggests that they probably could have lower hematotoxicity and higher oral bioavailability than primaquine—hence higher therapeutic indexes [140].

The primacenes are ferrocene–primaquine metallocenes in which a ferrocenyl unit is covalently coupled to the amino functionalities in the primaquine scaffolds as shown in Figure 11.6. Their design was inspired by the observation that insertion of a ferrocenyl moiety within the side chain of chloroquine enhanced antimalarial potency of resultant drug conjugates—the ferroquines [141]. In biological activity studies, both transmission-blocking and blood schizontocidal activities of the parent drug were conserved in the primacenes bearing a basic aliphatic amino group, while hypnozoitocidal activity was either conserved or enhanced in all the metallocenes. Notably, a metallocene obtained by replacement of the primaquine's side chain with a hexylferrocene produced a 45-fold increase in hypnozoitocidal activity [142]. Thus, both imidazoquines and primacenes represent a novel class of 8-aminoquinoline-based antimalarial hybrids.

11.7.3.2 Antileishmanial Hybrid Compounds Several academic research groups have engaged in antileishmanial drug discovery efforts through molecular hybridization. One such effort reported by Torres-Santos *et al.* involved hybridization of a naphthoquinone core with a pterocarpan moiety to obtain a novel hybrid compound christened naphthopterocarpanquinone (Fig. 11.7). Naturally occurring *para*-quinones and their synthetic analogues are an important class of biologically active compounds, and some of them

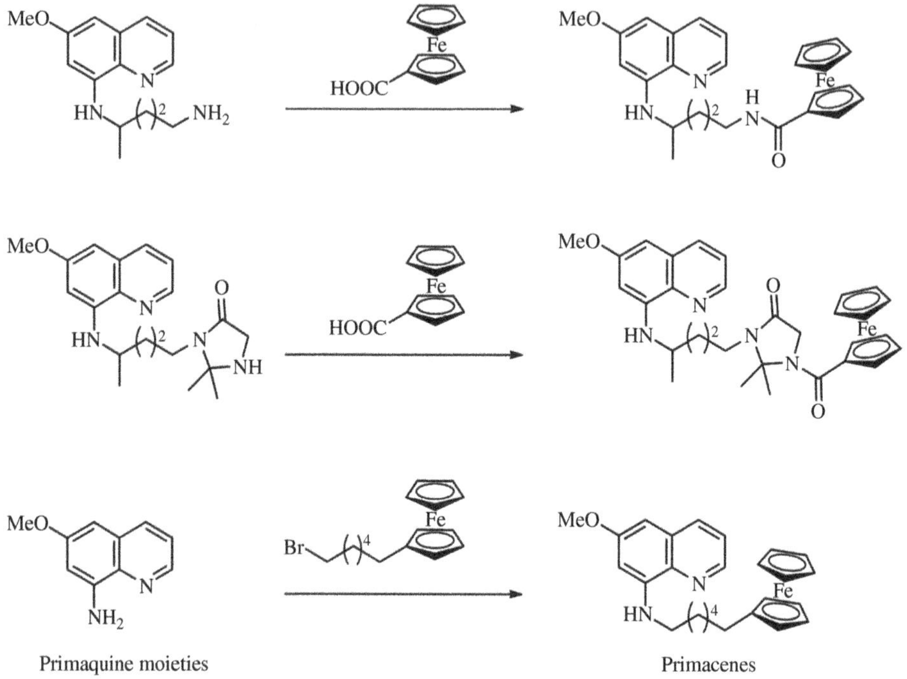

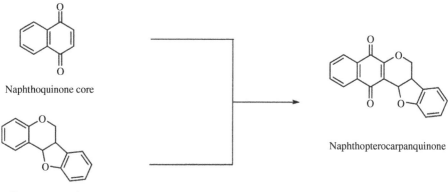

Primaquine moieties   Primacenes

FIGURE 11.6 Some bioactive primacenes obtained from primaquine-like scaffolds [142].

Naphthoquinone core

Naphthopterocarpanquinone

Pterocarpan moiety

FIGURE 11.7 Designed antileishmanial naphthopterocarpanquinone hybrid molecule [144].

have been used in the clinic as antineoplastics and antiparasitics [143]. These include the anticancer anthraquinones daunorubicin and doxorubicin and the synthetic antiparasitic atovaquone, which, in addition to its known antimalarial activity, is used in the treatment of *Pneumocystis jirovecii* infections and has shown *in vitro* activity against *Leishmania infantum* [144]. On the other hand, pterocarpans are a class of isoflavonoids with various biological activities [145]. The naphthopterocarpanquinone hybrid was locally and orally efficacious in an experimental cutaneous leishmaniasis model. It had moderate selectivity

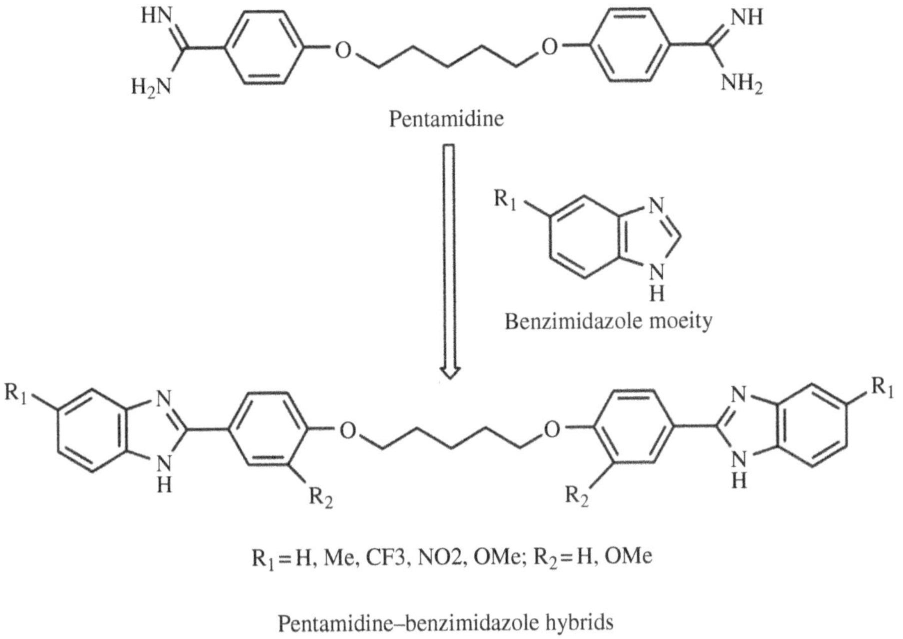

Pentamidine

Benzimidazole moeity

$R_1 =$ H, Me, CF3, NO2, OMe; $R_2 =$ H, OMe

Pentamidine–benzimidazole hybrids

FIGURE 11.8   Design strategy for the pentamidine–benzimidazole hybrids [146].

for intracellular amastigotes of *L. amazonensis* with an IC_{50} value of 1.4 µM, which was significantly less compared to the activity against the macrophages (IC_{50} 18.5 µM).

In 2008, Torres-Gomez *et al.* designed and synthesized ten hybrids based on pentamidine and benzimidazole [146]. Pentamidine—an aromatic diamidine—is used primarily for first stage of Gambian sleeping sickness, as second-line treatment for antimony-resistant leishmaniasis, and for opportunistic *P. jirovecii* pneumonia [147]. The benzimidazole motif is a recognized privileged scaffold in medicinal chemistry due to its diverse biological activities [148]. As shown in Figure 11.8, the design strategy retained the central pentyldioxyphenyl spacer in pentamidine but replaced the terminal diamidine groups with a 5-substituted benzimidazole scaffold. The hybrids were tested against five protozoa: *Leishmania mexicana*, *Plasmodium berghei*, *Trichomonas vaginalis*, *Giardia lamblia*, and *Entamoeba histolytica*. The bioactivity observed against the five protozoal parasites indicated that the introduction of benzimidazole into the pentamidine structure, and the inclusion of electron-donating moieties, enhanced the antiprotozoal activity [146].

Another recent hybrid antileishmanial drug discovery was reported in 2009 by Porwal *et al.* whose work was based on pentamidine and aplysinopsin. Aplysinopsins are a class of natural products possessing a cyclic guanidine function and have been shown to have affinity for certain biological targets including plasmepsin II and serotonin receptors [149]. The design strategy envisaged a molecular hybrid in which one amidinophenoxy function of pentamidine is replaced with aplysinopsin, thus preserving the dicationic character of pentamidine that is necessary for interaction with its molecular target [150]. However, as shown in Figure 11.9, their synthetic attempts did not deliver the target hybrid, but led to the discovery of a new class of antileishmanial agents. Through optimization, it was noted that incorporation of pentamidine substructure to the 2-thio analogue of aplysinopsin

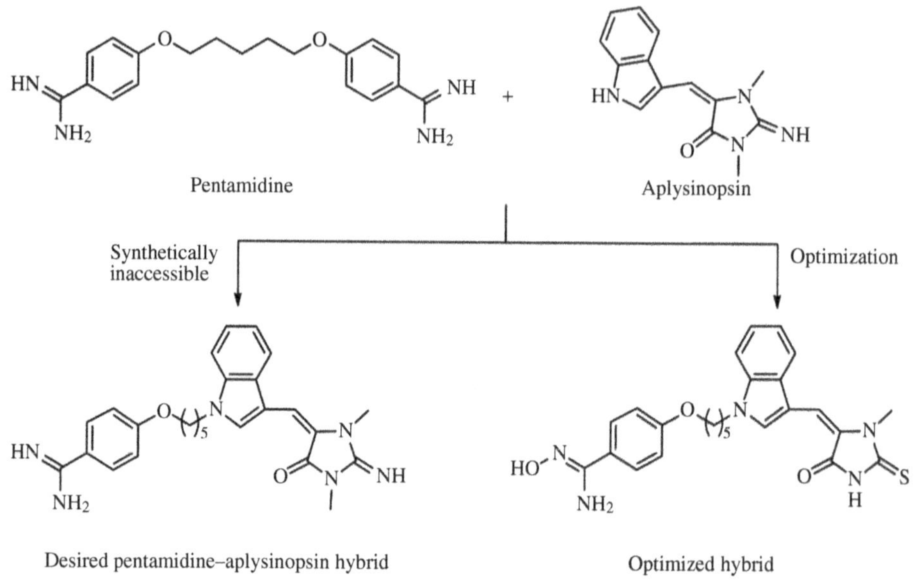

Pentamidine   Aplysinopsin

Synthetically inaccessible   Optimization

Desired pentamidine–aplysinopsin hybrid   Optimized hybrid

FIGURE 11.9   Design of pentamidine–aplysinopsin hybrid and subsequent optimization of newly discovered antileishmanial hybrid compound [149].

increased both the selectivity and activity of the hybrid compound. The optimized hybrid possessed 10 times more activity and 401-fold less toxicity than the parent drug pentamidine in cell-based assays [149].

11.7.3.3 Antitrypanosomal Hybrid Compounds   A number of research groups have reported multitargeted drug discovery efforts for HAT. Inspired by the antiparasitic activity of quinone-based natural products such as lapachol, plumbagin, aloe emodin, and diospyrin, Bolognesi *et al.* designed and synthesized 16 naphtho- and anthraquinones as captured in Figure 11.10. The framework combination design strategy involved incorporation of a selection of aryl groups mimicking the structural elements in the general biocide triclosan at the 2-position of the quinone motif [151]. A number of compounds had good potency against *Trypanosoma* and *Leishmania* at low concentration. One lapachol derivative had an *in vitro* IC_{50} value of 80 nM against *T. b. rhodesiense* and a safety index of 74, which are very close to the WHO/TDR specifications for a promising antitrypanosomatid hit for HAT [61].

Chalcone–benzoxaborole hybrid molecules represent the most recent antitrypanosomal hybrid drug discovery efforts. Benzoxaboroles, characterized by a unique five-membered oxaborole ring fused with a phenyl ring, are effective antitrypanosomal agents [152]. Due to their numerous pharmacological activities including antiparasitic effects, chalcones are versatile scaffolds in antiprotozoal drug discovery research [153, 154]. Previous studies have reported growth inhibitory effect of chalcone-like compounds against *T. brucei* and *T. cruzi* [140]. Inspired by these observations, Zhou and coworkers designed and synthesized a series of hybrid molecules bearing chalcone and benzoxaborole motifs as captured in Figure 11.11. A number of hybrids showed good *in vitro* antitrypanosomal efficacy against *T.b. brucei*, accompanied by satisfactory cytotoxicity profiles. Further, two compounds from the series demonstrated excellent *in vivo* efficacy in a murine infection model

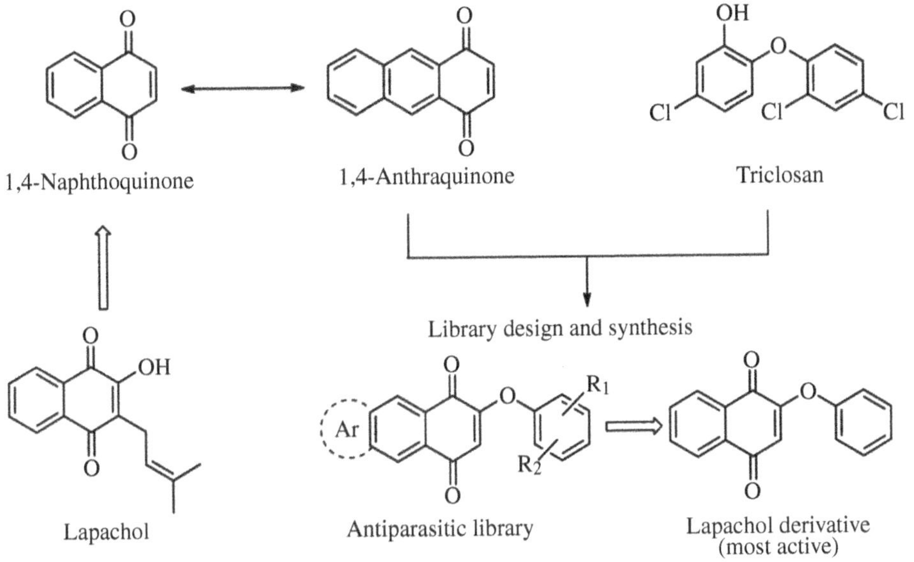

1,4-Naphthoquinone   1,4-Anthraquinone   Triclosan

Library design and synthesis

Lapachol   Antiparasitic library   Lapachol derivative (most active)

FIGURE 11.10 Designed multitarget antitrypanosomal hybrid compounds based on quinone motifs.

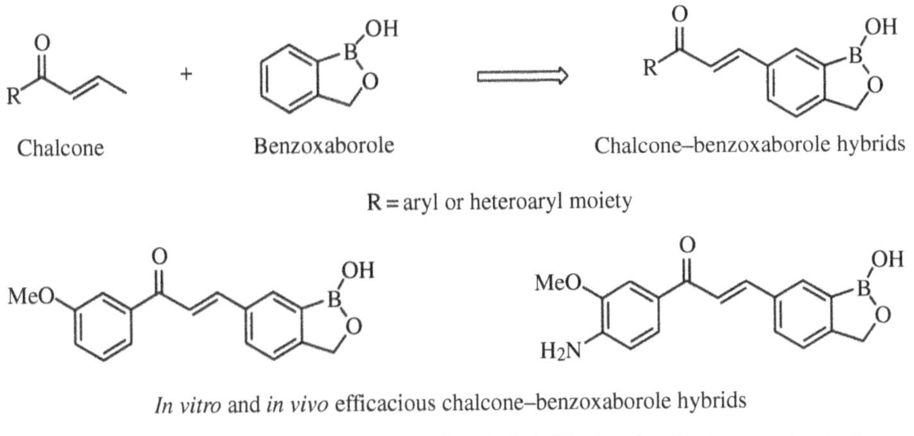

Chalcone   Benzoxaborole   Chalcone–benzoxaborole hybrids

R = aryl or heteroaryl moiety

In vitro and in vivo efficacious chalcone–benzoxaborole hybrids

FIGURE 11.11 Design of chalcone–benzoxaborole hybrids showing the two most potent compounds [153].

characterized by complete elimination of *T.b. brucei* parasites 30 days' postexposure and 100% survival rate of the assay mice [153].

11.7.3.4 Antichagasic Hybrid Compounds Several research groups have reported application of molecular hybridization in antichagasic drug discovery. Selected examples of hybrid compounds from these efforts are shown in Figure 11.12. Carvalho *et al.* designed and synthesized a new class of 1,3,4-thiadiazole-2-arylhydrazones based on megazol and guanylhydrazone—both of which have independent trypanocidal activity [155]. One hybrid, brazilizone A, showed high trypanocidal effect with IC_{50} values two-fold more potent than the parent megazol [156]. Independently, Gonzalez *et al.* designed

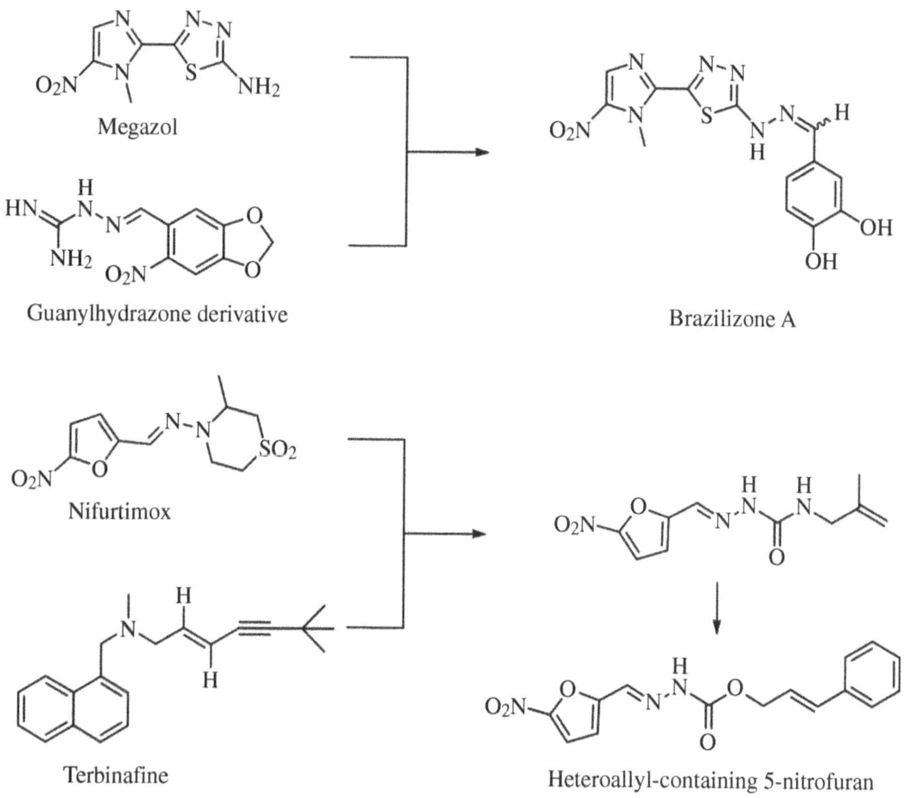

Megazol

Guanylhydrazone derivative

Brazilizone A

Nifurtimox

Terbinafine

Heteroallyl-containing 5-nitrofuran

FIGURE 11.12   Selected examples of antichagasic hybrid compounds.

and synthesized two series of thiosemicarbazone hybrid compounds by combining structural features of trypanocidal hydrazone moieties and the benzofuran pharmacophore [157]. The resultant thiosemicarbazone hybrid, though with a slight toxicity toward mammalian cells, had high trypanocidal activity and provided a useful starting point for further investigation. The same group also pursued hybrid compounds based on nifurtimox and terbinafine [158]. The design strategy involved framework combination of the nitrofuran moiety, which is responsible for the antichagasic oxidative stress of nifurtimox with the heteroallyl group in terbinafine that could potentially confer activity against squalene epoxidase [158]. The resultant heteroallyl-containing 5-nitrofurans were more active than either nifurtimox or terbinafine against *T. cruzi.*

The most recently described antitrypanosomal hybrids are the benzaldehyde-thiosemicarbazone derivatives of kaurenoic acid, a kaurane derivative. Kauranes and their biosynthetic analogues are naturally occurring diterpenes isolated from several plant species [159]. Diverse bioactivities have been described for the kaurane diterpenes, including antimicrobial, antiparasitic, antitumor, and anti-inflammatory properties, among others. On the other hand, thiosemicarbazones are an established class of synthetic compounds with a wide variety of biological properties including antiparasitic activity [160]. The design strategy involved covalent linkage of derivatized benzaldehyde-thiosemicarbazone analogues to kaurenoic acid via the terminal nitrogen as shown in Figure 11.13. The results of

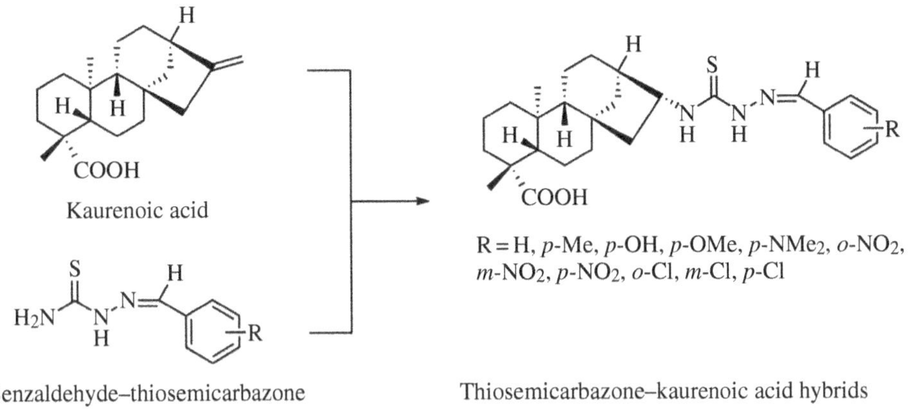

R = H, *p*-Me, *p*-OH, *p*-OMe, *p*-NMe2, *o*-NO2, *m*-NO2, *p*-NO2, *o*-Cl, *m*-Cl, *p*-Cl

Kaurenoic acid

Benzaldehyde–thiosemicarbazone

Thiosemicarbazone–kaurenoic acid hybrids

FIGURE 11.13 Design strategy for thiosemicarbazone–kaurenoic acid hybrids [161].

biological evaluation studies indicated that the molecular hybridization approach was successful since most of the hybrid compounds had greater activity than kaurenoic acid against *T. cruzi* in the *in vitro* assay. The *o*-nitrobenzaldehyde-thiosemicarbazone-based hybrid was the most active compound (IC_{50} = 2.0 μM), way much higher compared to kaurenoic acid (IC_{50} = 101.7 μM). It also had the best selectivity index (SI = 9.0) in the cytotoxicity assay, much better than that obtained with kaurenoic acid (SI = 0.6) [161].

11.8 DRUG REPOSITIONING AND REPURPOSING

11.8.1 Introduction

In spite of major advancement in human medicine, the balance sheet is still indelibly stained by unmet medical needs, either due to total absence of treatment options or severe inadequacy of existing therapies. The discovery of completely new drug classes remains a difficult, laborious, long (10–15 years), and costly (US$ 500–800 million) affair [15, 162]. Therefore, in the short term, innovative drug discovery is not going to plug the treatment deficit. One efficient way to fast-track drug discovery is to identify new indications for existing drugs. The process of finding new uses for the existing drugs outside the scope of the original indication is variably referred to as drug repositioning, drug repurposing, drug reprofiling, and drug redirecting [163]. Strictly speaking, drug repurposing is the investigation of new uses for the existing drugs, whether in clinical use or not, without any structural modification of the drug. On the other hand, drug repositioning differs from repurposing in that the drug in question undergoes structural modifications to improve the desired activity. As James Black, a pharmacologist and Nobel laureate, once remarked, "the most fruitful basis for the discovery of new drug is to start with an old drug", several pharma companies and start-up ventures have utilized this approach. As shown in Table 11.6, there already are excellent drug repositioning successes in the clinical market in various therapeutic areas.

Drug repositioning and repurposing offer a better risk-versus-reward trade-off compared with other drug development strategies in a number of ways [163]. First, drug repositioning provides label extension for a drug prolonging its clinical utility, thus

TABLE 11.6 Repositioned Drugs in Various Therapeutic Areas[a]

Drug	Original indication	New indication
Celecoxib	Arthritis	Familial adenomatous polyposis, colon and breast cancers
Finasteride	Benign prostatic hyperplasia	Hair loss
Mifepristone	Pregnancy termination	Psychotic major depression
Minoxidil	Hypertension	Hair loss
Paclitaxel	Cancer	Restenosis
Raloxifene	Breast and prostate cancer	Osteoporosis
Sildenafil	Angina	Male erectile dysfunction
Phentolamine	Hypertension	Impaired night vision
Zidovudine	Cancer	HIV/AIDS
Topiramate	Epilepsy	Obesity
Thalidomide	Sedation, nausea, and insomnia	Leprosy and multiple myeloma

[a] From Ref. 163.

maintaining a revenue base for the company. Second, it poses reduced financial risk because a repositioning candidate will usually have been through several stages of clinical development and therefore has well-known safety and pharmacokinetic profiles. Third, the route to the clinic is much shorter since *in vitro* and *in vivo* screening, chemical optimization, toxicology, formulation development, and even early clinical development have, in many cases, already been completed and can therefore be bypassed [163, 164]. Fourth, a repositioning candidate can offer a useful starting point for lead optimization especially where its molecular target is known.

In the field of NTDs, the current R&D pipeline for most tropical infections remains weak and thus unlikely to yield novel drug classes that meet desired target product profiles in the short term. Drug repositioning and repurposing strategies would have a dramatic effect in the discovery and development of novel drugs for neglected diseases [165]. This approach is fast gaining currency and is credited with recent drug development successes among the PPPs involved in NTD drug discovery [109, 166]. The use of this strategy is informed by the fact that due to lack of commercial markets for drugs against NTDs, many drugs against tropical infections were first developed for other human and veterinary indications [39, 167]. A few classical examples of drugs that owe their entry into human antiparasitic artillery through drug repurposing include the antitrypanosomal eflornithine first applied as an anticancer agent, the antimalarial fosmidomycin first used as an antibacterial agent, and the antileishmanial miltefosine initially developed for breast cancer [59].

The availability of many diverse chemical libraries in pharmaceutical companies, virtual discovery organizations, and academic institutions makes drug repurposing and/or repositioning a very attractive venture for efficient drug discovery for NTDs. The literature describes several approaches to drug repositioning and repurposing such as cell biology, exploitation of genome information, compound screening studies, exploitation of coinfection drug efficacy, and *in silico* computational technologies [109, 163, 167–169]. Selected examples of repositioned drugs and drug candidates for NTDs discovered through these approaches are presented in Table 11.7 and described in greater detail in the following subsections.

TABLE 11.7   Selected Repositioned Drugs and Drug Candidates for Neglected Diseases

Drug	Original indication	New indication	Repositioning approach
Methotrexate	Antiproliferative	Antimalarial	Cell biology
Posaconazole	Antifungal	Antichagasic	Cell biology
Fosmidomycin	Antibacterial	Antimalarial	Genome information
Astemizole	Antihistamine	Antimalarial	Phenotypic HTS
Closantel	Veterinary anthelmintic	Onchocerciasis	Target-focused HTS
Tinidazole	Antiamebic	Antimalarial	Coinfection drug efficacy

11.8.2   Cell Biology Approach

In this approach, a repositioned drug is discovered through exploitation of the similarity in biological processes of different diseases. Infectious microorganisms such as bacteria, viruses and plasmodia, and tumorous cells have rapid proliferative character. Thus, drugs developed against one of these cells could also potentially block division of the other cell types. In fact, the sulfa-based antimalarials such as sulfadoxine and sulfalene owe their origin from antibacterial sulfonamides. These compounds achieve their therapeutic efficacy by inhibiting the same target, the dihydropteroate synthase, involved in folate biosynthesis in both plasmodia and bacteria [168]. Another enzyme in the folate biosynthetic pathway, the dihydrofolate reductase (DHFR), is an established drug target for antibacterial, antimalarial, and antineoplastic chemotherapy. Studies have shown that the antifolate antineoplastics methotrexate and trimetrexate have clinically significant antimalarial efficacy at a much lower dose than used for cancer chemotherapy, indicating their repurposing potential for malaria treatment [170]. A related but independent study showed that trimetrexate is a potent inhibitor of *T. cruzi* DHFR activity with concomitant effectiveness in killing *T. cruzi* parasites *in vitro* [171].

As a class of enzymes, proteases are validated druggable targets that could offer repurposing and/or repositioning opportunities for discovery of lead compounds against NTDs. Research studies have demonstrated that the antiretroviral protease inhibitors such as indinavir, saquinavir, ritonavir, nelfinavir, lopinavir, amprenavir, and atazanavir used in the treatment of HIV/AIDS have excellent *in vitro* inhibitory effects against *P. falciparum* in which they block plasmodial aspartyl proteases (plasmepsins) [172]. It is plausible to speculate that the APIs could produce therapeutically significant inhibitory effects against aspartyl proteases in other tropical parasites such as *Trypanosoma*, *Leishmania*, and *Schistosoma*.

Sterol biosynthesis is an essential pathway in eukaryotes and is a target of various chemotherapeutic agents including the sterol 14α-demethylase inhibitor azole antifungals and the HMG-CoA reductase inhibitor anticholesterolemic statins [173]. Previous incremental studies revealed that the triazole antifungals posaconazole and ravuconazole have *in vitro* and *in vivo* antitrypanosomal and antileishmanial efficacy arising from inhibition of the protozoal sterol 14α-demethylase [174]. Thus, posaconazole and E1224 (a prodrug of ravuconazole) are currently in phase II clinical trials as repurposing candidates for Chagas disease [175].

11.8.3 Exploitation of Genome Information

This drug repositioning strategy involves identification of drug targets using genome homology between a potential target and a validated target in another therapeutic area. It is based on the recognition that a single biological target may be important in multiple disease pathologies. Exploitation of genome information to drug repurposing and/or repositioning is best exemplified by the discovery of the antimalarial efficacy of fosmidomycin, a natural antibiotic originally developed for the treatment of bacterial infections [176]. Previous research had shown that fosmidomycin exerts its antibacterial effects by inhibiting reductoisomerase, a key enzyme in the mevalonate-independent 1-deoxy-D-xylulose-5-phosphate (DOXP) pathway for isoprenoid biosynthesis. Subsequent genome sequencing studies revealed that the DOXP pathway is conserved in both bacteria and plasmodia, and the testing of fosmidomycin, which had already been developed as an antibacterial, led to its discovery as an antimalarial [168].

As argued in great detail in Section 11.5.3, protein kinases have proved a valuable druggable target class in cancer chemotherapy. Protein kinases are key regulators of many biochemical processes in eukaryotes, not excluding tropical parasites. In fact, several protein kinases involved in various plasmodial life-cycle processes have been described [88]. Given the similarities in interrupting the cell cycle of a rapidly replicating tumor cell and that of a rapidly replicating parasite, identifying cyclin-dependent cell cycle kinase targets in parasites and screening these against existing mammalian cell cycle kinase inhibitors is a promising opportunity for drug repurposing and/or repositioning for the NTDs.

11.8.4 Compound Screening Studies

This approach involves screening of compound libraries consisting of drugs already approved for human use for potential "off-label" antiparasitic efficacy using *in vitro* target-based and/or phenotypic screening. The antiparasitic efficacy of a number of drugs has been revealed through this approach. A notable example is the H_1-histamine receptor antagonist astemizole whose antimalarial activity against *P. falciparum* was discovered through *in vitro* parasite-based HTS of a library of 2,687 existing drugs. Further experiments revealed that astemizole and its principal human metabolite, desmethylastemizole, had *in vivo* efficacy in two mouse models of malaria [177].

Most recently, a target-based HTS of a 4,000-compound library led to identification of antifungal acrisorcin, anticancer harmine, and metabotropic glutamate receptor agonist, (±)-2-amino-3-phosphonopropionic acid, as synergistic inhibitors of *P. falciparum* heat shock protein 90 [178]. Likewise, a target-focused HTS of 1,514 FDA-approved drugs against the *Onchocerca volvulus* chitinase OvCHT1 discovered potential application of closantel, a veterinary anthelmintic, in chemotherapy of onchocerciasis [179]. In a more focused study, tamoxifen, an established breast cancer antineoplastic, demonstrated *in vitro* and *in vivo* efficacy against promastigotes and amastigotes of *Leishmania*, implying its possible repositioning for the chemotherapy of leishmaniasis [180].

11.8.5 Exploitation of Coinfection Drug Efficacy

Nzila *et al.* have rightly commented that drug repositioning can arise from careful observation of coinfection treatment [168]. Studies have shown that most NTDs occur as coinfections such as malaria and leishmaniasis, malaria and trypanosomiasis, and malaria and helminthiasis. Thus, the treatment of one infection could also clear the concurrent disease. For instance, the antimalarial effect of antiamebic drugs was first hypothesized during studies on the treatment of the protozoan amebiasis in patients coinfected with malaria. As a result, 5-nitroimidazole derivative tinidazole has been evaluated in a phase II clinical trial as a potential repositioning antimalarial candidate for radical cure of *P. vivax* malaria [181].

11.8.6 *In Silico* Computational Technologies

Several computational approaches that enable discovery of new uses for existing drugs have been described, including similarity of side effects, similarity of gene expression profiles of different diseases, and structural similarity of binding sites [182]. The latter approach is by far the most exploited. This computational drug repositioning approach is analogous to structure-based drug discovery (SBDD) in which virtual screening is carried out using libraries of established drugs. As is the case with SBDD, this approach requires knowledge of the 3D structure of both the drugs and the molecular targets and pharmacophore modeling [183]. Although there is currently no repositioned drug for NTDs discovered through this approach, there are a few successful stories in other therapeutic areas. For instance, *in silico* approaches have been used to identify inhibitors of transporter proteins such as P-glycoprotein, human peptide transporter, and human apical sodium-dependent bile acid transporter, among others [109]. Thus, computational approaches can be used complementarily with experimental techniques in repositioning studies for NTDs.

11.9 FUTURE OUTLOOK

It is our contention that expert exploitation of niche areas hitherto underexploited, such as NTDs, is the next frontier in drug discovery. The NTDs present far much better prospects for innovative drug discovery than other therapeutic areas already saturated with both originator products and their generic versions. Application of strategic approaches, a few of which have been the subject of this chapter, will reinvigorate innovative research and go a long way in ensuring efficiency in drug discovery.

REFERENCES

1 World Health Organization, The World Health Report 1999, *Making a difference*. WHO Press, Geneva, 1999.

2 United Nations Children's Fund, *The progress of nations*, Division of Communication, New York, 1998.

3 World Health Organization, The World Health Report 2007, A safer future: Global public health security in the 21st century, WHO Press, Geneva, *2007*.

4 Lichtenberg, F.R. The impact of new drug launches on longevity: evidence from longitudinal, disease-level data from 52 countries, 1982-2001. Available online at www.nber.org/papers/w9754. Accessed October 9, 2013.

5 Trist, D., Ratti, E., da Ros, L. The apparent declining efficiency in drug discovery. In: Taylor, J.B., Triggle, D.J. (Eds.), *Comprehensive Medicinal Chemistry II, Vol. 1: Global Perspectives*, Elsevier, Amsterdam, 2007, pp. 615–625.

6   Verdine, G.L. The combinatorial chemistry of nature. *Nature*, 1996, *384*, 11–13.

7   Thomas, G., *Fundamentals of Medicinal Chemistry*, John Wiley & Sons Inc., West Sussex, 2003, pp. 113–130.

8   Müller, G. (2003). Medicinal chemistry of target family-directed masterkeys. *Drug Discov. Today*, *8*, 681–691.

9   Koehn, F.E., Carter, G.T. (2005). The evolving role of natural products in drug discovery. *Nat. Rev. Drug Discov.*, *4*, 206–220.

10   Kennedy, J.P., Williams, L., Bridges, T.M., Daniels, R.N., Weaver, D., Lindsley, C.W. (2008). Application of combinatorial chemistry science on modern drug discovery. *J. Comb. Chem.*, *10*, 345–354.

11   Mullard, A. (2011). 2010 FDA drug approvals. *Nat. Rev. Drug Discov.*, *10*, 82–85.

12   Munos, B. (2009). Lessons from 60 years of pharmaceutical innovation. *Nat. Rev. Drug Discov.*, *8*, 959–968.

13   Kola, I., Landis, J. (2004). Can the pharmaceutical industry reduce attrition rates? *Nat. Rev. Drug Discov.*, *3*, 711–715.

14   Schmid, E.F., Smith, D.A. (2005). Is declining innovation in the pharmaceutical industry a myth? *Drug Discov. Today*, *10*, 1031–1039.

15   Schmid, E.F., Smith, D.A. (2004). Is pharmaceutical R&D just a game of chance or can strategy make a difference? *Drug Discov. Today*, *9*, 18–26.

16   Drews, J. (2003). Strategic trends in the drug industry. *Drug Discov. Today*, *8*, 411–420.

17   Booth, B., Zemmel, R. (2004). Prospects for productivity. *Nat. Rev. Drug Discov.*, *3*, 451–456.

18   Ishikawa, M., Hashimoto, Y. (2011). Improvement in aqueous solubility in small molecule drug discovery programs by disruption of molecular planarity and symmetry. *J. Med. Chem.*, *54*, 1539–1554.

19   Lipinski, C.A. (2000). Drug-like properties and the causes of poor solubility and poor permeability. *J. Pharmacol. Toxicol. Methods*, *44*, 235–249.

20   Kerns, E.H., Di, L. (2003). Pharmaceutical profiling in drug discovery. *Drug Discov. Today*, *8*, 316–323.

21   Walters, W.P., Green, J., Weiss, J.R., Murcko, M.A. (2011). What do medicinal chemists actually make? *J. Med. Chem.*, *54*, 6405–6416.

22   Paul, S.M., Mytelka, D.S., Dunwiddie, C.T., Persinger, C.C., Munos, B.H., Lindborg, S.R., Schacht, A.L. (2010). How to improve R&D productivity: the pharmaceutical industry's grand challenge. *Nat. Rev. Drug Discov.*, *9*, 203–214.

23   Hotez, P., Ottesen, E., Fenwick, A., Molyneux, D. In: Pollard, A.J., Finn, A. (Eds.), *Hot topics in infection and immunity in children*, Springer, New York, 2006, pp. 23–33.

24   Hotez, P.J., Fenwick, A., Savioli, L., Molyneux, D.H. (2009). Rescuing the bottom billion through control of neglected tropical diseases. *Lancet*, *373*, 1570–1575.

25   Burrows, J.N., Chibale, K., Wells, T.N.C. (2011). The state of the art in antimalarial drug discovery and development. *Curr. Top. Med. Chem.*, *11*, 1226–1254.

26   Sachs, J., Malaney, P. (2002). The economic and social burden of malaria. *Nature*, *415*, 680–685.

27   World Health Organization, First WHO report on neglected tropical diseases 2010, Working to overcome the global impact of neglected tropical diseases, Available online at: http://www.gsk.com/media/downloads/WHO-report-on-NTD.PDF. Accessed October 24, 2013.

28   Knox, D.P., Redmond, D.L. (2006). Parasite vaccines—Recent progress and problems associated with their development. *Parasitology*, *133*, S1–S8.

29   Richie, T. (2006). High road, low road? Choices and challenges on the pathway to a malaria vaccine. *Parasitology*, *133*, S113–S144.

30 Hill, A.V.S. (2011). Vaccines against malaria. *Phil. Trans. R. Soc. B*, *366*, 2806–2814.

31 Birkett, A.J., Moorthy, V.S., Loucq, C., Chitnis, C.E., Kaslow, D.C. (2013). Malaria vaccine R&D in the decade of vaccines: Breakthroughs, challenges and opportunities. *Vaccine, 31S*, B233–B243.

32 Magez, S., Caljon, G., Tran, T., Stijlemans, B., Radwanska, M. (2010). Current status of vaccination against African trypanosomiasis. *Parasitology, 137*, 2017–2027.

33 Handman, E. (2001). Leishmaniasis: current status of vaccine development. *Clin. Microbiol. Rev., 14*, 229–243.

34 Kelland, K. (2015). World's first malaria vaccine gets regulatory go-ahead, faces WHO review. Reuters, Available at: http://www.reuters.com/article/2015/07/24/gsk-malaria-vaccine-idUSL5N1041K120150724. Accessed August 23, 2015.

35 Manokaren, K. (2013). First malaria vaccine en route for 2015. The McGill Tribune. Available at: http://mcgilltribune.com/first-malaria-vaccine-en-route-for-2015/. Accessed December 12, 2013.

36 Singh, S., Sivakumar, R. (2004). Challenges and new discoveries in the treatment of leishmaniasis. *J. Infect. Chemother., 10*, 307–315.

37 Dias, J.C.P., Silveira, A.C., Schofield, C.J. (2002). The impact of Chagas disease control in Latin America—A review. *Mem. Inst. Oswaldo Cruz, 97*, 603–612.

38 Klausner, R., Alonso, P. (2004). An attack on all fronts. *Nature, 430*, 930–931.

39 Campbell-Lendrum, D., Molyneux, D. In: Epstein, P., Githeko, A., Rabinovich, J., Weinstein, P. (Eds.), *Ecosystems and vector-borne disease control.* Available at: http://www.maweb.org/documents/document.317.aspx.pdf. Accessed December 24, 2013.

40 Caffrey, R.C., Steverding, D. (2008). Recent initiatives and strategies to developing new drugs for tropical parasitic diseases. *Expert Opin. Drug Discov., 3*, 173–186.

41 Pink, R., Hudson, A., Mouries, M.-A.., Bendig, M. (2005). Opportunities and challenges in antiparasitic drug discovery. *Nat. Rev. Drug Discov., 4*, 727–740.

42 Towse, A., Kettler, H. (2005). Advance price or purchase commitments to create markets for treatments for diseases of poverty: Lessons from three policies. *Bull. WHO, 83*, 301–307.

43 Trouiller, P., Olliaro, P., Torreele, E., Orbinski, J., Laing, R., Ford, N. (2002). Drug development for neglected diseases: A deficient market and a public-health policy failure. *Lancet, 359*, 2188–2194.

44 Chirac, P., Torreele, E. (2006). Global framework on essential health R&D. *Lancet, 367*, 1560–1561.

45 Hotez, P.J., Molyneux, D.H., Fenwick, A., Ottesen, E., Sachs, S.E., Sachs, J.D. (2006). Incorporating a rapid-impact package for neglected tropical diseases with programs for HIV/AIDS, tuberculosis, and malaria. *PLoS Med., 3*, e102.

46 Renslo, A.R., McKerrow, J.H. (2006). Drug discovery and development for neglected parasitic diseases. *Nat. Chem. Biol., 2*, 701–710.

47 Nwaka, S., Hudson, A. (2006). Innovative lead discovery strategies for tropical diseases. *Nat. Rev. Drug Discov., 5*, 941–955.

48 Weigelt, J. (2009). The case for open-access chemical biology. *EMBO Reports, 10*, 941–945.

49 Woodcock, J. (2010). Precompetitive research: A new prescription for drug development? *Clin. Pharmacol. Therapeut., 87*, 521–523.

50 Esserman, L.J., Woodcock, J. (2011). Accelerating identification and regulatory approval of investigational cancer drugs. *JAMA, 306*, 2608–2609.

51 Baxter, K., Horn, E., Gal-Edd, N., Zonno, K., O'Leary, J., Terry, P.F., Terry, S.F. (2013). An end to the myth: There is no drug development pipeline. *Sci. Transl. Med., 5*, 171cm1.

52 Khanna, I. (2012). Drug discovery in pharmaceutical industry: productivity challenges and trends. *Drug Discov. Today, 17*, 1088–1102.

53 Nwaka, S., Ramirez, B., Brun, R., Maes, L., Douglas, F., Ridley, R. (2009). Advancing drug innovation for neglected diseases—criteria for lead progression. *PLoS Negl. Trop. Dis.*, *3*, e440.

54 Tralau-Stewart, C.J., Wyatt, C.A., Kleyn, D.E., Ayad, A. (2009). Drug discovery: new models for industry-academic partnerships. *Drug Discov. Today*, *14*, 95–101.

55 Gutteridge, W.E. (2006). TDR collaboration with the pharmaceutical industry. *Trans. Roy. Soc. Trop. Med. Hyg.*, *100*, S21–S25.

56 Twenty-eighth World Health Assembly Resolution 28.71, WHO's role in the development and coordination of research in tropical diseases. Available at: http://www.who.int/tdr/about/governance/documents/wha28-71/en/index.html. Accessed December 16, 2013.

57 Nwaka, S., Ridley, R.G. (2003). Virtual drug discovery and developments for neglected diseases through public-private partnerships. *Nat. Rev. Drug Discov.*, *2*, 919–928.

58 Hotez, P.J., Molyneux, D.H., Fenwick, A., Kumaresan, J., Sachs, S.E., Sachs, J.D., Savioli, L. (2007). Control of neglected tropical diseases. *N. Engl. J. Med.*, *357*, 1018–1027.

59 Butler, D. (2007). Lost in translation. *Nature*, *449*, 158–159.

60 Croft, S.L. (2005). Public-private partnership: from there to here. *Trans. Roy. Soc. Trop. Med. Hyg.*, *99S*, S9–S14.

61 Fidock, D.A., Rosenthal, P.J., Croft, S.L., Brun, R., Nwaka, S. (2004). Antimalarial drug discovery: efficacy models for compound screening. *Nat. Rev. Drug Discov.*, *3*, 509–520.

62 Fishburn, C.S. (2013). Translational research: the changing landscape of drug discovery. *Drug Discov. Today*, *18*, 487–494.

63 Hughes, B. (2008). Pharma pursues novel models for academic collaboration. *Nat. Rev. Drug Discov.*, *7*, 631–632.

64 Ma, P., Zemmel, R. (2002). Value of novelty. *Nat. Rev. Drug Discov.*, *1*, 571–572.

65 Brewer, G.J. (2006). Fundamental problems lie ahead in the drug discovery and commercialization process: restructuring of the pharmaceutical industry and an improved partnership with academia are required. *J. Investig. Med.*, *54*, 291–302.

66 Breman, J.G., Alilio, M.S., Mills, A. (2004). Conquering the intolerable burden of malaria: What's new, what's needed: A summary. *Am. J. Trop. Med. Hyg.*, *71* (Suppl. 2), 1–15.

67 Mboya-Okeyo, T., Ridley, R.G., Nwaka, S. (2009). The African Network for Drugs and Diagnostics Innovation. *The Lancet*, *373*, 1507–1508.

68 Aksoy, S. (2010). Solutions to neglected tropical diseases require vibrant local scientific communities. *PLoS Negl. Trop. Dis.*, *4*, e662.

69 Nordling, L. (2013). Made in Africa. *Nat. Med.*, *19*, 803–806. Available at: http://www.nature.com/nm/journal/v19/n7/full/nm0713-803.html. Accessed December 12, 2013.

70 Gamo, F.-J., Sanz, L.M., Vidal, J., de Cozar, C., Alvarez, E., Lavandera, J.-L., Vanderwall, D.E., Green, D.V.S., Kumar, V., Hasan, S., Brown, J.R., Peishoff, C.E., Cardon, L.R., Garcia-Bustos, J.F. (2010). Thousands of chemical starting points for antimalarial lead identification. *Nature*, *465*, 311–315.

71 Plouffe, D., Brinker, A., McNamara, C., Henson, K., Kato, N., Kuhen, K., Nagle, A., Adrián, F., Matzen, J.T., Anderson, P., Nam, T., Gray, N.S., Chatterjee, A., Janes, J., Yan, S.F., Trager, R., Caldwell, J.S., Schultz, P.G., Zhou, Y., Winzeler, E.A. (2008). In silico activity profiling reveals the mechanism of action of antimalarials discovered in a high-throughput screen. *Proc. Natl. Acad. Sci. U.S.A.*, *105*, 9059–9064.

72 Boulton, I.C., Nwaka, S., Bathurst, I., Lanzer, M., Taramelli, D., Vial, H., Doerig, C., Chibale, K., Ward, S.A. (2010). CRIMALDDI: a co-ordinated, rational, and integrated effort to set logical priorities in anti-malarial drug discovery initiatives. *Malar. J.*, *9*, 202.

73 Hopkins, A.L., Gloom, C.R. (2002). The druggable genome. *Nat. Rev. Drug Discov.*, *1*, 727–730.

74 Zambrowicz, B.P., Sands, A.T. (2003). Knockouts model the 100 best-selling drugs—Will they model the next 100? *Nat. Rev. Drug Discov.*, *2*, 38–51.

75 Schmid, E.F., Smith, D.A. (2006). R&D technology investments: misguided and expensive or a better way to discover medicines? *Drug Discov. Today*, *11*, 775–784.

76 Arrowsmith, J. (2013). Phase II and phase III attrition rates 2011-2012. *Nat. Rev. Drug Discov.*, *12*, 569.

77 Arrowsmith, J. (2011). Phase II failures: 2008-2010. *Nat. Rev. Drug Discov.*, *10*, 328–329.

78 Sams-Dodd, F. (2005). Target-based drug discovery: is something wrong? *Drug Discov. Today*, *10*, 139–146.

79 Editorial. (2011). *Nat. Rev. Drug Discov.*, *10*, 883.

80 Chatterjee, A.K., Yeung, B.K.S. (2012). Back to the future: lessons learned in modern target-based and whole-cell lead optimization of antimalarials. *Curr. Top. Med. Chem.*, *12*, 473–483.

81 Lee, J.A., Uhlik, M.T., Moxham, C.M., Tomandl, D., Sall, D.J. (2012). Modern phenotypic drug discovery is a viable, neoclassic pharma strategy. *J. Med. Chem.*, *55*, 4527–4538.

82 Butera, J.A. (2013). Phenotypic screening as a strategic component of drug discovery programs targeting novel antiparasitic and antimycobacterial agents: An editorial miniperspectives series on phenotypic screening for antiinfective targets. *J. Med. Chem.*, *56*, 7715–7718.

83 Gilbert, I.H. (2013). Drug discovery for neglected diseases: molecular target-based and phenotypic approaches. *J. Med. Chem.*, *56*, 7719–7726.

84 Brown, D. (2007). Unfinished business: target-based drug discovery. *Drug Discov. Today*, *12*, 1007–1012.

85 Prüll, C.-R. (2003). Part of a scientific master plan? Paul Ehrlich and the origins of his receptor concept. *Medical History*, *47*, 332–356.

86 Wells, T.N.C., Alonso, P.L., Gutteridge, W.E. (2009). New medicines to improve control and contribute to the eradication of malaria. *Nat. Rev. Drug Discov.*, *8*, 879–891.

87 Johnson, L.N. (2009). Protein kinase inhibitors: Contributions from structure to clinical compounds. *Quarterly Rev. Biophys.*, *42*, 1–40.

88 Kappes, B., Doerig, C.D., Graeser, R. (1999). An overview of Plasmodium protein kinases. *Parasitol. Today*, *15*, 449–454.

89 Vyas, V.K., Ghate, M. (2011). Recent developments in the medicinal chemistry and therapeutic potential of dihydroorotate dehydrogenase (DHODH) inhibitors. *Mini Rev. Med. Chem.*, *11*, 1039–1055.

90 Phillips, M.A., Rathod, P.K. (2010). Plasmodium dihydroorotate dehydrogenase: a promising target for novel anti-malarial chemotherapy. *Infect. Disord. Drug Targets*, *10*, 226–239.

91 Arakaki, T.L., Buckner, F.S., Gillespie, J.R., Malmquist, N.A., Phillips, M.A., Kalyuzhniy, O., Luft, J.R., DeTitta, G.T., Verlinde, C.L.M.J., van Voorhis, W.C., Hol, W.G.J., Merritt, E.A. (2008). Characterization of *Trypanosoma brucei* dihydroorotate dehydrogenase as a possible drug target; structural, kinetic and RNAi studies. *Mol. Microbiol.* 68, 37–55.

92 Cordeiro, A.T., Feliciano, P.R., Pinheiro, M.P., Nonato, M.C. (2012). Crystal structure of dihydroorotate dehydrogenase from *Leishmania major*. *Biochemie*, *94*, 1739–1748.

93 Hast, M.A., Beese, L.S., Structural biochemistry of CaaX protein prenyltransferase. In: Tamanoi, F., Hrycyna, C.A., Bergo, M.O. (Eds.), *The enzymes: protein prenylation, Part A*, Vol. XXIX, Academic Press, San Diego, 2011, pp. 235–257.

94 Kelland, L.R. (2003). Farnesyl transferase inhibitors in the treatment of breast cancer. *Expert Opin. Investig. Drugs*, *12*, 413–421.

95 Hast, M.A., Fletcher, S., Cummings, C.G., Pusateri, E.E., Blaskovich, M.A., Rivas, K., Gelb, M.H., van Voorhis, W.C., Sebti, S.M., Hamilton, A.D., Beese, L.S. (2009). Structural basis for binding and selectivity of antimalarial and anticancer ethylenediamine inhibitors to protein farnesyltransferase. *Chem. Biol.*, *16*, 181–192.

96   Agüero, F., Al-Lazikani, B., Aslett, M., Berriman, M., Buckner, F.S., Campbell, R.K., Carmona, S., Carruthers, I.M., Chan, A.W.E., Chen, F., Crowther, G.J., Doyle, M.A., Hertz-Fowler, C., Hopkins, A.L., McAllister, G., Nwaka, S., Overington, J.P., Pain, A., Paolini, G.V., Pieper, U., Ralph, S.A., Riechers, A., Roos, D.S., Sali, A., Shanmugam, D., Suzuki, T., Van Voorhis, W.C., Verlinde, C.L.M.J. (2008). Genomic-scale prioritization of drug targets: the TDR Targets database. *Nat. Rev. Drug Discov.*, *7*, 900–907.

97   Crowther, G.J., Shanmugam, D., Carmona, S.J., Doyle, M.A., Hertz-Fowler, C., Berriman, M., Nwaka, S., Ralph, S.A., Roos, D.S., van Voorhis, W.C., Agüero, F. (2010). Identification of attractive drug targets in neglected disease pathogens using an *in silico* approach. *PLoS Negl. Trop. Dis.*, *4*, e804.

98   Sykes, M.L., Avery, V.M. (2013). Approaches to protozoan drug discovery: phenotypic screening. *J. Med. Chem.*, *56*, 7727–7740.

99   Swinney, D.C., Anthony, J. (2011). How were new medicines discovered? *Nat. Rev. Drug Discov.*, *10*, 507–519.

100   Hellerstein, M.K. (2008). Exploiting complexity and the robustness of network architecture for drug discovery. *J. Pharmacol. Exp. Ther.*, *325*, 1–9.

101   Younis, Y., Street, L.J., Waterson, D., Witty, M. J., Chibale, K. (2013). Cell-based medicinal chemistry optimization of high throughput screening (HTS) hits for orally active antimalarials-Part 2: Hits from SoftFocus kinase and other libraries. *J. Med. Chem.*, *56*, 7750–7754.

102   Chatterjee, A.K. (2013). Cell-based medicinal chemistry optimization of high-throughput screening (HTS) hits for orally active antimalarials. Part 1: Challenges in potency and absorption, distribution, metabolism, excretion/pharmacokinetics (ADME/PK). *J. Med. Chem.*, *56*, 7741–7749.

103   Paquet, T., Gordon, R., Waterson, D., Witty, M.J., Chibale, K. (2012). Antimalarial aminothiazoles and aminopyridines from phenotypic whole cell screening of a SoftFocus library. *Future Med. Chem.*, *4*, 2265–2277.

104   United States Food and Drug Administration, Paving the way for personalized medicine: FDA's role in a new era of medical product development. Available at: http://www.fda.gov/downloads/ScienceResearch/SpecialTopics/PersonalizedMedicine/UCM372421.pdf. Accessed January 10, 2014.

105   McCormick, S.R., Lillemoe, T.J., Beneke, J., Schrauth, J., Reinartz, J. (2002). HER2 assessment by immunohistochemical analysis and fluorescence in situ hybridization. *Am. J. Clin. Pathol.*, *117*, 935–943.

106   Davis, P.B., Yasothan, U., Kirkpatrick, P. (2012). Ivacaftor. *Nat. Rev. Drug Discov.*, *11*, 349–350.

107   Pene, F., Courtine, E., Cariou, A., Mira, J-P. (2009). Toward theragnostics. *Crit. Care Med.*, *37*, S50–S58.

108   Espino-Rodriguez, T.F. (2006). A review of outsourcing from the resource-based view of the firm. *Int. J. Management Rev.*, *8*, 49–70.

109   Ekins, S., Williams, A.J., Krasowski, M.D., Freundlich, J.S. (2011). *In silico* repositioning of approved drugs for rare and neglected diseases. *Drug Discov. Today*, *16*, 298–310.

110   Association of Clinical Research Organizations, CRO market. Available at: www.acrohealth.org/cro-market. Accessed December 14, 2013.

111   Kitano, H. (2007). A robustness-based approach to systems-oriented drug design. *Nat. Rev. Drug Discov.*, *6*, 202–210.

112   Morphy, R., Rankovic, Z. (2007). Fragments, network biology and designing multiple ligands. *Drug Discov. Today*, *12*, 156–160.

113   Frantz, S. (2005). Playing dirty. *Nature*, *437*, 942–943.

114 Medina-Franco, J.L. (2013). Shifting from the single to the multitarget paradigm in drug discovery. *Drug Discov. Today*, *18*, 495–501.

115 Csermely, P., Agoston, V., Pongor, S. (2005). The efficiency of multi-target drugs: The network approach might help drug design. *Trends Pharmacol. Sci.*, *26*, 178–182.

116 Morphy, R., Rankovic, Z. (2009). Designing multiple ligands—Medicinal chemistry strategies and challenges. *Curr. Pharm. Des.*, *15*, 587–600.

117 Donawho, C.K., Shoemaker, A.R., Palma, J.P., Principles of chemotherapy and pharmacology. In: Taylor, J.B., Triggle, D.J. (Eds.), *Comprehensive medicinal chemistry II, Vol. 7, Therapeutic areas II: Cancer, infectious diseases, inflammation, immunology and dermatology*, Elsevier, Amsterdam, 2007, pp. 33–53.

118 Chung, M.C., Ferreira, E.I., Santos, J.L., Giarolla, J., Rando, D.G., Almeida, A.E., Bosquesi, P.L., Menegan, R.F., Blau, L. (2008). Prodrugs for the treatment of neglected diseases. *Molecules*, *13*, 616–677.

119 White, N.J. (2004). Antimalarial drug resistance. *J. Clin. Invest.*, *113*, 1084–1092.

120 Morphy, R., Rankovic, Z. (2005). Designed multiple ligands: An emerging drug discovery paradigm. *J. Med. Chem.*, *48*, 6523–6543.

121 Bell, D., Winstanley, P. (2004). Current issues in the treatment of uncomplicated malaria in Africa. *Br. Med. Bull.*, *71*, 29–43.

122 Barrett, M.P., Vincent, I.M., Burchmore, R.J.S., Kazibwe, A.J.N., Matovu, E. (2011). Drug resistance in human African trypanosomiasis. *Future Microbiol.*, *6*, 1037–1047.

123 Griensven, J., Balasegaram, M., Meheus, F., Alvar, J., Lynen, L., Boelaert, M. (2010). Combination therapy for visceral leishmaniasis. *Lancet Infect. Dis.*, *10*, 184–194.

124 Fraga, C.A.M. (2009). Drug hybridization strategies: Before or after lead identification? *Expert Opin. Drug Discov.*, *4*, 605–609.

125 Morphy, R., Kay, C., Rankovic, Z. (2004). From magic bullets to designed multiple ligands. *Drug Discov. Today*, *9*, 641–651.

126 Viegas-Junior, C., Danuello, A., Bolzani, V.d.S., Barreiro, E.J., Fraga, C.A.M. (2007). Molecular hybridization: A useful tool in the design of new drug prototypes. *Curr. Med. Chem.*, *14*, 1829–1852.

127 Muñoz-Torrero, D. (2013). Editorial, complexity against complexity: multitarget drugs. *Curr. Med. Chem.*, *20*, 1621–1622.

128 Njogu, P.M., Chibale, K. (2013). Recent developments in rationally designed multitarget antiprotozoan agents. *Curr. Med. Chem.*, *20*, 1715–1742.

129 Meunier, B. (2008). Hybrid molecules with a dual mode of action: Dream or reality? *Acc. Chem. Res.*, *41*, 69–77.

130 Cosledan, F., Fraisse, L., Pellet, A., Guillou, F., Mordmuller, B., Kremsner, P.G., Moreno, A., Mazier, D., Maffrand, J.-P., Meunier, B. (2008). Selection of trioxaquine as an antimalarial drug candidate. *Proc. Natl. Acad. Sci. U.S.A.*, *105*, 17579–17584.

131 Egan, T.J. (2008). Recent advances in understanding the mechanism of haemozoin (malaria pigment) formation. *J. Inorg. Biochem.*, *102*, 1288–1299.

132 Egan, T.J., Kaschula, C.H. (2007). Strategies to reverse drug resistance in malaria. *Curr. Opin. Infect. Dis.*, *20*, 598–604.

133 Guantai, E., Chibale, K. (2010). Chloroquine resistance: Proposed mechanisms and countermeasures. *Curr. Drug Deliv.*, *7*, 312–323.

134 Burgess, S.J., Selzer, A., Kelly, J.X., Smilkstein, M.J., Riscoe, M.K., Peyton D.H. (2006). A chloroquine-like molecule designed to reverse resistance in *Plasmodium falciparum*. *J. Med. Chem.*, *49*, 5623–5625.

135   Kelly, J.X., Smilkstein, M.J., Cooper, R.A., Lane, K.D., Johnson, R.A., Janowsky, A., Dodean, R.A., Hinrichs, D.J., Winter, R., Riscoe, M. (2007). Design, synthesis, and evaluation of 10-N-substituted acridones as novel chemosensitizers in *Plasmodium falciparum*. *Antimicrob. Agents Chemother.*, *51*, 4133–4140.

136   Wells, T.N.C., Burrows, J.N., Baird, J.K. (2010). Targeting the hypnozoite reservoir of *Plasmodium vivax*: The hidden obstacle to malaria elimination. *Trends Parasitol.*, *26*, 145–151.

137   Baird, J.K. (2009). Resistance therapies for infection by *Plasmodium vivax*. *Clin. Microbiol. Res.*, *22*, 508–534.

138   Philip, A., Kepler, J.A., Johnson, B.H., Carroll, F.I. (1988). Peptide derivatives of primaquine as potential antimalarial agents. *J. Med. Chem.*, *31*, 870–874.

139   Araujo, M.J., Bom, J., Capela, R., Casimiro, C., Chambel, P., Gomes, P., Iley, J., Lopes, F., Morais, J., Moreira, R., de Oliveira, E., do Rosario, V., Vale, N. (2005). Imidazolidin-4-one derivatives of primaquine as novel transmission-blocking antimalarials. *J. Med. Chem.*, *48*, 888–892.

140   Vale, N., Prudencio, M., Marques, C.A., Collins, M.S., Gut, J., Nogueira, F., Matos, J., Rosenthal, P.J., Cushion, M.T., do Rosario, V.E., Mota, M.M., Moreira, R., Gomes, P. (2009). Imidazoquines as antimalarial and antipneumocystic agents. *J. Med. Chem.*, *52*, 7800–7807.

141   Biot, C., Nosten, F., Fraisse, L., Ter-Minassian, D., Khalife, J., Dive, D. (2011). The antimalarial ferroquine: From bench to clinic. *Parasite*, *11*, 207–214.

142   Matos, J., da Cruz, F.P., Cabrita, E., Gut, J., Nogueira, F., do Rosario, V.E., Moreira, R., Rosenthal, P.J., Prudencio, M., Gomes, P. (2012). Novel potent metallocenes against liver stage malaria. *Antimicrob. Agents Chemother.*, *56*, 1564–1570.

143   Netto, C.D., da Silva, A.J.M., Salustiano, E.J.S., Bacelar, T.S., Rica, I.G., Cavalcante, M.C.M., Rumjanek, V.M., Costa, P.R.R. (2010). New pterocarpanquinones: Synthesis, antineoplastic activity on cultured human malignant cell lines and TNF-α modulation in human PBMC cells. *Bioorg. Med. Chem.*, *18*, 1610–1616.

144   Cunha-Junior, E.F., Pacienza-Lima, W., Ribeiro, G.A., Netto, C.D., Canto-Cavalheiro, M.M., Silva, A.J.M., Costa, P.R.R., Rossi-Bergmann, B., Torres-Santos, E.C. (2011). Effectiveness of the local or oral delivery of the novel naphthopterocarpanquinone LQB-118 against cutaneous leishmaniasis. *J. Antimicrob. Chemother.*, *66*, 1555–1559.

145   Netto, C.D., Santos, E.S.J., Castro, C.P., da Silva, A.J.M., Rumjanek, V.M., Costa, P.R.R. (2009). (±)-3,4-Dihydroxy-8,9-methylenedioxypterocarpan and derivatives: Cytotoxic effect on human leukaemia cell lines. *Eur. J. Med. Chem.*, *44*, 920–925.

146   Torres-Gomez, H., Hernandez-Nunez, E., Leon-Rivera, I., Guerrero-Alvarez, J., Cedillo-Rivera, R., Moo-Puc, R., Argotte-Ramos, R., Rodriguez-Gutierrez, M.d.C., Chan-Bacab, M.J., Navarrete-Vazquez, G. (2008). Design, synthesis and in vitro antiprotozoal activity of benzimidazole-pentamidine hybrids. *Bioorg. Med. Chem. Lett.*, *18*, 3147–3151.

147   Barrett, M.P., Boykin, D.W., Brun, R., Tidwell, R.R. (2007). Human African trypanosomiasis: Pharmacological re-engagement with a neglected disease. *Br. J. Pharmacol.*, *152*, 1155–1171.

148   Horton, D.A., Bourne, G.T., Smythe, M.L. (2003). The combinatorial synthesis of bicyclic privileged structures or privileged substructures. *Chem. Rev.*, *103*, 893–930.

149   Porwal, S., Chauhan, S.S., Chauhan, P.M.S., Shakya, N., Verma, A., Gupta, S. (2009). Discovery of novel antileishmanial agents in an attempt to synthesize pentamidine-aplysinopsin hybrid molecule. *J. Med. Chem.*, *52*, 5793–5802.

150   Nussbaum, K., Honek, J., Cadmus, C.M.C.v.C., Efferth, T. (2010). Trypanosomatid parasites causing neglected diseases. *Curr. Med. Chem.*, *17*, 1594–1617.

151 Bolognesi, M.L., Lizzi, F., Perozzo, R., Brun R., Cavalli, A. (2008). Synthesis of a small library of 2-phenoxy-1,4-naphthoquinone and 2-phenoxy-1,4-anthraquinone derivatives bearing anti-trypanosomal and anti-leishmanial activity. *Bioorg. Med. Chem. Lett.*, *18*, 2272–2276.

152 Ding, D., Zhao, Y., Meng, Q., Xie, D., Nare, B., Chen, D., Bacchi, C., Yarlett, N., Zhang, Y-K., Hernandez, V., Xia, Y., Freund, Y., Abdulla, M., Ang, K-H., Ratnam, J., McKerrow, J.H., Jacobs, R.T., Zhou, H., Plattner, J.J. (2010). Discovery of novel benzoxaborole-based potent antitrypanosomal agents. *ACS Med. Chem. Lett.*, *1*, 165–169.

153 Qiao, Z., Wang, Q., Zhang, F., Wang, Z., Bowling, T., Nare, B., Jacobs, R.T., Zhang, J., Ding, D., Liu, Y., Zhou, H. (2012). Chalcone-benzoxaborole hybrid molecules as potent antitry-panosomal agents. *J. Med. Chem.*, *55*, 3553–3557.

154 Rahman, M.A. (2011). Chalcone: A valuable insight into the recent advances and potential pharmacological activities. *Chem. Sci. Jour.*, *29*, 1–16.

155 Carvalho, S.A., da Silva, E.F., Santa-Rita, R.M., de Castro S.L., Fraga, C.A.M. (2004). Synthesis and antitrypanosomal profile of new functionalized 1,3,4-thiadiazole-2-arylhydrazone derivatives designed as non-mutagenic megazol analogues. *Bioorg. Med. Chem. Lett.*, *14*, 5967–5970.

156 Carvalho, S.A., Lopes, F.A.S., Salomao, K., Romeiro, N.C., Wardell, S.M.S.V., da Silva, E.F., Santa-Rita, R.M., de Castro, S.L., Fraga, C.A.M. (2008). Studies toward the structural optimization of new brazilizone-related trypanocidal 1,3,4-thiadiazole-2-arylhydrazone derivatives. *Bioorg. Med. Chem.*, *16*, 413–421.

157 Porcal, W., Hernandez, P., Boiani, L., Boiani, M., Ferreira, A., Chidichimo, A., Cazzulo, J.J., Olea-Azar, C., Gonzalez, M., Cerecetto, H. (2008). New trypanocidal hybrid compounds from the association of hydrazone moieties and benzofuran heterocycle. *Bioorg. Med. Chem.*, *16*, 6995–7004.

158 Gerpe, A., Odreman-Nunez, I., Draper, P., Boiani, L., Urbina, J.A., Gonzalez, M., Cerecetto, H. (2008). Heteroallyl-containing 5-nitrofuranes as new anti-*Trypanosoma cruzi* agents with a dual mechanism of action. *Bioorg. Med. Chem.*, *16*, 569–577.

159 Garcia, P.A., de Oliveira, A.B., Batista, R. (2007). Occurrence, biological activities and synthesis of kaurane diterpenes and their glycosides. *Molecules*, *12*, 455–483.

160 Chipeleme, A., Gut, J., Rosenthal, P.J., Chibale, K. (2007). Synthesis and biological evalua-tion of phenolic Mannich bases as benzaldehyde and (thio)semicarbazone derivatives against the cysteine protease falcipains-2 and a chloroquine resistant strain of *Plasmodium falci-parum*. *Bioorg. Med. Chem.*, *15*, 273–282.

161 Haraguchi, S.K., Silva, A.A., Vidotti, G.J., dos Santos, P.V., Garcia, F.P., Pedroso, R.B., Nakamura, C.V., de Oliveira, C.M.A., da Silva, C.C. (2011). Antitrypanosomal activity of novel benzaldehyde-thiosemicarbazone derivatives from kaurenoic acid. *Molecules*, *16*, 1166–1180.

162 O'Connor, K.A., Roth, B.L. (2005). Finding new tricks for old drugs: an efficient route for public-sector drug discovery. *Nat. Rev. Drug Discov.*, *4*, 1005–1014.

163 Ashburn, T.T., Thor, K.B. (2004). Drug repositioning: identifying and developing new uses for existing drugs. *Nat. Rev. Drug Discov.*, *3*, 673–683.

164 Cavalla, D. (2009). APT drug R&D: the right active ingredient in the right presentation for the right therapeutic use. *Nat. Rev. Drug Discov.*, *8*, 849–853.

165 Uliana, S.R., Barcinski, M.A. (2009). Repurposing for neglected diseases. *Science*, *326*, 935.

166 Mullard, A. (2011). Could pharma open its drug freezers? *Nat. Rev. Drug Discov.*, *10*, 399–400.

167 Chong, C.R., Sullivan Jr., D.J. (2007). New uses for old drugs. *Nature*, *448*, 645–646.

168 Nzila, A., Ma, Z., Chibale, K. (2011). Drug repositioning in the treatment of malaria and TB. *Future Med. Chem.*, *3*, 1413–1426.

169   Lipinski, C.A. (2011). Drug repositioning. *Drug Discov. Today*, *8*, 57–59.

170   Kiara, S.M., Okombo, J., Masseno, V., Mwai, L., Ochola, I., Borrmann, S., Nzila, A. (2009). *In vitro* activity of antifolate and polymorphism in dihydrofolate reductase of *Plasmodium falciparum* isolates from the Kenyan coast: emergence of parasites with Ile-164-Leu mutation. *Antimicrob. Agents Chemother.*, *53*, 3793–3798.

171   Senkovich, O., Bhatia, V., Garg, N., Chattopadhyay, D. (2005). Lipophilic antifolate trimetrexate is a potent inhibitor of Trypanosoma cruzi: prospect for chemotherapy of Chagas' disease. *Antimicrob. Agents Chemother.*, *49*, 3234–3238.

172   Parikh, S., Gut, J., Istvan, E., Goldberg, D.E., Havlir, D.V., Rosenthal, P.J. (2005). Antimalarial activity of human immunodeficiency virus type 1 protease inhibitors. *Antimicrob. Agents Chemother.*, *49*, 2983–2985.

173   Lepesheva, G.I., Villalta, F., Waterman, M.R. (2011). Targeting *Trypanosoma cruzi* sterol 14α-demethylase (CYP51). *Adv. Parasitol.*, *75*, 65–87.

174   Lepesheva, G.I., Waterman, M.R. (2011). Sterol 14alpha-demethylase (CYP51) as a therapeutic target for human trypanosomiasis and leishmaniasis. *Curr. Top. Med. Chem.*, *11*, 2060–2071.

175   BIO Ventures for Global Health, Chagas Disease, Washington DC, 2013. Available at: http://www.bvgh.org/Biopharmaceutical-Solutions/Global-Health-Primer/Diseases/cid/ViewDetails/ItemID/1.aspx. Accessed on June 21, 2015.

176   Wiesner, J., Borrmann, S., Jomaa, H. (2003). Fosmidomycin for the treatment of malaria. *Parasitol. Res.*, *90*, S71–S76.

177   Chong, C.R., Chen, X., Shi, L., Liu, J.O., Sullivan, Jr. D.J. (2006). A clinical drug library screen identifies astemizole as an antimalarial agent. *Nat. Chem. Biol.*, *2*, 415–416.

178   Shahinas, D., Liang, M., Datti, A., Pillai, D.R. (2010). A repurposing strategy identifies novel synergistic inhibitors of *Plasmodium falciparum* heat shock protein 90. *J. Med. Chem.*, *53*, 3552–3557.

179   Gloeckner, C., Garner, A.L., Mersha, F., Oksov, Y., Tricoche, N., Eubanks, L.M., Lustigman, S., Kaufmann, G.F., Janda, K.D. (2010). Repositioning of an existing drug for the neglected tropical disease Onchocerciasis. *Proc. Natl. Acad. Sci. U.S.A.*, *107*, 3424–3429.

180   Miguel, D.C., Zauli-Nascimento, R.C., Yokoyama-Yasunaka, J.K.U., Katz, S., Barbieri, C.L., Uliana, S.R.B. (2009). Tamoxifen is effective against *Leishmania* and induces a rapid alkalinization of parasitophorous vacuoles harbouring *Leishmania* (*Leishmania*) *amazonensis* amastigotes. *J. Antimicrob. Chemother.*, *63*, 365–368.

181   Macareo, L., Lwin, K.M., Cheah, P.Y., Yuentrakul, P., Miller, R.S., Nosten, F. (2013). Triangular test design to evaluate tinidazole in the prevention of *Plasmodium vivax* relapse. *Malar. J.*, *12*, 173.

182   Haupt, V.J., Schroeder, M. (2011). Old friends in new guise: repositioning of known drugs with structural bioinformatics. *Briefings in Bioinformatics*, *12*, 312–326.

183   Dudley, J.T., Deshpande, T., Butte, A.J. (2011). Exploiting drug-disease relationships for computational drug repositioning. *Briefings in Bioinformatics*, *12*, 303–311.

12

IMPACT OF INVESTMENT STRATEGIES, ORGANIZATIONAL STRUCTURE AND CORPORATE ENVIRONMENT ON ATTRITION, AND FUTURE INVESTMENT STRATEGIES TO REDUCE ATTRITION

GEOFF LAWTON

Garden Fields, Hertfordshire, UK

12.1 ATTRITION

Despite the title of this book, perhaps there is too much emphasis on attrition in drug discovery. Successful treatment of patients and the economic health of the biopharmaceutical industry are driven by our successes, not by our failures. Failures are a necessary evil. By definition, research moves into unknown territory and cannot be certain of a successful outcome. If there were no attrition, it would not be research. Human biology, in contrast to the physical world, is by no means fully understood. Creating a new medicine cannot appropriately be compared (as it sometimes has been) to engineering a new aircraft or to getting to the moon. In both of these latter cases, the underpinning physics has been understood for centuries and designs, although complex and expensive to put into operation, have little risk of complete operational failure. In contrast, modulation of disease processes by new drug candidates is a research and subsequent development operation rather than a purely technical development problem.

Our attention has been focused on the failures because it is now widely understood that the current ratio of new medicine outputs compared to the financial input is not sustainable into the future. Both the cost structure and the output rate need to be corrected to allow the continuous stream of new medicines required by the future growing and aging population and by the people of the emerging economies, who have so far been inadequately served by pharmaceutical research.

Attrition in the Pharmaceutical Industry: Reasons, Implications, and Pathways Forward,
First Edition. Edited by Alexander Alex, C. John Harris and Dennis A. Smith.
© 2016 John Wiley & Sons, Inc. Published 2016 by John Wiley & Sons, Inc.

Marginal gains can be vitally important. There is a 95% failure rate from candidate selection to launch of a drug. Simple mathematics means that converting 1% of these failures into successes would give an almost 20% increase in output.

We have seen in earlier chapters how the incomplete knowledge/understanding of biology or of medicinal chemistry can lead to technical failures that account for a great part of the attrition. Nevertheless, in all of the published analyses of reasons for failure, a significant proportion of the failures are due to "commercial," "strategic," or "financial" reasons. My experience teaches that reasons for failure are not always captured or reported accurately. For example, when a company restructuring results in a change in therapeutic focus, it is sometimes politically unacceptable to blame project failure on deliberate management policy, and so commercial/strategic reasons are frequently underreported. Perfectly good assets may fail to achieve efficacy because the patient population in the clinical trial was chosen to satisfy a commercial goal rather than a medical one. The recent, admirably honest, review of portfolio failures in AstraZeneca refers to "projects (that) were pushed to investigate what were perceived as higher value, more commercially attractive indications, but the scientific and medical basis for these indications was weak. Three quarters of these projects failed owing to efficacy issues later in development" [1].

This paper also discusses the "strategic" failures: "The remaining project failures were categorized as 'strategic' and represented decisions to close projects for non-technical or non-scientific reasons. For example, AstraZeneca exited osteoarthritis R&D during the period analysed, and consequently closed a number of projects in this indication. Strategic closures accounted for 7% of project closures, and four clinical projects were terminated for strategic reasons during this assessment period … it should be noted that repeated strategic shifts in disease area focus can also contribute to increased attrition and reduced productivity."

When a company has a major drug in a particular indication, there is some advantage to working on other approaches to the same indication. The marketing function will be very keen to develop a "franchise" in this field. Forcing research in this way is rarely efficient, despite the potential leverage in some aspects, and encourages the researchers to work in areas that they know have a lower probability of success. This deliberate acceptance of higher risk of failure in order to satisfy a marketing aspiration should be ascribed to commercial attrition rather than to poor research performance.

The critical economic issues in drug discovery are the cost of the successes and the risk to the capital employed in the process. Undoubtedly, the major cost of creating the successful drugs lies in paying for the failures. A key part of the management of the risk to capital is in the construction of a portfolio of research and development (R&D) projects that can continuously generate a proportion of successful outcomes. The revenues from these successes are able to service the costs of the capital.

Learning from past failures can help in the construction of the portfolio, but care must be taken in handling these historical examples. No two research projects are the same and applying "lessons" in the wrong context could be the worst approach. This is a common failure in large organizations, which often have a large historical dataset, and the "hit to lead" and "lead to candidate" phases are often driven by a rules-based tick-box culture rather than by assessment of each new situation on its merits.

The "herd behavior" of the drug discovery companies plays a big part in the costs and creates significant systemic risk. Simultaneous initiation of projects by several companies often follows publication on a new "target" in the latest edition of *Nature*. The resulting duplication (multiplication) of costs has heavy consequences for the overall economics of the system. Before the embryonic stirrings of "open innovation," we were not able to learn from the failures of others and were condemned to replicate their mistakes. If the economic arguments against this "herd behavior" were not enough to avoid it, we have recently learned that much that has been reported in the serious scientific literature could not be reproduced by others, and so the starting premise for the project may be flawed [2, 3].

It has been estimated that one-third of proteins under investigation in pharmaceutical research belong to the kinase family [4]. So far, only 13 NMEs from this class have reached patients. Clearly, more drugs will emerge in the coming years, but nevertheless, the returns on the costs sunk to date have been spectacularly poor. In this area, the "herd" underwent radical changes of direction. In the 1980s, when the first kinase inhibitor programs were begun by a few 'mavericks', the consensus view in the industry was that messing with these ubiquitous signaling processes would have a devastating effect on normal physiology and unavoidably lead to safety issues. As biological knowledge of the field developed, the full extent of the kinome revealed, and signaling pathways better understood, the 'herd' rapidly changed its collective mind and the resulting 'stampede' led to very large volumes of research projects in many organizations. Despite the extent of the kinome, more than 500 proteins, most of these projects have been directed at the same few kinase proteins [4].

Good scientific method and excellent experimental design should at least ensure that a hypothesis can be clearly invalidated. The problem of the simultaneous reaction to the latest publication is that the 'invalidation' costs are replicated in multiple organizations. When the published target proves to be 'valid', then we have the problem referred to as the 'me-too' drugs. Because a number of players began similar projects at the same time, a number of them (perhaps 10–15 years down the road) may get to successful medicines. In the past, it was likely that many of these would reach the market and provide adequate returns even with the resulting market fragmentation. Today's regulators are likely to approve only members of the same drug class that are adequately differentiated from each other in terms of effects on patients. This differentiation can usually be achieved only through extensive, expensive phase III clinical trials. Failure at this late stage has a devastating impact on the economics of the portfolio.

The future will create better understanding of the sensible boundary between precompetitive and competitive situations. This will usually be at the stage when chemical lead structures have been identified. One extreme view [5] is that the precompetitive stage should reach through to the proof of clinical efficacy when evaluating a new approach. Recent moves to outlaw the patenting of gene sequences in the United States [6] have given a big push in the right direction. The precompetitive, knowledge generating, stage can be funded either by consortia of companies or by public funds (or a mix of the two). Only at the stage of commercially valuable IP generation would private investors take over. In one recent example [7], Baxter took over the development role for a university/NIH-funded sickle cell disease drug.

12.2 COSTS

12.2.1 The Costs of Creating a New Medicine

There are many estimates of the R&D costs of discovering a new drug. The truth is that no one knows for sure. Some commentators think they know the answer and many more pretend to know because they have a particular point to prove or interest to protect.

Estimates range from the widely criticized academic viewpoint [8], which considers $43mio to be a good guess (but seems to ignore the costs of the failures), to $10s of billions, which is arrived at by taking the whole R&D expenditure of the industry and dividing it by the number of new medicines approved each year. This high number is further increased if spending by charities and non-governmental organizations (NGOs) is added and further still if the funding for relevant academic activity that often underpins drug discovery projects is included. Matthew Herper has calculated [9] the number for each pharmaceutical company by dividing the total published R&D spend over 15 years by the number of drugs launched. Twenty of the large companies by this calculation spent more than $4bio for each drug launched.

For many years, the figure derived by researchers at the Tufts Institute from 2003 has been widely quoted [10]. This $802mio figure is now considered to have inflated to >$1.5bio. The Tufts' numbers were reinforced by Adams and Brantner [11]. In their paper, it was shown that costs vary significantly between different operating companies and also depend heavily on the disease area being addressed.

Recently, as the public perception of pharmaceutical companies has become more hostile, commentators point out that the Tufts research is paid for by the industry. In the past, it was seen to be in the industry's interest that a high number should emerge in order to justify the significant tax breaks for pharmaceutical research in many countries and the high prices charged for the medicines. More recently, the commentariat uses the high number to point out the very poor 'productivity' of the industry's R&D and resulting poor return on investment (RoI). As this point of view gains currency, future investment in the pharmaceutical companies is threatened.

R&D costs are heavily biased toward D. The cost of a single clinical trial can be $10s of millions. Considerable costs are incurred in 'life-cycle management' of a newly launched medicine, and these are frequently assigned to 'R&D'. These costs would more appropriately be allocated to the marketing function, as their intention is to increase revenues from an existing product. The R&D tax credit systems operating in various countries of course encourage this practice.

12.2.2   The Costs of *Not* Creating a New Medicine

The lack of available treatments for some diseases has a very large societal impact.

12.2.2.1   The War on Cancer   After the successful Apollo missions to the moon in the 1960s, US President Nixon turned the nation's attention to the next big challenge: The War on Cancer. The National Cancer Act of 1971 was intended to eradicate cancer as a major cause of death. Cancer survival rates in the United Kingdom have doubled in the last 40 years [12], but although many cancers are now successfully treated, the outcomes of the first 40 years of the 'war' are somewhat chastening and often considered by commentators to be a failure.

Indeed, cancer is still the number two cause of death in the United States. Many hundreds of billions of dollars have been spent by governments and charities, supplemented by billions more by the pharmaceutical industry. In this case, it is reasonable to assume that the cost per new medicine is extremely high. However, the alternative would have been a much higher proportion of deaths due to cancer than we have today because of the considerably greater life expectation now (76.3 years in 2011) than in 1971 (67.9 years) and the corresponding increase that this would have inevitably caused in cancer-related deaths. One of the societal benefits of the enormous investment is the vastly greater biological knowledge that we now have. The study of cancers has spilled over into knowledge creation in a whole range of biology. The costs are visible and more or less measureable. However, the benefits are much less tangible even when considering only direct benefit to today's patients.

12.2.2.2 *Bacterial Resistance*

In the 1990s, most pharmaceutical companies stopped their research on new antibacterial drugs. It was considered that the problem was solved, although no new classes of antibacterial drugs had emerged for several decades, and that new drugs would achieve poor financial return. Even if a drug with a new mechanism of action were to be produced, it was argued that this would be kept in reserve and not allowed to achieve a large market.

To understand the mismatch between commercial judgment and the societal need, it is interesting to reflect on 2013 newspaper headlines such as:

'Antibiotic-resistant diseases pose apocalyptic threat, top expert says'.

The 'top expert' is the UK's chief medical officer quoted in *The Guardian* January 23, 2013. Many such headlines have appeared and there are several reasons for the antibacterial crisis. In addition to the cessation of most research into new antibacterials almost 20 years ago, the old drugs have been more widely, and often inappropriately, prescribed resulting in rapid resistance development. Increased global travel has allowed rapid spread of resistant strains from their original point of appearance.

The commercial assessment by companies may not have changed significantly. It is likely that for a new antibacterial with a new mechanism of action, effective against resistant strains, regulators would restrict wide market use and thus curtail the profitability for the company that invented it. As a counter to these commercial complexities, probability of successful transition through the development phases is found to be greater for antibacterials than for other disease indications [13].

Irrespective of the risk/financial reward analysis, society clearly needs these new agents, and mechanisms must be found to finance their R&D.

One of the interesting new drug approvals from the US FDA in 2012 was raxibacumab. This is a treatment for anthrax, a disease that has been eliminated and therefore the only potential patients are those who have been exposed to the agent used in biological warfare. GSK has a $200mio deal with the US government to supply 60,000 doses over 4 years.

12.2.2.3 *The Dementia Crisis*

As life expectancy inexorably increases, age-related diseases become more prevalent. For some, such as macular degeneration, the pharmaceutical companies have spotted the market opportunity and importantly have noted that the status of biological understanding of the disease provides opportunities for drug discovery projects. For neurodegenerative diseases, including Alzheimer's disease, the situation is less clear. Market opportunity is enormous, but the biological understanding is less good, and the need for extensive clinical trials to prove efficacy is likely to be unavoidable. Thus, many pharmaceutical companies have considered the high costs and high risk of failure for drug discovery projects in this field to be too great for inclusion in the project portfolio even taking into account the potentially enormous financial rewards.

It seems that, as in the case of the war against cancer, society's agents in the form of governments and charities will need to make the required investments.

12.3   INVESTMENT STRATEGIES

12.3.1   RoI

Different funders have differing requirements for the type, the timing, and the level of the RoI. They also greatly differ in their tolerance of risk. A discussion of risk is included in Section 12.5.

12.3.1.1   Different Types of RoI   There is much discussion today of RoI. Return on R&D investment is a current favorite topic with the major consultancy groups, and so the pharma company 'herd' have rapidly adopted it as a top priority. Many commentators consider only the financial return. For some investors, including venture capital funds, this may indeed be the full story. But for many other stakeholders, other returns are important. When a charity funds a drug discovery project, it might not be seeking a financial return at all. Its only purpose might be to grant fund for the creation of a new medicine to benefit a particular group of patients. The expected return is in improved health. Charities and venture philanthropists increasingly look for some financial return to allow them to fund future projects on a continuing basis. The subject of 'impact investing' is widely discussed and captures this mix of social and financial return. Civil society through governments and NGOs may have other social returns in mind in addition to the improvement in individual patient's health. In those countries where government pays the costs for healthcare, the impact of new medicines in the whole cost structure of the health service is an objective. In emerging economies, addressing diseases neglected by developed world researchers might warrant societal investment. Local governments will have local economic uplift high on their agenda. Local employment maybe their desired return. Other intangible social returns come from the knowledge creation and reinforcement of the positive feedback loop for wider academic education in biology, chemistry, and medicine, which spills over into other business sectors.

12.3.1.2   Timing of the Return   In the twentieth century, return on pharmaceutical research investment was typically delayed until revenues were achieved from marketed drugs. In present times, the beginning of the vertical disintegration of the pharmaceutical companies [14] has allowed financial returns to be achieved at earlier value-inflection points along the lengthy pathway.

Relying on the sales revenues from drug reaching patients requires a great deal of patience from the investor. Timing varies depending on the target disease. The HIV drug saquinavir achieved one of the fastest times on record from project initiation to launch of the drug, achieving this in 9 years. But average time spent in just the clinical stages of development for drugs for CNS diseases was 102.1 months based on launches between 1996 and 2010 [15].

Many VC-funded companies advance to a point where they are acquired by a bigger company releasing returns to the initial funders. This ideal situation (from the investor point of view) in fact occurs too infrequently to create a sustainable system.

Now, it is common for a project to begin in one organization that passes it on to another in return for an upfront payment and the potential of further financial return as downstream milestones are met. This increasing trade in assets, rather than in companies, offers a better chance of reaching sustainability. As we shall see later, the transfer of an individual asset can increase the possibilities for a successful outcome provided that the asset can always be in the hands of the most appropriate stakeholder.

12.3.2 Investment in a Portfolio of R&D Projects

Whichever investment model is adopted, the high failure rate of projects has to be taken into consideration. The only way to have sustained success is to have a stake in a critical mass of projects to ensure a high probability of continuous outputs.

In the successful days of the Fully Integrated Pharmaceutical Company (FIPCO), this is exactly what many companies managed to achieve. In the 1990s, it became possible for a short time, for venture capital funds to put together a sufficiently large project portfolio by funding a panel of companies that in aggregate achieved a good probability of sustained output.

The current very high failure rates and high costs have made the operation of a sufficiently large portfolio unfeasibly capital intensive. An emerging solution is to access 'soft' capital (grants, loans on favorable terms, etc.) to complement the mainstream capital provided by investors and also by pharmaceutical partners. There is a growing number of different business models in which this mixed stakeholder environment can be accommodated.

12.3.3 Asset-Centered Investment

Asset-centered investment is efficient and flexible and can be suitable for many different stakeholder types. In this system, capital is applied to progressing individual projects rather than to building fully fledged companies. Asset-centered investment allows the asset to be taken to go/no go point with a smaller investment than is typical in the company-funding model, thereby reducing the average cost of drug development.

The traditional venture approach is to build a company with a portfolio of assets to reduce risk. This is inefficient and is particularly wasteful for early-stage opportunities, which have very high failure rates. Companies often waste energy and resources moving forward with some assets that are unlikely to succeed and building infrastructure that proves unnecessary when product candidates fail. Conversely, in order to get a rapid return on their investment, the funders frequently force all of the resources of a company into the most advanced project to the detriment of other valuable entities in the portfolio.

This asset-centered model works for all parties. The investor gets to spend their money only on the work that is directly related to their individual objective. They can manage their investment within the familiar portfolio and project management framework, and when the project is completed, there is no need for a clean-up of assets or messy disposals. The talent gets to work on projects in small teams of people free from the overhead and complexity of corporate constraints. Originators get their new medicine discovery ideas off the ground faster as they need to raise less money and can focus on resourcing only the essence of the project.

Investment in a single asset (drug discovery project) has many advantages for this high-risk sector. In contrast to the company-based investment system, funding a collection of individual projects allows fine-tuning of the risk profile of the individual investor's portfolio. Capital required to develop a single compound to a data package that can be partnered with a pharmaceutical company is modest. Revenues coming from licensing deals can be distributed at an early time point to the investor(s). Ownership structures can be tailored to the requirement of individual investors.

In the current economic environment, the probability of a licensing transaction on an attractive asset with early clinical data is much greater than the probability of an 'exit' for a company. The structure is also very motivating to the originators and operators of the individual project as they are more likely to achieve their desired return in a reasonable time frame. Complex unwinding of company structures at the time of the exit is avoided. The asset-centered funding model efficiently uses capital and talent.

12.3.4 Sources of Funds

The economic failure of the biopharma system has encouraged innovative approaches. Profits from existing products and equity funding from the public markets or venture capital funds remain the important mainstays. Other sources of funds are becoming increasingly important. Charities and venture philanthropists offer grant funding for some projects, but increasingly they are using convertible bonds and other creative forms of debt. Charitable bonds are essentially loans that are repaid with a coupon if the project is successful.

Drug discovery projects that are funded by charities generally are not fundable by the commercial route either because a higher than average failure rate is expected or the financial return is predicted to be low. For these funders, the higher risks can be tolerated. The goal is a social dividend rather than a financial one. Nevertheless, some of the charity-funded projects are indeed commercially successful (e.g., current cystic fibrosis drugs are very promising). A recent *Nature* commentary by Heidi Ledford [16] points out that more philanthropic investors are beginning to capture the financial gains from the successful projects in order to allow greater investment in the next generation of "non-commercial" projects.

Crowdfunding has been successful in other business sectors and is increasingly talked about for drug discovery. The Jobs Act will make this more accessible in the United States by removing restrictions for nonprofessional investors. Antabio, a French antibacterial company, has already successfully used this route [17]. Many individuals are comfortable with the idea of using their available capital to contribute to society while generating a reasonable return. Some are prepared to put the capital at risk in return for the possibility of a significant social impact. Thus, in the crowdfunding field, there is a range of possibility from charitable donation to full equity investment.

Debt funding including charity bonds and securitization of future revenues are also under evaluation. Better returns than from banks in today's low interest world might be achievable by appropriate portfolio selection.

Monetization (securitization) of royalties on marketed products has been a possibility for some time. Royalty Pharma has been operating successfully since 1996. We are now seeing this approach move into earlier phases with future potential milestone payments being monetized for earlier stage projects. Andrew Lo has proposed [18] a $30bio fund to finance cancer projects and Cancer Research Technology's (CRT's) Pioneer Fund operates in this way.

Some of these new approaches might be more patient than current equity markets. Effectively, the investors buy in to a long-term revenue stream resulting from the successful projects in the portfolio. Increasing contributions of these types of funding model will counterbalance to some extent the rapid decline in research investment by the major pharmaceutical companies. The growth in diversity of funding mechanisms together with the diversification of business operating models will allow evolution to take its course and the fittest processes to expand.

12.4 BUSINESS MODELS

Historically, the operator of a project portfolio would also be the funder/investor. As the system continues to disaggregate, a range of different investor types can be accommodated within one operational portfolio, and conversely, a single funder can carry out its projects in different operational business structures.

The financial portfolio does not have to be the same as the operational portfolio. At the two extremes, the funded portfolio of projects can each be managed virtually and operated by one or more contract research organizations or, in the FIPCO model, the project portfolio is funded and operated by the pharma company.

The operator manages the laboratories in which the experimental processes of drug discovery are carried out. This is where the experimental data underpinning the intellectual property residing in the drug itself is created.

A range of different investor types can be accommodated within one operational portfolio. Funders can include stakeholders whose goal is a simple financial return, philanthropists aiming purely at public benefit, and those, such as patient advocacy groups, who want to direct their philanthropic "investment" towards specific disease indications. Some examples of funders of virtual portfolios include Wellcome Trust, Bill & Melinda Gates Foundation (BMGF), Medicines for Malaria Venture (MMV), Drugs for Neglected Diseases (DNDi), and the Global TB Alliance.

12.4.1 FIPCO

This is the traditional form of the pharmaceutical company. All operations are owned and managed by a single company. The FIPCO model was very effective for 50 years. Indeed, it was so profitable in the 1990s that more consolidation was promoted by the consultants. Size is good for incremental advances and for managing the life cycle to extract maximum bang for the buck. Large organizations are generally less good in achieving innovation. The real disrupters are shunned by the bureaucracies necessary to operate a very large operation.

FIPCOs are today encouraged by the management consultancies to "exit research to create shareholder value" and to change R&D to "search and development" [19]. Few of these consultancies seem concerned that recommendation to eat seed corn will lead to short-term gains at the expense of long-term survival.

The creation of a new FIPCO is extremely rare. It will take a minimum of 20 years to build such a company along with a great deal of capital and lots of luck. Some recent examples are Actelion (started with ready-made development candidates discarded by Roche), Gilead (the star performer), and Vertex (very long time from start-up to the market, greatly facilitated by "social" capital from the Cystic Fibrosis Foundation). In the early days of new biological entities as drugs, Amgen, Biogen, and Genentech (now owned by Roche) were created.

12.4.2 Fully Integrated Pharmaceutical Network (FIPNET)

It is now very rare for a company to carry out all operations internally. Many pieces of work are outsourced. This might be to access external skill sets not available within the company, to manage peaks and troughs in demand on resources, or to achieve cost savings.

Some of the FIPCOs have broken their research function into more or less autonomous separate small units. Despite the increase in accountability and "agility" that this process engenders, it has the potential to achieve the worst of both worlds with too much energy spent on raising funds and servicing debt from a single provider (the parent company). At the same time, the high operating overhead cost of the big company is retained.

The "herd behaviour" described in Section 12.1 with respect to the selection of projects also applies to business strategy. Pharmaceutical companies have a tendency to behave like lemmings led by consulting groups and stock analysts. The changing technical fashions through the past three decades have ranged through rational design, target-centered discovery based on the genomic revolution, high-throughput technologies, and systems biology (phenotype). In the company strategy field, we have seen the following:

Consolidate to realize the benefits of size

Fragment research functions to gain the agility of small companies

Exit research to create value

Locate in Cambridge (United States or United Kingdom)

Increase RoI

Build franchise

Get out of difficult areas

None of these has been successful in terms of reducing attrition or increasing output or cost-effectiveness and, unfortunately for patients, the last of the strategic examples includes ceasing research in the areas of highest medical need. This collective behavior severely reduces diversity and markedly increases the risk of extinction of the whole "herd."

The leaders of many of the large pharma companies have poorly managed the expectations of the owners (shareholders). It is difficult to blame them. They are faced with a barrage of publications from analysts and consultants giving them arbitrary financial targets. This has provided another demonstration that if you set the wrong targets and slavishly follow them, system damage on a grand scale can result. Other examples include the banking industry and NHS in the United Kingdom. For pharma companies, the financial marketplace has decreed a return on research investment, which is clearly far away from what the system can deliver overall. Even the good leaders are therefore forced into the observed high-risk (for long-term sustainability) behavior: focus on late-stage projects, however poor they are; blockbuster mania; exit research to create value; and so on. Good leaders can select, motivate, and organize teams of good scientists to innovate even in difficult areas of science. The shareholders must be convinced that the goal is improvement in patient well-being and that this is a necessary precursor of financial return in the long (and even the medium) term.

As economic pressures become truly unsurvivable, some genuine differences in business strategy are beginning to emerge: increase/dispose of diagnostics/generics/veterinary medicines/over-the-counter (OTC) medicines. Recent examples include asset exchanges between GSK and Novartis and Abbott's creation of AbbVie, Merck's disposal of its OTC business, and Novartis exit from animal health.

12.4.3   Venture-Funded Biotech

Venture capital funds have achieved some significant successes in the past two decades and to some extent have sustained the pipelines of the FIPCOs who have acquired either the assets of the VC-funded companies or the companies themselves. In recent times, the VC funds have followed the FIPCOs in attempting to reduce risk and have increased emphasis on later-stage drug discovery projects. This replication of strategic direction of their customers, as opposed to complementation by providing what the customers have decreased internally, seems likely to further compromise future sustainability of the system.

There seem to be more late-stage failures in projects operated by small companies than in those that are partnered with pharma companies [20]. The authors speculate that this is because underfunding leads to less than optimal staffing and clinical trial design. It may also be that the small company avoids the "killer experiment" resulting in negative impact on valuation, that is, the project is not stopped when it should be.

12.4.4   Fee-for-Service CRO

A pure fee-for-service company generates experimental data, and the rights to this IP are wholly owned by the client. The range of providers includes those that only carry out experiments to the client's design through to those that provide all of the design, the client simply providing the funding and cost control. All of the many facets of drug discovery from target discovery through candidate identification and clinical development can be sourced from this type of business structure. Some providers can offer a "one-stop shop" preferred by some clients, while others provide only single skill sets and the client puts together a supply chain of providers.

12.4.5   Hybrids

Many of today's interesting companies fall into the hybrid category with a sliding scale between the two extremes of low risk/low reward to high risk/high reward.

At the low end, we have CROs who adopt "shared risk" with clients. They accept a lower fee and receive a bonus for success. They have no rights to the products.

Next, we have the "Alliance" example. This type of company has different types of collaboration, but broadly they agree a disease area with a client and generate project ideas for a relatively small access fee. If the client selects a project, that triggers a payment that is used to progress the project through a series of milestones, each of which triggers a payment that more than covers the cost of advancing to the next milestone. If at any time the client declines to take the project on to the next stage, rights to the IP revert to "Alliance."

Many of the hybrid companies are built on a particular technical approach (e.g., Addex, Heptares, Astex) where their ability to create the alliances derives from their particular technical skill.

A different hybrid version (at various times operated by Evotec among many others) is to operate largely as a fee-for-service company and to use some of the revenues to fund the generation of fully owned IP. This almost invariably causes problems among the shareholders, some of whom invested for regular dividends rather than growth.

Some companies mainly do their own drug discovery and occasionally, to reduce the cash burn, accept an alliance fee for particular projects with milestone payments (with or without some retained IP rights).

12.4.6 Academic Institute

All of the operator types described earlier are regular companies limited by shares aiming to deliver financial returns to their shareholders.

More recently, a different type of operator has emerged. Here, the main aim is more altruistic. Scientists frustrated by lack of take-up of their ideas, and governments unhappy with the economic output from their investments in academic research, encourage the early-stage translation of ideas through the first stages of derisking in order to make them more attractive to investors.

As the large pharmaceutical companies wind down their discovery operations, there is a burst of growth in academic drug discovery operations. A recent survey [21] identified 78 such organizations in the United States working in small-molecule drug discovery. Many of these have joined a newly established Academic Drug Discovery Consortium (ADDC) that has more than 80 members [22]. This situation is rapidly being replicated around the world. The UK academic drug discovery operations were recently reviewed [23].

Is this the most effective way to harness social (in this case government) finance to create new medicines for future patients? In one successful example, Princeton received $524mio in royalties from Alimta, which depended on a license to IP owned by the university. There is an argument that with very few exceptions, the failure rate will be higher in these academic institutes because they deliberately select unproven (and therefore higher-risk) targets. The cost of these failures would then be loaded onto the education system. Some pros and cons of the academic business model are as follows:

Pros:

> Investment decisions are more likely to be driven by scientific opportunity and social impact rather than simple financial return.
>
> Provides a mechanism to bridge the "valley of death" to attain positions where the large capital requirements of drug development can be accessed with reduced risk for the investor.
>
> Likely to follow many high-risk targets which the committees of big pharma will not. Likely to increase attrition but will allow a small number of breakthrough medicines that otherwise would not be achieved.
>
> It is often argued that target validation is the greatest cause of attrition. There is great potential for academic units to produce effective tool compounds to contribute to target validation.
>
> There are already signs of new venture-fundable companies emerging.

Cons:

A system with so many organizations operating below critical mass is not efficient.

Few academic institutions can accommodate the multidisciplinary team culture essential for efficient drug discovery.

Many of the operating units do not have access to excellent medicinal chemistry or to *in vivo* pharmacology. These are two essential pillars of successful drug discovery.

Portfolio management is likely to be inefficient. The academic funding process rarely encourages the well-timed transfer of resources from failing projects to succeeding ones.

Huryn [24] has summarized well the strengths of the academic system

12.4.7 Social Enterprise

In my view, a better solution than the university drug discovery centers is to establish autonomous institutes. This avoids the potential financial risk to the education system and removes the problem of conflict of purpose. The primary education and knowledge generation missions of an academic institute cannot be fully aligned with that of a drug discovery enterprise. The disadvantages noted earlier in the academic system can be avoided. Six such autonomous drug discovery institutes from around the world have formed a global alliance and no doubt more will join (http://drugdevelopmentalliance. com/index.php). Some of these are configured as charitable foundations where profits are used to grant fund more projects and others are set up as social enterprises in which all profits are reinvested. It is very common for governments and academic institutions to be important stakeholders in such institutes. Various legal structures are possible, but the overall aim is to achieve focused high-quality drug discovery in an enterprise that reinvests profits into sustaining and growing the institute rather than for the financial benefit of shareholders.

12.5 PORTFOLIO MANAGEMENT

12.5.1 Portfolio Construction

Each individual project in drug discovery and development has a high probability of failure. The only way for a business to sustainably succeed is by ownership of a portfolio in which overall risk is managed by balancing the individual risks of the projects in the portfolio.

The dimensions of portfolio risk are as follows:

Timing of the delivery of output

Potential return from the individual output

Scale (can you buy enough tickets to be sure of winning the lottery?)

Scope of the portfolio (narrow focus vs diversity)—leverage of technology/knowledge common to multiple projects (but avoiding all eggs in one basket)

Innovation style (incremental vs. disruptive)

Risk management will depend on which stakeholder "owns" the portfolio.

A charity or patient advocacy group for a disease with no current effective treatment might be prepared to invest heavily in "high"-risk projects. The few or even single successes will more than compensate in social terms for the higher number of failures. In this situation, the higher attrition could be balanced by a lower cost of capital.

A VC fund is also likely to invest in a high-risk portfolio. In this case, the high attrition is balanced by a very high financial return from the successes.

Pharma companies usually aim for a balanced portfolio funded by the profits from earlier successes. The time frame of delivery from the portfolio is crucial. Revenues from the earlier successes are perishable. Patent life is limited and, when it reaches the "cliff," the revenue income from the asset decreases markedly and is not sufficient to cover the costs of a high-risk R&D portfolio. Different companies therefore have to adopt different approaches to risk management in their portfolio depending on the time profile of the expiry of their product patents.

Franchise development is a reason for a company to accept a potentially high risk of failure for a particular project. Disease area franchise is an obvious example where adding projects to an already well-established portfolio can have the effect of raising the barrier to entry for competitors. Technical skill and target family expertise can also provide important leverage and may provide good reasons for accepting higher risk.

There is much talk and analysis from the business consultancies on individual company strategy. A one-size-fits-all solution is usually proposed despite the very different shapes of the companies' product portfolios. The long elapsed time between research investment and revenue generation from launched drug means that the quality of management decisions is not capable of assessment in real time and confidence in the leaders has to be based on trust. Trust in business is in very short supply in the days of banking collapse and economic fragility. The resulting paralysis of pharmaceutical company management and investment has the potential to destroy the whole industry.

Biopharma 1.0 consisted almost entirely of FIPCOs and venture-funded biotechs. In the partnerships of Biopharma 2.0, the individual members of the partnership are likely to be members of multiple partnerships and are therefore each managing a portfolio of activities. There are thus multiple dimensions to portfolio management. The members of partnerships have different motivations and therefore are aiming for different output measures. Examples of the goal of an individual portfolio "integrator" could be:

Maximize financial return (if an investor)

Maximize fit with strategic plan (if a pharma company)

Maximize patient impact (if a medical research charity)

However large the business structure, innovation capacity in a single organization is limited. Unnecessary attrition sometimes results. The portfolio managers are likely to apply a forced ranking system that can lead to stopping projects in one organization that, in another organization's portfolio, would have a relatively high probability of success compared to that organization's existing projects.

12.5.2 Project Progression

The target product profile is an important tool in management of a project portfolio. Each project has specific goals identified at the beginning of the process. Criteria will include the size of the patient population, the factors that will differentiate the new medicine from current therapies, the route and frequency of administration, the acceptable safety margin, cost of goods, and so on. Progress toward these goals will be continuously monitored. An individual project will progress through a series of "milestones." Typically, these will include identification of a "target," generation of a "lead" compound, identification of a "candidate" medicine, entry into the "first-in-human" study, proof of clinical efficacy, proof of differentiation from competitors, and finally approval from the regulatory authorities.

There is lots of scope for this process to go wrong and to contribute to poor productivity. In some organizations, the milestones are treated very inflexibly in a "tick-box" culture. In this system, "millstone" could be a more apposite term. No project proceeds perfectly smoothly and lack of flexibility can result in "killing" projects that with creative management would be capable of survival. On the other hand, particularly in organizations where the "milestone" triggers a reward for one group of people, pushing poor quality projects through milestones will increase the probability of later and more expensive failure.

The milestones serve as a mechanism for identifying value-inflection points and opportunities to move a project from one owner to another.

Value-inflection points and timelines for project progression are project specific. A common mistake in portfolio management is to apply universal definitions and generic expectations. Different disease areas may follow different paths.

Reducing overall project failure rate is critically dependent on high-quality decision making through the milestone transitions. In a small company with few assets, it is tempting to continue projects with poor prospects. Indeed, this may be necessary for the survival of the company. The progression of poor quality compounds through the phases leads to more late-stage (expensive) failures. Even in large companies, attaching undue importance to the sunk costs in a project can lead to continuation of poor projects.

12.5.3 The Risk Transition Point

At the stage of selection of a "candidate drug," the innovation process undergoes a radical change. From here on the work is highly regulated. The success rate is still low. Approximately 5% of "candidates" become drugs available to patients. Importantly from an investor's perspective, at this point, the risk profile of an "asset" changes. Before this point the risk is proportional; a failure does not necessarily mean that the "target" has failed and more work to generate new and better molecules may lead to project success. After this point the die is cast, the risk is binary, the candidate either succeeds or fails. In this latter situation, it is often easier for potential investors to take evidence-based rational decisions. Before this point, much reliance is placed on the opinions of experts. These expert-led decisions are the critical success factor in the first phase of pharmaceutical innovation.

Since many of the activities in the development phase are defined by regulation, the tasks are very amenable to outsourcing by the owner. A group of highly efficient CROs have become the dominant operators in this part of the process.

12.6   PEOPLE

Creating a new medicine requires a very wide range of scientific and business skills. Reducing attrition depends critically on the people and ensuring their engagement in the process. Since individuals have differing motivations, the proliferation of business models and resulting increased diversity of employment opportunities offers a greater chance for an individual to find the best-fit employer. Productivity in the whole system will markedly improve.

12.6.1   Motivation

Different individuals require different environments for maximum performance. Continuous high output from an organization can be maintained only with motivated staff. Proximity to and ability to influence the decision-making process (autonomy), development of expertise (mastery), and working toward a worthwhile goal (purpose) are important components of job satisfaction. It is generally accepted that true motivation requires at least two of these three areas to be satisfied to ensure full engagement of the individual [25].

Inventions are made by individuals. The inventors are often eccentric, not easily managed in rules-based cultures, and indeed frequently considered by "the management" to be mavericks. The creative people are often motivated by scientific achievement/status and continuity of employment more than by financial reward. They are easily distracted by managers and by "strategic initiatives."

Experts in the research phase respond differently to those operating in the development stages of the process. In the latter, following the regulations is the only way to deliver the product.

12.6.2   Culture and Leadership

The multitude of skills required in creating new medicines necessitates establishment of teams of individuals with complementary skills. Increased probability of success is likely when organizational culture promotes team-working of the disparate experts. This is reinforced through the reward and recognition systems. Those cultures where one group of people provide "services" to others in the same organisation are rarely optimal.

Very few individuals can effectively meet the demands imposed on the leaders of pharmaceutical companies. In addition to meeting the technical demands of the R&D and marketing parts of the organization, creating a vision for the future direction of the company and convincing the staff to follow a clear direction through the very chaotic current business environment are essential.

The frequent changes in leadership and direction during the past decade have been a significant contributor to the current poor success rate of projects in large pharmaceutical companies. One of the drivers for the present fragmentation of the industry is to allow it to operate in manageable pieces. Of course, this then demands a new style of leadership capable of integrating appropriate fragments.

12.6.3   Sustainability

Fragmentation into smaller operating units compromises the capacity for organizational learning. Reducing future attrition depends on continuous development of the knowledge base underpinning the decision-making process. Some of this knowledge is explicit and can be captured in scientific publications or company operating procedures, but the vast proportion of valuable knowledge is tacit and resides in the heads of the employees. Maintaining and improving the learning process is a key role for company managers. Individual organizations that aim to be long-lived must develop excellent processes for recruitment and personal development of their staff. The risk arising from high staff turnover can be mitigated to some extent by capturing, and making explicit, as much of the tacit knowledge as possible. Smaller organizations frequently access a wider knowledge base through use of expert technical consultants or advisory boards. This approach increases agility and allows the company to change its technical area without wholesale employee changes but usually relies on experts who have been trained in the previous world of large organizations. Continuous supply of such individuals in a fragmented system is not assured.

12.7 FUTURE

Aging populations and increased access of developing economies to medical education ensure that the need for new medicines will continue. The business environment as a whole will find ways to achieve sustainability.

12.7.1 Business Structures

Not all business entities in the system need to sustain. Indeed, there is space for investors who find an asset, exploit it aggressively using operators, and partner businesses that are then discarded when their role in the R&D is complete.

But in order for this type of ephemeral business to be capable of existence in the future, other business structures that can provide a continuous stream of embryonic assets and skilled people must be present in the system. In the past, the integrated pharmaceutical companies have fulfilled both of these roles. Today's short-term shareholders do not have the patience for these activities, and unless attitudes change in at least some of the pharma companies, other types of enterprises must grow to fill the roles.

The number of different types of business engaged in drug discovery is expanding. The biopharma business environment is under great stress at present, and the increased diversity in business models will promote survival of the fittest and of the system as a whole.

We have seen in earlier chapters that improvements in science and technology and in decision making will contribute to reduced attrition in the future. An increased focus on identifying the most appropriate patients for clinical trials and treatment will reduce the failure rate. Integrated diagnostic/drug combination and disease/treatment monitoring devices will improve the performance of those medicines that make it to the marketplace. More emphasis on preventative medicine will also increase the opportunities (with a corresponding increased safety hurdle).

From a business perspective, and stripping out the details, the overall economics of discovering and developing a new medicine in Biopharma 1.0 do not look encouraging. The average overall cost for each new drug is probably in the region of $1.2bio. Most launched drugs do not achieve this sum in total revenue. The probability of failure of an individual project is extremely high. The time frame for the overall process from project initiation to launch is typically 15–20 years. The total output each year from the entire industry is small (the 39 new approvals from the FDA in 2012 were an exceptionally high number) and not replicated in 2013. For the new drugs that do make large returns in the future world of stratified (or "personalized") medicines, the cost per patient is often unfeasibly high and precludes widespread accessibility for the worldwide population.

Lack of diversity in the businesses has been a significant contributor to poor performance of the pharmaceutical industry as measured by NME output in the past 20 years. We have seen a period of consolidation. Mergers and acquisitions have reduced the number of major participants and correspondingly reduced the innovation capacity. At the same time, many small businesses have been created, but often these have a short lifetime funded by venture investors seeking a rapid exit. These companies either succeed and are eaten by the big fish or fail and die. Very few have grown into large self-sustaining companies. In the late twentieth century, Genentech (now acquired by Roche), Amgen, and Biogen, more recently joined by Gilead and Vertex, grew from discovery of new drugs through to launch and management of a portfolio of drugs. Vertex achieved this only after 23 years and through use of significant funding from the Cystic Fibrosis Foundation.

Many observers believe that the consolidation phase will be followed by increased disaggregation. The megasized companies will increasingly "outsource" more of their operations and will focus on the later stages of proving efficacy of the new medicine in large patient populations and marketing the new drugs.

The emerging biopharma "ecosystem" is one of partnerships between an increasingly diverse set of stakeholders. Greater involvement of charitable foundations and venture philanthropy will allow a better balance between social and financial aims, movement away from short-term financial motivation, and a return to focusing on the customers. The apparent customers for new medicines are patients, and indeed their voice will be increasingly heard. The HIV story is a very clear demonstration that patient advocacy groups can change the entrenched view of the marketing functions within big business. Medical research charities representing the patient are important funders of drug discovery and are raising their profile in the decision-making processes in order to achieve impact from their investments. The real customer, who pays for the medicine, is only occasionally the patient. Usually, governments or insurance companies are the customers. It is indeed surprising that these stakeholders have for decades ignored the process of generating new medicines and this situation is likely to change. They will increasingly become stakeholders in the new partnership structures. It is interesting to note that Roche has recently partnered with Swiss Re to offer insurance policies to Chinese people that will pay for cancer therapeutics if they should need them. Biopharma 2.0 is likely to include partnerships in various forms between the providers and the payers and the patients.

It is very likely that some pharmaceutical companies operating the FIPCO model will survive. It will be interesting to see if these survivors are dominated by those that are owned by families or by charitable foundations. This ownership structure has hitherto adopted a more patient approach to RoI than those owned by equity investors.

Diversity in sources of funding (including the crowd) is now complemented by diverse operating organizational structures. The herd is being broken up. Increased adoption of asset-centered investment, where individual drug discovery projects are separately owned and operated, encourages success by placing the right asset in the right partnership. Importantly, when individual stakeholders want to change strategic direction, they can achieve this by manipulating their portfolio of assets and not necessarily causing widespread systemic destruction as has happened in the past decade. The increasing separation between the funders of drug discovery and the operators allows for each of the two types to separately build their project portfolios. Risk (attrition) is thus managed in more than one dimension, and each of these can put mitigation procedures in place. This business restructuring is very likely to significantly reduce or even remove entirely the 'commercial' failures.

Early-stage drug discovery projects are increasingly taken on by universities. In the past, adding this medical/biological knowledge to society was considered to be part of the academic funding contract. Today and in the foreseeable future, the institutes are required to achieve a financial return on these assets. In the early days of this transition, technology transfer functions were created from scratch. The inexperienced practitioners were unable to set appropriate valuations on immature assets and frequently acted as a barrier to the development of these assets in the commercial world. This situation is rapidly evolving into an efficient system with a variety of operating modes. One such mode involves the academic institute itself carrying out the early development of the asset. Optimum ways to further develop these projects are under experimental development and include direct to pharmaco, direct to investor, or via an incubator such as a social enterprise.

There is much debate under the heading 'Open Innovation' of where lies the boundary between 'pre-competitive' and 'competitive' innovation. In one extreme form, organizations such as Sage and SGC [26] are creating consortia of companies, which will manage a portfolio of projects that remain 'pre-competitive' until after the phase II studies that prove (or disprove) efficacy in the clinic.

There is much talk these days about the virtues of virtual drug discovery businesses. For investors requiring a short-term return, it is clearly attractive (even essential) to avoid funding the overheads required for 'wet' drug discovery. Most players agree, however, that the core requirement for the virtual solution is a small team of experienced drug hunters. These individuals have the tacit technical and process skills that are essential for making the critical decisions—which project, which technology, which outsource provider, which lead, which candidate, and when to stop—and can provide effective management of the outsourced provisions.

12.7.2 Skilled Practitioners

The vocational training of drug discoverers can be achieved only in an environment of active drug discovery projects. With very few exceptions, this is not best achieved in universities. In the past, it was very much seen as the pharma company's role, and they were very successful in producing almost all of the current leaders in today's diverse world of new medicine discovery.

Much of the present-day drug discovery is carried out at CROs. These include specialist organizations (e.g., in clinical development) and one-stop shops providing all the skills of drug discovery and development. Present fee structures do not allow in most CROs much scope for specific training budgets. Instead, they rely on recruiting skilled people from the shrinking R&D departments of pharma companies. This is clearly a temporary situation, and when this source dries up, either with time or more optimistically with the end of the shrinkage, the CROs will have to incorporate more training and employee development in order to survive. Early adopters are likely to use the ready-made learning opportunity of working with a wide variety of clients on a wide variety of projects to promote internally trained highly skilled staff as a market differentiator.

Today, there are many individuals who learned the skills in the fully integrated pharmaceutical companies where they were exposed to all of the very wide-ranging components of successful drug hunting. Many of these individuals are managing the virtual portfolios of investors and charities/NGOs/venture philanthropists. But where is the training ground for the next generation of drug hunters? Exposure to many successful and unsuccessful projects is a key component of the learning. Will the breed die out?

Will those CROs providing 'integrated' drug discovery 'services' be able to take on enough of the decision-making responsibilities for projects to effectively develop the high-level skills or will a heavy focus on cost containment, coupled with the need for firewalls between projects, impede the learning process in these organizations? In the new world of multidimensional partnerships, a key skill required for success is relationship management.

12.7.3 Partnerships

We have argued [14] that the complete process of drug discovery, development, and commercialization does not fit well into a single vertically integrated business but each phase is best handled in a different business structure. In this way, each business can easily access multiple providers of inputs and multiple potential customers and can flexibly tailor its portfolio and operational strategy to prevailing market conditions.

The supply chain concept implies that there is a 'controller' of the system. There is now increasing recognition that partnerships between different stakeholders are essential in the new landscape of Biopharma 2.0. At present, the partnerships are usually formed and controlled by large pharma companies (FIPNET). Almost all big pharma companies now advertise themselves as the hub of a network of partnerships and collaborations. It is likely that the drivers of future partnerships will include a wider spectrum of stakeholders. These will include venture philanthropists, medical insurance companies, and governments. We are already seeing the emergence of, for example BMGF, Michael J. Fox Foundation, and Cystic Fibrosis Foundation using their power to bring together the components of the supply chain. The members of the partnerships will be drawn from biopharma companies (pharma companies, biotechs, generic drug companies), academic institutes and their varied forms of technology transfer operations, other not-for-profit drug discovery units/consortia, contract research organizations, patient advocacy groups and other medical charities, various types of financial institution, and increasingly 'the crowd' (through a variety of crowdsourcing engines).

The presence of multiple stakeholder types within each of these partnerships will help to ensure that all of the value of all assets produced within the group will be effectively exploited. In Biopharma 1.0, it is commonplace to discard potentially valuable projects because they fail to fit the strategy (often newly defined) of the single current owner.

One clear advantage of the partnership model is that each useful 'asset' has more than one possible successful outlet. The medicine is much more likely to reach the patient if it does not have to rely on a consistent coherent strategy of a single owner during the 10–15-year development process.

When patient benefit becomes again the central focus in the creation of new therapeutics, there is opportunity for productive social output (health and economic benefit) when direct financial return is relatively small (or even negative in the case of charities).

New business models are emerging in which other stakeholders collaborate with pharmaceutical companies [27]. In rare, but likely to become more frequent, cases, these alternative agencies carry out (or pay for) the whole translation process from discovery to development to delivery to patients. These other stakeholders ("social investors") today include governments, medical charities, patient advocacy groups, and venture philanthropists. In collaboration mode, these stakeholders provide funding that allows pharmaceutical companies to reduce their costs and also their risks. However, in cases where the patient population is unable to pay realistic prices for new medicines, the whole process will need to be covered by the "social investors." These alternative stakeholders will increasingly emphasize social return along with the straightforward financial return on the investment in a new medicine.

As the pharmaceutical companies search for different economically viable ways of introducing new medicines, they are increasingly supporting "open innovation" and entering into a large number of collaborations with other players. So far, most of these "collaborations" have usually involved the pharma company trawling the outside world for ideas that it can support relatively cheaply until the point at which the asset has been sufficiently derisked to justify bringing it into the internal portfolio. This one-way traffic will increasingly be superceded by a two-way traffic system. Assets in the internal portfolio, which either no longer fit a changed strategic direction for the company or are forecast to give insufficient financial return, will be out-placed to be further "translated" by other more suitable stakeholders.

12.7.4 A Personal View of the Future

Diversity in the drug discovery ecosystem is vital to its long-term survival. The present apocalypse is clearly a result of the "herd behavior" of the pharma companies in the past decade.

Society greatly needs new medicines to treat the diseases of an aging population. Social enterprises are a key component of the new landscape. These businesses, with less emphasis on short-term profitability, and more emphasis on maximizing scientific advances for the benefit of future patients, can also provide the "experienced" drug hunters who will be required for the next phase of the business cycle.

MRCT, CRT, and the Wellcome Trust's Seeding Drug Discovery Initiative (SDDI) are three successful UK examples of alternative business models. In each case, a portfolio of drug discovery projects is funded, the project's initiators are able to share their risk, and additional partners are brought in to carry out the development and commercialization phases. The retained stakes in commercially successful outcomes can be recycled to fund further innovative discoveries. In the cases of MRCT and CRT, the projects are operated in dedicated business-focused in-house laboratories (separate from funded academic laboratories). In the case of the SDDI, the portfolio is managed virtually by the Trust with the work done mostly by CROs.

This social enterprise concept is rapidly developing around the world (http://drugde velopmentalliance.com/index.php) and can provide a focus for more trusting partnerships. These organizations are under the control of social stakeholders and may give confidence in a more truly "open" motive and allow "honest-broker" behavior to provide lubrication for the partnerships. They might also be able to effectively manage the elusive boundaries between precompetitive and competitive technologies.

Another important development is the IMI Lead Factory. This public–private partnership aims to provide wide access to hit finding technologies used in the first stages of drug discovery (http://www.imi.europa.eu/content/european-lead-factory). I look forward with great anticipation to this blossoming and becoming an important focus for collaborations.

The availability of alternative drug development organizations will make it socially and economically unacceptable for companies to simply discard high-quality drug discovery projects that fail to fit the strategic or commercial goals of their owner. It will become normal practice for these to be "out-placed." In one embodiment of this, the NIH (through NCATS) is promoting repurposing of potential drugs shelved by pharma companies, and in another, MRCT has access to a portfolio of cancer drug candidates unwanted by AstraZeneca.

Biopharma 2.0 must have as its focus the patients of the future (15–20 years if an individual project is in its early stage). It is vital to consider the future marketplace. Who will be paying for these new medicine products? Today's payers are mostly in the developed world and are usually government departments or insurance companies with some revenues from wealthy individuals and increasing amounts from NGOs and charities (including venture philanthropists and patient advocacy groups). The payers are likely to be the same categories in the future, but the global spread will be very different and the balance between the categories is likely to alter. It is important now to debate: How much of the costs of the R&D should be borne by the payers and how much allocated to RoI capital?

Attrition in the R&D of new drugs will always be high. Continued improvement in our understanding of the complexities of human pathology will allow incremental reduction in the failures due to lack of efficacy and will allow us to reduce the safety failures. The commercial failures, however, offer an immediate opportunity for a step change in increasing the proportion of successful outcomes.

A recipe for change in the commercial system might look like this:

1. Ensure that outcomes from "precompetitive" research are widely shared
2. Promote the formation of diverse business structures
3. Encourage partnerships between different stakeholders including those with social goals
4. Foster the easy transfer of individual project assets between stakeholders
5. Ensure that there are at least some organizations with the critical mass to nurture the necessary skills and to train the next generation of drug hunters

The recent decrease in outputs coupled with increased costs has produced a crisis of confidence in researchers and in investors in drug discovery. Unless this is reversed, there is a danger of complete market failure. The recipe mentioned before has the potential to rapidly improve overall success rates, which in turn would allow a return in confidence in the system as a whole.

Some of the best brains of our generation are working in the biopharma sector, and if we can remove the blinkers of tradition, we are well placed to identify the gaps in the evolving ecosystem and to generate new business structures to complement the old models.

REFERENCES

1   Cook, D., Brown, D., Alexander, R.M., Morgan, P., Satterthwaite, G., Pangalos, M.N. (2014) Lessons learned from the fate of AstraZeneca's drug pipeline. *Nat. Rev. Drug Disc.*, *13*, 419–431.

2   Begley, C.G., Ellis, L.M. (2012) Drug development: Raise standards for preclinical cancer research. *Nature*, *483*, 531–533.

3   Prinz, F., Schlange, T., Asudullah, K. (2011) Drug development: Raise standards for preclinical cancer research. *Nat. Rev. Drug Disc. 10*, 712.

4   Fabbro, D., Cowan-Jacob, S.W., Mobitz, H., Martiny-Baron G. (2012) Targeting cancer with small-molecular-weight kinase inhibitors. *Methods Mol. Biol.*, *795*, 1–34.

5   Bountra, C. (2011) Traditional drug-discovery model ripe for reform. *Nature*, *471*, 17–18.

6   Check-Hayden, E. (2014) Biotech reels over patent ruling. *Nature*, *511*, 138.

7   NIH (2014) First drug candidate from NIH program acquired by biopharmaceutical company developed to treat sickle cell disease. Available at http://www.nih.gov/news/health/jul2014/ncats-09.htm (accessed on June 20, 2015).

8   Light, D.W., Warburton, R. (2011) Demythologizing the high costs of pharmaceutical research. *BioSocieties*, *6*, 34–50.

9   Herper,M.(2013).Availableatwww.forbes.com/sites/matthewherper/2013/08/11/how-the-staggering-cost-of-inventing-new-drugs-is-shaping-the-future-of-medicine

10 DiMasi, J.A., Hansen, R.W., Grabowski, H.G. (2003) The price of innovation: new estimates of drug development costs. *J. Health Econ.*, *22*, 151–185.

11 Adams, C.P., Brantner, V. (2006) Estimating the cost of new drug development: Is it really $802 million? *Health Aff.*, *25*, 420–428

12 CRUK (2014). Available at http://www.cancerresearchuk.org/cancer-info/cancerstats/survival/ (accessed on June 20, 2015).

13 Hwang, T.J., Carpenter, D., Kesselheim, A.S., (2014) Target small firms for antibiotic innovation. *Science*, *344*, 967–8.

14 Dixon, J., Lawton, G. and Machin, P.J. (2009) Vertical disintegration: a strategy for pharmaceutical businesses in 2009? *Nat. Rev. Drug Disc.*, *8*, 435.

15 DiMasi, J.A. (2012) Pace of CNS drug development and FDA approval lags other drug classes. *Tufts CSDD Impact Report*, *14*, 1.

16 Ledford, H. (2011) Charities seek cut of drug royalties. *Nature*, *475*, 275–276.

17 Powers, M. (2012) Crowdfunding carries Antabio across the valley of death. *Bioworld Today*, October 3.

18 Boissel, F. (2013) The cancer megafund: mathematical modeling needed to gauge risk. *Nat. Biotechnol.*, *31*, 494.

19 Baum, A., Verdult, P., Chugbo, C.C., Abraham, L., Mather, S., Bradshaw, K., Nieland, N. (2010) Pharmaceuticals: exit research to create value, Morgan Stanley Report.

20 Czerepak, E.A., Ryser, S. (2008) Drug approvals and failures: implications for alliances. *Nat. Rev. Drug Disc.*, *7*, 197–8.

21 Frye, S., Crosby, M., Edwards, T., Juliano, R. (2011) US academic drug discovery. *Nat. Rev. Drug Disc.*, *10*, 409–10.

22 Slusher, B., Conn, J., Frye, S., Glicksman, M., Arkin, M. (2013) Bringing together the academic drug discovery community. *Nat. Rev. Drug Disc.*, *12*, 811–812.

23 Tralau-Stewart, C., Low, C.M., Marlin, N. (2014) UK academic drug discovery. *Nat. Rev. Drug Discov.*, *13*, 15–16.

24 Huryn, D.M., (2013) Drug discovery in an academic setting: Playing to the strengths. *ACS Med. Chem. Lett.*, *4*, 313–315.

25 Pink, D.H. *Drive: The Surprising truth about What Motivates Us*, Riverhead Books, New York, 2011.

26 Norman, T., Edwards, A., Bountra, C, Friend, S. (2011) The precompetitive space: time to move the yardsticks. *Sci. Transl. Med.*, *3*, 76.

27 Dixon, J.D., England, P., Lawton, G., Machin, P., Palmer, A. (2010) Medicines Discovery in the 21st Century: The Case for a Stakeholder Corporation. *Drug Discov. Today*, *15*, 700.

INDEX

Attrition in the Pharmaceutical Industry: Reasons, Implications, and Pathways Forward,
First Edition. Edited by Alexander Alex, C. John Harris and Dennis A. Smith.
© 2016 John Wiley & Sons, Inc. Published 2016 by John Wiley & Sons, Inc.